Sustainable Ecological Engineering Design

Lloyd Scott • Mohammad Dastbaz
Christopher Gorse
Editors

Sustainable Ecological Engineering Design

Selected Proceedings from the International Conference of Sustainable Ecological Engineering Design for Society (SEEDS) 2019

Springer

Editors
Lloyd Scott
Technological University Dublin
Dunboyne, Ireland

Mohammad Dastbaz
University of Suffolk
Ipswich, Suffolk, UK

Christopher Gorse
Leeds Sustainability Institute
Leeds Beckett University
Leeds, UK

ISBN 978-3-030-44380-1 ISBN 978-3-030-44381-8 (eBook)
https://doi.org/10.1007/978-3-030-44381-8

Cover credit: Photograph taken outside the Waterfront building, University of Suffolk, Neptune Quay Ipswich

This Springer imprint is published by the registered company Springer Nature Switzerland AG
The registered company address is: Gewerbestrasse 11, 6330 Cham, Switzerland

Preface

We held the Inaugural Sustainable Ecological, Engineering Design for Society Conference in 2015 at Leeds to try and bring together researchers from across the world to exchange ideas about our common problems and the challenges our planet face. Three weeks after our first SEEDS conference, 150 world leaders attended the "UN Sustainable Development Summit" in New York to discuss the challenges facing our plant, the fast disappearing natural resources. The conference set a vision for 2025–2030 to develop a "plan of action for people, planet and prosperity". Five years on it is both ironic and dangerous that we have influential voices around the world that there is a climate disaster happening and that we have significant responsibility to try and reverse our history of deliberate and destructive impact on our environment over the past centuries. It is also refreshing that the weight of public opinion has forced significant changes in government behaviours across the world.

Through research and proven practice, the aim of the SEEDS conference, each year, is to foster ideas on how to reduce negative impacts on the environment while providing for the health and well being of society. The professions and fields of research required to ensure buildings meet user demands and provide healthy enclosures are many and diverse. In 2019, the SEEDS conference addressed the interdependence of people, the built and natural environments, and recognized the interdisciplinary and international themes necessary to assemble the knowledge required for positive change.

The selected proceedings of SEEDS 2019 presented here is organized into six sections covering: Sustainable Development and Urban Spaces; Sustainability Education; Project Management in Sustainability; Energy and Energy Efficiencies; Sustainable Retrofit and Lifecycle Assessment and Ecology and General Sustainability.

We hope that SEEDS 2019 selected proceedings provide a platform for interested policy makers, researchers, practitioners and educators to discuss the recognized and important problems affecting sustainable built environment.

Dunboyne, Ireland — Lloyd Scott
Ipswich, Suffolk, UK — Mohammad Dastbaz
Leeds, UK — Christopher Gorse
February 2020

Contents

About the Editors

Mohammad Dastbaz graduated in Electrical and Electronic Engineering and went on to do a PhD in "Design, Development and Evaluation of Multimedia Systems" at Kingston University. In 1989, he established one of UK's first multimedia computer companies called "Systems 2000 Ltd." and was one of the only companies alongside Philips Corporation to participate in the UK's first Multimedia Systems Exhibition at London Olympia in 1990. He Joined Kingston University in 1994 as a research lead for developing multimedia-aided learning packages and has since worked in number of UK Universities, progressing to become the Dean of School of Computing and Engineering at the University of East London. In 2011, Professor Dastbaz joined Leeds Metropolitan University (now Leeds Beckett University) as their dean and pro-vice-chancellor for Faculty of Arts, Environment and Technology.

Professor Dastbaz remains research active and has published over 60 refereed journal, and conference papers, books and book chapters. He is on a number of editorial boards of international journals, has been chair of a number of international conferences (including IEEE's Information Visualisation) and remains a member on a number of international conference committees. His latest publications include an edited volume on "Green Information Technologies"; "Building Sustainable Futures—Design and the Built Environment" and a series of four edited volumes on "Technology and Sustainable Futures" published by Springer.

Professor Dastbaz is a Fellow of the BCS—The Chartered Institute for IT, Fellow of the Higher Education Academy, and a Fellow of Royal Society of Arts (RSA).

Christopher Gorse is the director for the Leeds Sustainability Institute, located at Leeds Beckett University. The building performance and sustainability research unit that Professor Gorse leads has amassed one of the most comprehensive sets of actual building thermal performance data in the UK. The work has informed government policy and regulation, is used extensively by industry and is now embedded in the work of the International Energy Agency's programme to inform whole building testing and performance measures. The research team has extensive knowledge and expertise in whole building performance tests in the field and laboratories, hygrothermal and thermal modelling, building simulation, elemental building component testing, energy and behaviour monitoring.

Work extends across domestic and industrial developments, refurbishment and retrofit projects and looks at the impact and effectiveness of interventions and changes. The efficiencies of building fabric, services and renewable energy technologies are a major focus of the group's work. The more extensive work of the Sustainability Institute includes energy, waste, business and social governance, information and communication technology, corporate social responsibility, planning, ecology, process management and project management, all engaged in reducing negative impact on the natural environment.

Professor Gorse leads a sub-task group for the International Energy Agency, Annex 58, on whole-scale building testing, working with over 30 international partners. More recently, Professor Gorse led a major contract project funded by the Department of Energy and Climate Change to undertake research on the Core Cities Green Deal Go Early project. The work involved the intensive and extensive monitoring and evaluation of buildings that are benefiting from Green Deal and Eco funding in the Leeds area.

Lloyd Scott is a professor of Academic Advisor and Partnership Co-Ordinator in the School of Surveying and Construction Management at Dublin Institute of Technology (DIT). He joined the DIT as a lecturer in Construction Management and Technology in 2000. He is currently supervising 7 PhD researchers. Apart from his lecturing, supervision, research, and academic administrative duties, Lloyd has completed a PhD in the field of Built Environment Education and has developed a framework for assessment led learning strategies for Built Environment education. In 2014 he accepted the position of 'Professor of Practice' at the Haskell and Irene Lemon Construction Science Division in the College of Architecture at the University of Oklahoma. Along with this he has produced many peer-reviewed conference and journal papers. He serves as the Associate

Editor of the International Journal of Construction Education and Research. His research interests include modern approaches to thermal performance in domestic construction, development of sustainable energy sources and their practical application in Ireland and project delivery methods for a sustainable environment in Ireland. He also serves on the editorial board of Structural Survey an academic journal that publishes contemporary and original research in building pathology and building forensics, refurbishment, and adaptation. In 2016 Lloyd accepted the position of Research Fellow at the Sustainability Institute at Leeds Beckett University.

About the Authors

Mohsen M. Aboulnaga has 35 years of vast experience in higher education, high-level government posts and consultancy in sustainable cities, climate change, sustainable development and sustainable energy. He is a qualified architect in Egypt, graduated from Cairo University with BSc Arch (1979) and MSc in Housing Economic (1985), and holds a PhD from The University of Leeds in the UK (1991), https://www.leeds.ac.uk/. Dr. Aboulnaga is a professor of Sustainable Built Environments at Cairo University, https://scholar.cu.edu.eg/mohsen_aboulnaga/. He is a registered senior expert on Urban Energy Policy at European Union–NEAR, ENPI South, https://www.ces-med.eu/. He is an associate partner of European Sustainable Development Network in Vienna, https://www.sd-network.eu/; a permanent member of World Renewable Energy Congress/Network in the UK, https://www.wrenuk.co.uk/; a senior expert at Union for Mediterranean in Barcelona, https://ufmsecretariat.org/first-platform-sustainable-urban-development/, and UNESCO Inclusive Policy Lab in Paris, https://en.unesco.org/inclusivepolicylab/users/mohsen-aboulnaga; as well as a member of UrbanFarm 2020 International Jury in Italy, https://site.unibo.it/urban-farm/en. He is an invited speaker to 425 international conferences and has 200 international publications and presentations. He trained 520 government officials of Ministry of Petroleum—Egypt on strategy planning, SDGs and Energy Efficiency as well as 168 government officials in Kazakhstan and Azerbaijan on green cities strategies. Professor Aboulnaga is the author of a book *Urban Climate Change Adaptation*, https://www.springer.com/gp/book/9783030054045, plus ten book chapters (Springer and Elsevier).

Clinton Aigbavboa is vice-dean for the Postgraduate Studies, Research and Innovation and head of the Sustainable Human Settlement and Construction Research Center. He is also an editor for the *Journal of Construction Project Management and Innovation* (JCPMI). He has written a number of conference papers and journal articles and published a book on human settlement. He also holds a PhD from the University of Johannesburg.

Ghizlane Ben Baha graduated with distinction from the MSc Strategic Project Management course at Leeds Beckett University, after previously reading Architecture at Cambridge. Her research into Programme Management challenges assumptions held by the contemporary theory of Programme Management.

Colin A. Booth is associate head of Research and Scholarship in the Faculty of Environment and Technology at the University of the West of England, Bristol (UK). He holds the distinguished titles of Visiting Professor of Civil Engineering and Visiting Professor of Sustainability at prestigious international universities. He is the author/co-author of 8 books and ~180 scientific papers. His research interests include sustainability in the built environment, environmental management in construction, property level flooding, sustainable drainage systems (SuDS), climate change adaptation strategies, built environment studies and urban pollution.

John Bruen trained and qualified as Charted Architect in the UK and Ireland. He holds an additional Masters in Construction Project Management and a PhD in Architecture. He has trained and qualified as a Certified Passive House designer with the Passive House Institute in Darmstadt, Germany. He runs his own architectural practice focusing on contemporary energy-efficient design in a variety of contexts and has contributed to the Construction Project Management Masters in Queens University, Belfast.

Alan Bugg joined the faculty of the McWhorter School of Building Science at Auburn University in 2016. Prior to joining the faculty at Auburn, he worked for the U.S Army Corps of Engineers in a variety of positions for over 33 years. Mr. Bugg earned a bachelor's degree in agricultural engineering in 1983, a master's degree in Business Administration in 2003 and a master's degree in Building Construction in 2011, all from Auburn University. Mr. Bugg is a registered Professional Engineer in the State of Alabama, a certified Project Management Professional (PMP) and a certified DBIA Design-Build Professional.

Richard Burt trained and qualified as a Chartered Building Surveyor in the UK. He holds a master's degree in Construction Management and a PhD in Architecture from Texas A&M University. He is currently the McWhorter Endowed Chair & Head of the McWhorter School of Building Science at Auburn University in Alabama.

He currently serves on the Board of Directors of the AGC Education and Research Foundation and the Board of Trustees of the American Council for Construction Education. Dr. Burt also serves as the co-coordinator of the International Council for Building (CIB) Working Group WG089—Education in the Built Environment.

Colin Butlin graduated with distinction from the MSc Strategic Project Management course at Leeds Beckett University, and is a senior practitioner with extensive project practice experience in the development and delivery of capital projects in the UK.

John Lester Clarke, BSc (Hons), PGCE, MSc, PhD is currently deputy module leader at University of Estate Management in Reading. He has worked as a public and private sector building surveyor and on several low environmental impact buildings in the UK and overseas, utilizing strawbale, adobe, rammed earth and timber framed structures and other sustainable materials and has taught a broad range of built environment topics in schools, colleges and universities to a wide variety of learners. His teaching and research interests are in developing and promoting economic, social and environmental sustainability related to the built environment.

He has an honours degree in Building Surveying, a master's degree in Renewable Energy & Architecture and a Postgraduate Certificate in Education. He completed his PhD entitled "Sustainable Buildings: Sustainable Behaviour?" in 2013. This focused on how sustainable buildings, throughout their design, construction, operation and use, impact on sustainable construction practices and the behaviours of key stakeholders.

Fidelis Emuze, PhD is professor and head of the Department of Built Environment and head of the Unit for Lean Construction and Sustainability at the Central University of Technology, Free State (CUT), South Africa. Lean construction, health, safety, and well-being and sustainability constitute the primary research interest of Dr. Emuze, who is a National Research Foundation (NRF) funded researcher that have published over 200 research outputs in the last 7 years. Dr. Emuze is the editor of *Value and Waste in Lean Construction* (published by Routledge), *Valuing People in Construction* (published by Routledge), and co-editor of *Construction Health and Safety in Developing Countries* (published by Routledge). Dr. Emuze is a member of editorial advisory boards of international journals including the ISI indexed Proceedings of the Institution of Civil Engineering–Municipal Engineers. He is a member of the Association of Researchers in Construction Management, and the Board of Directors of the Engineering, Project, and Production Management association (EPPM-Association). Dr. Emuze is the international coordinator of CIB TG59—People in Construction task group.

Thomas Franzen has 11 years' experience in engineering design and recently graduated with a Bachelor of Engineering (Civil) with First Class Honours at the University of Southern Queensland (USQ) and was awarded the University Medal. Tom also holds a Bachelor of Business Administration and a Diploma of Civil Construction Design. He was awarded the 2018 IPWEA (NSW) David Abbott Young Public Works Leader of the Year. David Thorpe is associate professor (Engineering/Technology Management) at USQ, where he has a strong interest in sustainable engineering management. Prior to joining the university, he worked in engineering design, construction and research management.

Mark Gazzard is a Chartered Building Surveyor and Chartered Building Engineer and is a former postgraduate student in the Faculty of Environment and Technology at the University of the West of England, Bristol (UK). His research interests include construction technology, project design and implementation, sustainable infrastructure and environmental management.

Peter Gudde is a researcher at the University of Suffolk looking at the sustainability of the UK energy transition. He has been involved professionally in sustainability for over 25 years, currently working in the public sector supporting energy infrastructure projects. He was a director of a solar company and in the due diligence team during its acquisition. He was a founder of the Suffolk Climate Change Partnership and in 2006 wrote one of the first local authority Climate Action Plans in England. He is a Chartered Environmentalist and Waste Manager and holds degrees from the Universities of Durham and Cranfield.

Rajat Gupta is director of the Oxford Institute for Sustainable Development at Oxford Brookes University. Rajat's research interests lie in evaluating building performance from a socio-technical perspective, local energy mapping, smart energy systems and scaling up energy retrofits. Rajat is currently PI of a £1.5 million UK-India EPSRC-DST project on residential energy reduction in India (RESIDE). He is also Co-I of £8 million EPSRC Energy Revolution Research Consortium (EnergyREV: Core), PI of EnergyRev Plus project on *smart energy tools* and lead academic in the £13.8 million Innovate UK Local Energy Oxfordshire (LEO), a smart local energy system demonstrator. He is a member of EPSRC and ESRC peer review colleges.

John Heathcote trained as a civil engineer and delivered a broad range of engineering and business improvement projects in a 22-year career in practice. He holds an HNC in civil engineering, and an honours degree in project management and an MBA from Sheffield Hallam University's Sheffield Business School. He now lectures and researchers into the management of projects, based at Leeds Beckett University, where he has been for 17 years. John has been a visiting lecturer at the University of Applied Sciences Wuerzburg-Schweinfurt. Chairing the APM's Value Management Specific Interest Group for 5 years promoting an interest in projects as an open system concept, John's research takes an interest in reconceptualizing normative project management to exploit their latent value, something that might be key to creating a more sustainable future.

Samaa E. Helmy is trained and qualified architect in Egypt. She graduated from Cairo University with BSc Arch in 2014 and holds a MSc in Architecture–Biomimicry Architecture in 2018 from the same university. She is registered in the PhD degree programme at Department of Architecture, Faculty of Engineering at Cairo University since March 2019. Ms. Helmy is currently teaching assistant at Architectural Engineering and Technology Programme (AET) at Cairo University and also teaching assistant at Architectural Department, Misr University for Science and Technology (2017–Present). In addition, she is assisting Professor Aboulnaga in many projects and courses, namely Smart Building (2016–present). She received a 3-week summer school training on "Sustainability and Energy Conservation in Architecture" at the University of Lincoln in the UK (2012).

Siyabonga Jikeka currently works for the National Department of Public Works, South Africa, as a director: Construction Project Management.

He obtained a National Diploma in Building Science from the then Eastern Cape Technikon in April 2004, a B-Tech in Quantity Surveying in April 2009, and a Master of Science in Built Environment: Property Economics and Valuations in April 2014 from the then Nelson Mandela Metropolitan University.

Siyabonga is a registered Professional Quantity Surveyor with the South African Council of Quantity Surveying Profession.

Shaherah Jordan is a third year doctoral candidate at the University of East London. Her work focuses on different factors that could impact electric vehicle adoption in London including the built environment and socio-economic demographics. Her thesis focuses on cultural attitudes to money and finance as a potential factor in electric vehicle adoption. She holds an MBA from the University of East London.

Hadi Kazemi is a senior lecturer and researcher into project and construction management at Leeds Beckett University. He holds Civil Engineering and Construction Management degrees from the University of Surrey and Sheffield Hallam University.

Nokulunga Mashwama is an assistant lecturer in the Department of Construction Management and Quantity Survey at the University of Johannesburg. She earns Masters in Construction Management from the University of Johannesburg. She is a PhD candidate in Quantity Surveying with the University of Johannesburg. She has published a number of conference papers. She has worked for the industry for 9 years as quantity surveyor with Owen Thindwa and Associate Chartered Quantity Surveyors and Construction Companies as a site quantity surveyor.

Barry McAuley is a Chartered Construction Project Manager and full-time lecturer in Digital Construction and Engineering within the School of Multidisciplinary Technologies at Technological University (TU) Dublin. This role involves the teaching and development of new academic programmes within architecture, engineering, and construction (AEC) analytics, Building Information Modelling, and Digital Construction related courses. He is also the current programme chair for the MSc in applied BIM and Management course. Prior to his current position, Barry spent a number of years working in the construction and facilities management sector, which enabled him to develop his managerial skills through employment in a number of diverse roles. He completed a PhD in 2016, which focused on using BIM to demonstrate how early integration of Facilities Management professionals into the design team can result in reducing life cycle costs. On completion of his PhD, Barry spent 2 years working as the primary postdoctoral researcher on the CitA Lead Enterprise Ireland funded BIM Innovation Capability Programme of Ireland. As a result of his research to date, he has had a significant body of work published through a combination of industry reports, conference proceedings and journal papers. This has resulted in Barry been named one of "Autodesk 40 under 40 Construction Champions of 2019" which acknowledges and recognizes extraordinary professionals under 40 years of age within the global AEC industry.

Michela Menconi holds a master's degree in architecture, from Florence University, in Italy, where she specialized in Building Conservation and Restoration Technology. She has worked in different practices, in Italy and in Germany, on a range of projects, from new design to adaptation and repair of heritage buildings, for the public and private sector. She is currently doing her PhD on retrofit strategies for traditional listed dwellings (TLDs), in School of Environment and Technology, University of Brighton, where she has also worked as a part-time lecturer for the past 7 years. Her main interests include modelling and energy simulation for analysis of energy performance of TLDs and heritage conservation.

Paul Mundy trained and qualified as a building surveyor, working in the UK construction sector for 10 years. Having undertaken a master's degree in Sustainable Development in Practice at the University of the West of England, he is currently researching sustainable construction within the UK, to bring research and experience together. Working in the role of senior lecturer in Building Surveying at the University of the West of England and using his research to influence his teaching provides graduates with sustainability literacy to take forward and shape the industry in their practice.

Niamh Murtagh is a Senior Research Fellow at the Bartlett School of Construction and Project Management, University College London (UCL). An environmental psychologist by discipline, Niamh's focus has been on applying insights from psychological theory to a sustainable built environment, investigating the psychological underpinnings of pro-environmental behaviour in construction professionals including planners, architects and builders. Niamh was managing guest editor for a Virtual Special Issue of the *Journal of Cleaner Production on Sustainable and Resilient Construction*, and has published in, and is a regular reviewer for, journals including *Journal of Environmental Psychology, Frontiers, and Environment and Behavior*.

Dean Myers My name is Dean Myers and I am a Sustainability Officer with a Registered Provider of social housing in West Yorkshire. My interests are energy efficiency and renewable energy technologies. I am a 5th Year PhD student with Leeds Sustainability Institute at Leeds Beckett University studying overheating risk in the UK's first large-scale Passivhaus retrofit. The aim of my research is to inform improvements in design and installation of future low-energy buildings. I was awarded the Building Performance Evaluation award at the SEEDS Conference 2019 for my latest conference paper.

Mpho Ndou is a trained and qualified professional quantity surveyor. He holds a master's degree in Construction Management and is currently a PhD candidate at the University of Johannesburg, South Africa. He is affiliated with both the local and international Professional Counsels of Quantity Surveying with a membership status of Pr.QS and MRICS, respectively. Similarly, he also serves as an academic assessor at the Royal Institution of Chartered Surveyors (RICS).

His research interests are in Sustainable Development, Environmental Science, and Ecological Economics.

Chibuisi Chinasaokwu Okorieimoh is a trained teacher and a doctoral research student in the School of Electrical and Electronic Engineering at Technological University Dublin, Ireland, under the supervisions of Professors Brian Norton and Michael Conlon. He holds a Diploma in Computer and Basic Physical Sciences, Bachelor of Science Education in Physics, master's degree in Solar Energy Physics and Material Science (all from University of Nigeria) and Master (Mobility) in the master's degree programme of Physics from University of Valladolid, Spain. He was among the 2019 Global Teacher Awardee.

He is currently a part-time assistant lecturer in the School of Multidisciplinary Technologies at Technological University Dublin, Ireland.

Gloria Osei is a PhD student at Anglia Ruskin University, Engineering and the Built Environment, and has a master's degree in civil engineering from City University, London. Gloria has also obtained an associate lecturer position teaching undergraduate Architect students.

Alice Owen trained as an engineer and worked in policy development before completing her PhD; she is now associate professor of Business, Sustainability and Stakeholder Engagement at the Sustainability Research Institute in the University of Leeds; Dr. Kathryn Janda trained in architecture and is a Principal Research Fellow in Organisations and Non-Domestic Buildings in the Bartlett School of Environment, Energy and Resources at UCL. Dr. Kate Simpson trained as a building surveyor and worked in further education before becoming a researcher at the Turing Institute, as part of Imperial College London. All three authors work in pragmatic, interdisciplinary and high impact ways.

Noel Painting, BSc Grad Dip Arch MRICS has been at the University of Brighton since 1992 and is currently course leader of the BSc Quantity Surveying degree. His research is in design, procurement and cost management on which he lectures at undergraduate and postgraduate levels. Noel was a member of the ARCOM scientific committee from 2006 to 2012 and is currently an external examiner at Leeds Beckett University. He has validated courses internationally and has been part of six CIAT accreditation panels for BSc Architectural Technology courses.

He also carries out consultancy for a major music festival promoter with a focus on health and safety.

Federica Pascale is a senior lecturer in Architectural Technology, Anglia Ruskin University. Federica was involved in Phase 2 of The Health and Care Infrastructure Research and Innovation Centre (HaCIRIC), which was a collaboration between existing research centres at Imperial College London and the Universities of Loughborough, Reading and Salford.

Mehmet Koray Pekeriçli, MSc, PhD has received his Bachelors in Architecture at Middle East Technical University in year 2000. He received British Council Chevening TEV Scholarship and received a master's degree in Construction Management at Reading University, UK, in the year 2000. With the support of NSF and Turkish Higher Education Scholarships he pursued his PhD studies at Carnegie Mellon University, PA/USA between 2003 and 2006. He took part in KIM-Grand Challenge Project as a researcher at Reading University between 2006 and 2011, and completed his PhD thesis there. He has been teaching and conducting research at Middle East Technical University since 2011. His interest areas cover Construction Management, Building Information Management, Sustainable Buildings, Intelligent Buildings, Smart Cities, Responsive Environments, and Building Informatics.

Neil Pickavance is a staff Senior Construction Planner (Scheduler) in the UK construction industry, currently at the UK construction arm of the world's largest non-state-owned construction and engineering company. With 25 years' full-time experience at Tier 1 contractors Neil has held senior and lead planner/scheduler roles on numerous major UK projects. Technically qualified in building and construction, a full member of the Chartered Institute of Building (MCIOB), and holding a master's degree in Project Management (Construction) from Nottingham Trent University, Neil is currently undertaking a part-time MPhil/PhD Built Environment at Liverpool John Moores University, investigating construction project management cultures.

Poorang Piroozfar is a Principal lecturer (associate professor) of Architectural Technology, the academic program lead for BSc (Hons) Architectural Technology and MSc Digital Construction, and the director of Digital Construction Lab at School of Environment and Technology, University of Brighton. People, environment, information and technology are at the heart of Poorang's research where different methods are triangulated to formulate the most viable solutions. He has numerous research outputs, and together with his research team, he has won several research grants and awards for his research on building performance and evaluation, building energy simulation, facility management of complex heritage buildings, responsible and future-proof energy retrofits, life cycle assessment of buildings, double-skin and PV integrated façade systems as well as values and frames in design for sustainability.

Alison Pooley is currently the deputy head for the School of Engineering and the Built Environment, Anglia Ruskin University. Alison has worked as a housing officer in East London and was in architectural practice for several years before going on to teach at the Centre for Alternative Technology, Wales, where she taught on postgraduate programmes for nearly 10 years.

Kenneth M. Rampou is an assistant director—Real Estate Investment Services at the Department of Public Works and Infrastructure in the Republic of South Africa and serves as a member of the Regional Green Building Technical Committee. The author has an avid interest in green buildings and the circular economy and is currently studying towards and MSc in Project Management, having recently completed a BSc Honours in Quantity Surveying at the Nelson Mandela University, with research co-supervision under the guidance of Ms Sharon Dent and Professor Geritt Crafford.

Josie Rothera is the course director for MSc Civil Engineering at Leeds Beckett University. Josie is a Chartered Engineer with the Institution of Civil Engineers, an Associate of the Association of Project Management, a Fellow of the Higher Education Academy and an Early Career Researcher. Josie has also recently been appointed as a Panel Member for the Institute of Apprenticeships and Technical Education. Josie has over 10 years' experience of project site management on contracts over £10M, often the sole female representation on the delivery team. Josie is a company director of STEER Support and Mentoring CIC, which supports undergraduates in transitioning into Built Environment careers.

Lloyd Scott is a professor of Sustainable Construction Technologies and head of the Knowledge in Sustainable Construction Research Group at the Technological University Dublin (TU Dublin), Ireland. He is also vice-chairman of the Board of the Association of Researchers in Construction Management (ARCOM) and chair of the Sustainable Environmental Engineering in Design for Society (SEEDS) Conference (2018–2020).

He is a visiting professor of Practice in Sustainable Construction at the College of Architecture, Oklahoma University, USA (2014–), and an active member of the editorial board of Mass Timber Construction journal (2018–). His research focuses on building performance, mass timber construction and sustainable knowledge and expertise in the sAEC sector. He has collaborated on planning/design build student competitions and studies with the University of Oklahoma and the Milwaukee School of Engineering. The competitions have been part of the Associated Schools of Construction (ASC) program, where he served as a board member of the ASC (2011–2018).

Eric Simpeh is currently a lecturer and the acting head of Department of Construction Management and Quantity Surveying at Cape Peninsula University of Technology. He holds a PhD degree in Construction Management from Nelson Mandela University, a master's degree in Construction Management and Bachelor of Technology in Quantity Surveying from Cape Peninsula University of Technology. Eric's first venture into research was in 2012 when he developed a rework probability model for assessing rework occurrence in construction projects. His areas of research interest include sustainable construction and resource optimization in construction.

John Smallwood is the professor of Construction Management in the Department of Construction Management, Nelson Mandela University, and the principal, Construction Research Education and Training Enterprises (CREATE). Both his MSc and PhD (Construction Management) addressed construction health and safety (H&S). He has conducted extensive research and published in the areas of construction H&S, ergonomics, and occupational health (OH), but also in the areas of health and well-being, primary health promotion, quality management, and risk management.

Lilian Nwanyisonde Smart is currently a PhD research student in the School of Art, Design and Architecture, Global Disaster and Resilience Centre, University of Huddersfield, United Kingdom. She holds an MRes in Government and International Affairs, Durham University, and a master's degree in International Relations, University of York, all in the United Kingdom. Lilian has worked as a teaching assistant as well as research assistant, Global Disaster and Resilience Centre, University of Huddersfield, United Kingdom. She was the chairperson of a Postgraduate Research conference on Sustainable Development held in the University of Huddersfield, UK. Has won several awards including the prestigious presidential honours award. She is currently working with the Student Union, University of Huddersfield, on a project under the Inspiring Future Leaders program as the project manager of BAME students' mental health.

John P. Spillane is an experienced research focused lecturer in Construction Management at the University of Limerick, with a demonstrated history of working in both higher education and the construction industry. As a Chartered Construction

Manager, working within the remit of Digital Construction, Sustainability, Dispute Resolution and Confined Site Construction, John is currently working on a number of research projects throughout Ireland and the UK.

John L. Sturges began work in the Foundry Industry in the West Midlands, before going to Aston University, where he obtained a Junior Research Fellowship in the Department of Metallurgy. From there he took up a research post in the Department of Mechanical Engineering at the University of Leeds, pursuing work on the dynamic flow and fracture of materials. He moved to Leeds Beckett University as senior lecturer in Materials Science in the School of Civil Engineering and Construction. While there, he pioneered the teaching of environmental impact and sustainability to final year courses across the school. He presently serves as visiting professor in the Leeds Sustainability Institute.

Ali Murat Tanyer, MSc, PhD, LEED AP BD+C; O+M After graduating from the Department of Architecture of Middle East Technical University (METU), Ankara—Turkey, in 1997, he started his master studies at the same department. He completed his master's degree in 1999 in construction management field. He started his graduate studies at the Salford University, United Kingdom, in 2000 and obtained his PhD in 2005. He worked as a post-doc researcher in European Union (EU) Framework 6 (FP6) funded INTELCITIES research project.

His main areas of expertise are construction information technology and sustainability and how they support built environment. He has been a full-time instructor since 2006 and currently working as a full professor at METU. He has been lecturing in *Construction Management, Professional Practice for Architects*, *Construction Informatics*, and *Sustainability in Construction*. He has several publications in refereed conferences, journals and books.

Didi Thwala is a professor with the university of Johannesburg, and A-rated researcher. He is a SARCHI chair holder, and an HOD for the Department of Construction Management and Quantity Surveying. He has published a number of conference papers and journal articles and a book on Health and Safety. He has a PhD from the University of Johannesburg. He is an Editor in Chief of the Journal of Construction Project Management and Innovation (JCPMI).

Su Taylor is the Dean of Research for the Faculty of Engineering and Physical Sciences, a Professor of Structural Engineering and lead the Intelligent infastructure group at Queen's University Belfast.

Blake Wentz is chair of MSOE's Civil and Architectural Engineering and Construction Management Department. He is vice-president of the Associated Schools of Construction (ASC), the national professional association of educators and industry practitioners working together for the development and advancement of construction education. Wentz joined the MSOE faculty in 2005. He holds a bachelor's degree in finance and master's degree in engineering, both from the

University of Nebraska-Lincoln, and a PhD in technology management from Indiana State University. From 2009 to 2014, he served as MSOE's construction management program director. More recently he has served as owner's representative for MSOE's Grohmann Tower renovation project and for the university's new Dwight and Dian Diercks Computational Science Hall project, which will break ground this spring. He is a LEED Accredited Professional and American Institute of Contractors (AIC) Certified Professional Constructor (CPC).

Stephen Wilkinson has been in higher education for 34 years, having been a principal lecturer at Leeds Beckett University. His interests range from Robotics and Automation to serious VR, including hand operations for carpal tunnel syndrome; refuse recycling autoclaves and WW1 prisoner camps.

He has co-authored books on Engineering and has presented many papers. His qualifications are BSc (Hons), PGCE, MSc, PhD and CEng MIET. He is a Senior Fellow of the HE academy.

He is currently a fellow of the Leeds Sustainability Institute at Leeds Beckett University and has presented four papers at recent SEEDS conferences on sustainability.

Müge Yüksel is a project architect and research specialist at Yüksel Proje Int. Co. with over 4 years of experience in planning and detailed design of Railway Systems, including international projects. She has completed her bachelor's degree at İzmir Institute of Technology with interests in BIM, parametric design and sustainability. Projects she has involved in professionally have been processed with BIM. She is currently an MSc student at the Building Science Programme of Middle East Technical University, studying on "Sustainability Assessments of Railway Systems".

Part I
Sustainable Development and Urban Spaces

The Significance of Social Sustainability

J. L. Sturges

Introduction

Sustainability first emerged as a matter of concern at least five centuries ago when people expressed concern about the impact of mankind's consumption of natural materials and its impact on the environment. These concerns centred on timber, a very visible resource. Some of the earliest literature on silviculture was produced in Germany 500 years ago (Caradona, 2014). In England, the Woodlands Act was passed by parliament in the reign of Henry VIII, as there was widespread concern about the depletion of forests. Timber was being cut for building, for ship-building, for fuel and for charcoal-burning for iron production. In Japan Tokugawa Ieyasu founded the last Shogunate in 1603, and an early measure instituted was one to protect Japan's forests which were rapidly becoming depleted at that time. These concerns show us that sustainability first had an environmental and resource-depletion element, and this is still the case today.

We have come to realise that there are two other equally important dimensions to sustainability, namely the economic and social ones. Elkington (1997), for example, refers to the 'triple bottom line', where economic prosperity, environmental quality and social justice are the three elements. The need for social sustainability is logical, as it is people and their economic activities that impact the world's environment. Elkington points out that conventional economics places no monetary value on natural materials and resources such as air and water, which are assumed to be free and freely available. He makes the point that sustainability cannot be achieved unless social sustainability is also achieved. In describing social sustainability, he quotes from Gladwin (1996), who calls for a paradigm shift in our current attitudes. This will involve a transformation of human values, its political values, and normal behaviour to:

J. L. Sturges (✉)
Leeds Sustainability Institute, Leeds Beckett University, Leeds, UK
e-mail: j.sturges@leedsbeckett.ac.uk

L. Scott et al. (eds.), *Sustainable Ecological Engineering Design*,
https://doi.org/10.1007/978-3-030-44381-8_1

> *economic efficiency towards social equity, from individual rights to collective obligations, from selfishness to community, from quantity to quality, from separation to interdependence, from exclusion to equality of opportunity, from men to women, from luxury to necessity, from repression to freedom, from today to tomorrow, and from growth that benefits a few to genuine human development that benefits us all.*

These words set out a vision of society moving in the opposite direction from the one we live in at present. We can compare them with the findings from the case studies.

After sharing the Earth with other, earlier hominins, *Homo sapiens* became the dominant species around 30,000–40,000 years ago. Emerging from Africa, *Homo sapiens* spread to Europe, Asia and America by 15,000 BC, eventually even colonising small islands in the Pacific Ocean. Human societies are complex adaptive systems, and form in response to the local conditions where they take root. The earliest societies evolved sustainable lifestyles, generally in harmony with their local conditions. If so, these societies must have achieved Social Sustainability, in a way that has been lost, or at least no longer exists. Today, in 2019, the fact that we no longer live sustainably is a matter of great concern, and the question arises; do we have any evidence of societies that lived sustainably among all the diverse societies in the world? How do these societies relate to the present global situation and what can we learn from them?

Writers on sustainability have recognised the three elements of environment, economy and society, and represented a sustainable situation in the form of a Venn diagram with three overlapping circles as shown in Fig. 1. Our present world is on a non-sustainable trajectory, and this could similarly be represented by a second Venn diagram where the circles do not overlap, as depicted in Fig. 2.

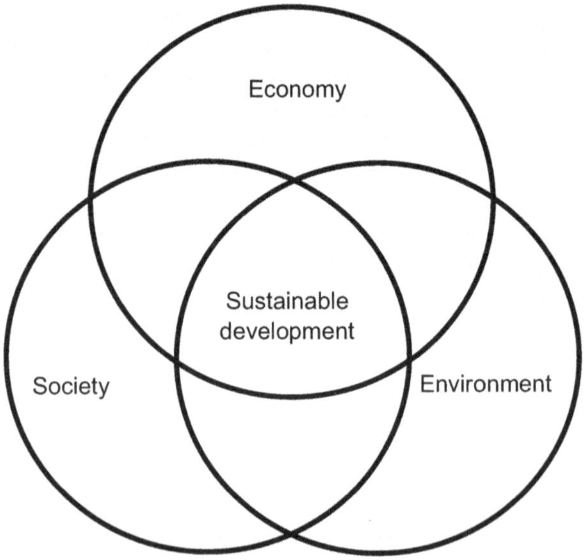

Fig. 1 Idealised picture of sustainability, with the environment, the economy and society being in harmonious relationship

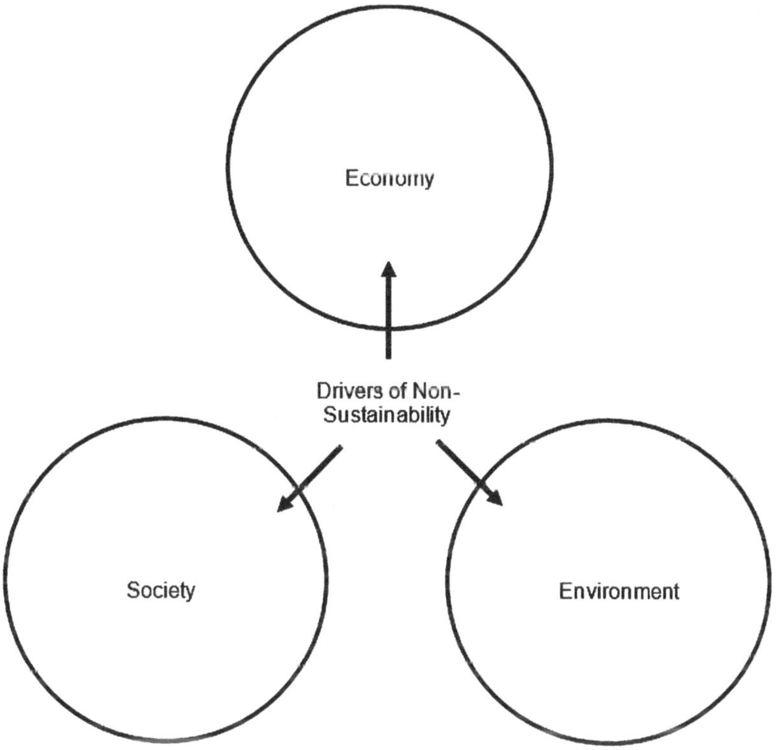

Fig. 2 Picture of our current non-sustainable situation where the environment, the economy and society are seriously out of harmony

Development of Human Society

The evolution of human society has been punctuated by a series of events that have changed its course. When *Homo sapiens* emerged as potentially the dominant life form on Earth around 200,000 years ago, he adopted the role of hunter-gatherer to survive. Then around 10,500 years ago, he began to adopt Agriculture. Just over 500 years ago, Spain and Portugal carved out empires for themselves in Central and South America and in Asia, and adopted the role of mercantile capitalists. Three hundred years later came the Industrial Revolution when industrial capitalism became the human role. Finally, following World War II came the evolution of consumer capitalism. Throughout history, man has always sought to maximise his access to energy, in whatever form. Table 1 below summarises the stages in human development, giving approximate dates, population sizes and *per capita* energy consumption. This table summarises the situation for most, but not all, of the human population of our world. When much of the world had reached the industrial and consumer stage, there were still a few remote societies living in the agricultural and hunter-gatherer modes.

Table 1 Showing transitions in mode of living, including dates, global population and per capita energy consumption

Mode of living	Date	Global population (millions)	Per capita energy consumption (W)
Hunter-gatherer	200,000 years ago	Ancestor group	300
Farmer	10,500 years ago	5	2000
Mercantile capitalist	1500 AD	500	2200
Industrial capitalist	1800 AD	950	4000
Consumer capitalist	1950 AD	2500	8000

Source: Lewis and Maslin (2018)

Each transition from one mode of living to the next is brought about by an increase in the amount of available energy, an increase in information available and an increase in the scope for human action. Although dates are quoted in Table 1, these transitions were not overnight changes, but took some time to become established. In addition, not all peoples and societies made all the transitions; this was determined by their remoteness or proximity to the great centres of population. However, at each stage, the adoption of the next stage gave rise to an increase in population, *per capita* energy consumption and an increase in CO_2 in the atmosphere. The transition from hunter-gatherer to farmer involved forest clearance to make way for fields to grow crops. The total number of trees on Earth at the hunter-gatherer to agriculture transition was around 6 trillion, whereas the number on Earth today is around 3 trillion (Crowther et al., 2015).

Having outlined the development of human society since the dawn of history, it is interesting to examine the scale of human energy and power consumption at various points over the past three millennia. Table 2 below taken from the work of Smil (2017) clearly illustrates the exponential rise in energy use by mankind during this time.

Newcomen's engine is an interesting item in this list as it represented the first time in history that someone had devised a way of converting fossilised sunlight in the form of coal into mechanical power and Lovelock (2019) suggests this event as the start of the industrial revolution. The engine was installed to pump water out of a deep coal mine near Dudley in Worcestershire. This event really should be linked with another development that occurred only about 20 miles away in the Severn gorge at Coalbrookdale when in 1708 Abraham Darby first smelted iron using coke (non-renewable) instead of charcoal (a renewable material). The industrial revolution could not have happened without a plentiful supply of iron, and iron production had been severely restricted by the limits placed on the cutting of timber to make charcoal. The ability to use coke 'liberated' iron-making from dependence on

Table 2 Power ratings from a candle to global power consumption

Examples and dates	Power (W)
Small wax candle burning (800 BCE)	5
Egyptian boy turning Archimedean screw (500 BCE)	25
Dutch Tread-wheel powered by 8 men (1500)	800
Newcomen's atmospheric engine pumping water (1712)	3750
Large Dutch windmill draining a polder (1750)	12,000
Calder Hall nuclear reactor (1956)	202,000,000
Rocket engine launching Saturn C5 rocket (1969)	2,600,000,000
Global commercial energy consumption (2015)	17,530,000,000,000

Source: Smil (2017)

timber and helped stimulate the demand for coal and the need for pumps to keep mines clear of flooding.

We shall now examine various societies that have achieved sustainability and where we have the evidence, what effect contact with western civilisation subsequently had on them.

Examples of Long-Term Sustainability

In finding case studies of sustainable societies, the work of Diamond (2006) and Norberg-Hodge (2000) has been most useful. These authors' treatment of the case studies was quite discursive, and so the main points will be summarised here. These societies were in various ways cut off from the main centres of population (Tikopia, New Guinea, Ladakh) or decided to cut themselves off from external contact (Japan) or never adopted agriculture and reliance on the land at all (NE Pacific Indian tribes of Alaska).

Tikopia

- Small Pacific Island, near Vanuatu and Solomon Islands, area 1.8 square miles.
- Continuously inhabited for 3000 years.
- Population kept at around 1200 by various birth control measures.
- The people lived by agriculture and fishing.
- Society organised on a bottom-up basis.
- They have exercised excellent stewardship of their environment.

New Guinea

- Large island in East Indies, first visited by Europeans four or five centuries ago.
- People have lived there sustainably for around 45,000 years.
- Until 1930s the interior was never explored and was believed to be uninhabited. Interior inhabited for thousands of years, with people living sustainably and independently from the rest of the world.
- Society organised on a bottom-up basis.
- Very sophisticated agricultural techniques developed.
- They have exercised excellent stewardship of their environment.

Japan

- Large island archipelago east of China, completely cut off from the rest of the world during the Tokugawa Shogunate from 1603 until its end in 1868.
- Christian missionaries ejected and foreign traders only allowed access to small island in Nagasaki Bay by order of the Shogun.
- Society organised on a top-down basis, and it enjoyed peace and prosperity for the duration of the Shogunate.
- No external wars were fought during the Shogunate.
- Population very stable during the Shogunate.

Ladakh

- Land-locked area under the Karakoram, in the trans-Himalayan region of Kashmir.
- It is a 2000-year-old kingdom, an area of Tibetan Buddhism, originally of Tartar herders who have adopted agriculture.
- People live in villages, largely organised on a bottom-up basis.
- They have adopted policies to limit and control their population size.
- They have adopted a very collaborative culture.
- They have exercised excellent stewardship of their environment.

Indian Tribes of North East Pacific in Alaska (Studied by Dr. Jago Cooper)

- These tribes are interesting because they never adopted agriculture, rather they have always depended upon fishing.

- They never developed a sense of ownership of land in the way that the previous four peoples mentioned above did, and they survived the impact of Europeans very much better than the Indian tribes living to the South and East in the USA and Canada.
- They have retained their traditional skills, for example, building wooden houses and making boats from tree trunks.
- They have exercised excellent stewardship of their environment.

Discussion

We have examined several societies that have lived sustainably for periods of time ranging from a couple of centuries to many millennia. Some were island communities, some located inland, all living by farming and just one set of coastal communities living by fishing. Because of their remoteness and inaccessibility they remained untouched by western civilisation until well into the twentieth century. The detailed information that we have is a result of prolonged contact made by a few people who took a great interest in them, who lived with them, befriended them and recorded their observations in detail. The work of Firth in Tikopia (1936, 1939) and Norberg-Hodge in Ladakh (2000) typify this approach.

Most of these societies achieved sustainability by a bottom-up approach, i.e. they did it by collective decisions. Tokugawa Japan was the exception, and this was due to the remarkable character of Tokugawa Ieyasu, who combined a certain military genius with an outstanding capacity for wise statesmanship. The leadership that he gave and the decisions that he took ensured that the Japanese enjoyed a good standard of life and freedom from warfare for over two centuries, and Japan benefits from his legacy to this day.

Much of the literature on sustainability and the future of human civilisation today makes frequent reference to the problems posed by the Earth's burgeoning population. It is the multiplier that exacerbates all our problems. Remarkably, all the societies described in this study solved this problem, taking steps to ensure that there were never too many mouths to feed. Furthermore, following contact with western civilisation, these societies became non-sustainable, and they suffered rapid population growth. Norberg-Hodge (2000) describes this process very well. We have seen that human society passed through various historical transitions to reach its present form. Drawing on what we have learned from the case studies outlined above, we can see that the transition from hunter-gatherer to agriculture did not lead to a non-sustainable society though it did give rise to a modest increase in population. The next transition from agriculture to mercantile capitalism did produce a society that was ultimately non-sustainable. It involved an increasingly intensive agriculture and the growing of cash-crops using slave labour. To make way for this, forests were cleared on an increasing scale, leading to an increase in atmospheric CO_2. The initial impact of the Spanish and Portuguese conquest of their American empires was to cause a population crash among the native Indians who succumbed to the European diseases against which they had no immunity. The number who died is not known but median estimates of around 50 million have been quoted (Lewis & Maslin, 2018). The immediate result of this was that Indian farmland reverted to forest bringing about a reduction in atmospheric CO_2. This may have helped bring about the mini 'ice age' noted in Europe. However, once the Spanish and Portuguese began growing cash-crops with slave labour, the situation was reversed again. These events illustrate just how sensitive our world is to changes that in former years were thought to be insignificant.

The transition to industrial capitalism was enabled by the increasing use of non-renewable sources of materials and energy. Agriculture became increasingly mechanised, leading to migration of redundant farm workers to the towns which became centres of manufacturing industry. The overall social cohesion in the form that could exist in rural village communities was completely lost, and people could no longer take responsibility for their lives. We have now reached the point where over half of the world's population live in cities. Cities are covered in concrete and asphalt, which act as solar-powered storage heaters, leading to the 'heat island' effect. In this age of consumer capitalism, millions of tonnes of manufactured goods are transported around the globe in huge container ships, causing more atmospheric and marine pollution. Each development leads us further down the path of non-sustainability. Globalisation means that most nations around the world are now following the western development model, but the chances of achieving global agreement on steps to reverse global warming are vanishingly small. We have entered a 'development trap' and it is not clear how we can escape.

The work on Ladakh reported by Norberg-Hodge (2000) is in general agreement with the descriptions of Tikopia, New Guinea and Japan reported by Diamond (2006), but it has the advantage of being a primary source. In it, she gives a very balanced and finely nuanced account of the impact of the west, pointing out all the advantages and drawbacks of each society. Both authors refer to the fact that western men automatically assumed a position of cultural superiority to the 'undeveloped' societies they discovered without taking time to study and understand them. This was usually unjustified, and in a few cases fatal. In New Guinea, westerners saw the vertical drainage channels used on the terraces where yams and sweet potatoes were grown. They 'knew' this was wrong and persuaded a few of the islanders to use horizontal channels. These retained water and during the next heavy rainfall the whole terraced system, crops and all were washed down the valley and into the river. Flannery (2019) also reports a conversation with New Guinea islanders. They were observed planting fruit and nut trees which took several decades to produce food, but which were known to attract game animals. The New Guinea people were asked why they did this when they would not live to obtain the benefit, and their reply was that they did it to ensure that their grandchildren had something to eat. This is inter-generational thinking, part of sustainable living. They were wiser than the westerners asking the questions. In 1845, Sir John Franklin led a well-resourced expedition to find the North West passage with 134 men. The crew had some contact with the Inuit whom they looked upon as savages, and they all perished because of this attitude. The Norwegian, Roald Amundsen, the man who succeeded, took the trouble to live with the Inuit for some months prior to his attempt and learned how to survive the Arctic conditions.

Clearly these apparently primitive societies had developed highly sophisticated agricultural practices. Their techniques were not wasteful and did not rely on huge tonnages of artificial fertilisers as are employed by western farmers. Soil fertility was maintained in a non-polluting way without the degradation produced by western 'industrial agriculture'. Modern, industrial agriculture is thermodynamically less efficient than traditional farming, as illustrated in Table 3 below.

The data is from the USA and relates to the growing of maize, a high-energy food crop. It shows a comparison between the total energy inputs and food energy outputs for the years 1950 and 1970. This is particularly interesting because 1950 was the year that industrial agriculture took off. The first thing to notice is that labour input was halved, while food energy output more than doubled, i.e. productivity has been quadrupled. This is the basis for claiming that this type of farming is more 'efficient' than the traditional methods. However, the figures also show that the ratio of energy input to output is lower than the traditional route (2.82 as against 3.18). This represents lower thermodynamic efficiency. The other factor to notice is the tremendous increase in the use of artificial fertilisers (figures for nitrogen, phosphorus and potassium) and insecticides and weed killers. It must be remembered that these figures relate to 1970, nearly 50 years ago, and industrial agriculture is pursued more intensely today. Another factor is run off from agricultural land during rainfall. The excess fertilisers end up in streams and water courses, rivers and eventually the seas, where they cause algal blooms and eventually dead zones. The cost in biodiversity loss and loss of fish stocks is never taken into account, because

Table 3 Production of maize

Energy input type	1950 (MJ/acre)	1970 (MJ/acre)	% Change
Labour	41	21	−49
Machinery	1047	1758	+67
Gasoline	2578	3336	+32
Nitrogen	527	3938	+647
Phosphorus	64	197	+208
Potassium	44	285	+548
Seeds	169	264	+56
Irrigation	93	142	+52
Insecticides	5	46	+820
Weed killers	3	46	+1433
Drying	126	502	+298
Electricity	226	1298	+474
Transport	126	293	+132
Total energy input	5049	12,126	+140
Maize—food energy—Output	16,034	34,177	+113
Ratio output:input	**3.18**	**2.82**	

Energy inputs and outputs for the years 1950 and 1970
Data supplied by Prof. D. Bradley (2005), F.R.S., University of Leeds

neither the fertiliser manufacturer nor the farmer sees the results of this type of agriculture. The manufacturer of industrial chemicals just looks at his bottom line. The farmer just looks at the costs of operating his farm. There is no overall oversight, no one can take responsibility for the system, and this illustrates the disconnection between action and results that typifies the western world. This situation did not exist in sustainable societies; we have entered a 'development trap' of global proportions and therefore we do not have social sustainability.

This example is drawn from agriculture and not from construction and building, but it illustrates very clearly the importance of social sustainability. The evolution of mercantile capitalism, industrialisation and the consumer society leading to a human population explosion have combined to drive the economy, society and the environment out of alignment.

Summary of Conclusions

Certain conclusions can be drawn from the case studies of sustainable societies, as follows:

- These societies were free from outside influences.
- They had a finite area of land available to them.
- They knew that there was an upper limit to the amount of food they could produce, and this knowledge placed an effective upper limit on their population size.
- They devised strategies to keep their populations at a stable size.
- They adopted cultural attitudes that emphasised collaboration rather than the exercise and assertion of individual rights to ensure the survival of their societies.
- They exercised good stewardship of the environments in which they lived.

In the western, modern world we have placed too much emphasis on the economic bottom line, over-emphasised individual freedom and rights, and placed no emphasis on responsibilities.

- These societies have adopted a culture based on individual liberty and the exercise and assertion of individual rights.
- Little or no emphasis has been placed upon individual responsibilities or obligations to society.
- People are unable to take responsibility for their own lives. They increasingly rely on what others provide and cannot know where things come from, or whether they are obtained sustainably or in ways damaging to the environment.
- Because of this detachment, blame for the ills of society is often placed on the wrong people and money and resources are devoted to symptoms rather than causes.
- Western societies have exercised 'freedom to pollute' and did not exercise good stewardship of their environments.
- Adoption of western ways of living has always led to a large population increase.

The main conclusion to be drawn from this must be that without the achievement of social sustainability, overall sustainability cannot be achieved. Everyone in society must be involved, and this involvement must ultimately be global. This is often referred to as the age of 'globalisation', but ironically, we have globalised the drivers of non-sustainability, but not the social sustainability which is of such vital importance.

Acknowledgements The author acknowledges the helpful discussions and encouragement given by his colleagues Mike Bates and Chris Gorse.

References

Bradley, D. (2005) Energy inputs and outputs for the years 1950 and 1970, Data obtained and adapted through private communication, University of Leeds, accessed 2005, unpublished.

Caradona, J. L. (2014). *Sustainability. A history*. Oxford: Oxford University Press.

Crowther, T. W., Glick, H. B., Covey, K. R., Bettigole, C., Maynard, D. S., Thomas, S. M., … Bradford, M. A. (2015). Mapping tree density at a global scale. *Nature, 525*, 201–205.

Diamond, J. (2006). *Collapse. How societies choose to fail or survive*. London: Penguin Books.

Elkington, J. (1997). *Cannibals with Forks. The triple bottom line of 21st century business*. Oxford: Capstone Publishing Ltd..

Firth, R. (1936). *We, the Tikopia*. London: George Allan and Unwin.

Firth, R. (1939). *Primitive Polynesian economy*. London: George Routledge and Sons.

Flannery, T. (2019). *Europe. A natural history*. London: Allen Lane.

Gladwin, T. N. (1996). Socio-economic challenges of social sustainability. In *Third Senior Executives Seminar, "Sustainability and Profitability. Conflict or Convergence?" HRH Prince of Wales' Business & the Environment Programme, September 15 19, 1996*. Cambridge: Cambridge University.

Lewis, S. L., & Maslin, M. A. (2018). *The human planet. How we created the anthropocene*. London: Pelican Books.

Lovelock, J. E. (2019). *The Novacene*. London: Allen Lane.

Norberg-Hodge, H. (2000). *Ancient futures* (Revised ed.). London: Rider, Ebury Press.

Smil, V. (2017). *Energy and civilization. A history*. Cambridge, MA: MIT Press.

Urbanisation and the Built Environment: Exploring How the Built Environment Can Enhance the Health and Wellbeing of the People Living in Urban Slums

Lilian Nwanyisonde Smart, Dilanthi Amaratunga, and Richard Haigh

Introduction

According to the report from UN-Habitat (2015), one in eight people live in slums resulting in a total of 1 billion people living in slums conditions around the world today. Despite the argued progress in improving slums and preventing their formation which led to a decrease from 39% to 30% of urban slum dwellers in developing countries between 2000 and 2014, the absolute numbers have continued to grow. Shockingly, projections also show that this number could rise to about 2 billion over the next 30 years, and 3 billion by 2050 (Moir, Moonen, & Clark, 2014; UN-Habitat, 2003). The UN-Habitat (2015) sees the slum issue as a huge challenge that remains a critical factor for the persistence of poverty in the world. These are crucial revelations on the persistent growth of slums, a growth that has continued to result to an increased health and wellbeing challenges. Unfortunately, the population of the people residing in urban slums increases with the increased rate of urbanisation. According to Krefis, Augustin, Schlünzen, Oßenbrügge, and Augustin (2018), rapid urbanisation itself contributes to making the health and wellbeing of the people an increasingly distinguishable challenge.

The report from the World Health Organisation (WHO, 2017), however, sees high rate of urbanisation as a welcomed development, stating that growing cities play a crucial role in promoting and protecting the health and wellbeing of the people. While Bai, Nath, Capon, Hasan, and Jaron (2012) acknowledge that health and wellbeing status is better in urban areas than the rural areas, they assert that cities contribute hugely to human health and wellbeing challenges instead of promoting and protecting it. Following the review of some studies Bai et al. (2012) argue that most advantages of the cities especially those that have to do with the health and wellbeing of the people can be windswept by the adverse challenges emanating

L. N. Smart (✉) · D. Amaratunga · R. Haigh
The University of Huddersfield, Huddersfield, UK
e-mail: lilian.smart@hud.ac.uk; d.amaratunga@hud.ac.uk; r.haigh@hud.ac.uk

L. Scott et al. (eds.), *Sustainable Ecological Engineering Design*,
https://doi.org/10.1007/978-3-030-44381-8_2

from the urban environment. Pineo and Rydin (2018) argue that the most difficult of these challenges facing the world today is how to accommodate the teeming urban population in a way to ensure long-term health and wellbeing of the people and to provide them the opportunity to live a fulfilled life.

With this, Pineo and Rydin (2018) have identified the lack of adequate accommodation as well as the inability to provide the residents with opportunities to live a fulfilled life as part of the factors that escalates the health and wellbeing challenges in the urban centres. It is arguable that some of the immediate effects of this lack of adequate accommodation in these cities could be a high cost of accommodation and an increased population in urban slums environments where the accommodation cost is presumably affordable. Accordingly, the need to conduct detailed research on the health and wellbeing of this increased number of people living in such deplorable environment called slums emerges.

Krefis et al. (2018) note that several disciplines, including public health, urban planning, and natural sciences, have conducted remarkable research on the link between urban areas and health and wellbeing of the people. Different institutions such as World Health Organization (WHO), the London School of Economics and Political Science (LSE), and the Healthy People 2020 Organisation that monitors health-related issues researched the health and wellbeing of the urban residents. Additionally, the International Council for Science (ICSU) came up with plan on facilitating health and wellbeing in the changing urban environment, focusing on the importance for shaping cities for health (ICSU, 2011). Moreover, Krefis et al. (2018), who researched most of these body of literature, point out that the major recommendations from the above institutions basically centre on improving urban health and wellbeing in order to reduce health inequalities and to build capacity on national and regional levels (see also WHO, 2017). These elaborate studies show the importance of providing a health and wellbeing-enabling environment for the people residing in the cities. However, we observed from the review that these studies do not cover issues on the health and wellbeing of people who reside in urban slums built environments, meaning that the researchers seem silent on this crucial aspect of research. Accordingly, this study focuses on the health and wellbeing of the slum dwellers and their built environment. The aim is to demonstrate how to design and structure the urban slum built environments to enhance the health and wellbeing of the residents. The question emanating from the aim of research is: how can an urban built environment be structured and designed to enhance the health and wellbeing of the residents? With this question, we will be able to identify the exact ways the built environment can be restructured and redesigned to enhance the health and wellbeing of the people in urban slums.

Therefore, to be able to answer this question, this study is divided into sections; the section after this introduction is the methodology in brief. Followed by the section on the concept of urbanisation, this will immediately be followed by definitional perspectives on the term urban slums and the concept of the built environment. The section following will be the review of literature centring on the health, wellbeing, and the built environment. This will be followed by the discussion of the characteristics of the health and wellbeing-enhancing built environment. After this, there will be a section on how to structure the urban slums built environment to enhance the health and wellbeing of the residents. Then the conclusion of the study.

Methodology

This methodology section describes actions taken to investigate the research problem, detailing the processes of data generation. Instead of the use of the qualitative, quantitative, or mixed-method research design, which uses such methods as interviews, focus group discussion, experiments, and surveys, an integrative literature review method was employed in this research. To address the topic which centres on exploring how the built environment can enhance the health and wellbeing of the people living in urban slums, the researcher reviewed various relevant pieces of literature on the topic area, mainly to find out the current debate in this area and identify the research gap. The literature was sort through different search engines. Some of these search engines include the university summons, google scholars, Business Source Premier (EBSCO), Emerald Management e-Journals, and Science Direct (Elsevier) electronic databases.

After searching and gathering the literature materials, about 50 different literature, they were read intensely, analysed, and critiqued, and about 27 of the work that are most relevant to the study were synthesised and used to be able to come up with an argument in this study. The reason for the use of this method is because, first, there is elaborate recent literature on the built environment, urbanisation, urban slums, and other keywords of the research which is capable of providing the researcher with recent on-going debates in the area. This is important because, to be able to find the gap in the literature, there is need to review literature massively to understand the trending arguments around the topic as well as ascertain the areas that are less researched. Secondly, the researcher has no intention to do a fieldwork data collection for the research presently but solely depended on the literature-based data. Moreover, the extensive literature review was able to help the researcher to achieve the research aim and answer the research questions posed, thereby addressing the identified research gap.

The Concept of Urbanisation

There is no stereotype definition of the concept of urbanisation, as some scholars define it from their perspectives. According to McGranahan and Satterthwaite (2014), urbanisation generally involves the shift in population from rural to urban settlements. To make this very clear, McGranahan and Satterthwaite (2014) state that urbanisation does not occur when urban and rural populations grow at the same pace, that for it to take place, the urban population of the cities must outgrow that of the rural areas. This explanation entails that urbanisation can only occur when the share of a country's population that resides in the cities outweighs its rural counterparts. Based on this, urbanisation is defined as the increased total number of a country's population that resides in the cities as against their rural counterparts.

According to Pineo and Rydin (2018), city residents have access to parks, public transport, healthy food and other amenities that support health and wellbeing, and the concentration of opportunities and services in urban areas is advantageous to the health and wellbeing of urban residents. Pineo and Rydin (2018), however, argue that the densely populated living and working conditions in cities equally create conditions for the spread of pollution and diseases in the cities. The last line of argument shows that the same urban built environment equipped with facilities that can help to improve the health and wellbeing of the occupants can equally generate health and wellbeing challenges especially if the urbanisation is rapid and unplanned. According to Krefis et al. (2018), rapid urbanisation aggravates the already complicated and prevalent health and wellbeing challenges in our urban centres. This scenario implies that as the urban population grows, the urban built environment becomes unconducive for the enhancement of the health and wellbeing of the people due to density and pressure in the built environment. It could be on this basis that Burdett and Taylor (2011) and Krefis et al. (2018) conclude that the urban environment provides some of the best as well as some of the worst environments for health and wellbeing. Perhaps, this is because, according to Pineo and Rydin (2018), there are different life-threatening communicable diseases present in the urban centres which pace of transfer from one person to another increase with the increase in the urban population.

Bai et al. (2012) also identified that apart from the prevalent communicable diseases in fast-growing cities, there is also the rise of other chronic and non-communicable diseases in the cities. The latter diseases result from unhealthy urban lifestyles such as physical inactivity, unhealthy diets, tobacco smoking, and harmful use of alcohol. The physical inactivity and unhealthy diet can be said to be mostly a consequent of the features, designs, and structure of the built environment. According to Bai et al. (2012), other health issues ravage the urban environment, which also increase with the pace of the urban population growth, and they include the following:

- Prevalent infectious diseases resulting from overcrowding in substandard living conditions and urban squalors—this one is very rampant in urban slums built environment.
- Diseases associated with industrial pollution—acute and chronic diseases such as respiratory disease and pulmonary cancer.
- Some injuries that result from motor vehicle collisions, violence, and crime.

From the above, it is clear that rapid urbanisation hugely contributes to most of the health and wellbeing challenges that bedevil both the main cities and the urban slum areas. The argument is that this rapid urbanisation mounts significant pressure on the built environment resulting to such issues like overcrowding, urban congestion, and increased population in urban slums. This pressure on the concerned built environments can be seen as perhaps the most spectacular of the consequences of rapid urbanisation. No wonder Bai et al. (2012) argue that one of the effective ways to resolve these problems can more probably be through addressing the urban environmental issues rather than focussing narrowly on healthcare facilities. Indeed, this a good suggestion which further justifies the reason why this study focuses on the built environment, the health and wellbeing of the people in urban slums. Therefore, how can we make sense of the concept of urban slums?

What Is Urban Slum?

Ezeh et al. (2017) statistically demonstrates that enormous slums characterise most of the cities in many developing countries today, with the population of slums increased massively in the past 60 years. The United Nations Educational Scientific and Cultural Organisation (UNESCO) defines urban slums as urban spaces characterised with inadequate housing and essential services (UNESCO as quoted in Ezeh et al., 2017). This explanation may mean that any urban area that lacks both adequate housing and basic amenities is classifiable as urban slums. This, however, may not be a sufficient explanation of what an urban slum entails. Ezeh et al. (2017) state that the most used and pertinent definition of slums is the one given by the United Nations Human Settlements Programme (UN-Habitat), a definition based on households. According to UN-Habitat, the slum household is defined as a group of individuals that live under the same roof with the lack of one or more of the following conditions: access to improved water, access to improved sanitation, sufficient living space, the durability of housing, and secure tenure. According to Ezeh et al. (2017), the health of the slum dwellers is mostly jeopardised by such factors as inadequate water supply, sanitation, and drainage and the lack of rubbish collection in a crowded environment which influences obstinate diarrhoea and diseases such as typhoid, hookworm, and cholera that characterises slum settlements.

Ezeh et al. (2017) advance that an unfortunate aspect of the slum settlements is that children are specifically vulnerable because of such factors as low breastfeeding rates, under-nutrition, and poor sanitation, which predispose children to chronic diarrhoea, impaired growth, and cognitive development (please see Ezeh et al., 2017 for more details). According to UN-Habitat (2003), slums are the physical and spatial manifestation of urban poverty and intra-city inequality. Regarding abject poverty in urban slums, Ezeh et al. (2017) state that the health and life of the people in urban slums are usually under threat whenever they get ill because of the lack of extra cash to take care of their health. This is a clear indication of the fact that slums are poverty-stricken places with poverty being one of the primary reasons why people reside there. In fact, Ezeh et al. (2017) raise the unfortunate historical case of the marginalisation and stigmatisation that urban slum dwellers face, and how they experience displacement, expropriation of property, and the denial of access to essential services for a long time.

Arguably, the definition of urban slums provided above is grounded in the condition of the slums built environment and the health and wellbeing challenges of the people resulting from the said environments. However, it is important to state here that urban slums are not to be seen as an informal settlement because even in Sustainable Development Goals (SDGs), informal settlements and slums are treated

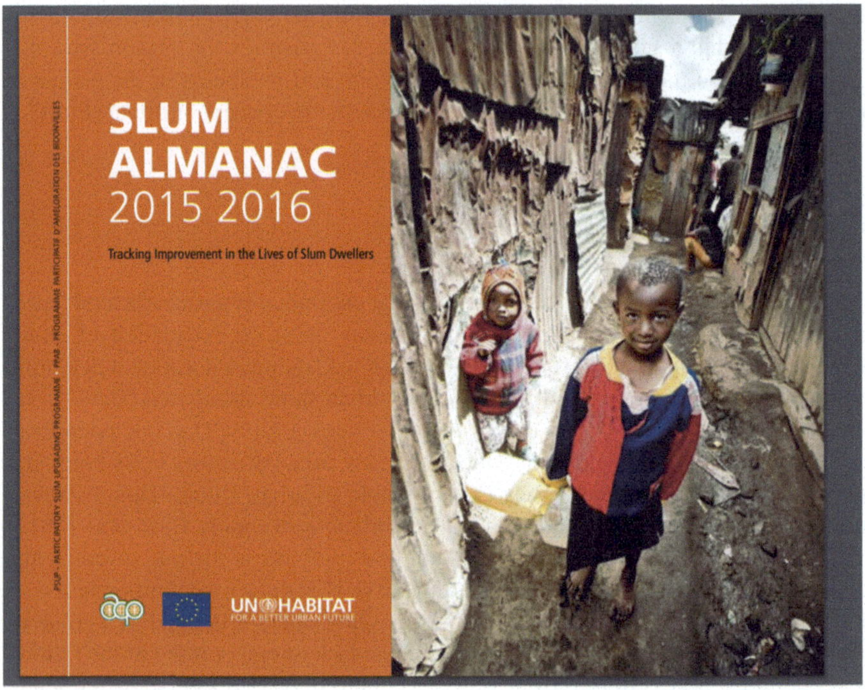

Pictorial representation of urban slums. (Adapted from UN-Habitat, 2015)

as two different terms (Ezeh et al., 2017). However, like informal settlements, urban slum increases mostly with an increasingly urban population, which means that as the population in the cities increases, the population in urban slums increases even more rapidly, despite the deplorable conditions of slums.

Indeed, the condition of the slum dwellers is pathetic, and this may be why the UN-Habitat (2003) advised that tackling this situation hugely depends on the rate at which urban policies should continue to aim at using better housing policies to create safer cities for the urban low-income population, especially slum dwellers. Urban planning and management policies should be designed to prevent the emergence of slums, to create cities without slums, and to even resort to slum upgrades with new urban planning strategies. This suggestion is essential considering the longstanding consequences of living in urban slums environment. However, it is worrisome that despite these strong recommendations given in 2003, recent studies still demonstrate that adequate attention has not been paid on creating an urban slum built environment that can facilitate the health and wellbeing of the people (Ezeh et al., 2017). It is worrisome mostly because according to the review, it is detrimental for people to continue to live in urban slum built environments. That gets one wondering why the government, policy makers, and building professionals seem to neglect enacting effective policies that will help to create habitable cities that are also void of any manner of slums. For the sake of clarity, the factors that characterises urban slums built environment are represented in Fig. 1.

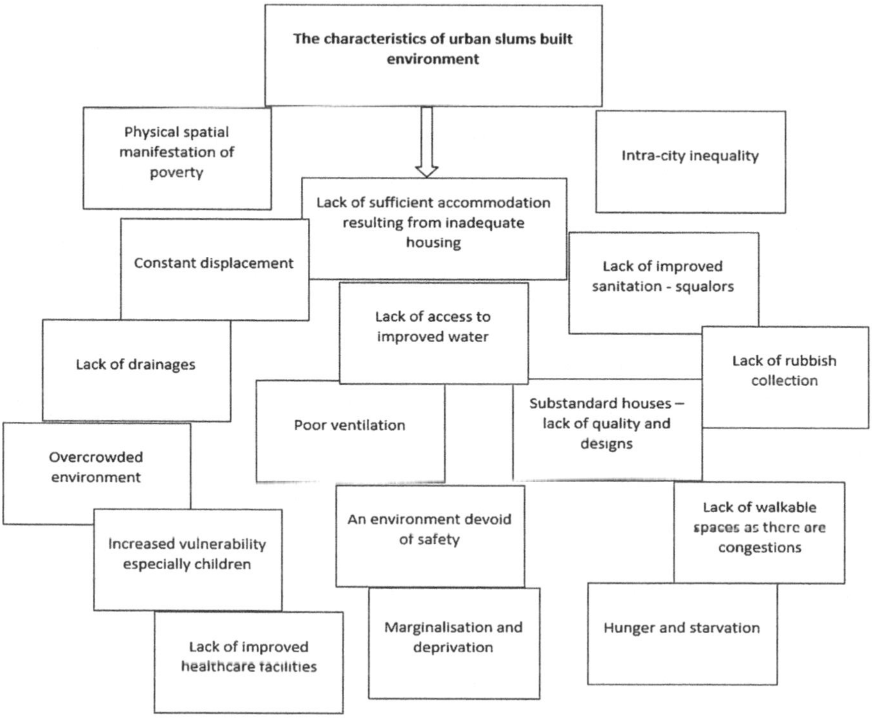

Fig. 1 A framework demonstrating most of the factors characterising the urban slum built environments

A built environment characterised by these factors will continue to be inimical to the health and wellbeing of the people. However, not all built environments have the full blast of these characteristics; it is more prevalent in urban slums. No wonder Ezeh et al. (2017) assert that there is a need to differentiate between the health of the people who reside in a well-built and structured built environment from the health of the people who reside in urban slums. According to them, this distinction should be mainstreamed in the implementation of the Sustainable Development Goals and the New Urban Agenda. In fact, the above framework as well as the pictorial representation of an instance of urban slum is a pointer of the significant differences between the slum and some of the well-designed cities we know of—the major differences is in their built environment. At this point, it becomes important to discuss the concept of the built environment.

The Concept of the Built Environment

According to Barton (2009), a built environment has to do with the planned and structured aspects of our surroundings, which include buildings, transit routes, and parks. Frank and Engelke (2005) define the built environment as an environment with all the physical structures in which we live, work, travel, and play such as houses, apartments, offices, parks, streets, shopping centres, parking lots, factories, superhighways, transit stations, and so on. To Glasgow Centre for Population Health (2013), the built environment is the physical structures engineered and designed by people to live, play, and socialise. While Williams (2013) acknowledges these definitional perspectives of the built environment which according to him comprises hard infrastructure like houses; he argues that there is no need for the neglect of the role of soft infrastructure like walkable routes in the built environment. According to him, these routes encourage interactions as well as make goods and services accessible. The inference from the assertion of Williams is that there should be a connection between the natural environment and the built environment to create a more conducive environment. Writing about this Glasgow Centre for Population Health (2013) adds that the connection between the built infrastructural spaces and a range of natural features should be an integral part of the built environment. Indeed, the above has called our attention to the fact that the built environment is not just the place we live and work, it is much more than that and should comprise carefully styled hard and soft infrastructure, which is a combination of the built and the natural spaces.

The importance of designing a carefully styled built environment that has the above qualities is paramount because it will positively influence people's lives (Bergman, 2018). Bergman (2018) advance that the built environment, on the whole, plays a vital role in influencing people's lives and their overall performances. Williams (2013) states that the components of the built environment affect our daily decisions and the way we live our lives. Further to this, Frank and Engelke (2005) explains that the technique used to design and build our environments has significant impacts on the decisions we make, our health, and quality of life. Moreover, Williams (2013) confirms that the design and layout of the built environment can significantly contribute to our psychological and physiological health and wellbeing. Thus, these authors have reminded us of the connection between the built environment, the health and wellbeing of the people as well as our daily decisions and possible productivity. From this, one can conclude that most health and wellbeing challenges that people face are traceable to the environment they found themselves.

These scholarly definitions above show that the built environment is supposed to be a well-planned, duly structured, conscientiously engineered, and nicely designed quality environment where people can comfortably live, work, play, socialise/interact, travel, walk, and spend their entire lives. In addition, it is arguable that every built environment is made by the people and for the people's habitations and comforts, and its designs and features have a considerable impact on the health and wellbeing of the occupants. In all, the built environment possesses specific attributes that can either hamper or enhance the health and wellbeing of the people depending on some factors. Moreover, the definitions show that building professionals do have a significant role to play in getting the built environment right because it is more dangerous not to get it right. Getting it right here means ensuring that there are not only places to live, but also places to work, play, socialise/interact,

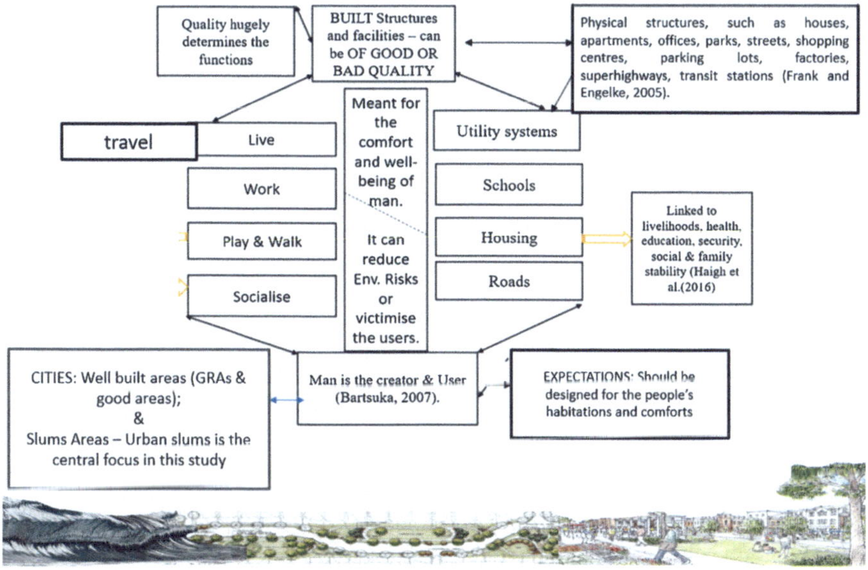

Fig. 2 Framework that demonstrates the meaning of the built environment

walk, and travel as stipulated by the scholars above. Below is a framework demonstrating the meaning of the concept of the built environment (Fig. 2).

Unfortunately, the discussion on urban slums above seems not to reflect the meaning of the built environment spelled out in the above framework, no wonder the health and wellbeing of the slum dwellers remain at stake more than that of the people residing in a more conducive built environment. With these enumerations, it is essential to discuss the health and wellbeing of the people in the context of the built environment. This section will help us to understand correctly the connection that exists between health, wellbeing, and the built environment.

Health, Wellbeing, and the Built Environment

The World Health Organization (WHO) defines health as a state of complete physical, mental, and social wellbeing of the people, not just the absence of disease. According to Barton (2009), this definition explained health in its wholesome perspective, associating it with the social, economic, and the environmental aspect of life. Pineo and Rydin (2018) show that the urban environment has long been recognised as an essential determinant of the health and wellbeing of the residents. A situation where the more significant part of the health and wellbeing of the people hugely depends on their environment. The definitions have shown us that health may be seen as physical, mental, and social completeness, which is associated with the social, economic, and environmental aspects of life. From this, we can conclude that social, economic, and environmental factors could trigger health and wellbeing-related challenges. However, this study looks at environmentally induced health and wellbeing challenges in urban slums. According to Dodge, Daly, Huyton, and Sanders (2012), the term wellbeing is a growing area of research, and the question of how it should be defined remains unanswered. Notwithstanding, Dodge et al. (2012) propose that wellbeing is the balance point between an individual's resource pool and the challenges they faced. To them, the individual resources that determine his/her wellbeing includes psychological, social, and physical forces. Some other scholars said wellbeing of such factors as happiness, the satisfaction of life, and quality of life explains the wellbeing of an individual (Bai et al., 2012; Kjellstrom et al., 2007).

The above has provided us with some of the vital concepts with which to explain the health and wellbeing of the people. Perhaps the most striking of it all is the fact that what affects the health of an individual may equally affect their wellbeing and vice versa. For instance, people with mental, social, and environmental issues may as well lose their happiness, satisfaction of life, and quality of life and vice versa. With this, it is arguable that the best way to enhance the wellbeing of the people is to equally enhance their health and quality of life and vice versa.

Therefore, wellbeing in this study is defined as a state of mind enveloped with peace, happiness, satisfaction emanating from psychological soundness and impressive social, physical, and or environmental factors. This, in the real sense, means that psychological, social, as well as environmental factors have huge effects on people's wellbeing. Our interest in this study, however, is the effect of the environmental factors (built environment) on the health and wellbeing of the people. The recent research conducted by Pineo and Rydin (2018) demonstrates that chronic health conditions that impose high costs on countries are rising globally, that in the UK alone, these health conditions accounts for about 70% of the spending on health and social care. According to Pineo and Rydin (2018), this alarming rate of chronic health conditions in the UK, which are strongly influenced by the built environment, is preventable. Pineo and Rydin (2018) also suggest that this uncalled-for expenses on health issues is preventable if the built environment is redesigned and reconstructed in such a way it can help facilitate the health and wellbeing of the people. The absence of this may result in the inability to avert this rapid health issues. Indeed, this is a strong recommendation that needs implementation without any form of delay to at least prevent the health and wellbeing challenges associated with the built environment that are preventable according to Pineo and Rydin (2018).

Some scholars are clear on the fact that the built environment has a strong influence on our health and majorly determines the wellbeing of the people (Barton, 2009; Freeman, Thompson, & Jalaludin, 2011). The relationship between the built environment, health, and wellbeing of the people is becoming clear through these various arguments, unlike the argument of Bai et al. (2012), which states that the relationship between wellbeing and health and the urban built environment is unclear and needs detailed explanation. The supportive role of the built environment for human health is a fast-growing area of interdisciplinary research, evidence-based policy development, and other related practices (Kent & Thompson, 2012).

In addition, the increasing link between the built environment, physical, mental, and social wellbeing of residents has caught the attention of public health professionals and land use planners (Majid, Cox, & Wu, n.d.). The reason is not farfetched, and it is perhaps because human beings spend their lives in the built environment, like their houses, offices, and so on. Haigh, Hettige, Sakalasuriya, Vickneswaran, and Weerasena (2016), writing about housing, clearly show that housing is an essential and complex asset linked to livelihoods, health, education, security, and social and family stability. Barakat (2003) equally shows that housing is essential to the well-being and development of societies; that it is a complex asset, which has links to all of those listed by Haigh et al. (2016).

From the preceding, it is clear that there is a link between the built environment, health, and wellbeing of the occupants. It is also clear that the built environment can affect the health and wellbeing of the people positively or negatively. The inference is that in order to enhance the health and wellbeing of the urban residents, the role of built environment is paramount. This is not to pretend that there are no other factors that affect people's health and wellbeing negatively; the focus in this study is on the built environment. While it is crucial to acknowledge the clarity and consensus of these scholars on the impact of the built environment on the health and wellbeing of the people, what seems missing is a clear indication of how the built environments can be designed and structured to enhance the health and wellbeing of the occupants in urban slum environments. Indeed, this should be an essential part of a study of this sort. Thus, it is vital to discuss the possible factors that characterises a health and wellbeing-enhancing built environment before proceeding to discuss how these factors can help to facilitate the health and wellbeing of the people.

The Supposed Characteristics of the Health and Wellbeing-Enhancing Built Environment

It is crucial to start this section by stating that the built environment is everywhere both in the rural areas, main cities, informal settlements, and urban slums. Although there may not be any perfect built environment, its impacts on the health and wellbeing of the occupants depend hugely on how developed the area is and the features, designs, and structure of the built environment. This means that there may be a level of health and wellbeing issues connected to every built environment no matter the location. However, this study does not pretend to cover all these areas. Its focus is on the urban slums built environment. Without a doubt, understanding the impacts of the built environment on the health and wellbeing of the people who reside in urban slums should be of paramount importance to the policymakers, governing authorities, as well as the building professionals. The reason is that there are several conditions in urban slums built environment that can threaten the health and wellbeing of the people, and some of these conditions require urgent eradication (please see above for the description of the conditions in urban slums). Moreover, it is essential to state here that the conditions in an urban slum built environment can be said to be much more deplorable when compared to the built environment in most of the main cities.

At this point, it is necessary to discuss the possible factors that can be associated with the urban built environment. Majid et al. (n.d.) assert that there are numerous factors in the built environment which can hugely affect the quality of life and wellbeing of an individual and these factors include pollution—air and water, the natural areas, and public green spaces. Pineo and Rydin (2018) confirm that lack of air pollution, the presence of green, and walkable spaces for increased physical activity are among the factors that should be found in a built environment. Writing on the need for physical activity in a built environment, Williams (2013) suggests that for a built environment to be capable of promoting physical activity it should make provision for safe pedestrian routes, connected street networks, ample street lighting, dynamic land-use mix, and recreational centres.

In confirmation of the need for the provision of spaces for physical activity in the built environment and the need for designing street networks, some other scholars add that there should also be walking and cycling routes (Freeman et al., 2011; Kent & Thompson, 2012). That also, there should be the creation of an environment for social cohesion. Thus, the expression of social cohesion has emerged in this study, and it is vital to explain it before moving forward, albeit in brief. According to Dempsey (2008), the study of social cohesion or 'the social glue of a society' is a long-standing study which centres in examining society and social relations in a variety of social settings. To Mannakkara and Wilkinson (2013), a cohesive society is a society that 'hang together', partly through social interaction. According to them, this integration of individuals can partly be achieved through residents interacting with one another and getting to know neighbours, and the absence of these results in people being stuck in loneliness and living their separate lives in the same area (this is social isolation).

These elaborations have shown that it is good for the built environment to enhance social cohesion instead of social isolation because social isolation can have devastating effects on health and wellbeing. For instance, Kent and Thompson (2012) found out that social isolation and obesity are among the significant risk factors for many of the chronic diseases facing contemporary society. Therefore, instead of social isolation, the built environment must connect to enhance social cohesion amongst communities, to strengthen communities, neighbourhood, social relations, and social networking (Freeman et al., 2011; Kent & Thompson, 2012). Glasgow Centre for Population Health (2013) adds that the designs and the quality of the built environment affects social connections, accessibility, and physical activity levels. Thus, the emphasis on the importance of getting the features, designs, and quality of the built environment right has been laid. Moreover, Pineo and Rydin (2018) state that the urban built environment needs to have reasonable access to healthy food through the reduction of fast-food centres around school environments. There should be the retention of peri-urban agricultural lands to enable the ease assessment of affordable healthy food as well as put measures in place to encourage the establishment of community and farmers gardens. The essence of this is to ensure that people live in a suitable environment as well as have easy access to the required diet from their environment.

Indeed, the above has provided us with some of the crucial factors that characterises health and wellbeing-enhancing built environment. The review has also shown that a built environment is not only the houses we live, the offices or places we work, or the shopping malls, but it is also an environment where we do such things as play, socialise, walk, and cycle. It is an environment that should have natural areas, greenery areas, walkable spaces, reasonable access to healthy foods, and spaces for social cohesion. It can be added that an urban built environment needs to have an improved water supply, improved sanitation, waste management techniques, and enhanced health care facilities. These factors are fully noted in this study. Accordingly, the next section will be the discussion of how the built environment can help to facilitate the health and wellbeing of the people in urban slums and it will further be represented in a framework (Fig. 3).

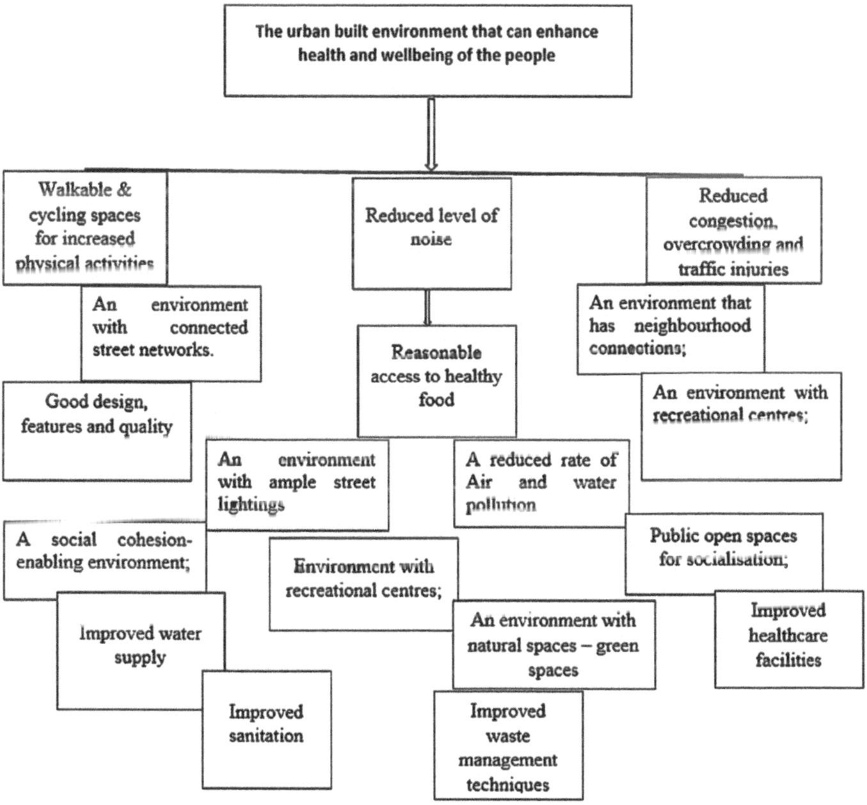

Fig. 3 A framework demonstrating the features of an urban slum health and wellbeing-enhancing built environment

How to Enhance the Health and Wellbeing of Urban Slum Dwellers Using the Built Environment

So far, the literature review has shown the factors that characterise the built environment. Accordingly, this section will centre on how to enhance the health and wellbeing of urban slum dwellers using the built environment. For a smooth line of discussion, most of the points raised in the review above on the characteristics of a built environment will be discussed one after the other, albeit in brief.

Air and Water Pollution

The review above reveals that most urban environments are characterised with air and water pollution, which grossly affects the health and wellbeing of the residents. This situation could even be worse in urban slums characterise by issues such as minimal waste management, poor sanitation, and stagnant water. These pollute the air people breath and the water they drink and as such capable of endangering the health and wellbeing of the people. Therefore, it is strongly recommended that the building professionals and even the government look into the reconstruction, redesigning, and restructuring of the urban slums built environment in such a way that there will be a reduced rate of these high rate of pollution.

The Integration of Natural or Green Spaces to the Built Environment

According to Williams (2013), for a built environment to enhance the health and wellbeing of the people, there is a need to integrate the natural environment to the built environment to make the environment more conducive and stress-reducing instead of a boring and a stress-generating environment. Glasgow Centre for Population Health (2013) adds that the built infrastructural spaces and a range of natural features should be an integral part of the built environment. However, a look at the section that deals with the meaning of urban slums reveals the lack of the integration of the natural spaces to the built environment. The resultant effect is the generation of different health and wellbeing issues. We, therefore, call the attention of the building professionals to revisit the built environment and ensure that it comprises a carefully styled hard and soft infrastructure, built spaces and the natural spaces to produce a habitable environment.

Public Open Spaces for Physical Activities

Scholars above state that for a built environment to enhance the health and wellbeing of the people, there needs to be walkable and cycling spaces for enhanced physical activities. These spaces should also include safe pedestrian routes, connected street networks, ample street lighting, and recreational centres (Williams, 2013). According to Williams, any built environment with the provision of places meant for physical activities can facilitate the health and wellbeing of the people, and this can be vice versa. This is because physical acivities can facilitate the development of muscles and increases physical strength, thereby enhancing the health and mental wellbeing of the people. Therefore, it is arguable, however, that such health and wellbeing-enhancing factors of a built environment are lacking in urban slums. The reason is that the level of congestion and overcrowding of the area with perhaps no enough space to create open spaces, walkable, and cycling routes affects the availability of such facilities. Therefore, it is advisable for the building professionals to put these in place to help eradicate some of the built environment-induced health and wellbeing challenges that are prevalent in slums.

The Features, Designs, and Quality of the Built Environment

The features, designs, and quality of the built environment play a significant role in either enhancing the health and wellbeing of the people or endangering it. Glasgow Centre for Population Health (2013) affirms that it is the design and the quality of the built environment that affects social connections, accessibility, and physical activity levels. According to the report of the Chartered Institute of Building (CIOB), quality is critical and about the greater public good expected from buildings to promote human health, safety, and wellbeing as well as addressing the many social, cultural, environmental, and economic concerns of today. It is arguable from the definition of the urban slums above that its environment lacks no designs, excellent features, and quality and this could raise the height of their health and wellbeing challenges. Therefore, it is recommended for the government and the building professionals to look into providing a built environment that is of high quality, feature, and design to the urban slum dwellers.

Congestion, Overcrowding, and Traffic Injuries

A built environment that is characterised by congestion, overcrowding, and increased traffic injuries such that are found in the urban slums can worsen the health and wellbeing of the inhabitants. Most scholars writing about the meaning of urban slums enlisted some of these factors as the characteristics of the slum areas, a situation that requires the attention of the building professionals to use their capabilities to restructure the area into a more habitable, less congested, and less crowded area so that people's health and wellbeing will be enhanced.

Environment for Social Cohesion

Some scholars above state that there is a need for the creation of an environment for social cohesion, an environment that glue the society together in social relationships and neighbourhood (Freeman et al., 2011; Kent & Thompson, 2012). Social cohesion is achieved when residents interact with one another, getting to know their neighbours and do so many things together and in unity (Mannakkara & Wilkinson, 2013). These interactions and relationships can reduce the risks associated with people suffering from depression and stress resulting from being stuck in loneliness. Therefore, for the built environment in urban slums to be enhance the health and wellbeing of the people, the building professionals should try to put in place an environment that supports social cohesion.

Reasonable Access to Healthy Food

According to Williams (2013), the combination of physical activity with a healthy diet has been found to reduce some chronic diseases like hypertension, delay the onset of partial or total disability, and even reduce mortality levels. He advances that for the built environment to support the health and wellbeing of the people, there is a need for reasonable access to healthy foods and the retention of peri-urban agricultural lands to enable the ease assessment of affordable healthy food (Pineo & Rydin, 2018).

A closer look at the framework representing urban slums built environment above will show that the environment lacks most of these health and wellbeing facilitating factors, no wonder there seems to be increased health and wellbeing challenges in urban slums than in the other urban areas. Below is a representation of most of the factors that can enhance the health and wellbeing of the urban slums dwellers when put in place:

From the above, it is clear that urban slum built environments need to possess the above health and wellbeing-facilitating factors to eradicate the health and wellbeing challenges that people face in urban slums.

Conclusion

This research focuses on exploring the impact of the built environment on the health and wellbeing of the people who reside in urban slums. The essence of this study is to increase recognition and understanding of the built environment as an essential determinant factor in people's health and wellbeing. To also highlight that appropriate decision-making of the building professionals and others can help to ensure the health and wellbeing of the people in slums. The study reviewed the literature on some of the vital concepts such as urbanisation, urban slums, built environment, health, and wellbeing in order to understand the trending argument on the subject area. In the study, the literature on the built environment as well as the health and wellbeing in the context of the built environment were reviewed. The findings show that there is a relationship between the health and wellbeing of people and their built environment. In addition, the results from the review show the various ways the built environment can help to facilitate the health and the wellbeing of the people who live in slums. Most of this literature reveals that urban slums, unlike the rest of the cities, lack excellent infrastructural facilities and almost all the necessary amenities such as suitable housing, drinkable water, and healthcare facilities. It shows that there is a need to redesign and restructure the built environment to include at least most of the enlisted health and wellbeing-facilitating factors (see Fig. 2 above). This mandate is achievable only if the building professionals, urban planners, and policymakers should see this need and work together towards achieving it.

Therefore, this study is a wake-up call to both the built environment professionals and urban policymakers to make policies that centres on redesigning the urban slums and turn the environment into a habitable area capable of enhancing the health and wellbeing of the residents. This study also suggests that we draw from the recommendation of Pineo and Rydin (2018) that the issues of health and wellbeing in cities should neither be a private finance issue nor a public finance issue; it is a far more critical matter that requires collaboration for effective and lasting solutions. Moreover, Bai et al. (2012) affirm that the effective interventions to improve urban health and wellbeing require action that is beyond the services that the formal health sector offers.

According to them, there is need for the engagement of many different societal sectors such as water supply, sanitation, housing, transport, education, and all levels of government—local, provincial, and national to be able to create a built environment that can facilitate the health and wellbeing of people who resides in urban slums. The achievement of the kind of built environment that will enhance the health and wellbeing of the people should be based on collaborative efforts, and like Jakab (2011) rightly summarised, we must all work towards the realisation of health and wellbeing. The implication is that it may be challenging to create an environment that is habitable and enhances health and wellbeing without proper planning, designing, and structuring. This is where the work of the building professionals comes in handy because they require enough time and knowledge to plan, design, and structure the built environment. In fact, the role of professional knowledge when it comes to building construction and delivering a liveable and life-enhancing built environment cannot be overemphasised.

References

Bai, X., Nath, I., Capon, A., Hasan, N., & Jaron, D. (2012). Health and wellbeing in the changing urban environment: Complex challenges, scientific responses, and the way forward. *Current Opinion in Environmental Sustainability, 4*(4), 465–472.

Barakat, S. (2003) *Housing reconstruction after conflict and disaster*, Network Papers (Vol. 43, pp. 1–40). Humanitarian Policy Group: London.

Barton, H. (2009). Land use planning and health and well-being. *Land Use Policy, 26*, S115–S123.

Bergman, J. (2018). *Social cohesion and everyday built environments.*

Burdett, R., & Taylor, M. (2011). *Can cities be good for you.* Paper presented at the Proceedings of the Cities, Health and Well-Being Conference, London School of Economics and Alfred Herrhausen Society in Partnership with the University of Hong Kong, Hong Kong.

Dempsey, N. (2008). Does quality of the built environment affect social cohesion? *Proceedings of the Institution of Civil Engineers—Urban Design Planning, 161*(3), 105–114.

Dodge, R., Daly, A. P., Huyton, J., & Sanders, L. D. (2012). The challenge of defining wellbeing. *International Journal of Wellbeing, 2*(3), 222–235.

Ezeh, A., Oyebode, O., Satterthwaite, D., Chen, Y.-F., Ndugwa, R., Sartori, J., … Watson, S. I. (2017). The history, geography, and sociology of slums and the health problems of people who live in slums. *Lancet, 389*(10068), 547–558.

Frank, L. D., & Engelke, P. (2005). Multiple impacts of the built environment on public health: Walkable places and the exposure to air pollution. *International Regional Science Review, 28*(2), 193–216.

Freeman, E., Thompson, S., & Jalaludin, B. (2011). *Healthy built environments: Stakeholder engagement in evidence based policy making.* Paper presented at the Proceedings of the State of Australian Cities Conference.

Glasgow Centre for Population Health. (2013). *The built environment and health: An evidence review, Briefing paper, 11 concepts series.* Retrieved from https://www.gcph.co.uk/assets/0000/4174/BP_11_-_Built_environment_and_health_-_updated.pdf

Haigh, R., Hettige, S., Sakalasuriya, M., Vickneswaran, G., & Weerasena, L. N. (2016). A study of housing reconstruction and social cohesion among conflict and tsunami affected communities in Sri Lanka. *Disaster Prevention and Management, 25*(5), 566–580.

ICSU. (2011). *Health and wellbeing in the changing urban environment: A systems analysis approach: An interdisciplinary science plan.* Paris: International Council for Science. Retrieved from https://council.science/cms/2017/05/Health-and-wellbeing-in-the-changing-urban-environment-web.pdf

Jakab, Z. (2011). *Presentation: Designing the road to better health and wellbeing in Europe, at the 14th European Health Forum Gastein, October 7, Bad Hofgastein, Austria.* Retrieved from http://www.euro.who.int/__data/assets/pdf_file/0003/152184/RD_Dastein_speech_wellbeing_07Oct.pdf

Kent, J., & Thompson, S. (2012). Health and the built environment: Exploring foundations for a new interdisciplinary profession. *Journal of Environmental and Public Health, 2012*, 95817.

Kjellstrom, T., Friel, S., Dixon, J., Corvalan, C., Rehfuess, E., Campbell-Lendrum, D., … Bartram, J. (2007). Urban environmental health hazards and health equity. *Journal of Urban Health, 84*(1), 86–97.

Krefis, A., Augustin, M., Schlünzen, K., Oßenbrügge, J., & Augustin, J. (2018). How does the urban environment affect health and well-being? A systematic review. *Urban Science, 2*(1), 21.

Majid, M., Cox, S., & Wu, E. (n.d.). Discussion paper: The impact of the built environment on mental wellness.

Mannakkara, S., & Wilkinson, S. (2013). Build back better principles for post-disaster structural improvements. *Journal of Structural Survey, 31*(4), 314–327.

McGranahan, G., & Satterthwaite, D. (2014). *Urbanisation concepts and trends*. London: IIED.

Moir, E., Moonen, T., & Clark, G. (2014). *What are future cities? Origins, meanings and uses*. Compiled by The Business of Cities for the Foresight, Future of Cities Project and the Future Cities Catapult.

Pineo, H., & Rydin, Y. (2018). *Cities, health and well-being*. London: Royal Institution of Chartered Surveyors (RICS). Retrieved from https://www.rics.org/globalassets/rics-website/media/knowledge/research/insights/cities-health-and-well-being-rics.pdf

UN-Habitat. (2003). *The challenge of the slums. Global report on human settlements 2003*. London: Earthscan Publications Ltd. Retrieved from https://www.un.org/ruleoflaw/files/Challenge%20of%20Slums.pdf

UN-Habitat. (2015). *Slum Almanac 2015–2016; Tracking improvement in the life of slum dwellers*. Retrieved from https://www.worldurbancampaign.org/psup/slum-alamanac-2015-2016

WHO. (2017). *Environment and health for European cities in the 21st century: Making a difference* by Laurence Carmichael, Francesca Racioppi, Thomas Calvert, Danielle Sinnett. Retrieved from http://www.euro.who.int/__data/assets/pdf_file/0020/341615/bookletdef.pdf?ua=1

Williams, L. (2013). *Getting to know the built environment as a complex system*. Toronto: Wellesley Institute.

Future Cities: The Role of Biomimicry Architecture in Improving Livability in Megacities and Mitigating Climate Change Risks

Samaa E. Helmy and Mohsen M. Aboulnaga

Introduction

"The best way to predict the future is to design it" (Buckminster Fuller). Megacities face colossal challenges, mainly high energy use, urban air pollution due to transport, large number of inhabitants' activities and natural resources in all sectors. In cities, air is the main component of atmosphere all humans, animals, plants and other living organisms that depend on it for survival. According to the World Health Organisation (WHO), at least 96% of the populations in large cities are exposed to PM2.5 that exceeds the WHO air quality guidelines levels (World Health Organization, 2019). In addition, 70% of the world population will be living in cities by 2050 (UN DESA | United Nations Department of Economic and Social Affairs, 2019). Thus, improving city's infrastructure, mainly buildings, is one of the major steps needed to enhance livability in cities. Biomimicry architrave has been adopted in many cities worldwide as a model of interaction between building and its surrounding environments. Since the early starting of universe and creation, man and creatures were enclitic by nature and well organized in harmony. In the meantime, history is the master source of inspiration when it comes to nature that inspired human in several aspects in the life. Architecture is considered one of the noteworthy countenances in communities that cannot be separated or detached from nature (Aanuoluwapo & Ohis, 2017).

In addition, the built environment is subject to the worldwide matrix of energy and water; energy emergency and crisis, clean water deficiency, and ecological contamination are overall issues (Ragheb, Ayad, & Galil, 2017; United Nations University, 2014). Understanding urban communities as powerful and regularly advancing ecosystems can assist in defining the methodologies for a reasonable urban future and liveable cities. This could also be seen as decent shapes in nature

S. E. Helmy (✉) · M. M. Aboulnaga
Department of Architecture, Faculty of Engineering, Cairo University, Giza, Egypt
e-mail: maboulnaga@eng.cu.edu.eg

L. Scott et al. (eds.), *Sustainable Ecological Engineering Design*,
https://doi.org/10.1007/978-3-030-44381-8_3

and comprehends the standards behind them; it can motivate numerous thoughts that can prompt new inventive arrangements that are profoundly more asset effective where biomimicry is a rich source of inspiration to create green and sustainable cities (Vincent, Bogatyreva, Bogatyrev, Bowyer, & Pahl, 2006). The importance of this chapter is lying on the highlights of the analytical comparative analysis of several biomimetic architectural buildings worldwide to identify their sustainability elements and addresses how biomimicry contributes to climate change mitigation and adaptation, with the approach of narrative method and qualitative strategy. It is imperative to think of the research question: "*does biomimicry approach have an impact on creating a sustainable architecture in livable cities and adapt to climate change?*" This chapter presents an assessment of the value of the adopting biomimicry concept as a sustainable design tool in architecture based on its potential to create regenerative built environment.

Structure of the Study

This study defines how to get inspiration from nature's behaviours, and the way that nature can influence on theoretical or conceptual design in any project's decisions. The structure of the chapter is divided into two main folds: the first fold presents a literature review and theoretical comprehension of biomimicry concept, and the second fold highlights the global case studies and the analytical comparison between global architectural design examples and explores the relationship between biomimicry, sustainability and climate change as shown in Fig. 1.

Aim and Objectives

The aim of this study is to address the balance between the built and natural environment through assessing the value of nature and corresponding biomimicry design concept as a sustainable design tool in architecture due to its potential in creating

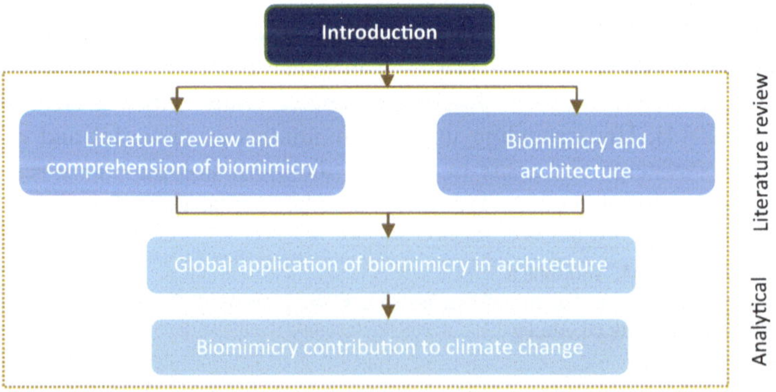

Fig. 1 Structure of the study

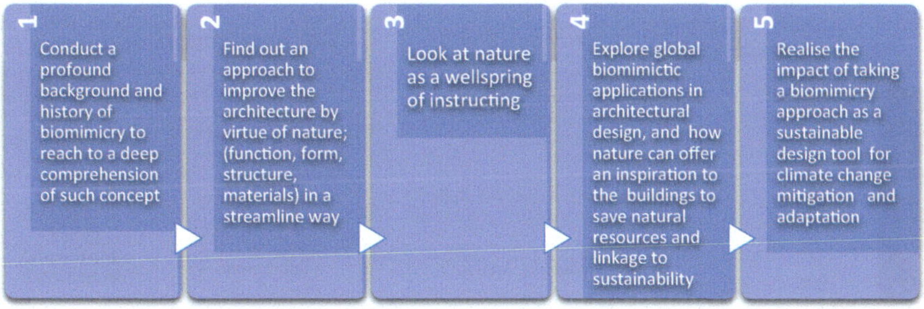

Fig. 2 The objectives of the study

regenerative, sustainable and environmentally friendly built environment. Focus on the relevance of enhancing nature appreciation and the level of awareness and knowledge of biomimicry concept, as a sustainable design tool in architecture is examined by looking into nature for solutions, followed by a development in the way we think of nature from a source of extraction of materials to a model, a measure and a mentor that leads to a comprehensive understanding of nature's genius principals of design. The study's objectives are shown in Fig. 2.

Methodology

A research methodology has been carried out to accomplish research objectives. First, a literature review about biomimicry and a framework of application of biomimicry in architectural design, case studies and how biomimetic buildings worldwide conserve the environment, natural resources, human health, energy and adapt to climate change. The methodology of the research work encompasses many steps: (a) research strategy; (b) research design; (c) methods of data collection and analyses; (d) research approach; and (e) the type of the research problem are the key components of any scientific research method (Fig. 3).

Overview

The term biomimicry is derived from the Greek word "bios", which means life, and "mimesis", which means emulation or imitation; it is the testing of nature through its models, systems, processes and elements, which suggests solutions to human problems Rao (2014). The ability of adaptation found in natural organisms is one of the impressive biological processes. Flora and fauna offers a lot of examples of adaptation methods. Architects nowadays are continuously looking for inventive thoughts and brilliant ideas for reduction, recycling and reusing energy and

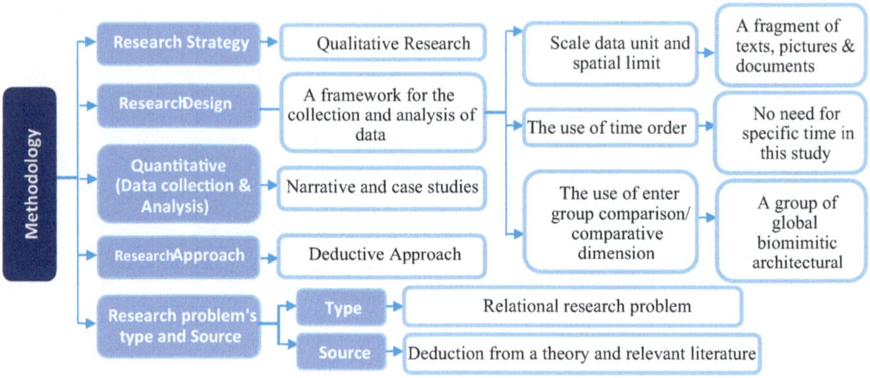

Fig. 3 Methodology outlines

minimizing carbon footprint in architectural field. Nevertheless, biomimicry can give elective answers for different basic proficiency, e.g. water effectiveness, zero waste frameworks, comfort condition and vitality supply. These are considered a basic for any sustainable building design in the thirty-first century (Al-Ahmar, 2011).

Origin and History of Biomimicry

Architects, engineers, planners and designers have sought biology for motivation since the beginnings of the science in the mid-nineteenth century. They didn't copy the types of plants and creatures, yet in addition to discover techniques in design analogous to the process of development and advancement in nature (Al-Ahmar, 2011). Human and nature relationship has evolved and improved since the beginning of creation. In the early ancient Egyptian and Greeks civilization, engineers and inventors had turned and looked to nature for inspiration and ideas (Zari, 2007). Truth be told, it might be contended that biomimicry isn't something new, yet a "*Revival and Return*" to our most punctual motivations. For example, famous architects such as Alvar Alto, Frank Lloyd Wright and Le Corbusier have laid accentuation on this association. Architecture is being impacted by few subjects of natural and sociologies (Rust, 2011). Looking back at history, there were some historical examples of problem solving inspired by nature for creating a flying machine. Da Vinci found that it is critical to study the anatomy of the birds and the flying techniques (Fig. 4). Although Da Vinci's machine was never completed, the abstract principle of taking the inspiration from nature let him to be pioneer in biomimicry and who got their inspiration to construct the first air plane from flying pigeon.

Also, Gaudi had looked to nature when he designed his buildings: (a) the tree structure in Sagrada Familia church in Barcelona, sequoia trees were used as an inspiration; (b) The Casa Batlo where the inspiration of exterior structure was drawn

Fig. 4 Drawings trials for flying machine by Leonardo Da Vinci's. *Image source*: http://www.sciencedirect.com/science/article/pii/S187802961630055X

Fig. 5 Sagrada Familia Church and Casa Batlo building—Barcelona, Spain. (**a**) Interior of Sagrada Familia and (**b**) exterior structure of Casa Batlo. *Image source*: Authors/http://www.projects.science.uu.nl/urbanbiology/articlepagebiomim.html

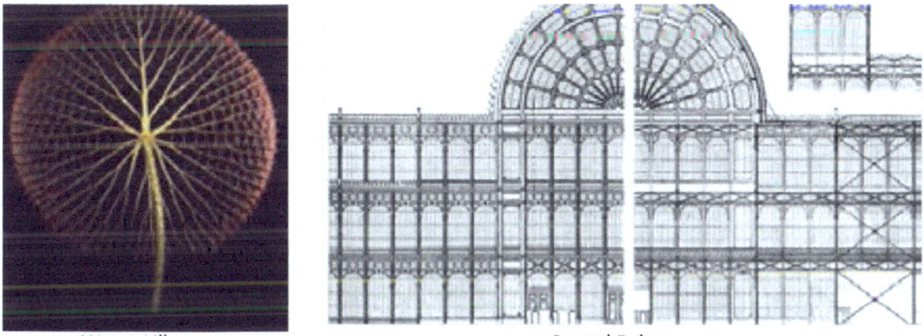

Fig. 6 Crystal Palace Building, London, UK. (**a**) Water lily and (**b**) Crystal Palace. *Image source*: http://www.sciencedirect.com/science/article/pii/S187802961630055X

from the human pelvic bones as shown in Fig. 5. In 1851, James Paxton designed the structural system of the Crystal Palace from his observation of giant water lilies' leaves of Amazonian water lily plant as illustrated in Fig. 6. The leaves of this water lily look brittle but can support a weight. The Crystal Palace's structure is founded on the ribs and trunk of the lily leaves (Arslan, 2014; ElDin, Abdou, & ElGawad, 2016).

Overlap Between Architecture and Nature

The importance of the connection between nature and architecture is investigated through few trials, where the focus is on the morphological aspect. Paolo Portoghesi worked on the combination of both nature and architecture in the encyclopaedic reference "*Nature and Architecture*". This book is a comprehensive aggregate of identification between nature and architecture, and examines both structure and function (Snijders & Pronk, 2019). Biomimicry and nature relationship could by clarified by taking nature as mentor, measure and model as shown in Fig. 7.

Sustainability and Global Application of Biomimicry in Architecture

Many countries worldwide apply the concept of biomimicry in architecture and build several mimicking buildings. The more the country is aware of environmental impact, the better the sustainability outcomes.

Biomimicry Application in Line with Sustainability

The core of biomimicry approach is engaging the design with the function, transferred by particular natural adaptation. This is the secret key to fruitful executing biomimicry principles. Architecture's practical applications exist in biomimicry, which are reflected into many levels that need to be achieved in the design process in order to mitigate and adapt to climate change and create better built environments (Fig. 8) (Rankouhi, 2012).

Fig. 7 Biomimicry and the vision to the nature

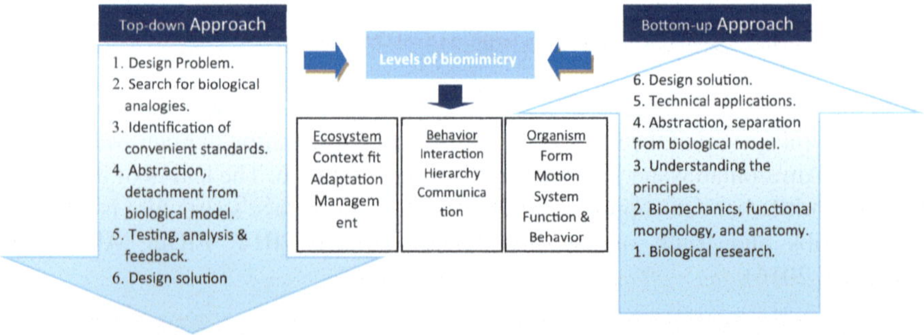

Fig. 8 A framework for understanding the application of biomimicry in architectural design

Architectural Design Example

Several global architectural design examples, with different levels and approaches in applying biomimicry thinking, are presented and assessed. Some of the thirteen examples are already built and others are hypothetical futuristic vision, including: (1) Waterloo International Terminal in London, UK, that mimic the pangolin organism; (2) Hydrological Centre for the University of Namibia that mimic Stenocara beetle; (3) East Gate Building in Harare, Zimbabwe, that mimic termite mound behaviour in ventilating the building; (4) Beijing Olympic Stadium and National Aquatic Centre in China that mimic birds' nest and water bubble; (5) Swiss Re–Gherkin Tower in London, England, that got inspiration from Venus flower form and air craft inspiration; (6) Habitat 2020, China, a future sustainable housing for large cities (Yoneda & Yoneda, 2008), (7) Qatar's giant cactus building (MMAA) that mimic cactus plant behaviour in surviving in harsh climate conditions; (8) Esplanade theatre in Singapore, that mimic Durian plant in keeping the users inside in comfort condition as this plant keeps the seeds inside it; (9) Sagrada Familia Church in Barcelona—Spain, that mimic tree trunk where using the same structure systems for carrying the loads of the vaults; (10) The lotus temple in Delhi—India, that mimic the form and behaviour of lotus flower (Lotus Temple in India, 2014); (11) Residential building in New Songdo City, South Korea, that mimic human spine and honeycomb structure to reduce structural cost of the building; (12) Tree scraper tower of tomorrow in the USA, that mimic the tree as a living system and (13) CH2 Melbourne City Council House 2 in Australia, and many other buildings and examples of biomimicry architectural design buildings.

By looking at all these examples and based on the analytical comparison between the best four case studies (existing buildings and hypothetical ones), it can be deduced that in order to find out the best biomimetic buildings' design application to sustainable architecture evaluation elements should be assessed (Spaho, 2011). Table 1 lists the analysis of the buildings' performance and their ranking according to sustainability evaluation elements (WikiArquitectura, n.d.; Maglic, 2012; Rust, 2011). According to the comparison between the four buildings, it is found that CH2 Melbourne City Council House 2 is the optimum example for biomimicry approach that proved the concept could be environmentally sustainable approach in architecture. But nevertheless, the question is that "Does the CH2 building with all sustainability features it has could contribute and assist in reducing greenhouse gas (GHG) emissions?" If so, would it contribute to climate change; most probably the answer is yes (Nunn, 2007; Roger, Yoon, & Malek, 2007; Snijders & Pronk, 2019) (Table 2).

Biomimicry and Climate Change

David King expressed the seriousness of global climate change stating that it is the number one threat facing us and its severity overcomes the threat of terrorism. As the effects of climate change increment, arrangements and activities must be applied

Table 1 Biomimicry and worldwide buildings and its inspiration from nature

Building name	Photo of the building	Inspiration	Biomimicry level
Waterloo International Terminal in London, UK *Image source*: http://buildipedia.com			Organism level
Hydrological Centre for the University of Namibia *Image source*: http://scholarworks.umass.edu/cgi/viewcontent.cgi?article=1984&co ntexttheses	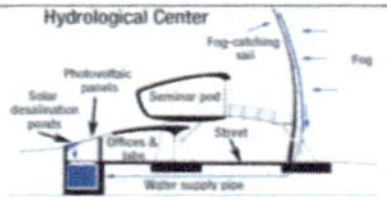		Organism level
East Gate Building in Harare, Zimbabwe *Image source*: http://meioinfo.eco.br/ biomimetica-design-arquitetura-complexa-cupinzeiros			Behaviour level
Beijing Olympic Stadium and National Aquatic Centre, China *Image source*: http://daily.swarthmore.edu, https://www.mcgill.ca/architecture/files/ architecture/BiomimicrySSEFessay2007.pdf			Behaviour level
Swiss Re–Gherkin Tower in London, England *Image source*: http://www.greatbuildings.com, http://www. isrjournals.org/journals/civil_environmental_journals/ biomimicryinarchitecture1401273613.pdf		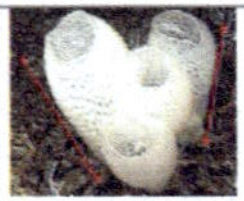	Behaviour level
Habitat 2020, China *Image source*: http://inhabitat.com/habitat-2020-off-the-grid-future-abode/		Living organism	Ecosystem level
Qatar's giant cactus building (MMAA) *Image source*: http://newatlas.com/qatars-giant-cactus-biomimicry/10993/#gallery, http://www.ucreative.com/inspiration/ you-be-inspired-10-nature-inspired-architectural-designs/			Behaviour level

Sagrada Familia church in Barcelona, Spain *Image source*: http://www.Pixbay.com, https://www.quora.com/What-are-some-of-the-best-examples-of-biomimicry-in-architecture			Organism level
The lotus temple in Delhi, India *Image source*: https://bashny.net/t/en/105425			Organism level
Esplanade theatre in Singapore *Images source*: MakeMyTrip/http://www.garykent.com, http://www.sciencedirect.com/science/article/pii/S1876610217308603#			Behaviour level
Residential building in New Songdo City *Image source*: http://docs.rwu.edu/cgi/viewcontent.cgi?article=1072&context=archthese	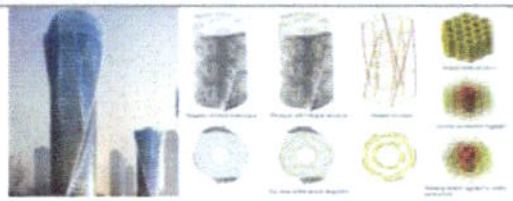	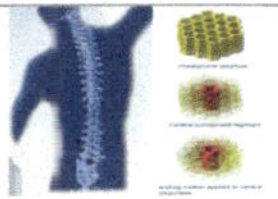	Organism level
Tree scraper tower of tomorrow in USA *Image source*: http://www.biomimeticarchitecture.com/2010/william-mcdonoughs-treescraper/		Tree in the ecosystem	Ecosystem level
CH2 Melbourne City Council House 2 in Australia *Image source*: https://architectureau.com/articles/ch2/		Inspired by the fundamentals nature	Ecosystem level

Table 2 Comparison between best four biomimetic buildings' cases worldwide

Building's name		East Gate building	CH2 Melbourne City Council House 2	Tree-scraper tower	Habitat 2020
Assessment elements					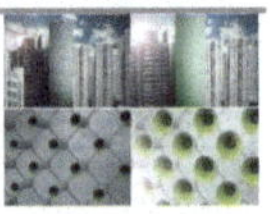
Architect		Mick Pearce	Advanced Environmental Concepts, Lincoln Scott, Design Inc.	William McDonough	Philips Design
Country		Zimbabwe	Australia	USA	China
Building type		Commercial office building	Commercial office building	High-rise office building	Residential or office building, not recognized yet
Existence		Existing	Existing	Proposal	Proposal
Energy efficiency	Energy saving	Use 10% less energy used in similar size conventional building	82%	√	√
	Natural lighting	√	Saved by 65%	√	√
	Natural ventilation	Use about 35% of energy required for temperature regulation	Saved by 65%, provide nearby 100% fresh air	√	√
	Usage of solar PV panels	√	√	√	√
Water recycling		√	72% reduction water consumption	√	√
Waste management		√	Recycling facilities for office waste	√	√
Biomimetic level		Behaviour	Organism and behaviour	Behaviour	Ecosystem
Ranking		2	1	3	4

without delaying to mitigate GHG emissions (Zari, 2016). In this regard, promoting and adopting biomimicry architecture is one of the solutions to curb the climate change impact due to low carbon generated in such buildings.

Climate Change Risks

Thinking about living organisms and their way of adapting similar problems like climate change and gas emissions will help us to find a solution to overcome the problem. The adaptation and survival of some plants and animals in the extreme weather or climate change will give us a perspective of how buildings and constructions should or could work. Biomimicry can create a comfortable and sustainable urban environment. By looking at the natural organisms, it is found that their survival depended on their ability to adapt to the change. The utilization of such science in the designing process remains unnoticed and not spread enough. There are several effects of climate change that are already existing such as continuous rise in temperature, frost-free seasons, changes in precipitation modules, drought and heat waves, and stronger and more intense hurricanes; by 2100 sea level will rise 1–4 ft and Arctic is likely to become ice free (Climate Change 2013).

The proper utilization of biomimetic methods will result not only in reaching sustainable solutions for the urban environment, but it can also regenerate and restore the natural ecosystem, and the differences between the meaning of mitigation and adaptation as illustrated in Fig. 9. The building sector has the most chance to reduce the effect of climate change with the least amount of expenses unlike other sectors, and biomimicry case studies demonstrate the approach's ability to guide eco integration. As such, architects are uniquely positioned to embrace biomimicry

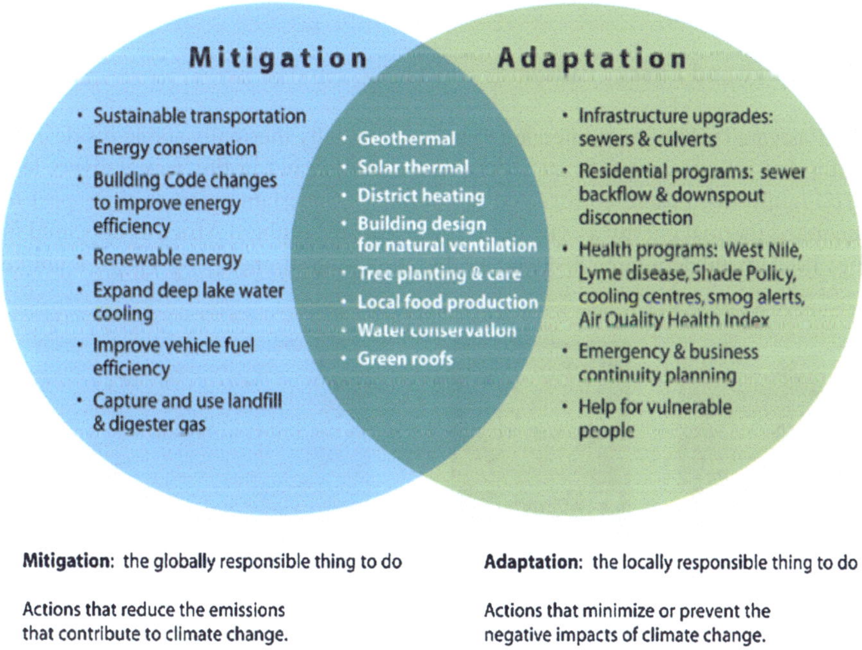

Fig. 9 Mitigation and adaptation to climate change. *Image source*: http://www.fewresources.org/adaptation%2D%2Dmitigation.html

and become leaders of a sustainability transition. Biomimicry has proven ecologically and economically viable in the face of critics who hang on to a careless view of humans as the champion life form.

Biomimicry: Energy Efficiency and Restoring Carbon Applications

Numerous cases of new advancements represent cantered on moderating GHG outflows, and can be sectioned into two methodologies:

1. The first approach is mimicking the energy utilization of living biological systems and organisms. The main target of this approach is reducing the amount of fossil fuels energy which will decrease the GHG outflow levels.
2. The second approach is to reach better methods for generating another form of energy with non-GHG emissions and decreasing our dependence on fossil fuel energy. These methods are the most widely recognized ways to deal with decreasing the reasons for environmental change related with the urban regions. Deal with reducing GHG discharges is examining living biological systems to discover cases and behaviours that enabled them to isolate and store carbon. Mimicking this accompanies the aim of zero GHG emissions that are radiated through human exercises and causing extra climate change (Zari, 2010).

Biomimicry and Energy Efficiency

The Eastgate building was intended to have a generally thermally stable inside condition that utilizes insignificant mechanical cooling (and in this way delivers less GHG outflows). Pearce constructed the outline to a certain extent with respect to generated thermal energy as saw in (termite) hills of southern Africa as appeared in Fig. 10. This made a huge decrease in the utilization of energy by 17–52% unlike

Fig. 10 Mitigation and adaptation to climate change. (**a**) Termite mound, (**b**) Eastgate building and (**c**) multiple deployed bio-wave units. *Image source*: https://www.researchgate.net/publication/268507389_Can_Biomimicry_Be_a_Useful_Tool_for_Design_for_Climate_Change_Adaptation_and_Mitigation and http://www.biomimicrynorway.com/blog/7/2/2015/wave-power-and-seaweed

other similar buildings in Harare. Discovering techniques to supplant fossil energy with renewable power sources is a way to reduce the reasons for climate change resulting into a long-term arrangement. Biomimicry likewise offers methods to deal with abundance carbon dioxide (CO_2) effectively in the environment and to reduce high GHG emissions, e.g. sea kelp technology as shown in Fig. 11. Sea kelp is great at withstanding strong and severe water waves.

CH2 Building: Best Biomimicry in Architecture Case Study

The CH2 building (Fig. 12) provides a sound and a gainful working environments to its users while diminishing the building's effect on nature through brilliance in design and development. More beneficial structures will enhance tenant well-being and prosperity. This building helped in decreasing GHG emission, and it does not harm the environment. In addition, it was the superb case of biomimicry approach in design that accomplished supportability, exceptionally in ventilation as biomimicking termite hill as indicated earlier.

Fig. 11 Interior and exterior of the CH2 building, Melbourne, Australia. *Image source*: https://www.researchgate.net/publication/262857243_Biomimetic_design_for_climate_change_adaptation_and_mitigation

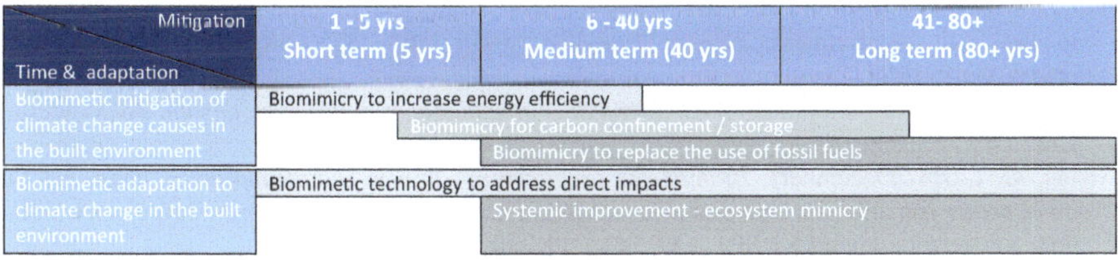

Fig. 12 Timeline for biomimicry approaches to address climate change. *Image source*: https://www.researchgate.net/publication/262857243_Biomimetic_design_for_climate_change_adaptation_and_mitigation

Biomimicry, Climate Change and Time Line

The responses of biomimicry to climate change in short, medium and long terms are shown in Fig. 9; some of them are more effective than others. But in the short and medium term existing technologies and techniques will be crucial; long-term solutions addressing climate change impacts could be obtained from biomimicry. Numerous researchers expressed the connection between bio-affected designs and enhanced mental and physical well-being. Regarding financial advantages, there is proof of economical sustainability for the structures unlike the traditional ones. As GHG discharges are progressively being managed, structures that don't meet lawful or execution expectations might be harder to rent, market, and secure resulting in high monetary life cycle expenses (Zari, 2010).

Conclusions

An assessment of the value of biomimicry architecture coupled with a comparative analysis has been conducted and global examples were reviewed. It is found that biomimicry concept could contribute in growing sustainability and it is a way to address the balance between built and natural environment. Four buildings based on biomimicry concepts were evaluated in terms of 8 sustainability elements and ranked. The study indicates that building CH2 Melbourne City Council House 2 is the optimum example for biomimicry approach. It becomes a way of thinking in achieving sustainability in architecture, yet mitigates climate change mitigation and contributes to adaptation actions. Biomimicry and how it mitigates GHG emissions in the built environments by virtue of energy efficiency, energy generation can offset climate change risks. In addition, the biomimetic sequestering and storing carbon are the ways forward to improve the livability and nature conservation in future cities.

References

Aanuoluwapo, O., & Ohis, A. (2017). Biomimetic strategies for climate change mitigation in the built environment. *Energy Procedia, 105*, 3868–3875. https://doi.org/10.1016/j.egypro.2017.03.792

Al-Ahmar, S. (2011). *Biomimicry as a tool for sustainable architectural design*. Alexanderia University, Alexanderia.

Arslan, G. (2014, December). *Biomimetic architecture—A new interdisciplinary approach to architecture* (pp. 29–35). Universiti Putra Malaysia. [Online]. Retrieved from http://www.frsb.upm.edu.my/dokumen/FKRSE1_157-542-1-PB.pdf

Church, J.A., P.U. Clark, A. Cazenave, J.M. Gregory, S. Jevrejeva, A. Levermann, M.A. Merrifield, G.A. Milne, R.S. Nerem, P.D. Nunn, A.J. Payne, W.T. Pfeffer, D. Stammer and A.S. Unnikrishnan, 2013: Sea Level Change. In: Climate Change 2013: The Physical Science Basis. Contribution of Working Group I to the Fifth Assessment Report of the Intergovernmental Panel on Climate

Change [Stocker, T.F., D. Qin, G.-K. Plattner, M. Tignor, S.K. Allen, J. Boschung, A. Nauels, Y. Xia, V. Bex and P.M. Midgley (eds.)]. Cambridge University Press, Cambridge, United Kingdom and New York, NY, USA.

ElDin, N., Abdou, A., & ElGawad, I. (2016). Biomimetic potentials for building envelope adaptation in Egypt. *Procedia Environmental Sciences, 34*, 375–386.

Lotus Temple in India. (2014). [Online]. Retrieved from https://www.india.sea7htravel.com/2014/01/lotus-temple.html

Maglic, M. (2012). *Biomimicry: Using nature as a model for design*. Master. University of Massachusetts Amherst, Amherst, MA.

Nunn, R. (2007). CH2. [Online]. ArchitectureAU. Retrieved from https://architectureau.com/articles/ch2/

Ragheb, S. A., Ayad, H. M., & Galil, R. A. (2017). An energy-resilient city, an appraisal matrix for the built environment. *WIT Transactions on Ecology and the Environment, 226*, 667–678. https://doi.org/10.2495/SDP170581

Rankouhi, A. R. (2012). *Naturally inspired design investigation into the application of biomimicry in architectural design*. The Pennsylvania State University, State College, PA.

Rao, R. (2014). *Biomimicry in architecture*. ISR Journals and Publications. [Online]. Retrieved from http://www.isrjournals.org/journals/civil_environmental_journals/biomimicryinarchitecture1401273613.pdf

Roger, A., Yoon, B., & Malek, C. (2007). *Beijing Olympic Stadium 2008 as Biomimicry of a Bird's Nest*. [Online]. Mcgill.ca. Retrieved from https://www.mcgill.ca/architecture/files/architecture/BiomimicrySSEFessay2007.pdf

Rust, W. (2011). *Blueprints from Nature: Investigating a biomimetic approach to architecture*. Master. University of Cape Town, Cape Town.

Snijders, A., & Pronk, L. (2019). *Urban biology*. [Online]. Projects.science.uu.nl. Retrieved from http://www.projects.science.uu.nl/urbanbiology/articlepagebiomim.html

Spaho, K. (2011). *Biomimicry: Architecture that imitates nature's functions, forms and parts*. Master. Roger Williams University, Bristol, RI.

UN DESA | United Nations Department of Economic and Social Affairs. (2019). *68% of the world population projected to live in urban areas by 2050, says UN*. UN DESA | United Nations Department of Economic and Social Affairs. [Online]. Retrieved April 14, 2019 from https://www.un.org/development/desa/en/news/population/2018-revision-of-world-urbanization-prospects.html/

United Nations University (UNY). (2014). *World Water Day 2014: UN Stresses Water and Energy Issues*. [Online]. Retrieved May 2, 2019 from https://unu.edu/media-relations/releases/wwd2014-un-stresses-water-energy-issues.html

Vincent, J., Bogatyreva, O., Bogatyrev, N., Bowyer, A., & Pahl, A. (2006). Biomimetics: Its practice and theory. *Journal of the Royal Society Interface, 3*(9), 471–482.

WikiArquitectura. (n.d.). *30 St Mary Axe (The Gherkin)—data, photos & plans—WikiArquitectura*. [Online]. Retrieved from https://en.wikiarquitectura.com/building/30-st-mary-axe-the-gherkin/

World Health Organization. (2019). *WHO Global Urban Ambient Air Pollution Database (update 2016)*. [Online]. Retrieved March 10, 2019 from https://www.who.int/phe/health_topics/outdoorair/databases/cities/en/

Yoneda, Y., & Yoneda, Y. (2008). *HABITAT 2020: Future Smart 'Living' Architecture*. [Online]. Inhabitat.com. Retrieved from https://inhabitat.com/habitat-2020-off-the-grid-future-abode/

Zari, M. (2007). *Biomimetic approaches to architectural design for increased sustainability*. New Zealand: School of Architecture, Victoria University. [Online]. Retrieved from http://www.cmnzl.co.nz/assets/sm/2256/61/033-PEDERSENZARI.pdf

Zari, M. (2010). Biomimetic design for climate change adaptation and mitigation. *Architectural Science Review, 53*(2), 172–183.

Zari, M. P. (2016). Can biomimicry be a useful tool for design for climate change adaptation and mitigation? In *Biotechnologies and biomimetics for civil engineering* (pp. 81–114). Cham: Springer.

Sustainable Development and Management of Low-Volume Road Networks in Australia

Thomas Franzen and David Thorpe

Introduction

Low-volume roads (LVRs) account for a significant proportion of world roads but can often have limited or insufficient funding. Depending upon the exact definition of a LVR, it has been claimed that between 75% (Zimmerman & Peshkin, 2003) and 90% (Irwin, 2003) of the worlds roads are classified as LVRs. Most of these roads are unsealed. In addition, around 85% of the sealed roads in the world have traffic volumes of less than 1000 vehicles per day (Douglas, 2016). These roads have a replacement value estimated at more than USD 7.6 trillion worldwide (Faiz, 2012). However, as world resources are finite, funding is often inadequate to cover their maintenance and rehabilitation cost (Zimmerman & Peshkin, 2003).

According to the Bureau of Infrastructure Transport and Regional Economics (BITRE), 73% of the total road length in Australia (over 873,000 km) are classified as local roads (BITRE, 2017), with low traffic volumes. These roads are important to Australia's economy as many rural products start and end their transportation on local roads. Most of the estimated 640,000 km of local roads in Australia are managed by local governments, with their net spending on roads accounting for 21% of total Australian road related expenditure (BITRE, 2017). It has been estimated that there is a shortfall in maintenance and renewal expenditure on local roads in Australia of AUD 17.6 billion for the period between 2010 and 2024 (Howard, Jeff Roorda and Associates, & Institute of Public Works Engineering Australia (IPWEA), 2013). Given the scale and importance of the local LVR network in Australia and

T. Franzen
Eurobodalla Shire Council, Moruya, NSW, Australia

D. Thorpe (✉)
School of Civil Engineering and Surveying, University of Southern Queensland, Toowoomba, QLD, Australia
e-mail: david.thorpe@usq.edu.au

L. Scott et al. (eds.), *Sustainable Ecological Engineering Design*,
https://doi.org/10.1007/978-3-030-44381-8_4

funding limitations, innovative network management practices are therefore required.

LVRs provide an essential service by enabling access to communities (ARRB Group, 2005). They require careful management to serve these communities well. However, authorities have often found it difficult to justify significant expenditure on their construction and maintenance. It has been argued that this attitude often costs society in the long run, with roads in poor condition resulting in higher costs to the road users (Irwin, 2003). It can be similarly argued that low-volume roads in poor condition pose a range of threats to their sustainable management, including safety concerns, excessive use of fuel for vehicles that use them, and lack of community social connectedness.

In Australia, which is governed at federal, state and local levels, state governments have the primary responsibility for managing their road networks and local governments manage the remaining roads including local roads, which include most LVRs (Austroads, 1998, 2016b). Funding is provided by all levels of government (Department of Infrastructure and Regional Development, 2017). In the 2015/2016 financial year, an amount of AUD 26.17 billion was made on road-related expenditure in Australia's Road Network (BITRE, 2017).

Because of the size of the LVR network and the challenges in Australia in the ability of this network to meet community requirements with limited funding, road authorities must manage it as efficiently and effectively as possible. This process requires a combination of best road network management practice and an innovative approach. Therefore, the objective of the research discussed in this paper has been to investigate and recommend improvements to network management strategies for Australian LVRs, in order to achieve the best options for achieving their sustainable development and management. In particular, this research has focused on the LVR network (sealed and unsealed) managed by local government authorities in regional and rural New South Wales (NSW). The main step in the research process, in terms of its inputs, processes and outputs, is summarised in the conceptual framework in Fig. 1.

Literature Review

Key Terms

There is no one accepted definition of the exact traffic volume that defines a low-volume road. For example, Faiz (2012) suggested that an Annual Average Daily Traffic (AADT) threshold of 1000 Vehicles Per Day (VPD) or less as defining a LVR. Alternatively, the American Association of State Highway and Transportation Officials (AASHTO) have used an AADT of 400 VPD or less (AASHTO, 2001). Other threshold values of AADT range from that of the Transportation Association of Canada value of 200 VPD (Douglas, 2016). Thus, there does not appear to be a

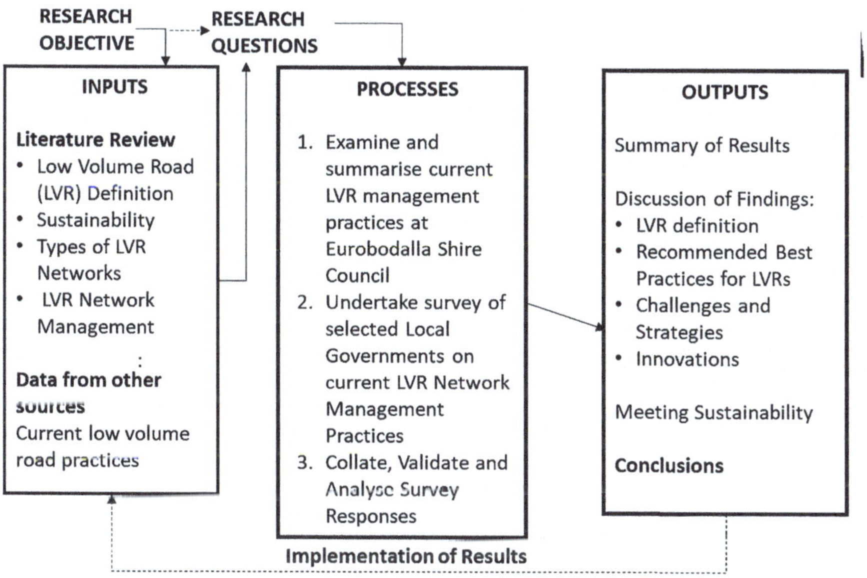

Fig. 1 Conceptual framework of research

universally accepted definition of the threshold traffic volume that defines a LVR. Similarly, in Australia a survey of state and local road authorities on sealed roads found that the commonly accepted equivalent AADT definition of LVR varied between 200 and 400 VPD, with the most common definition being 200 VPD (Austroads, 2015b). Another Australian survey indicated that LVRs had traffic volumes up to 1000 VPD (Austroads, 2000). Given these ranges in threshold values for the traffic volumes of LVRs, it is clear that there is no common definition of a LVR and that its identification is a function of items like location, land use, topography, traffic composition, population and other factors (Douglas, 2016).

Another aspect of defining a LVR is to link it with the functional classification (or hierarchy category) of roads (AASHTO, 2001). This approach determines appropriate road management and maintenance practices, and aligns engineering standards with the road's function (Giummarra, 2003), which is based on mobility and access. A road with a higher classification might focus on mobility and one with a lower classification more on providing access to properties. A road's function, which is typically used by road authorities to divide roads into different functional classifications (such as local roads), is reflected in its characteristics, such as like likely traffic volume (Austroads, 2015a). This classification aids in the allocation of resources (Giummarra, 2003). In this classification, LVRs tend to be associated with lower hierarchy roads such as local roads where the main function is to provide access to properties, farms and businesses, as opposed to meeting through traffic requirements (AASHTO, 2001). From this point of view, LVRs in NSW tend to have an AADT of less than 1000 VPD for urban roads and less than 200 VPD for rural roads (IPWEA NSW, 2015).

The other term that requires definition is "sustainable development", which is "development that meets the needs of the present without compromising the ability of future generations to meet their own needs" (Brundtland, 1987). This concept is further defined in terms of the three pillars of economic development, social equity and environmental protection (Drexhage & Murphy, 2010). From these three pillars, this research has proposed six sustainability goals with respect to LVR development and management. These goals are productivity, safety, usability, social equity, sustainable environmental management and resilience.

Types of Low-Volume Road Networks

The length of the unsealed LVR network in Australia is about 500,000 km (Austroads, 2006). The Unsealed Roads Manual of the Australian Road Research Board (ARRB) (Giummarra, 2009) is normally followed in designing them. Unsealed LVRs often do not require the same carriageway width as higher volume roads, and often have open table drains to drain water from the pavement (Giummarra, 2009). In addition, the pavement design and configuration of unsealed LVRs differs from that of sealed roads, as traffic volumes and relative benefits from them are usually lower. Additionally, LVRs have often been developed in stages, from a basic cleared unformed surface, to a formed road with no gravel, to a gravel paved road with sheeting material on its surface (ARRB, 1993). As a result, a number of them may not necessarily meet current design standards.

It is also estimated that there are 210,000 km of sealed local (or lower volume) roads in Australia. Many of their pavement structures have evolved over time and have therefore not been upgraded systematically (ARRB Group, 2005). Normally, they consist of a flexible pavement and a bitumen or asphaltic surface. The main function of the surfacing material is to protect the underlying courses from moisture and withstand loading and environmental effects. Certain surfacing materials, like asphaltic concrete, can aid the road's structural strength. It has been claimed that many of the current challenges facing their planning, design, construction and maintenance are focused on improving sustainability, equity of access and transport choice (Austroads, 2009b).

Management of Low-Volume Road Networks

It is important to operate and manage road networks using a strategic, planned approach that makes the best use of resources. This approach aims to optimise road infrastructure at a network level in order to meet the needs of users (Austroads, 2016a). It requires an interdisciplinary approach, and is embodied in the development of network level innovative solutions focused on adding value to the road network and increasing the efficiency of network operations (World Road Association, 2003). This process requires the integration of asset management, safe systems and road infrastructure strategies. The implementation of an organisation's asset management strategy will therefore directly influence the performance of the road network and how it meets community expectations (Austroads, 2016a). The important function of maintenance aims to ensure that adequate levels of service are provided, and that the road reaches its intended asset service life. Such maintenance can be divided into preventive maintenance, which aims to predict when defects will occur in advance and rectify them (ARRB, 1993) and corrective (or reactive) maintenance, which addresses defects as they arise, and is thus considered a less efficient use of limited maintenance resources. Reactive maintenance can also lead to higher costs for road users and more severe deterioration impacts if defects are left uncorrected.

Challenges in maintaining unsealed LVRs include rapid deterioration from rain and traffic (Giummarra, 2009), relatively high operating costs, limited access during and after high intensity rainfall events (thus impacting on their resilience), higher consumption of natural materials, environmental and heritage impacts, a higher demand for water due to frequent maintenance requirements, and often more risk of accidents (Austroads, 2009a). Some of these issues can be addressed through regular maintenance activities, which include routine grading and periodic reshaping and re-gravelling (ARRB, 1993; Giummarra, 2009).

Normally the maintenance of sealed LVRs is classed as routine or periodic (ARRB Group, 2005). While routine maintenance activities are normally minor, periodic maintenance, which aims to reinstate the condition or surface of the pavement, requires planning and tends to be more expensive and cyclic. Thus, the overall maintenance strategy for sealed LVRs is to undertake regular inspections and review, to identify defects and minimise deterioration.

Other issues in managing LVRs include having a duty of care to the safety of road users, maintaining community productivity, managing social equity (for example, providing access in remote areas), practising sustainable environmental management, providing resilience (ensuring that the network can withstand and bounce back from shocks), and using innovative practices to make the most effective use of funds.

Research Questions

In summary, the literature review has focused on definitions of the key terms of low-volume roads and sustainability and has considered issues in the management of LVRs. It has also indicated several questions with respect to the information available and the main principles by which LVRs are managed in Australia, including NSW, where the research described was undertaken. These questions, which underpin the overall research objective, include:

1. What is the commonly accepted definition of a LVR?
2. What level of service should be provided for LVRs and how is this achieved?
3. What should the design life be for a typical LVR pavement or surface treatment?
4. What standards are commonly adopted for planning, design, and construction of LVRs?
5. Are there any common activities that are being successfully used to extend the life or reduce the costs spent on the renewal of LVRs?
6. What are the most critical issues in the management of LVRs?
7. What strategies are being used to overcome these challenges?

Research Methodology

Background

The research discussed herein has reviewed and recommended key best practices for managing LVR networks within regional and rural areas of NSW. A first step was to investigate the current network management practices being used at the Eurobodalla Shire Council (ESC) in NSW, in which one of the authors works. This investigation formed the baseline for a survey questionnaire from practicing engineers and professionals representing a cross section of other local government authorities across different regions in NSW to investigate current LVR network management practices. Where possible, it was verified and validated against publicly available information and previous local government surveys.

The second component of the research involved a survey questionnaire that was distributed to selected local government organisations, primarily in NSW. This constituted a major source of data. Such data was critical in analysing current road network management practices that were used for both sealed and unsealed LVRs, particularly within rural and regional areas of NSW. The survey collected data on current network management practices related to levels of service and maintenance, future demand management, planning, design and construction, life cycle and asset management, asset renewal, and monitoring and performance. Ethics approval was obtained prior to carrying out the survey and ethical practices were implemented during its administration. A significant amount of data was obtained from primary information sources, where there may be some limitations and potential individual respondent bias. To minimise resulting errors in the survey, additional measures were taken, such as verifying data against that published independently in a range of government and professional organisation sources, and validation against previous survey findings by reputable organisations such as the Institute of Public Works Engineers of Australia (IPWEA) and Austroads, the association of Australasian road transport and traffic agencies.

The Survey Questionnaire

The survey followed a similar methodology as adopted by Austroads (2015b) in the LVR survey relating to seal design improvements, but covered broader aspects relating to the management of both sealed and unsealed LVRs at a network level. It was undertaken online using a proprietary platform, thus giving it a range of advantages including speed, timeliness, flexibility, convenience and ease of data entry and analysis. To overcome any concerns by participants, the survey was made completely voluntary with confidentially assured.

It comprised a set of structured questions with a mix of both multiple-choice questions (single answer and multiple answer responses), along with some open-ended questions that were designed to be as straightforward as possible. Questions were grouped as follows:

1. General questions on the size and definition of the local LVR network for which the council is responsible.
2. Questions related to levels of service and maintenance, in terms of community expectations and technical measures of performance for sealed and unsealed roads.
3. Questions related to the planning, design, and construction of LVRs.
4. Questions related to life cycle management and renewal of LVRs.
5. Questions related to current challenges and strategies for the LVR network.

Responses were obtained from 33 local government organisations throughout NSW, or 26% of all local governments in that state. A further five responses were received from other local governments in Australia and New Zealand. It is therefore considered that data was obtained from a good cross section of LVR Network Managers in NSW. Participation in the survey was encouraged with the offer of providing participating councils a summarised report of the findings at the completion of the study upon request, thus allowing them to review their current practices and highlight if there are any other innovative practices that could be implemented that would lead to improved outcomes.

Results

There were 30 questions in the survey. A summary of responses to them, listed by the number of the research question that they are addressing, is in Table 1.

Table 1 Summary of survey responses

RQ	Summary of question	Main response (percentages rounded)
	General questions	
NA	Name of Council	Provided by respondent
NA	Urban or rural	26% urban; 74% rural
NA	Estimated population	>20,000—52%; <20,000—48% (NSW only)
NA	Length of roads (km)	Average 1936 km. Most <2000 km
1	Traffic threshold unsealed LVR	<100 vpd—49%; <150 vpd—69%
1	Percentage unsealed LVRs of total	81% to 100%—38%; 61% to 80%—12%
1	Traffic threshold sealed LVR	<100 vpd—29%; <250 vpd—76%
1	Percentage sealed LVRs of total road network	0% to 20%—16%; 21% to 40%—34%; 41% to 60%—22%; 61% to 80%—19%: 80% to 100%—9%
	Levels of service	
2	Community satisfaction—unsealed road service	Neither satisfied or dissatisfied—50%; Satisfied—27%
2	Annual expenditure—unsealed	Other data (Verity, 2018)—AUD 1791/km[a]
2	Inspection frequency—unsealed	One per year—31%; twice per year—17%
2	Grading frequency—unsealed	Once per year—24%; Once per 2 years—17%
2	Community satisfaction—sealed road service	Satisfied—45%; neither satisfied nor dissatisfied—41%
2	Annual expenditure—sealed	Other data (Verity, 2018)—AUD 3832/km[a]
2	Inspection frequency—sealed	Once per year 45%, twice per year 24%
2	System to record/manage defects	Software based—78%, None—11%
2	Conduct road safety reviews or audits	Yes—29%; No—71%.
2	Extraordinary practices improving level of service	17 responses received: 11 nil; 2 using "Otta Seal"; 2 using polymer additives
	Planning, design, and construction	
3	Typical design life—unsealed pavements	0 to 5 years—24%; 5 to 10 years—17%; 10 to 15 years—21%; 15 to 20 years—17%
3	Typical design life—sealed pavements	10 to 20 years—31%; 20 to 30 years—31%; 40 to 50 years—17%
3	Design life—sealed surface treatment	Bitumen: 10 to 15 years—39%; 15 to 20 years—36%; 20 to 25 years—14%; Asphalt: 20 to 25 years—35%; 25 to 30 years 29%
4	Technical guidelines followed	Multiple responses—most used Austroads Guide to Road Design; Several did not indicate they were using LVR standards

(continued)

Table 1 (continued)

RQ	Summary of question	Main response (percentages rounded)
	Life cycle management and renewal	
3	Average pavement age—unsealed	5 to 10 years—11%; 10 to 15 years—11%; not recorded 64%
5	Renewal activities—unsealed	14 participants—resheeting with new gravel; 14 participants—resheeting with gravel nearby; 7 participants—stabilisation
3	Average pavement age—sealed	10 to 20 years—11%; 20 to 30 years—15%; 30 to 40 years—15%; not recorded 52%
3	Average age—bitumen sealed surface	5 to 10 years—7%; 10 to 15 years—18%; 15 to 20 years—25%; not recorded—32%
5	Renewal activities—sealed (number of response)	Bitumen resealing—26; replacement of drains/culverts—17; stabilisation—15.
5	Innovative or out of the ordinary activities—extending service life of LVR networks	6 participants—includes stabilisation; rubber S35E for reseals; widening road for larger trucks
	Challenges and strategies	
6	*Main challenges in managing LVR network—rank 1 to 5* Inadequate funding—53 responses; ageing network—42 responses; increasing traffic volume—31 responses; increasing construction cost—25 responses	
7	*Strategies for managing challenges* New assets to meet traffic demands—17 responses; leveraging funds—17 responses; investigate emerging techniques for asset preservation and management—15 responses. Set and deliver on preventative maintenance targets—14 responses.	

"RQ" = Research Question Number (1 to 7)
[a]Individual data varied. Thus used other sources

Discussion and Conclusion

Main Findings with Respect to Research Criteria

Overall, the above results, which have addressed all research questions, indicate that low-volume road practices vary significantly between local governments. The main findings of this research are summarised below.

Low-Volume Roads Definition

The definition of a LVR in the respondent group was typically less than 200 vehicles per day for both unsealed and sealed LVRs (this is particularly the case for rural and lower populated areas). Thus, a LVR could be defined as one with 200 vehicles per day or less.

Unsealed Low-Volume Roads

Most unsealed LVRs, particularly in rural areas, provided a level of service that communities were neither satisfied nor dissatisfied with. More frequent inspections of unsealed LVRs could be related to better community satisfaction.

Councils adopted a range of grading schedules. Generally, those with lower community level of service satisfaction undertook less frequent grading than with satisfied communities. However, higher grading frequencies did not always result in good community satisfaction.

The average design life for unsealed LVR pavements adopted by the survey participants was 17 years. Most survey participants did not know or record the current average age of unsealed LVR pavements that they managed.

Improvements in the awareness of technical guidelines specifically covering the planning, design, and construction of unsealed LVRs were possible.

Sealed Low-Volume Roads

Most survey participants did not know the current average age of the sealed LVR pavements that they managed. Responses from those participants which recorded this figure indicated that this average age was about 39 years.

Knowledge of the current average age of bitumen seals on LVRs was better than that for their associated road pavements, but was still not high. This figure was 23 years, which exceeded the average design age of new sealed road pavements by 6 years. This result suggests that further expenditure on the renewal of bitumen seals in the surveyed area is required to maintain adequate levels of service.

Challenges and Strategies for Future Development

The survey found that the most common system to record and manage defects on LVR networks was through specialised computer systems. While use of these systems did not guarantee community satisfaction, the use of a formal recording system aided it.

Although the literature suggests that undertaking road safety reviews or audits on LVRs could provide significant safety and economic benefits from reduced crash rates, only a small proportion of participants undertook road safety reviews or audits on LVRs.

Maintaining and renewing drainage systems was important for maximising LVR useful life.

The major challenge in the effective management of LVRs in the surveyed area were found to be inadequate funding and an ageing network. The most effective strategies to manage these challenges were leveraging grants and investigating emerging techniques for asset preservation and management.

Innovative Practices

The study found a number of innovative solutions for managing the life cycle of sealed LVRs. They included:

- Using rubber S35E (a polymer modified bitumen) for reseals and using cement-stabilised sand as a bridging layer over poor subgrades. These practices may only be practical in certain circumstances.
- Widening selected roads to improve transportation.

There were a number of recommendations, aimed at implementing the above findings.

Meeting Sustainability Goals

Productivity: Community productivity depends on good, safe roads that foster efficient transportation and property access. Therefore, enhancing the low-volume road network through developing and maintaining it in a good, safe condition is important. The survey indicated that a number of improvements could be made in these areas.

Safety: While road safety reviews and audits are important in maintaining community well-being, it was found that only a small proportion of respondent councils undertook them.

Usability: The road systems surveyed appeared to be usable overall. Improved maintenance, such as a higher frequency of grading on unsealed roads and maintaining drainage, is important in this process.

Social equity: The main factors in achieving social equity were having as many roads as possible sealed and having a good level of service. The decision to seal a road can however lead to significantly increased life cycle costs, which require consideration. Similarly, the proportion of councils that indicated that their communities were neither satisfied nor dissatisfied with their level of service was quite high. There is room for improvement in both of these areas.

Sustainable environmental management: Most unsealed road grading materials are taken from quarry sources, with the potential to damage the environment. Some councils were using stabilised materials for this purpose. Some were using polymer and other innovative seals. While questions on drainage were not asked, this is also significant environmentally.

Resilience: Resilience was not specifically reviewed. However, roads that are sealed and have good drainage could be expected to be more resilient than unsealed or poorly drained roads.

Conclusion

In conclusion, this study has contributed significantly to the body of knowledge that exists in network management in low-volume roads in Australia, and in particular in the state of NSW. It has been found that there is a significant opportunity to increase the level of road safety reviews for these roads. Leveraging funding, ensuring that new low-volume roads meet future traffic demands, and continuing to investigate best practices for life cycle based sustainable asset management, development and preservation are considered the most successful strategies to meet these challenges.

Future work could include extending the rigour of this research through more detailed participant surveys and assessment and extending the scope of a study of this type to assess in depth practices in the development and management of low-volume roads throughout Australia, and ultimately internationally. Other studies could focus in depth in other types of low-volume roads, and in-depth research into use and potential for innovative practices.

References

AASHTO. (2001). *Guidelines for geometric design of very low-volume local roads (ADT 400)*. Washington, DC: American Association of State Highway and Transportation Officials (AASHTO).

ARRB. (1993). *Unsealed roads manual: Guidelines to good practice*. Vermont South: Australian Road Research Board (ARRB) Ltd.

ARRB Group. (2005, July). *Sealed local roads manual—Guidelines to good practice: Design, construction, maintenance and rehabilitation of pavements* (Revised and expanded ed.). Vermont South: ARRB Group Ltd.

Austroads. (1998). *Responsibilities for local roads* (Publication no. AP-129-98). Sydney: Austroads Incorporated.

Austroads. (2000). *Use of recycled materials and the management of roadside vegetation on low trafficked roads* (Publication no. AP-R154-00). Sydney: Austroads Incorporated.

Austroads. (2006). *Asset management of unsealed roads: Literature review, lga survey and workshop (2000–2002)* (Publication no. AP-T46-06P). Sydney: Austroads Incorporated.

Austroads. (2009a). *Guide to pavement technology part 6: Unsealed pavements* (Publication no. AGPT06-09). Sydney: Austroads Ltd.

Austroads. (2009b). *Guide to road transport planning* (Publication no. AGRTP-09). Sydney: Austroads Incorporated.

Austroads. (2015a). *Guide to traffic management part 1: Introduction to traffic management* (Publication no. AGTM01-15) (3rd ed.). Sydney: Austroads Ltd.

Austroads. (2015b). *Seal design improvement for low volume roads: Technical report* (Publication no. AP-T302-15). Sydney: Austraods Ltd.

Austroads. (2016a). *Guide to traffic management part 4: Network management* (Publication no. AGTM04-16). Sydney: Austroads Ltd.

Austroads. (2016b). *Reforming remote and regional road funding in Australia* (Publication no. AP-R526-16). Sydney: Ausroads Ltd.

Brundtland, G. H. (1987). *Report of the World Commission on Environment and Development—Our common future*. New York, NY: United Nations General Assembly.

Bureau of Infrastructure Transport and Regional Economics. (2017). *Yearbook 2017: Australian infrastructure statistics, statistical report*. Canberra: BITRE.

Department of Infrastructure and Regional Development. (2017). *Annual report for the Department of Infrastructure and Regional Development 2016–17*, Canberra.

Douglas, R. A. (2016). *Low-volume road engineering: Design, construction, and maintenance* (1st ed.). Boca Raton, FL: CRC Press.

Drexhage, J., & Murphy, D. 2010. *Sustainable development: From Brundtland to Rio 2012*, Background, Paper prepared for consideration by the High Level Panel on Global Sustainability at its first meeting, September 19, 2010.

Faiz, A. (2012). *Transportation research circular e-c167: The promise of rural roads: Review of the role of low-volume roads in rural connectivity, poverty reduction, crisis management, and livability*. Washington, DC: Transportation Research Board of the National Academies.

Giummarra, G. (2003). Establishment of a road classification system and geometric design and maintenance standards for low-volume roads. *Transportation Research Record: Journal of the Transportation Research Board, 1819*, 132–140.

Giummarra, G. (2009). *Unsealed roads manual: Guidelines to good practice* (3rd ed.). ARRB Group: Vermont South.

Howard, J., Jeff Roorda and Associates, & Institute of Public Works Engineering Australia (IPWEA). (2013). *Road asset benchmarking project 2012: Road management report*. Sydney: IPWEA NSW Division Roads and Transport Directorate.

IPWEA NSW. (2015). *Local government functional road classification*. Sydney: IPWEA NSW Roads and Transport Directorate.

Irwin, L. H. (2003). Foreword to eighth international conference on low-volume roads 2003. *Transportation Research Record: Journal of the Transportation Research Board, 1819*, xvii–xviii.

Verity, S. (2018). *2017 Road management report: Road asset benchmarking project*/prepared by Steve Verity, Technologyone. Sydney: IPWEA NSW Division, Roads and Transport Directorate.

World Road Association. (2003). Technical Committee 16 Network Operations. *Road network operations handbook*. World Road Association.

Zimmerman, K., & Peshkin, D. (2003). Applying pavement preservation concepts to low-volume roads. *Transportation Research Record: Journal of the Transportation Research Board, 1819*, 81–87.

A Community-Driven Nature-Based Design Framework for the Regeneration of Neglected Urban Public Spaces

Gloria Osei, Alison Pooley, and Federica Pascale

Definitions The paper will have the corresponding word definitions, unless stated:

Urban: An environment of high population density and advanced infrastructure.
Sustainability: The built environment's ability to value social, economic and environmental factors in equal measure, enduring into the future.
Community: The local users of nature-based public spaces including residents and local workers.
Nature: Land containing green and/or blue spaces within the ability of human management.
Public space: Spaces designed for access to all the community, and to be utilised for leisure and respite, not just for passage, e.g. walkway.
Neglected: Public spaces that have been socially abandoned because of, or leading to, failure to accomplish its designed purpose.

Introduction

Nature-Based Solutions in Urban Environments

Nature-based solutions are approaches stimulated by nature through the use of green and/or blue spaces and interventions. Nature-based solutions are currently found in coastal, river, watershed, freshwater, forested, mountainous, rural and urban ecosystems (IUCN, 2019). The term nature-based solution is a recent concept and is currently being used for the UNFCCC (2018) Paris Agreement to help 'combat climate change and adapt to its effects', and the EU Horizon 2020

G. Osei (✉) · A. Pooley · F. Pascale
Faculty of Science and Engineering, School of Engineering and the Built Environment, Anglia Ruskin University, Chelmsford, UK

L. Scott et al. (eds.), *Sustainable Ecological Engineering Design*,
https://doi.org/10.1007/978-3-030-44381-8_5

(European Commission, 2017) in their innovation and research program for 'smart, sustainable and inclusive growth'. The implementation of nature-based solutions in urban environments has been considered for: climate change adaptation by aiding heat, energy and flood regulations (Kabisch, Korn, Stadler, & Bonn, 2017); natural disaster regeneration by allowing ecosystems to restore and increase (Asian Development Bank, 2016); agricultural sustainability by improved integrity of water management and food security (Sonneveld, Merbis, Alfarra, Unver, & Arnal, 2018).

Public Space

Public spaces such as coasts, canal fronts, walkways and parks are used daily as an accessible domain, vital to the day-to-day activities of human beings. The nineteenth century saw the beginning of public parks as a space for social control as well as a meaningful way to draw communities out of their busy lifestyle and into the appreciation of the opposing, natural environment (Holland, Clark, Katz, & Peace, 2007). Modern urban public spaces are becoming less individual in their national identity and increasingly completive with other nations as a factor of development. This competitive drive has therefore meant a movement from the traditional aims of community centred space, to achieving international value. As space is expensive in the urban environment (Nabarro & Smart, 1978), there is demand for commercial development over 'other' types of space, potentially squeezing out nature-based solution spaces. This has proven a strain on local councils in reduced funds for such developments. Nature-based public spaces have hence been seen as unprofitable developments, that need justification of cost effectiveness and value, demonstrated by the debate for a housing development to replace the Limehouse Triangle biodiversity scheme originally formed by the Tower Hamlet council's Locksley Environment project (Brooke, 2019).

An inevitable consequence on the contending attitude of public space development has been the effect on originality of design, steering designs from representing the individual location/community identity to attention on globally accepted design. Public spaces are now therefore designed with geometric shapes to satisfy a modern, clean, global design, such as that adopted by the design of Granary Square, King's Cross, contrary to the emphasis of unconfined nature with an assortment of reminders of the past. Kim and Kwon (2018) note that new urbanism design and sustainable urbanism design draw inspiration from traditional design and historical relevance whilst post urbanism designs are inspired by contemporary design influenced by globalisation and mediatisation (media shaped communication).

The increase in pseudo-public spaces has seen a reduction in the 'free', 'accessible' and 'all-are-welcome' mode of public spaces, being transformed into areas in which you can potentially be asked to leave if your presence is not wanted, e.g. the prevention of the Occupy London protesters in Paternoster square, London, due to its private ownership. As taxes are used to fund, to an extent, the production of public spaces, there have been calls for the need to look into the distribution of these funds to optimise the desired effects of improved health and well-being for the community. This distribution could potentially include neighbourhood percentage shares that allow for the promotion of regeneration amenities, even in underprivileged areas, for the encouragement of human and economical capital (Radoslav et al., 2012).

Summary of the Problem

The issue at the core of this research is that public spaces by their very definition are spaces for the public, yet development of these spaces as well as the literature around the subject is far more theoretical and professionally influenced, rather than being community influenced. Communities within many societies are very diverse, ranging in: culture, ethnicity, class, age and social preference. With the global increase in diversity within urban areas progressing rapidly, it has become necessary for leaders of regeneration to take these diversities into consideration, especially if sustainability is likely to be achieved. The complexity of needs from one region to the other gives an enhanced need for professionals and authorities to regenerate with the community, and not for them (Levy-Storms, Chen, & Loukaitou-Sideris, 2018).

Literature Review

The literature on urban regeneration gathers a vast range of fields of knowledge, including politics, social science, biodiversity and urban planning. The research divides into differing themes such as crime-prevention, human health and increased general economy. Within the capacity of this paper, a narrative literature review begins to draw information on community driven, nature-based public spaces in neglected urban sites.

Social Neglect of Spaces

Social neglect can lead to an area becoming a derelict site, locations that have fallen victim of disuse and been relinquished or abandoned (Bradshaw & Chadwick, 1980). These sites are not only found on built-up development, but could include parks, walkways and open spaces (Lans, Mulder, & van Rij, 2007). Socially neglected spaces are not used to their optimal potential as prescribed by the original design, and therefore fail to fulfil their design purpose. This type of neglect is brought about by more complex reasons than that of Brownfield sites, and often as a result of a combination of factors. Some possible factors are: lack of funding for needed facilities (Grimski & Ferber, 2001, p. 143; Kim, 2016), need for maintenance (Grimski & Ferber, 2001, p. 148; Rosol, 2010, p. 551), increase competition of groups within the public for differing uses of the space (Malone, 2002), as well as a simple lack of the 'right' facilities provided by the space (Phillips, Walford, Hockey, Foreman, & Lewis, 2013). For the neglect of public spaces to be averted, design must focus on sustainability through careful strategies.

Potential of Public Spaces to the Community

The availability of public spaces to every type of user means that design must be sensitive to all abilities and vulnerabilities. The scope of vulnerability includes, but is not limited to: elderly, physically disabled, young, mentally ill or homeless. Many vulnerable individuals identify little differentiation between their physical and psychological well-being, and speak of both as critical aspects when discussing open spaces (Levy-Storms et al., 2018). The concentration on the possible adaptations of urban regeneration for the inclusive ease of use by the more vulnerably members of society, have few articles addressing it, with reduced focus on possible design strategies, as spoken by Pani (2016) in *Improving the lives of those with dementia through urban design*.

The public natural environment is aimed at inviting all to use its space without the separation of class or identity. Modern facilities such as seating areas have, on the other hand, been criticised for being arranged in a way that promotes individuality rather than social cohesion. The popular isolated group seating with separating metal handle encourages 'atomised' public life, whilst the use of clearly defined narrow pathways promotes the brisk movement of people rather than strolling enjoyment (Byass, 2010).

The unstoppable rise of the use of social media presenting itself as a virtual public space (Floridi, 2015) poses as a competitor for physical public space but with potential of isolation from real-life interaction and tangible emotional connections. Albert Mehrabian (2017) in his extensive studies of communication complexities expresses the importance of non-verbal, physical communication, more so the limited ability of the dependency on purely words as a mode of interaction. There is therefore an increasing social need to provide effective public space in order to stimulate social cohesion.

Nature-Based Solutions

Nature-based solutions are designed to 'help societies address a variety of environmental, social and economic challenges in a sustainable way' (ECDG, 2015, p. 5). EU Horizon 2020 funded several projects demonstrating innovative nature-based solutions in cities in the 2017 call: Nature-based solutions for inclusive urban regeneration.

Research into the field of nature-based public spaces as a nature-based solution has declared a long list of benefits on the necessities of its presence within urban environments, including economic development (Buckley & Brough, 2017), improvement of environmental conditions and comfort (Boeri, Gaspari, Gianfrate, & Longo, 2017), aesthetics (Julier, 2005; Kim & Kwon, 2018), social cohesion (Ivanova, 2016) and human health and well-being (van den Bosch & Ode Sang, 2017). Systematic research in the immersion of humans with nature, however, shows that basic physical walks through a natural environment has little effect on a human's connectedness with nature; amplified contact with nature by emotional and sensory activities, on the other hand, provides a greater connection with nature, potentially resulting in increased human benefits (Lumber, Richardson, & Sheffield, 2017). The availability for these experiences within the natural environment is associated both with the size of the natural environment and the distance the individual must travel to access the amenities (Ekkel & de Vries, 2017; Russo & Cirella, 2018). Russo et al. (2017) suggest that residents should have 2 ha or more of natural green space within 5 min or less walking distance from their residence. It is however considered that even when there is availability, 'it is not just the physical environment that encourages or discourages people to come out in public, but also the opportunity to see something different' (Holland et al., 2007, p. 64). In order to create a space that encourages community participation, it is essential that the public are involved in the design process of their space.

Nature-Based Public Space Planning and the Community

The benefit of nature-based design has also been documented in its use for the improvement of environmental conditions and comfort (Boeri et al., 2017), and even as a means of reconnecting communities after national conflict (Ivanova, 2016). The sustainability of the urban environment as a whole has drawn out increased questions regarding whether more contemporary designed public spaces would benefit from a heightened vernacular architectural approach, focusing on traditional and indigenous design (Kim & Kwon, 2018).

Community input has further been assessed in the context of strategies to best involve the public; through open-ended, personal communication prior to development (Golden, 2014), post-psychological response from long-term and recent residents (Bélanger, Cameron, & de la Mora, 2012; Salone, Baraldi, & Pazzola, 2017), as well as the need of careful design aesthetics to incorporate culture, place identity and place attachment (Julier, 2005); Mihaylov & Perkins, 2014), aimed at recognising the individuality of the community. Quality of public space due to class prejudice (Wolch, Byrne, & Newell, 2014) is also recognised in literature, where it is acknowledged that more vulnerable communities face gaining a less valued public space to those in a more wealthy communities.

It has been argued that the built environment pays little attention to the community as relevant agents when it comes to planning and design (Ravetz, 1980). In *Remaking Cities*, Ravetz gives an analysis of the urban environment and the use of the built environment as a control mechanism for groups within society over others in the community. More recently advancements to produce methods of community involvement have been made in the planning process of public space regeneration (Cilliers & Nicolene, 2016; Cilliers & Timmermans, 2014; Stratigea, Kikidou, Patelida, & Somarakis, 2017).

There are many professional bodies that play as actors in the process of regeneration with some being called in the past to expand their roles to address the needs of the community. In the UK, Town Centre Management has been regarded as a tool for improving town centre environments. The findings of Otsuka and Reeve (2007) identify that Town Centre Managers are influential to local authorities in regeneration by providing information on local community needs and interests. However they are limited as their work does not see them directly communicating with the users for development purposes, but provide report through observation. They are further unable to perform as a tool for social equity in regeneration, as they are usually an under-resourced small team.

The use of public artists as a medium between the community and authorities within urban regeneration projects is increasingly being explored. UK City of Architecture and Design 1999, Glasgow, shows the success of community-driven regeneration from the offset. The use of the new genre public art approach allowed for the production of identity and ownership through engagement with community (Sharp, 2007). Sharp reflects that for urban regeneration to succeed, there should be 'processes through which communities are activated and stimulated into action' (Sharp, 2007, p. 288).

Research Review and Methodology

To explore current frameworks for nature-based public space design, this research uses a meta-synthesis literature review method (Walsh & Downe, 2005) to build and analyse secondary data on the available nature-based public space design frameworks to date. The research looks into a number of frameworks and analyses their focus and content, evaluating the extent in which community preference is considered. The methodological approach employed is a grounded theory analysis. This approach is used to gather data to the point of saturation, with the intention of analysing if current design frameworks are community-driven. For this purpose, an inductive research approach is necessary. Other alternative techniques using a deductive approach are not used. This is because the paper intends to explore the phenomena of design representation in nature-based public space frameworks. This search is best done by analysing and reanalysing different aspects of existing nature-based pubic space design frameworks. It will interpret and interrogate the theories adopted and evaluate the common consensus of representation.

Research Method

When looking at the available public space design frameworks, this paper considers a search into the available academic articles providing this service to designers and planners. The records were identified through database searching. With the exact phrases located anywhere within the article, '"public space" "nature based" "framework"' was placed into the search engine. The search originally brought up 728 results from the year 2015 onwards, patents and citations excluded. Those results that had a new design aiding 'framework' mentioned as part of the title, abstract or keyword was then further investigated. The frameworks were also limited to design for general multiuse spaces for the open use of the community, and restricted frameworks for specified areas were disregard.

There were found very little results between the years 2015 and 2016 and therefore the search was restricted to 2017 onwards, allowing for the evaluation of frameworks following a time where the concept of nature based solutions was more popular. After this date the authors would have the availability of a wide range of literature. In so doing, the thesis looks at 14 frameworks from 2017 focusing primarily on nature-based public space design. In reducing the search of design frameworks to that of nature-based public spaces only, the literature review focuses on analysing the frameworks on the basis of: (1) themes covers, (2) steps required to enforce the framework, (3) the source of information used to produce the framework and (4) testing of the framework. The source of information is generalised into two categories in this research: literature review and community participation. Community participation is considered as interacting with the community to gain their views, and therefore site observations are not considered a valid community participation method.

Research Results

The literature review on nature-based solutions shows significant potential in its use to benefit society and the urban environment, with nature-based public spaces providing a possible forum in which these solutions could be actualised. To allow for the optimum implementation of public spaces as a nature-based solution the paper looks into the articles providing frameworks that could aid the design of these spaces. As the frameworks would be considered as a protocol for design, the paper aims to evaluate the considerations given towards providing tangible user requirements in public space development.

To explore the extent community preference is considered in current nature-base public space design frameworks, the paper compares the existing studies in Table 1.

Table 1 Framework comparisons

Framework title	Steps to enforce	Source
NBS co-benefits FW [Raymond et al. (2017)]	1. Identify problem or opportunity 2. Select NBS and related actions 3. Design NBS implementation processes 4. Implement NBS 5. Frequently engage stakeholders and communicate co-benefits 6. Transfer and upscale NBS 7. Monitor and evaluate co-benefits	Literature review
Peri-Urban Agriculture (UPA) FW [Artmann and Sartison (2018)]	1. Vision definition 2. Implementation efficiency 3. Impact efficiency	Literature review (*166 Academic articles*)
Nexus FW [Chiabai, Quiroga, Martinez-Juarez, Higgins, and Taylor (2018)]	1. Identify 'driver'—climate change 2. Analyse 'pressure' on green space due to driver 3. Evaluate the 'state' needed—size or quality of green space and ecosystem services 4. Select 'exposure' to space—passive, consumptive, active 5. Evaluate health 'effects'—direct and indirect, positively or negatively 6. Apply any 'adaptations' throughout and enforce any 'mitigations'	Literature review (*117 study articles*)
Human centred FW [Swierad and Huang (2018)]	1. Identify options for park activities 2. Build action strategies: conventional planning + human-cantered design 3. Evaluate connection to: family and loved ones, community and neighbourhood, self, nature 4. Analyse improved health and well-being 5. Improve community engagement	Community participation (*20 residential interviews*)

(continued)

Table 1 (continued)

Framework title	Steps to enforce	Source
Urban resilience in food practices FW [Dezio and Marino (2018)]	1. Pressures—look at problems that need to be addressed 2. Action—plan and implement the individual actions needed to address pressures 3. Expected impact—note expected results, and the planned and/or performed monitoring 4. Indicator—evaluate impact indicators that verify effectiveness of actions	Community participation (*50 practices—activation of gardens and training for community*)
Human mental health FW [Buckley and Brough (2017)]	1. Quantify types of park users and park uses in a manageably small number of categories 2. Quantify proportional changes in mental health parameters, for different categories of people and experiences 3. Quantify economic values of mental health outcomes, using national economic statistics for public health	Literature review
Urban Green Space—3 stage FW (UGS-3CC) [Mukherjee and Takara (2018)]	1. Contextual concept—thinkers—corresponds to the pre-design stage 2. Core competency—makers—corresponds to the design and Implementation 3. Contribution calculation—traders—corresponds to monitoring, evaluation and improvement	Literature review (*Lit. review and site observations*)
Small park design FW [Currie (2017)]	'Not to construct a list…but identify foundational design principles'	Literature review and community participation (*site observations, Lit. Review, 8 professional and 7 community interview*)
Urban green equity FW [Nesbitt, Meitner, Sheppard, and Girling (2018)]	1. Consider temporality 2. Investigate condition and preference 3. Deal with ownership 4. Representation 5. Procedure 6. Analyse community desire to participate 7. Analyse community ability to participate	Literature review (*45 articles 9 studies from books 8 professional reports*)

(continued)

Table 1 (continued)

Framework title	Steps to enforce	Source
FW for Children's Human Nature Situations—(ACHUNAS) [Giusti, Svane, Raymond, and Beery (2018)]	1. Identify qualities of Significant Nature Situations with the potential to 'connect' 2. List abilities of Human Nature Connection: (a) Being IN Nature, (b) Being FOR Nature, (c) Being WITH Nature	Literature review (*26 Professional Interviews 275 Studies in Literature*)
An affordance FW for green space [Lennon, Douglas, and Scott (2017)]	1. Encourage the desire to actualise affordance 2. Enhance confidence among potential users in the use of that green space 3. Prompt greater frequency and range of uses of that green space 4. enhance potential of that green space to assist societal health and well-being	Literature review
An integrated green space FW-health and well-being [Douglas, Lennon, and Scott (2017)]	1. Maximise streetscape greenness and green space provision 2. Engage all users in planning 3. Provide differing: lengths of paths, environment and varying degrees of challenge 4. Institutionalise good maintenance and renovation regimes	Literature review
Conceptual FW for landscape design intended for frail elderly [Charras, Bébin, Laulier, Mabire, and Aquino (2018)]	1. Attract curiosity of users and secure the environment 2. Allow for social uses: occupancy opportunities of spaces 3. Evaluate way spaces invites curiosity of users 4. Provide functional and ergonomic characteristics of setting 5. Facilitate orientation by identification and structuring of landscapes 6. Insure maintenance of uses of gardens by integrating the evolution and growth of natural elements that compose them	Literature review (*7 Articles Secondary data: Quantitative and Qualitative*)

(continued)

Table 1 (continued)

Framework title	Steps to enforce	Source
Social cohesion and urban green space conceptual FW [Jennings and Bamkole (2019)]	1. Locate presence and/or access to urban green spaces 2. Identify social determinants of health: social cohesion/ social capital 3. Evaluate potential outcomes 4. Analyse physiological and behavioural responses 5. Note benefits to physical and psychological health	Literature review

Themes Covered by the Framework

The frameworks have all suggested the benefits of nature in public spaces to human health: whether physically or psychologically and either within the full framework or as part of the frameworks scope. They highlight the multi-beneficial aspects of nature towards human living, from social cohesion (Jennings & Bamkole, 2019) and the feeling of belonging or ownership, to the chemical benefits that are implied to the environment (Chiabai et al., 2018) for cleaner air and therefore health. Some frameworks where specifically designed with a set of people at its centre, for example, for the benefit of children (Giusti et al., 2018), or the elderly (Charras et al., 2018).

Clarity of Framework for Design Purposes

When considering the steps required in enforcing the framework, there are obvious variations in the number and clarity of steps needed to follow the framework. In terms of readability and ease of following the pathways given by the framework, those frameworks posing more steps tend to provide a clearer guide for assessment of design and planning of nature-based public spaces. Raymond et al. (2017) and Chiabai et al. (2018) are examples of frameworks that identify precise stages needed to achieve the necessary results from nature-based design as well as classifying a necessary stage of monitoring and adaptation throughout the design process relevant for a sustainable design.

Source of Information for Development of Framework

The source of information used to produce the framework shows that 11 frameworks out of the possible 14 favoured the use of academic studies as secondary supply, with limited interactions with human participants for the creation of the framework. However, Swierad and Huang (2018) and Dezio and Marino (2018) both provide frameworks from primary analysis of community preference, and Currie (2017) made use of both academic documentation and community interviews.

Tests for the Validity of the Framework

The testing of the framework by others showed that Raymond et al. (2017) have been used by academics such as Frantzeskaki (2019). Frantzeskaki used the framework to produce seven lessons for nature-based planning which combines the case studies of 15 European nature-based solution spaces. Other frameworks were internally tested by the author through case studies to verify its effect on public space design (Charras et al., 2018).

Discussion

For nature-based public space frameworks to provide an efficient guide for the design, incorporating the diverse benefits nature-based solutions offer, it is evident a broad theme must be covered so as the physical and physiological profits may be realised. Clear and precise steps that link the designer or planner to the tangible requirements of community needs are also vital for an effective protocol. A majority of the frameworks express the necessity for regular maintenance within the lifespan of nature-based public spaces to ensure sustainability of design purpose.

Considering the validity of current frameworks to provide considerations towards tangible user requirements in public space developments, those that used a community centred source of primary investigation also consequently provided a framework for the active participation of users within public spaces. Dezio and Marino (2018) further provide a step for the improvement of community engagement as a final stage in their Human Centred Framework. The high proportion of literature reviews for framework sources provide evidence of underrepresented community input within public space planning.

Very few frameworks had been tested for their validity of use, yet in the situations where these tests had taken place, it was necessary to use case studies to validate and draw conclusions; these points to the need of actualised representations to enable clear validity. The physical adaptations of the framework could be internally represented by community involvement at the initial stages of development. The need for actualised testing of frameworks using case studies shows that the use of only literature review is limiting, and therefore a bottom-up approach would prove more efficient. Hollander, Foster-Karim, and Wiley (2018) identify the need of literature review to draw general understanding when looking at public spaces and urban inclusivity, yet for in-depth analysis, there is a need for addressing community reactions.

Lessons to Be Learnt

Care needs to be taken in order to represent the whole community so as to accomplish a democratic and well-informed basis of design research. Ball (2004) explores the negative views associated with community-driven regeneration and expressed the concern of non-representative data, where middle class, elderly and Caucasian people are far greatly represented. Individuals with dominant personalities and/or unrealistic aims also contribute to un-representative data. Other negative experiences researched are that community-driven processes are time consuming, over democratic, and hindered by a lack of trust between community and authority. These concerns reflect the need to collect community insight in an inclusive and broad way; this could be by active outreach involvement.

Van den Bosch and Ode Sang (2017) supplied a review on literature focused on the health benefits that could be provided to the public by nature-based solutions. Their analysis showed there is limited research into the psychological effects exposure to nature-based solutions has compared to physical effects. Acquiring evidence on nature's effects can be difficult as other factors must be eliminated to ensure nature is the stimulus of human health, as well as the length of time needed to ensure significant results are achieved. The extent of psychological benefits of nature on the community's daily well-being need further investigation.

Conclusion

The literature review of nature-based solutions expresses significant findings on the diversity of benefits associated with the use of nature within urban environments. Nature shows great potential in enhancing environmental, economical and human progress, allowing sustainable development. The reviews on human enhancements have however been dominated by the physical benefits of nature, rather than psychological health and well-being. The research results show literature on nature-based design frameworks is skewed towards theoretical and professional influence, disregarding community involvement. For the research on nature-based solutions benefits to be realised in practice, design frameworks for public spaces must open dialogues with communities, unlocking the psychological profits that could be available. This paper brings to light some themes covered by current design frameworks, such as design tailored for specific vulnerable groups. Out of the 14 frameworks analysed, only Raymond et al. (2017), Artmann and Sartison (2018), Chiabai et al. (2018), and Mukherjee and Takara (2018) did not primarily focus on direct human benefits. Nevertheless this paper notes that though 10/14 framework themes identified humans as key beneficiaries to nature-based public spaces, only 3/14 frameworks actively incorporated communities in their process of framework production.

The Next Step

The progressing PhD research consequently focuses on the direct benefits nature provides to human daily well-being, not by replicating the research of its potential benefits, but by considering how design of nature-based public spaces could be used to increase the experience of humans within this environment. The PhD research focuses on primary qualitative data collection to place the community at the focal point of design and suggestive changes. This will look at how increased immersion of vulnerable people within neighbouring neglected nature-based public spaces could be improved. The contribution to knowledge in the field of the built environment is therefore the production of a new nature-based public spaces framework for redeveloping neglected sites in urban areas. In addressing community needs and therefore encouraging social cohesion the project will aim to gain an understanding of current and future needs of local communities in nature-based public space design and implement unobtrusive strategies to improve usability and safety of all users.

Limitations

The limitation of this research paper is the reduced timeframe in which potential frameworks were collected and analysed, between 2017 and 2019. Analysis of frameworks from a wider range of time is beneficial to correlated clear proportions between professional preference on the use of literature review to community interaction.

References

Artmann, M., & Sartison, K. (2018). The role of urban agriculture as a nature-based solution: A review for developing a systemic assessment framework. *Sustainability, 10*, 1937. https://doi.org/10.3390/su10061937

Asian Development Bank. (2016). *Nature-based solutions for building resilience in towns and cities: Case studies from the Greater Mekong Region*. Manila: Asian Development Bank Institute.

Ball, M. (2004). Co-operation with the community in property-led urban regeneration. *Journal of Property Research, 21*(2), 119–142.

Bélanger, H., Cameron, S., & de la Mora, C. (2012). Revitalization of public spaces in a working class neighborhood: Appropriation identity and urban imaginary, Chapter 4. In H. Casakin & F. Bernardo (Eds.), *The role of place identity in the perception, understanding, and design of built environments* (pp. 47–62). Sharjah: Bentham Science.

Boeri, A., Gaspari, J., Gianfrate, V., & Longo, D. (2017). Accelerating urban transition: An approach to greening the built environment. *WIT Transactions on Ecology and the Environment, 223*, 3–14.

Bradshaw, A. D., & Chadwick, M. J. (1980). *The restoration of land: The ecology and reclamation of derelict and degraded land*. Berkeley, CA: University of California Press.

Brooke, M. (2019). *Limehouse Triangle green space lost by Tower Hamlets Council vote for building scheme next to Regent's Canal*. [Online]. Retrieved April 6, 2019 from https://www.eastlondonadvertiser.co.uk/news/politics/limehouse-triangle-lost-by-council-vote-for-tower-block-1-5972204

Buckley, R. C., & Brough, P. (2017). Economic value of parks via human mental health: An analytical framework. *Frontiers in Ecology and Evolution, 5*(16), 1–9.

Byass, R. (2010). From public garden to corporate plaza: Piccadilly Gardens and the new civic landscape. *Journal of Landscape Architecture, 5*(1), 72–83.

Charras, K., Bébin, C., Laulier, V., Mabire, J., & Aquino, J. (2018). Designing dementia-friendly gardens: A workshop for landscape architects: Innovative practice. *Dementia (London, England)*, 1–9.

Chiabai, A., Quiroga, S., Martinez-Juarez, P., Higgins, S., & Taylor, T. (2018). The nexus between climate change, ecosystem services and human health: Towards a conceptual framework. *Science of the Total Environment, 635*, 1191–1204.

Cilliers, E. J., & Nicolene, D. F. (Eds.). (2016). *49th ISOCARP Congress, Brisbane, October 2014*. The Hague: ISOCARP.

Cilliers, E. J., & Timmermans, W. (2014). The importance of creative participatory planning in the public place-making process. *Environment and Planning B: Planning and Design, 41*(3), 413–429.

Currie, M. A. (2017). A design framework for small parks in ultra-urban, metropolitan, suburban and small town settings. *Journal of Urban Design, 22*(1), 76–95.

Dezio, C., & Marino, D. (2018). Towards an impact evaluation framework to measure urban resilience in food practices. *Sustainability, 10*, 2042. https://doi.org/10.3390/su10062042

Douglas, O., Lennon, M., & Scott, M. (2017). Green space benefits for health and well-being: A life-course approach for urban planning, design and management. *Cities, 66*, 53–62.

ECDG. (2015). *Towards an EU research and innovate policy agenda for nature-based solutions and re-naturing cities*. Luxembourg: European Commission.

Ekkel, E. D., & de Vries, S. (2017). Nearby green space and human health: Evaluating accessibility metrics. *Landscape and Urban Planning, 157*, 214–220.

European Commission. (2017). *H2020-SCC-NBS-2stage-2016*. [Online]. Retrieved February 11, 2019 from https://cordis.europa.eu/projects/result_en?q=(relatedProgramme/programme/code%3D%27SCC-02-2016-2017*%27%20OR%20relatedSubProgramme/programme/code%3D%27SCC-02-2016-2017*%27)%20AND%20contenttype%3D%27project%27

Floridi, L. (2015). *The online manifesto: Being human in a hyperconnected era*. Cham: Springer.

Frantzeskaki, N. (2019). Seven lessons for planning nature-based solutions in cities. *Environmental Science & Policy, 93*, 101–111.

Giusti, M., Svane, U., Raymond, C. M., & Beery, T. H. (2018). A framework to assess where and how children connect to nature. *Frontiers in Psychology, 8*, 21. https://doi.org/10.3389/fpsyg.2017.02283

Golden, S. M. (2014). Occupied by design: Evaluating performative tactics for more sustainable shared city space in private-led regeneration projects. *WIT Transactions on Ecology and the Environment, 191*, 441–452.

Grimski, D., & Ferber, U. (2001). Urban brownfields in Europe. *Land Contamination and Reclamation, 9*(1), 143–148.

Holland, C., Clark, A., Katz, J., & Peace, S. (2007). *Social interaction with urban spaces*. Bristol: The Policy Press.

Hollander, J. B., Foster-Karim, C., & Wiley, A. (2018). Urban inclusivity through a service design framework. *Journal of Place Management and Development, 12*(1), 71–87.

IUCN. (2019). *Nature-based solutions*. Gland: Commission on Ecosystem Management.

Ivanova, E. (2016). Public gardening and the challenges of neighbourhood regeneration in Moscow. *Critical Housing Analysis, 3*(2), 26–32.

Jennings, V., & Bamkole, O. (2019). The relationship between social cohesion and urban green space: An avenue for health promotion. *International Journal of Environmental Research and Public Health, 16*(3), 452.

Julier, G. (2005). Urban designscapes and the production of aesthetic consent. *Urban Studies, 42*(5–6), 869–887.

Kabisch, N., Korn, H., Stadler, J., & Bonn, A. (2017). *Nature-based solutions to climate change adaptation in urban areas: Linkages between science, policy and practice*. Cham: Springer.

Kim, G. (2016). The public value of urban vacant land: Social responses and ecological value. *Sustainability, 8*(5), 486. Retrieved from https://www.mdpi.com/2071-1050/8/5/486

Kim, S., & Kwon, H. (2018). Urban sustainability through public architecture. *Sustainability, 10*(4), 1249. Retrieved from https://www.mdpi.com/2071-1050/10/4/1249/pdf

Lans, W., Mulder, A., & van Rij, A. (2007). Open space and social contact. *International Journal for Housing Science and Its Applications, 31*(1), 13–24.

Lennon, M., Douglas, O., & Scott, M. (2017). Urban green space for health and well-being: Developing an 'affordances' framework for planning and design. *Journal of Urban Design, 22*(6), 778–795.

Levy-Storms, L., Chen, L., & Loukaitou-Sideris, A. (2018). Older adults' needs and preferences for open space and physical activity in and near parks: A systematic review. *Journal of Aging and Physical Activity, 26*(4), 682–696.

Lumber, R., Richardson, M., & Sheffield, D. (2017). Beyond knowing nature: Contact, emotion, compassion, meaning, and beauty are pathways to nature connection. *PLoS One, 12*(5), 1–24.

Malone, K. (2002). Street life: Youth, culture and competing uses of public space. *Environment and Urbanization, 14*(2), 157–168.

Mehrabian, A. (2017). *Nonverbal communication*. New York, NY: Routledge.

Mihaylov, N., & Perkins, D. D. (2014). Community place attachment and its role in social capital development, Chapter 5. In L. C. Manzo & P. Devine-Wright (Eds.), *Place attachment: Advances in theory, methods and applications* (pp. 61–74). New York, NY: Routledge.

Mukherjee, M., & Takara, K. (2018). Urban green space as a countermeasure to increasing urban risk and the UGS-3CC resilience framework. *International Journal of Disaster Risk Reduction, 28*, 854–861.

Nabarro, R., & Smart, G. (1978). High cost and low value in urban land. *Built Environment, 4*(3), 229–236.

Nesbitt, L., Meitner, M. J., Sheppard, S. R., & Girling, C. (2018). The dimensions of urban green equity: A framework for analysis. *Urban Forestry & Urban Greening, 34*, 240–248.

Otsuka, N., & Reeve, A. (2007). The contribution and potential of town centre management for regeneration: Shifting its focus from 'management' to 'regeneration'. *Town Planning Review, 78*(2), 225–250.

Pani, B. (2016). Improving the lives of people with dementia through urban design. *Journal of Urban Design and Mental Health, 1*, 9. Retrieved from https://www.ccities.org/improving-the-lives-of-people-with-dementia-through-urban-design/

Phillips, J., Walford, N., Hockey, A., Foreman, N., & Lewis, M. (2013). Older people and outdoor environments: Pedestrian anxieties and barriers in the use of familiar and unfamiliar spaces. *Geoforum, 47*, 113–124.

Radoslav, R., Găman, M. S., Morar, T., Bădescu, S., Branea, A., & Ghena, C. (Eds.). (2012). *Sustainable urban development through the empowering of local communities*. Rijeka: InTech.

Ravetz, A. (1980). *Remaking cities (Routledge revivals)* (1st ed.). London: Routledge Ltd.

Raymond, C. M., Frantzeskaki, N., Kabisch, N., Berry, P., Breil, M., Nita, M. R., … Calfapietra, C. (2017). A framework for assessing and implementing the co-benefits of nature-based solutions in urban areas. *Environmental Science and Policy, 77*, 15–24.

Rosol, M. (2010). Public participation in post-Fordist urban green space governance: The case of community gardens in Berlin. *International Journal of Urban and Regional Research, 34*(3), 548–563.

Russo, A., & Cirella, G. (2018). Modern compact cities: How much greenery do we need? *International Journal of Environmental Research and Public Health, 15*(10), E2180. Retrieved from https://www.ncbi.nlm.nih.gov/pubmed/30301177

Russo, A., Ignatieva, M., Cirella, G., Marchesini, L., Krestov, P., Korzhov, E., … Escobedo, F. (2017). Biophilia: Nature-based solutions for sustainable cities.

Salone, C., Baraldi, S. B., & Pazzola, G. (2017). Cultural production in peripheral urban spaces: Lessons from Barriera, Turin (Italy). *European Planning Studies, 25*(12), 2117–2137.

Sharp, J. (2007). The life and death of five spaces: Public art and community regeneration in Glasgow. *Cultural Geographies, 14*(2), 274–292.

Sonneveld, B., Merbis, M. D., Alfarra, A., Unver, O., & Arnal, M. F. (2018). *Nature-based solutions for agricultural water management and food security 12 nature-based solutions for agricultural water management and food security*. Unpublished, Food and Agriculture Organization of the United Nations, Rome.

Stratigea, A., Kikidou, M., Patelida, M., & Somarakis, G. (2017). Engaging citizens in planning open public space regeneration: Pedio Agora framework. *Journal of Urban Planning and Development, 144*.

Swierad, E. M., & Huang, T. T. K. (2018). An exploration of psychosocial pathways of parks' effects on health: A qualitative study. *International Journal of Environmental Research and Public Health, 15*(8), E1693. https://doi.org/10.3390/ijerph15081693

UNFCCC. (2018). *The Paris Agreement*. Bonn: United Nations Climate Change.

van den Bosch, M., & Ode Sang, Å. (2017). Urban natural environments as nature-based solutions for improved public health—A systematic review of reviews. *Environmental Research, 158*, 373–384.

Walsh, D., & Downe, S. (2005). Meta-synthesis method for qualitative research: A literature review. *Journal of Advanced Nursing, 50*(2), 204–211.

Wolch, J. R., Byrne, J., & Newell, J. P. (2014). Urban green space, public health, and environmental justice: The challenge of making cities 'just green enough'. *Landscape and Urban Planning, 125*, 234–244.

Does the Planning System in England Deliver a Sustainable and Resilient Built Environment? A Study of the Experience of Town Planners

Niamh Murtagh, Nezhapi-Delle Odeleye, and Chris Maidment

Introduction

The latest report of the IPCC, detailing the difference between a global mean surface temperature rise of 1.5 °C versus 2 °C above pre-industrial levels, carries stark evidence for a world changed dramatically by global warming (IPCC, 2018). The term 'climate breakdown' has been suggested as an appropriate description of the degree of change underway. The level of risk depends, in part, on the extent of mitigation and adaptation that are pursued and both mitigation and adaptation are crucial. Even if global commitments to greenhouse gas emission reductions made as part of COP21 are achieved, global warming is still likely to surpass a 2 °C threshold (Rogelj et al., 2016). In the UK, risks have been identified from flooding, higher temperatures and extreme weather to people, business, infrastructure and buildings (DEFRA, 2017). A more resilient built environment has a key role to play in reducing vulnerability to risk and increasing capacity to recover in these domains (Lucon et al., 2014). Amongst the range of professionals involved in the delivery of the built environment, town (urban) planners are key (Jabareen, 2013). We examined the experience of planners in England in their role of delivering a sustainable and resilient built environment.

N. Murtagh (✉)
The Bartlett School of Construction and Project Management,
University College London (UCL), London, UK
e-mail: n.murtagh@ucl.ac.uk

N.-D. Odeleye
School of Engineering and the Built Environment, Anglia Ruskin University,
Chelmsford, UK

C. Maidment
School of Engineering and the Built Environment, Anglia Ruskin University,
Chelmsford, UK

Real Estate and Planning, The University of Reading, Reading RG6 6UD, UK

L. Scott et al. (eds.), *Sustainable Ecological Engineering Design*,
https://doi.org/10.1007/978-3-030-44381-8_6

Defining what is meant by sustainability remains a challenge (Johnston, Everard, Santillo, & Robert, 2007). Within the built environment, assessment systems such as BREEAM recognise differences between infrastructure and buildings, new build and refurbishment, and include requirements in aspects including energy, materials, waste, pollution and well-being. Resilience has been seen as an aspect of sustainability, in that a sustainable system will also be resilient (Quigley, Blair, & Davison, 2018). The two terms however have taken somewhat different paths in academic and policy discourse (Quigley et al., 2018). Definitions of resilience of the built environment are still evolving. Earlier conceptualisations referred to capacity of an urban system to maintain its functions under disturbance (Gunderson & Holling, 2001). More recently, Quigley et al. (2018) suggest that definitions to date can be viewed as dichotomous. Definitions with what they term as an engineering focus define urban resilience in terms of robustness, ability to predict and plan, capacity to withstand disturbance and to return to the status quo in an efficient manner. In contrast, socio-ecological approaches to resilience emphasise capacity to adapt, to maintain function after disturbance while evolving through feedback and learning. Of importance in this and other recent papers is the notion of resilience as an approach which recognises the essential interconnectedness of social and ecological systems, and of urban places as non-linear, dynamic, complex systems (Porter & Davoudi, 2013). While resilience is seen as a useful concept for climate change adaptation by scholars, there is evidence that implementation of resilience thinking is not yet happening in practice (Funfgeld & McEvoy, 2013; Poku-Boansi & Cobbinah, 2018). While a number of studies have explored practical advances in Ghana, African cities, Asian countries and the Netherlands, there is little current empirical work on progress in the UK.

In England, local authorities are responsible for determining what development may take place. In compliance with national policy, planners employed by local authorities gather evidence to develop local policy and manage its introduction, offer guidance on compliance, and provide judgement on applications for development. While they have 'delegated powers' to decide on typically small developments, their work is set within a wider context of local planning committees comprised of elected officials and a planning inspectorate which oversees appeals against planning judgements.

As the professionals who enact the planning regime, through application of national policy, formulation of local policy and judgements on development, planners have the potential to be critical mediators of sustainability and resilience and to influence the extent to which these concepts are enacted in daily decision-making in the planning system. We sought to explore the experience of planners in England in relation to the goals of a sustainable and resilient built environment. The research questions were: To what extent does the planning system in England deliver a sustainable and resilient built environment, and what is the role of the town planners in its delivery?

Method

Experienced urban planners were recruited through communications with alumni of accredited programmes in planning at the authors' institutions, and through the authors' professional networks. The recruitment advertisement sought planners with more than 7 years' experience working in England. Semi-structured interviews were conducted during 2018 with a total of 19 planners. Their experience ranged from 7 to 19 years. All but one worked in local authorities and eight were women. Job titles encompassed Senior Planning Officer or equivalent (4), Principal (5) and Manager (5) as well as employer-specific titles. Four participants worked primarily in development management (control), seven worked in policy and six combined both (this categorisation was not applicable for two participants). Interviews lasted approximately one hour, were audio recorded and transcribed verbatim, and had prior ethical approval. The interview schedule included sections on meanings of professionalism and of sustainability and resilience, which are analysed here. Using NVivo, thematic analysis was conducted (Braun & Clarke, 2006). The accounts were read and re-read, and phrases, sentences or short sections relating to sustainability and resilience were coded. The codes were clustered into more general themes and the most relevant of these to the research question are outlined below. All names are pseudonyms to protect confidentiality.

Findings

In this section, the theme of meaning of sustainability is presented, followed by responses to the concept of resilience.

Meanings of Sustainability

The participants in general understood sustainability to encompass environment, social and economic aspects and described the challenges in seeking outcomes that satisfied all three dimensions.

> *It's the economic, the social and the environment, and making sure it's all balanced. The trouble is, one will often not be in balance with the other one, but our job is to try and make sure it's all balanced.* [Heidi]

However there was recognition that balance may be an ideal and in reality, there may be pressure to achieve housing, economic or political targets:

> *I suppose you've always got- the political angle as well is that something might not be in the best location, say, environmentally. But if it brings a lot of jobs then... You know, it's a tricky one.* [Anne]

Examples were given of sustainability in the realm of transport, local economy and biodiversity. However, half of the participants questioned the usefulness of the term that has come to be seen as "*a form of tokenism…a buzzword*" [Gail]:

> *It's so elastic that it can mean anything to anyone… I would never do this, but you could almost write a committee report, or delegate a report sheet or whatever you were doing, and find and replace 'sustainable' with 'good', and it wouldn't actually make much difference, because that's how watered down the definition of sustainable has become.* [Kevin]

This lack of detail and precision in the term was linked to the participants' view of the core national policy document (National Planning Policy Framework, NPPF) as not useful, despite its stated objective of placing sustainable development as a core construct. The vagueness of critical concepts meant "*it's a lawyer's dream, because there's just so much you can interpret and fight the meaning of*" [Fliss]. Within a planning regime in which planners' decisions which are unacceptable to a stakeholder can be challenged in court, this implies that the policy is ultimately ineffective because "*it lacks teeth, and it means that, I would say, most planners are too scared to really rely on it as something they can resist [a scheme] on*" [Gail].

For the participants who described the NPPF in more positive terms, reference was made to local plans and it appeared that they used the NPPF to ensure compliance but relied more heavily on the local plan. Although one participant felt the NPPF was important and impactful in how it signalled a focus on sustainability, another pointed to an example where governmental policy direction was incompatible with climate change targets (support for gas extraction). In the absence of clear and strong national policy, "*it is just down to the planners, at the end of the day, to weigh everything up*" [Gail].

Several noted difficulties in implementation, including the contested definition and weak policy noted above, and one described the complexity of attempting to apply a high-level concept in practice on small projects: "*I think climate change is a really difficult thing to consider for an individual planning application*" [Charlotte].

Despite their dependence on inadequate policy, many of the participants saw planners as jointly responsible for the achievement of sustainable development. Planners were not seen as solely responsible: when asked who they viewed as responsible for delivering a sustainable built environment, the interviewees referred not only to central government but also to local government, council partners and specialist advisers such as Highways England. A few argued that everyone involved in the built environment shared responsibility, from citizens submitting a request for planning, to developers to advisers. The responses indicated a role for planners in sustainable development, alongside other stakeholders in the system.

The view that planners had an important role in achieving sustainability was echoed in a sense from many of the participants of a personal commitment to protecting the environment. For some, it was an overarching objective:

> *I'd like to think, as I go through my career, everything I do, every local plan I get through, every policy I help develop, every decision I input into will help me develop my ability to act as an advocate, really, for a sustainable built environment, for a resilient built environment, and natural environment.* [Gail]

The evidence was clear that most (though not all) of the participants pursued elements of a sustainability agenda from a sense of professional or personal commitment. They described fighting for greater biodiversity and protection of the green belt, water efficiency, more sustainable homes, fewer cars and eco-towns. This was evident in responses ranging from strong views on the need to reduce reliance on private cars to a more general aim to defend the natural environment where possible. In sum, "*you have to do your professional job in terms of you're trying to create sustainable places*" [Heidi]. For some, it was a goal linked to their professional identity: "*Because that's what I came into the profession to do. I wanted to make a difference.*" [Fliss]. This is consistent with theoretical understanding of internalised motivation being linked to self-identity (Ryan & Deci, 2000) and points to close alignment between professional identity as a planner and commitment to sustainability goals.

Responses to Resilience

When asked about the applicability of resilience in their work, there was no initial recognition of the concept by almost half of the participants, and in a few cases, there was acknowledgement of lack of knowledge on the topic:

> *Researcher: I was wondering is resilience something that you do talk about and what it might mean for the built environment.*
>
> *Participant: Not really. As a word we don't really use that, so I'm not sure how I would apply it to planning. Yes, it's not a word that appears in any of our policies, that I know of, and it's not something that really crops up.* [Ella]

However, more than half of the participants went on to provide some associations with the concept. Some confounded it with climate change mitigation and others referred to a wide range of factors including accessibility of homes, an ageing population, economic outcomes and biodiversity. Just over half connected the concept with a response to climate change, referring to planning for flooding, water stress and heatwaves. One talked about emergency planning and a number interpreted the concept in terms of future proofing, coping with adverse conditions and an ability to adapt: "*The ability for our developments to be able to mitigate, and adapt to, some of the issues is going to become more important.*" [Debbie]. Of note here is positioning the issues around resilience in the future tense ("is going to become"), in the same way as [Kevin] says: "*we don't yet know what it means properly…But it's early days.*" Understanding of resilience was markedly different to knowledge of sustainability, which all participants readily discussed and could explain how it was embedded in policy.

A small number of participants saw resilience as a concept with a broader meaning than preparing for climate breakdown, and incorporating flexibility to deal with change in market conditions and political contexts. One had attended the Royal

Town Planning Institute (RTPI; the professional body of planners) information sessions on resilient cities and another saw it as a goal:

> *Coming back to resilience, it's a big part of what we're trying to achieve and it's a very useful way to view some of those issues and a way to group things together and get people thinking about that longer term.* [Jack]

With the exception of this participant [Jack], there was little evidence of activity relating to resilience in planning work currently. Participants noted the absence of national policy and the corresponding gaps in local plans. Some also pointed to the challenge in development management of applying the overarching aims of resilience on a single development: "*We think very development specific, and the short term, in terms of from the point of receiving an application to granting consent, that's one of our thought processes really*" [Beth]. Where they offered a view on where responsibility lay for increasing resilience in the built environment, participants considered that planners had a role to play but that responsibility extended to all stakeholders:

> *I think it's a shared responsibility. I don't think that it resides in planners, though we have it. I don't think it's just a political thing though certainly politicians have a responsibility. I don't think there's anybody who's not touched by the issues of resilience.* [Jack]

Discussion

In the current study, 19 planners in England were interviewed, each with at least 7 years' experience, exploring the themes of sustainability, resilience and the planner's role.

The participants' accounts showed universal awareness of the ubiquitous definition of sustainability as requiring a balance between the 'triple bottom line' of environmental, social and economic sustainability (Elkington, 1997). However, there was acknowledgement of the challenges of delivering this balance in practice. These included political pressures such as the current focus on housing, the difficulty in applying the high-level concept of sustainability on specific planning applications, the vagueness of the term 'sustainability' and its amenability to differing interpretations, and the consequent ineffectiveness of national policy. Nonetheless, many of the participants positioned planners as sharing joint responsibility for delivery of a more sustainable built environment. A number of participants demonstrated strong personal commitment to the goals of sustainability and most, although not all, viewed delivery of a sustainable built environment as part of their professional identity. Identity theorists have argued that social structures, such as a profession, shape identities, and that an individual identity represents subjective processes or responses to such external institutions (Burke & Stryker, 2016). Thus, to identify as a particular type of professional is to act in a manner consistent with that identity.

In contrast to sustainability however, the notion of resilience was not familiar to nearly half of participants, echoing recent work with planners in Ghana

(Poku-Boansi & Cobbinah, 2018). When invited to consider the concept in interview, most could make some associations, more or less tenuously related to the term. Noticeably, the level of familiarity was markedly different from that of sustainability. There was little evidence of understanding of underlying principles of complex, non-linear, dynamic urban systems and of the relationship between social and ecological systems. The approach of 'resilience thinking', heralded as having the potential for a paradigm shift in planning (Shaw, 2013), was not evident. In a small number of cases, participants used the future tense, positioning resilience as something that will have to be addressed in the future. With strong current evidence for weather patterns affected by anthropogenic influence (Duffy et al., 2019) and the long-term impact of planners' decisions, as referenced by the participants themselves, this is a worrying finding. If planners today are not developing local policy and making judgements based on enhancing resilience, the impacts of the changing climate will have greater adverse impact into the future than could be the case.

In the UK, there has historically been a wide separation between town planning and health and emergency/disaster preparedness, with national policy being focused on civil emergencies (whether arising from accidents, natural hazards or human threats). Although the Planning and Compulsory Purchase Act of 2004 was meant to re-integrate health and wider concerns back into a new spatial planning approach, it would appear from the evidence here that this has had limited effect. It is possible that planning for climate change resilience may be being conducted within other functions of local authorities such as environmental health or emergency planning, and that there is greater knowledge and understanding in such areas. However, such an approach would indicate an engineering approach to resilience, that is, a reactive approach, planning to protect against change and aimed at returning a system to its original state with greatest efficiency. The potential benefits of an evolutionary approach, of developing learning, robustness, innovation and flexibility, are seemingly not being pursued. Resilient thinking, of necessity, should involve town planners, given the necessarily future-orientation of local development plans, which typically look 20–25 years ahead.

With few exceptions, there was little evidence of reference to urban resilience in national or local planning policies. However, the participants considered responsibility for sustainable and resilient environment to be shared, and to include planners, government and other stakeholders in the built environment. Although the evidence demonstrated lack of detailed knowledge and action, the theoretical framing of professional identity showed an underlying motivation is there.

Conclusion

The conclusion from the analysis is that while planners' professional identity drives a strong purpose of delivering social good, there are many challenges within the planning system to delivery of sustainability. On climate resilience, there is too little knowledge and awareness for the concept to inform day-to-day planning policy and decisions. This holds important implications for the profession, its professional body and policy makers.

Current national legislation is by and large unhelpful to planners in delivering sustainability. The legislation should address directly the difficulties of definition and the inherent tensions in aiming for environmental, economic and social sustainability. Consistency in policy is needed which transparently meets national commitments including to COP21 targets, the Climate Change Act (2008) and the UN Sustainable Development Goals.

Development of the built environment is guided by planners' expertise and long-term view. If planners nationally are unaware of the predicted impacts of climate breakdown on urban and other development, then the built environment now and into the future is not being prepared for future risks. Led by the professional body (RTPI), and drawing on experts within relevant economic sectors and from academe, it is imperative that the predicted risks to the built environment from climate breakdown and principles around urban resilience are disseminated. Resilience thinking must become embedded in day-to-day work of planners.

National planning policy requires urgent updating to address climate resilience but policy changes are not only top-down. Without a bottom-up drive from planners and their representatives, national policy will not develop in the direction needed.

With adequate legislation, and knowledge of principles and approaches to resilience, the evidence from this study suggests that planners have the motivation, through their professional identity to serve the public good, to deliver a sustainable and resilient built environment.

Further questions to consider include: given the political context of planning, are professional knowledge and motivation alone sufficient to ensure that best practice is pursued and implemented? Is the current structure of planning departments and processes in itself part of the problem? What would a planning system look like, which consistently delivered sustainable and resilient development?

References

Braun, V., & Clarke, V. (2006). Using thematic analysis in psychology. *Qualitative Research in Psychology, 3*(2), 77–101.

Burke, P. J., & Stryker, S. (2016). Identity theory: Progress in relating the two strands. In J. Stets & R. Serpe (Eds.). *New directions in identity theory and research* (pp. 658–682). New York: Oxford University Press.

DEFRA. (2017). *UK climate change risk assessment 2017*. Retrieved from https://www.gov.uk/government/publications/uk-climate-change-risk-assessment-2017

Duffy, P. B., Field, C. B., Diffenbaugh, N. S., Doney, S. C., Dutton, Z., Goodman, S., … Williams, A. P. (2019). Strengthened scientific support for the Endangerment Finding for atmospheric greenhouse gases. *Science, 363*(6427):eaat5982.

Elkington, J. (1997). *Cannibals with forks: the triple bottom line of 21st century business*. Oxford: Capstone.

Funfgeld, H., & McEvoy, D. (2013). Resilience as a useful concept for climate change adaptation? *Planning Theory and Practice, 13*(2), 324–328.

Gunderson, L., & Holling, C. S. (2001). *Panarchy: Understanding transformations in human and natural systems*. Washington, DC: Island Press.

IPCC. (2018). *Global warming of 1.5'C*. Retrieved from http://www.ipcc.ch/report/sr15/

Jabareen, Y. (2013). Planning the resilient city: Concepts and strategies for coping with climate change and environmental risks. *Cities, 31*, 220–229.

Johnston, P., Everard, M., Santillo, D., & Robert, K.-H. (2007). Reclaiming the definition of sustainability. *Environmental Science Pollution Research, 14*(1), 60–66.

Lucon, O., Ürge-Vorsatz, D., Zain Ahmed, A., Akbari, H., Bertoldi, P., Cabeza, L. F., ... Vilariño, M. V. (2014). Buildings. In *Climate change 2014: Mitigation of climate change. Contribution of working group III to the fifth assessment report of the IPCC*. Cambridge: Cambridge University Press.

Poku-Boansi, M., & Cobbinah, P. B. (2018). Are we planning for resilient cities in Ghana? An analysis of policy and planners' perspectives. *Cities, 72*, 252–260.

Porter, L., & Davoudi, S. (2013). The politics of resilience for planning: A cautionary note. *Planning Theory and Practice, 13*(2), 329–333.

Quigley, M., Blair, N., & Davison, K. (2018). Articulating a social-ecological resilience agenda for urban design. *Journal of Urban Design, 23*(4), 581–602.

Rogelj, J., den Elzen, M., Hohne, N., Fransen, T., Fekete, H., Winkler, H., ... Meinshausen, M. (2016). Paris Agreement climate proposals need a boost to keep warming well below 2'C. *Nature, 534*, 631–639.

Ryan, R. M., & Deci, E. L. (2000). Self-determination theory and the facilitation of intrinsic motivation, social development and well-being. *American Psychologist, 55*(1), 68–78.

Shaw, K. (2013). "Reframing" resilience: Challenges for planning theory and practice. *Planning Theory and Practice, 13*(2), 308–318.

Sustainability Assessments of Urban Railway Systems: Case Study Evaluations in Turkey

Müge Yüksel, Ali Murat Tanyer, and Mehmet Koray Pekeriçli

Introduction

The energy demand of society increases continuously due to technological improvements, population growth, and daily deficiencies. This energy demand is generally supplied from fossil fuels that damage the environment. Scientific evidence for warming of the climate system is unequivocal (IPCC, 2014). The significance of climate change has intensified after the industrial revolution due to human-induced activities. The condition is severe; the carbon footprint (CFP) of humans will reach over double of the earth's capacity by 2050. The Paris Agreement (2016) and the Kyoto Protocol (United Nations, 1998) clearly asserted that the built environment is a major contributor to climate change. According to the U.S. Green Building Council (USGBC), sustainability is a holistic concept that starts with the understanding of the built environment with its positive and negative effects on the natural environment, and evaluating the people who inhabit built environment every day. The Green Design Concept enables the mitigation of negative effects throughout the entire life cycle of buildings. In 1994, the Conseil International du Batiment (CIB, 1999), an international construction research networking organization, defined the goal of sustainable construction as "...creating and operating a healthy built environment based on resource efficiency and ecological design". Therefore, sustainability principles should be applied to resources, namely, land, materials, water, energy, and ecosystems, which are needed to create and operate the built environment during its entire life cycle.

M. Yüksel (✉)
Building Science Program, Middle East Technical University, Ankara, Turkey
e-mail: muge.yuksel@metu.edu.tr

A. M. Tanyer · M. K. Pekeriçli
Department of Architecture and Research Center for Built Environment (MATPUM), Middle East Technical University, Ankara, Turkey
e-mail: tanyer@metu.edu.tr; koray@metu.edu.tr

L. Scott et al. (eds.), *Sustainable Ecological Engineering Design*,
https://doi.org/10.1007/978-3-030-44381-8_7

Literature Review

With the increasing attention on the reduction of Global Warming Potential (GWP), numerous strategies have been developed by several countries. Especially for developing countries, sustainable development means more to accelerate improvements which can provide a competitive advantage on mitigation of greenhouse gas emissions. Many considerations turn out to be effective solutions for building level. On the contrary, the infrastructure level has a poor background despite its great return potential. The need for a more environment-friendly approach in infrastructure projects is an ever-growing demand throughout the world. Stripple and Uppenberg (2010) asserted that, to be able to internalize environmental problems and find innovative ways to solve them, a comprehensive advancement is needed. Typical life span of metro projects are more than 50 years, so CFP of them would have a great impact in the future. According to Andrade and D'Agosto (2016), infrastructure projects represent most of the emissions produced and energy used. Del Pero, Delogu, Pierini, and Bonaffini (2015) proved that, among the European Union, transportation industry is the second largest contributor to anthropogenic greenhouse gas emissions (GHG). Another crucial point is that up to now, most of the sustainability advancements indicate that Operational Carbon (OC) through operational energy performance has been succeeding. After the operation of energy use, embodied carbon in the built environment is a significant challenge. Yet, Embodied Carbon (EC) assessment and its reduction could not be completely tackled. Latest researches indicate that approximately 70% of green certifications and regulations address the carbon emissions from building materials. Therefore, materials play a significant role in sustainability practices in terms of life cycle impacts.

This paper basically explores the significance of the carbon footprint of metro projects. Due to the fact that metro stations are environmentally more preferable means of transportation, quantifying environmental impact data is crucial. In addition, the research provides researchers and practitioners new perspectives about optimizing carbon footprint. Spielmann and Scholz (2005) assume that the future's best option for evaluating the environmental impacts of the transport system is conducting LCA. Stripple and Uppenberg (2010) assume that LCA gives flexibility for working on complex projects.

This research highlights many of the critical considerations, but mainly focuses on two issues which are lifelong sustainability and configuration of complex project data sets. On the other hand, for metro projects, planners and other decision-makers must monitor the process in determining the benefits. The main aim of this study is to generalize the optimization of infrastructure projects, one type of which is metro stations. The principle is collecting and generating a database for a country and giving references by comparing each country's database. With these case studies, the aim is to identify potential reductions in carbon emission that might be achieved during the construction phase. The criteria within the research case studies are based on different construction-process types of metro stations. This study is planned for three cases of metro stations from İstanbul, Turkey.

Research Methodology

This study analyzes the EC of metro stations located in İstanbul, Turkey. Materials detrimental to carbon emissions are discovered in terms of environmental impacts. In order to implement LCA, ISO 14040 Standard is used at material level. Embodied carbon amount per m^2 (kg CO_2eq/m^2) for each metro station is calculated to evaluate carbon footprints of the metro stations. This study follows these questions: "What are the underlying reasons for them?" and "How would the carbon emission decrease at material level? What are the necessities for the frame of those studies?".

Nowadays, researchers, who contribute to the topic of sustainability, prove that a holistic approach to projects is obligatory. Life cycle assessment (LCA) is a well-defined and very detailed tool to observe evaluation phases of projects. Indeed, LCA is a framework which analyzes products and systems from cradle to grave. Therefore, with a broader perspective, this approach provides a logical methodology to solve GWP problems. CFP is one of the important criteria of the LCA study which supplies directly related results of embodied carbon. The method of the research includes performed International Organization for Standardization (ISO) assumptions by conducting LCA dealing with Product Category (PC) rules developed by the International Environmental Product Declarations (EPD) system.

According to Stripple and Uppenberg (2010), a significant part of the Environmental Product Declaration (EPD) comprises the description of quantities shaped by environmental restrictions. Chester and Horvath (2009) claim that the production of concrete is energy-intensive and releases CO_2, as well. Guidelines for planning and dimensioning infrastructure projects should emphasize the importance of optimizing the amounts of materials that are needed through the lifetime. It is possible to collect qualified data for all major flows of materials. The carbon footprint can be the method of implementation under ISO standards. Carbon Footprint is the quantitative measure of greenhouse gases that a human emits directly or indirectly through their actions. It is the pollution emitted through the burning of fuel, which influences climate change. This method includes manufacturing process, transportation, purchase of items that are manufactured, and have produced greenhouse gasses and the intake of food and amenity sources. For quantities, as it is mentioned before, EPD is needed for specialization of materials. There are several methods to quantify life cycle inventory burdens. For the development of EPD, subsystem analyses might be drawn to have a hierarchic material structure.

Case Studies

In the scope of this report, three types of metro stations were first quantified using the Revit software, and then analyzed with Tally software. The study displays the material level with a case study on three metro projects in İstanbul, Turkey. For future references, analyses on embodied carbon of these projects will be conducted in order to provide a reference model that may represent typical metro projects in Turkey.

Turkey is a developing country, having the potential to be a model for the following countries. İstanbul is not just a city, but a home to multicultural traditions and the cultural capital of Turkey. Therefore, authors think that a railway-system-based study for this city will provide a great opportunity for Turkey.

Case studies are designed to analyze the embodied carbon of the selected stations in this study. In the following parts of this report, stations are referred to as "Station 1," "Station 2," and "Station 3," and short descriptions of these are given.

All of three metro stations were first modelled in Revit, and in order to achieve this, required data for comparative analyses were provided by Yüksel Proje Intl. Co.

For the material level analysis, Tally software is used. It is easy to observe the behavior of materials due to the fact that all stations have approximately gross areas, and are constructed using similar materials. Therefore, comparing them with an LCA tool can provide similar results but it is a huge first step.

Sample Project, Station 1

The station is a mid-station with a platform width of 10.9 m and the platform size is 100 m. It is a four-story station with several facilities, namely, Platform, Mezzanine, and Concourse Level.

There exist one fixed stairway, three escalators, and two elevators that reach the ticket hall floor from the Platform floor level. While the surface is directly accessible with one emergency escape stairway at the beginning of the perimeter, circulation for each level is provided by the service stair at the end of the platform.

There are four fixed stairways, four escalators, and two elevators at the level of the ticket hall. Special doors have been arranged for passengers with the entry and exit on the cruise lines and for passengers with restricted movement.

Rooms for electrical and mechanical systems are located on the technical floor, on the mezzanine floor, and on the floor of the ticket hall. There are also rooms for the staff on the mezzanine floor (Fig. 1).

The first station is "cut and cover" type. It is located at the site as it is shown in Fig. 2. The gross built area is nearly 7500 m^2. For the calculations of environmental impacts, Gross Built Area (GBA) of this station is taken as 7.5K.

Sample Project, Station 2

The station is one with an edge platform, with a platform width of 4.5 m and a platform size of 100 m. The platform is designed as a tunnel. It is connected to the shaft structure by means of platform connection tunnels. There are four elevators and four escalators in the shaft structure (two of them are downhill), and concourses are

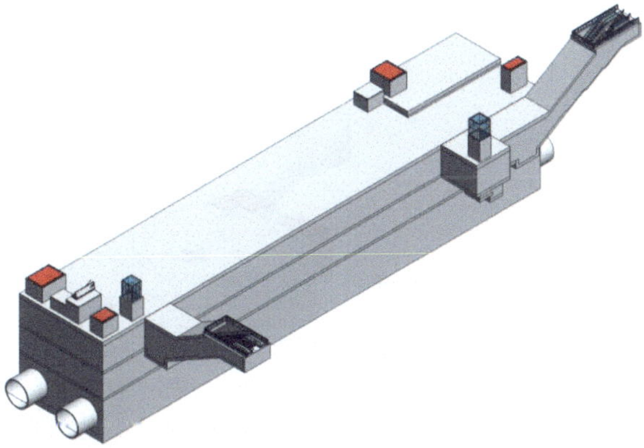

Fig. 1 Station 1—Revit model

Fig. 2 Station 1—site plan

transported. In addition, the surface can be reached directly by the emergency escape ladder located in the shaft structure (Fig. 3).

Another tunnel construction is joined at concourse floor; circulation areas, personal areas, and technical areas. Special doors have been arranged for passengers with the entry and exit on the cruise lines and for passengers with restricted movement. The concourse is used twice for technical spaces in the shaft above the floor. It is located at the site as it is shown in Fig. 4. The gross area is nearly 8000 m^2. For the calculations of environmental impacts, GBA of this station is taken as 8K.

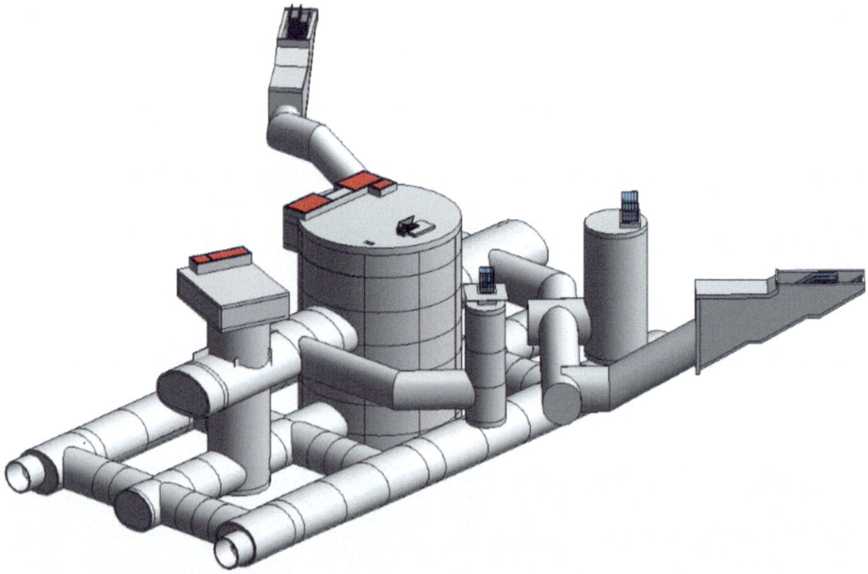

Fig. 3 Station 2—Revit model

Fig. 4 Station 2—site plan

Sample Project, Station 3

This sample is a station with an edge platform, with a net platform width of 3.9 m and the platform size is 100 m. The platform floor is designed as a tunnel. Access to the ticket hall floor is provided with the platform stairways of tunnels. There exist four escalators, two fixed stairs, and one elevator per passenger.

In the calculations, number of occupants, coefficients, capacities, etc., are evaluated.

The third station is a "tunnel type" one. It is located at the site as shown in Fig. 5. The GBA is nearly 7000 m^2. For the calculations of environmental impacts, GBA of the station is taken as 7K (Fig. 6).

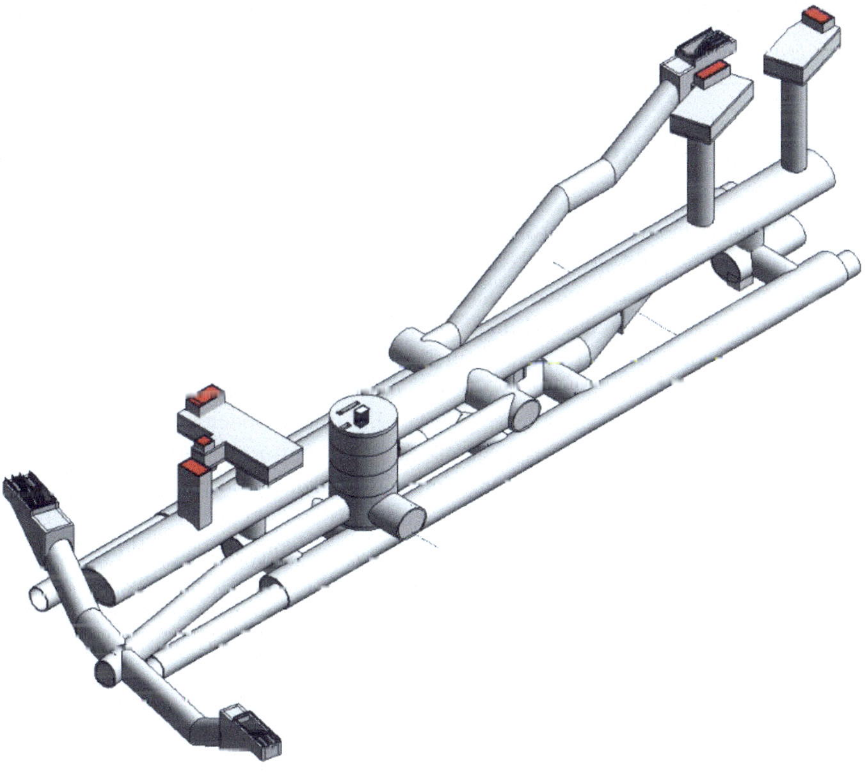

Fig. 5 Station 3—Revit model

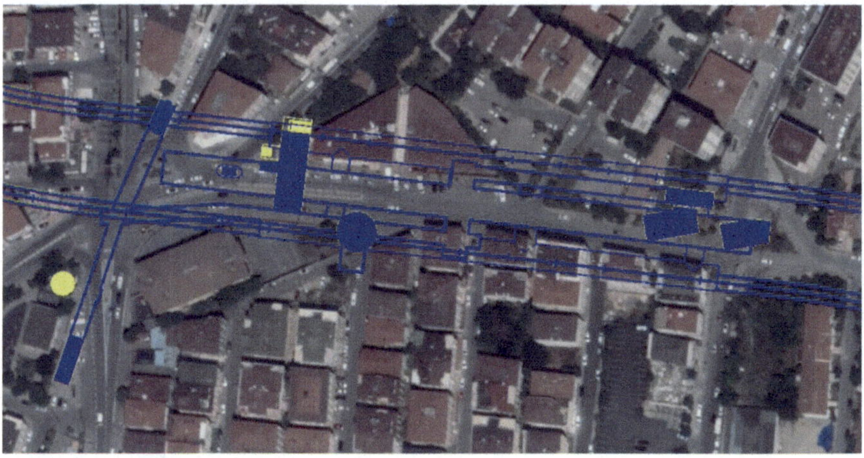

Fig. 6 Station 3—site plan

Analyses and Findings

The results are represented as raw data provided by the models, and their analyses are carried out by Tally software which is used for evaluating the GWP categories in terms of the U.S. database. By using the Tally software, environmental impacts of the three metro projects from İstanbul metro network according to the use of material are discussed. Analyses are prepared at material level; therefore, the bill of quantities are listed and quantified in the program. GWP values in $kgCO_2eq$ for total and per m^2 are calculated comparatively. The list of components for the embodied carbon analyses includes different components in metro stations and results are compared. Bill of Quantities for metro stations are listed for generating the benchmark of the study. Some of the materials are not calculated because of their uncertainty of the model reflection such as window frames, composites, and railings.

Embodied carbon of each component is calculated on a per kilogram basis ($kgCO_2eq/kg$) (Kayaçetin & Tanyer, 2020). The results can be seen in the following tables. For each station, Gross Built Area is assumed in order to reach embodied carbon per unit. The results are shown in Table 1.

Raw data provided by the Tally software are evaluated, discussed, and compared. Environmental impacts of metro stations are analyzed, and each impact category is calculated per m^2 for each material. In order to quantify CFP, embodied carbon per m^2 are calculated based on the total sum of GWP for each station.

It should be noted that selection and use of materials are similar on the same line of the İstanbul Metro. But to see deeper differences, all distinctive types are observed. Since the selection of materials and use are similar, categories are chosen as: materials by category and materials by the division of use for the analyses.

At the beginning of these calculations, it is assumed that all three stations have similar features, areas, materials, etc. In contrast, even the similar stations have a

Table 1 Raw data comparison

	Sum of acidification potential total ($kgSO_2eq$)	Sum of eutrophication potential total (kgNeq)	Sum of global warming potential total ($kgCO_2eq$)	Sum of ozone depletion potential total (CFC-11eq)	Sum of smog formation potential total (kgO_3eq)	Sum of primary energy demand total (MJ)	Sum of non-renewable energy demand total (MJ)	Sum of renewable energy demand total (MJ)	Sum of mass total (kg)	Gross build area (GBA)	$kgCO_2eq/m^2$
S1	72,152	3355	16,310,375	0	978,777	136,943,002	125,141,416	11,817,300	49,905,447	7500	2174.7
S2	55,167	2604	12,586,532	0	752,588	108,904,086	98,965,434	9,953,055	37,843,846	8000	1573.3
S3	36,038	1910	8,254,302	0	490,568	76,035,185	67,633,616	8,410,652	24,206,372	7000	1179.2

different weight of CFP. Moreover, to get the main purpose of the proposed study, a large number of stations must be analyzed and compared with different software analyses to have a holistic perspective. As indicated in Table 1, there are also other calculations related to environmental impacts.

Total CFP of stations per m^2 are as given below:

- Station 1 = 2,174.7 kg CO_2eq/m^2
- Station 2 = 1,573.3 kg CO_2eq/m^2
- Station 3 = 1,179.2 kg CO_2eq/m^2

Results for the embodied carbon of each component per kg were compared among the stations. In the literature, results are almost similar, but it is asserted that Turkey has a different issue from U.S. Lack of standards and using Tally software which makes use of the American standards might give uncertainty to the results. Furthermore, embodied carbon for the distributions of material categories are analyzed, and it is observed that the manufacturing phase has the biggest share in the chart of the global warming potential. For assessing the material level, with an enormous difference than other materials, concrete is the most detrimental material for the environment.

Taking everything into account, it can be concluded that, the main reason why concrete is the most detrimental material can be considered to be the impact of the manufacturing process. As it is mentioned in the literature survey, there are main impact categories of environment, and each impact is compared among the stations.

Regarding all the information above, since there is an abundance of parameters, there must be lots of questions and all of them must be tried for a proper benchmark. It is noticeable that results are consistent within themselves—with a few exceptions: Station 1 seems to have the biggest CFP, while Station 2 comes the second and Station 3 is the third one. It is shown that there is no relationship between GBA and GWP, while there are several parameters that need to be evaluated.

To conclude, it should be noted that, in case the calculations are based on the American database it would be early to conceive the exact reasons beyond embodied carbon in the metro projects in Turkey. Therefore, refined and detailed materials should be studied.

Conclusion and Discussion

In the literature, it is stated that the main issue of the environmental impact categories comes from the acquisition of raw materials and production of materials. The usage of materials such as steel, concrete, etc. that contributes to global warming should be analyzed. While steel is mostly related to the reinforcement process, cement is mostly used for shotcrete purpose in infrastructure systems. GHG emission reduction that comes from the material field can be managed by using lower CO_2 emitting materials. As it is proved in previous studies from literature, steel and concrete materials are top CO_2 emitters, so evaluating the use of these materials and reducing their use can be helpful. However, many aspects should be considered due to the complexity of both the processes and the projects.

This research aimed to generate an LCA framework for embodied carbon assessment of metro projects in Turkey. Therefore, the ultimate contribution of the research is that a sustainable framework will be presented for the metro projects in Turkey. In the light of the literature review, this study provides a benchmark for materials applied in metro stations throughout the world by generating Turkey's own database. It is important to generate the country's own database to take place in mitigation of GWP, all around the world. Up to now, there is no specialized work on the optimization of CFP of the metro stations, and this is the motivation of this research. In order to limit the scope of this study, all metro stations of three metro lines within the İstanbul Metro Network were analyzed as a case study to lead the way. This study aimed to widen the perspective for sustainable railway systems. Nevertheless, the findings from this research might prove to be valuable references for studies on sustainable metro stations. In order to improve the life cycle of the environmental performance of metro projects, several recommendations can be proposed by underlining the lessons learned from those analyzed case study experiences. Since this work is a progress report of a thesis, it can be said that this experience is probably a good preparation for facing real problems.

It seems that inclusive materials of the stations prove that, concrete has the major role because of its amount of use. Therefore, it can be said that just the right choice of concrete might cut of 50% of the CFP. Hence the target is searching on the effects of detrimental materials of metro stations, potential product improvements for future design strategies might be done accordingly. The reduction of carbon emission involved by "material" point of view is thought the potential of enhancement. As a result, the dominant start point is that up to now there is no comprehensive work about optimizing the carbon footprint of metro projects in Turkey. With this motivation, this research aims to examine the EC benchmark of the materials at the beginning and give a reference sustainability framework of the railway systems in Turkey.

With this study it is proved that innovative approaches and materials are urgently needed. Besides, using local resources can directly cut EC. Using local materials could also provide great profit for economy. Another important point is that a new vision should be introduced to the companies as new trends. The aim of this study is to be a joint point that addresses many problems of sustainability. With this intention, it is aimed to expand the studies to comparable levels and eventually to establish a range of EC estimation of materials. It is a great opportunity to capture the technologies of countries such as the UK and the USA to develop Sustainable Infrastructure Systems Strategies.

It is also important to remember that unlike operational carbon emissions the embodied energy and carbon cannot be reversed. Embodied energy and carbon is a topic of rising importance. In fact, it is normally possible to reduce the embodied energy and carbon of a building or construction project by 10–20% without adding to the build cost (Hammond & Jones, 2008). This study affirms that the consideration of EC in railway systems might be an important part of the aim of reducing Turkey's carbon emission to achieve the goals set by the Paris Agreement.

Acknowledgement The authors would like to express their gratitude to Yüksel Proje Intl. Co. for providing data for several railway system projects. In addition, for the Tally software, the official website of the software is used for gathering the information presented in this report and for doing analysis about research, and the student version of this software is used for the analyses.

References

Andrade, C. E. S., & D'Agosto, M. A. (2016). Energy use and carbon dioxide emissions assessment in the lifecycle of passenger rail systems: The case of the Rio de Janeiro Metro. *Journal of Cleaner Production, 126*, 526–536.

Chester, M. C., & Horvath, A. (2009). Environmental assessment of passenger transportation should include infrastructure and supply chains. *Environmental Research Letters, 4*, 024008.

CIB. (1999). *Agenda 21 on sustainable construction*. CIB Report Publication 237, Rotterdam, The Netherlands.

Del Pero, F., Delogu, M., Pierini, M., & Bonaffini, D. (2015). Life cycle assessment of a heavy metro train. *Journal of Cleaner Production, 87*, 787–799.

Hammond, G., & Jones, C. (2008). *Inventory of Carbon and Energy (ICE). Version 1.6a*. Bath: Sustainable Energy Research Team (SERT), Department of Mechanical Engineering, University of Bath.

IPCC. (2014). *Intergovernmental panel on climate change—Climate change 2014: Impacts, adaptation, and vulnerability*. Cambridge: Cambridge University Press.

Kayaçetin, N. C., & Tanyer, A. M. (2020). Embodied carbon assessment of residential housing at urban scale. *Journal of Renewable and Sustainable Energy Reviews, 117*, 109470. https://doi.org/10.1016/j.rser.2019.109470

Paris Agreement. (2016). United Nations Framework Convention on Climate Change (UNFCC).

Spielmann, M., & Scholz, R. W. (2005). Life cycle inventories of transport services. *International Journal of Life Cycle Assessment, 10*(1), 85–94. https://doi.org/10.1065/lca2004.10.181.10

Stripple, H., & Uppenberg, S. (2010). *Life cycle assessment of railways and rail transports*. Stockholm: IVL Swedish Environmental Research Institute.

United Nations (1998). Kyoto Protocol to the United Nations framework convention on climate change', New York: United Nations.

Perceptions of Teletubbyland: Public Opinions of SuDS Devices Installed at Eco-designed Motorway Service Areas

Mark Gazzard and Colin A. Booth

Introduction

The impact of urban development on a catchment (by reducing its permeability and increasing surface water runoff) reduces opportunities for water to be managed naturally on site and can lead to localised flooding, the affects of which are exacerbated when traditional piped drainage systems are unable to cope with heavy rainfall. Recent increases in instances of severe weather events (e.g. January 2014 was the wettest on record (Meteorological Office, 2014)) have led to flooding being identified as the most dangerous impact of climate change for the UK, with it hitting harder and faster than expected (Carrington, 2014).

Effective surface water management within the UK has become an important topic in recent years as the current shortfall in development land for housing has led to a marked increase in the number of areas at risk of flooding being developed and considered for development (Sayers, 2012). Sustainable drainage systems are becoming essential in managing effective stormwater infiltration (Thompson, 2014). The need to manage the capacity of surrounding land in dealing with unusual weather patterns and reducing the pressure on basic stormwater infrastructure is essential in providing a healthy, functioning natural environment to sustain economic growth, prospering communities, and personal well-being and reflect the critical value of nature (Natural England, 2009; The EU Sustainable Development Strategy, 2006; The UK Climate Change Act, 2008).

Traditional underground pipe infrastructure is no longer seen as a 'silver bullet' for dealing with flood issues and today it is recognised as something that must be integrated within a broader portfolio of non-structural and policy responses to help deliver 'multiple outcomes' (for ecosystems, the economy and society), while

M. Gazzard · C. A. Booth (✉)
Centre for Architecture and Built Environment Research (CABER), Faculty of Environment and Technology, University of the West of England (UWE), Bristol, UK
e-mail: colin.booth@uwe.ac.uk

L. Scott et al. (eds.), *Sustainable Ecological Engineering Design*,
https://doi.org/10.1007/978-3-030-44381-8_8

promoting resilience to changing and less predictable weather patterns (Cahill, 2012; Sayers, 2012). Linking non-structural drainage systems with future development and regeneration can deliver greener and more pleasant urban areas (Rossmiller, 2014), with trees and vegetation combining to define the scale of space, making large, unfriendly spaces smaller in parts and more conducive to a wider range of human activities (Ashley, Balmforth, Dignan, Glerum, & Stovin, 2012).

The concept of naturally inspired flood resilience has begun to be incorporated into sustainable design across a range of industries, including the relatively new concept of the Motorway Services Area (MSA), which is a direct response to the need to provide rest and refreshment to users of the expanding motorway network (Pegasus Group, 2010). There has been a recent shift in the archetype of contemporary MSAs in the UK, towards sustainably driven businesses, operating in eco-designed buildings that give attention to sustainable drainage (Charlesworth and Booth, 2017).

The M42 Hopwood Motorway Service Area in Worcestershire (completed in 1999) was the first MSA to incorporate sustainable drainage systems (SuDS). It utilises five principal SuDS components, with the aim of providing an affordable system, providing an attractive setting and added amenity (Robert Bray Associates, 2016). The system consists of filter trenches, a filter strip, swale, pond and wetland, forming two distinct management trains. Areas considered to pose a pollution risk to the environment have used the SuDS management train to ensure good water quality and deal with possible spillage events, with the HGV park, fuel filling area, coach park and service yard having an extended management train as they potentially pose a serious pollution risk (Dickie, McKay, Ions, & Shaffer, 2010). In general, water quality improves during passage through the more extended HGV and coach park SuDS, emphasising the importance of implementing SuDS units in management trains rather than in isolation (Heal et al., 2008).

Following the continuing success of 'sympathetically landscaped' Tebay Services (2016) on the Westmorland stretch of the M6 motorway in Cumbria, in 2010, Stroud District Council approved a detailed planning application for a new MSA to be known as Gloucestershire Gateway Services, located between junctions 11a and 12 of the M5 Motorway (Pegasus Group, 2016). The MSAs incorporate an extensive array of sustainable drainage processes within a carefully monitored management train, with features including green roofs, swales, bio-retention areas and filtration strips, ultimately discharging to a pond and wetland area (Water Active, 2015). The site was designed to provide improved landscape infrastructure, permeability, biodiversity and habitat creation, and to lessen flood risk (BWB Consulting, 2016), with the intention for storm events in exceedance of 100-years, plus 30% climate change events.

The Gloucestershire Gateway Services set a new standard for MSAs and are recognised nationally as a model for future MSA development (The Design Review Panel, 2013). This study explores public perceptions towards the sustainable drainage devices that have been incorporated into the eco-design of the amenity buildings and their surrounding landscapes at the southbound and northbound M5 Gloucestershire Gateway Services.

Background

In England and Wales, The Flood and Water Management Act (2010) requires new developments and refurbishments to have drainage plans for surface runoff approved by the SuDS Approval Board (SAB), where the construction work would have drainage implications. The SAB is responsible for adopting and maintaining new SuDS that serve more than one property and have been constructed as approved and function as designed (Defra, 2011). A proposed drainage system in England or Wales will not comply with these National Standards unless it is designed so that surface runoff is managed at its source where it is reasonably practicable to do so, with the Sustainable Drainage (Approval and Adoption) (England) Order 2012 defining exemptions to the requirement for approval. Furthermore, public space must also be used and integrated with the drainage system, where it serves more than one property, and the SuDS design must be cost-effective to operate and maintain over the design life of the development, in order to reduce the risk of the drainage system not functioning. The design of the drainage system will account for the likely impacts of climate change and changes in impermeable area over the design life of the development (Defra, 2011).

Sustainable Drainage Systems (SuDS) are technically regarded as a sequence of water management practices, control structures and strategies to drain surface water in a manner resilient to changes in climate, whilst minimising pollution and managing the entry of pollutants that could affect the quality of local water bodies (Susdrain, 2015a). 'Blue-Green' systems combine and protect the hydrological and ecological values of the urban landscape, while providing resilient and adaptive 'hard' and 'soft' measures to deal with flood events (The University of Nottingham, 2015). 'Soft' SuDS are naturally inspired landscaped measures that manage surface water typically above ground in a more sustainable fashion than traditional techniques. They often include trees and tree pits, swales, basins (dry, wet and infiltration), rain gardens and ponds. 'Hard' SuDS are of an engineered nature and include geocellular storage, proprietory products, separators, hydraulic controls and kerbside drainage. Soft SuDS can, however, create multiple benefits, including biodiversity, and are more flexible and adaptable to climate change (Ashley et al., 2012). A feature of retrofitting soft SuDS specifically is the introduction of more vegetation and trees into the urban landscape. This is a component of green infrastructure, helping to break down the harder appearance of constructed surfaces or unused green space and create a more pleasant vista (Balmforth et al., 2006).

The typical design of any SuDS system follows a stepwise hierarchy of various measures, commonly known as the 'Surface Water Management Train', which store and convey stormwater at different scales: (1) prevention (e.g. land use planning); (2) source control (e.g. green roofs, rainwater harvesting, permeable paving); (3) site control (e.g. vegetation or gravel filtration) and (4) regional control (e.g. retention ponds, wetlands) (Woods-Ballard et al., 2007a, 2007b, 2015). At the heart of SuDS is the 'SuDS square', which introduces the notion that to be successful, SuDS design should balance the desire to control water quantity, improve water quality, provide public amenity and encourage localised biodiversity benefits. Management of surface water on a local level, mimicking natural processes, provides an essential method of reducing the extent of pollution and contributing compliance with the EU Water Framework Directive by controlling both diffuse and point source pollution of surface water.

Public perceptions of sustainable drainage is an important indicator of the level of knowledge about these types of system; however, perceptions may not be easy to gauge, especially when the multiple benefits of SuDS components are often concealed as a result of their integration into a landscape or by being designed in a natural way (Sayers, 2012). All new and innovative drainage technologies applied in either residential or commercial developments, besides being technically and economically viable, must be designed to be accepted by the public. Poor public perception of SuDS may be a deterrent for developers in using them at new sites (Susdrain, 2015b). Further, more people are scantily informed about the behaviours required to ensure continued functioning and the development of the potential benefits from established devices. This will tend again to feed back to reduced functioning, worsened aesthetics, and further negative preferences (Everett, Lamond, Morzillo, Chan, & Matsler, 2016).

Implementing a shift away from hard 'grey' infrastructure requires the involvement of all stakeholders, including those local communities who will be affected, in developing new practices and behaviours to ensure functionality and sustainability (Everett et al., 2016). Public perceptions of SuDS are likely to be linked to several factors, including scheme performance, biodiversity issues, education strategies, aesthetics, perceived health and safety risks, water quality and respondent socio-economic status. Safety has already been proven to be one of the main concerns regarding SuDS application, for both developers and the public (Susdrain, 2015b).

Case Study

A partnership between Gloucestershire Gateway Trust and Westmorland Ltd. has facilitated the delivery of unique and visionary designed (BREEAM excellent) MSAs. Gloucester MSAs are unlike traditional services and appear almost seamless within their surroundings (Pegasus Planning Group, 2010). Costing £40 million to construct (by Buckingham Group Contracting Ltd.), the buildings on each side of the motorway display similar designs, layout and floor space, with a main facilities building (~3300 m^2), LGV drivers building (~30 m^2) and a petrol filling station (~230 m^2). The main facilities buildings are timber framed with lattice roof structures (~9 m high) convened between solid drystone walls made of locally sourced buff-coloured Cotswold limestone. The buildings are draped in a blanket of soil and turf to create bespoke green roofs supporting native grass covers, which contribute to the water management and biodiversity of the sites. Water conservation inside and outside the buildings incorporates rainwater harvesting, low flow sanitary appliances, which include dual flush toilets and aerated flow restricted taps, plus leak

Fig. 1 Photos of the main amenity building and surroundings at the M5 Gloucester motorway service areas

detection and control (using smart metering), and low water landscaping that reduces the quantity and improves the quality of site runoff (Fig. 1) (Booth & McLaughlin, 2017).

Drainage of the MSAs follows a SuDS management train and has been designed for storm events in exceedance of 100 years, plus 30% climate change events. Besides the source control measures already mentioned (e.g. green roofs and rainwater harvesting), a suite of devices have been incorporated into the landscape of the MSAs. The parking bays of the car parks (with kerb drains) and pedestrian walkways have permeable paving, which allow surface waters to percolate between the blocks to infiltrate and be stored in the stone beneath. The LGV parking and access roads are drained by shedding the flow to filter strips, before entering stone filled treatment trenches. A series of roadside swales and trenches (both wet and dry) provide an infiltration and conveyance network across the sites. These are joined by underground piped (various diameters) inlets/outlets, fitted with silt traps and flow control chambers. Excess waters eventually enter a series of ponds (permanent volume of 154.4 m^3; maximum attenuation volume 1004.7 m^3) and wetlands (permanent volume of 96.5 m^3; maximum attenuation volume 556.4 m^3) at the end of the MSA sites. As well as bestowing a visually attractive landscape, the sequence of site control devices improves the quality and runoff characteristics, which also contributes to the biodiversity and management of the MSAs (Booth & McLaughlin, 2017).

Research Design and Methodology

Data Collection

A questionnaire was chosen as an instrument for data collection because of the desire to consult as large a sample of MSA users as possible, without delaying their journey times or disrupting their experience of using the MSAs. For this reason, other methods were deemed to be impractical. The instrument was developed through an iterative process of literature reviews and consulting experts to refine the measurement items, before piloting with industry and academic professionals.

Feedback from the pilot exercise was used to amend and address possible issues before the final version of the questionnaire was distributed.

The main purpose of the questionnaire was to solicit public knowledge about 'blue-green' systems and their intended use and benefits, using the MSA as an exemplar case study. Data about individual views on the environment, climate change and extreme weather events were also collected, together with a range of questions to determine the demographics of the participants. A five-point Likert-type measurement scale (from strongly disagree (1) to strongly agree (5) responses) was adopted to gauge respondent ratings of statements. These were funnelled into a narrow frame to ensure they were specific, measurable, attainable, relevant and time-based (SMART). A small number of open-ended questions (dual approach) were also included to elicit rich qualitative data, alongside the quantitative approach.

Ethical approval was sought before the final questionnaire was shared. Approval meant all participants were informed in a participant information cover letter that their involvement was entirely voluntary and their decision to complete and return their completed questionnaire was their consent to take part in the study. As their responses would be anonymous, participants were also informed that there would be no opportunity to withdraw once the questionnaire had been returned.

Data Analysis

Data was entered into SPSS (version 20) before analysis and descriptive statistical information was derived. These are detailed and portrayed in the sections beneath.

Results

The findings of the data analyses are presented beneath, under four main sections: (1) background information about the participants; (2) their opinions of the environment; (3) their opinions of the Motorway Service Areas and (4) their opinions of Blue-Green Infrastructure.

Background Information

The questionnaire was distributed at Gloucester MSA (Northbound and Southbound), over 2 days during summer 2016. A weekday, including mainly morning and afternoon commuters (a cool and overcast day), and a weekend, including mainly families and leisure motorists (a hot and sunny day), were chosen. A total of 86 respondents (42 male and 44 female) between the ages of circa 18 and 80 years fully completed the questionnaire, which included a wide range of visitors, with purposes of travel ranging from leisure to commuting.

Opinions of the Environment

Of the participants, many (88%) believe climate change poses a major threat to the UK, but only 76% have a personal concern; however, there was consensus about the frequency of extreme weather events in the UK, with 84% agreeing that there has been an alarming increase, but only 70% recognising this as a personal concern. Flooding seems to be acknowledged as the major climate change consequence within the UK (agreed by 95% of participants); however, only 69% agree that they are worried about flooding.

Opinions of the Motorway Service Areas

Public opinion was very supportive of the design and presentation of the amenity building and surrounding landscape, with 100% of participants agreeing or strongly agreeing that the Gloucester MSAs blend effectively with the surrounding landscape, and the majority (68%) being aware that they have been designed to minimally impact the environment. A greater number of female participants (75%) confirm an awareness of the site's environmental attributes as opposed to only 60% of male participants. Furthermore, 85% of those who are aware of the site's aim of minimising impact on the environment live within a 50 mile radius, with only 48% of participants from outside a 50 mile radius having an awareness of the site's environmentally friendly design features. Overall, 94% of all participants believe that other motorway services areas should be designed or adapted to look like Gloucester MSAs.

The positive comments from participants about the aesthetics and environment were overwhelming, with comments including 'it's like a hobbit house' and 'an oasis compared to other MSAs'. One participant even noted that he loved it so much he had taken his wife there on a date, and several queried why there was no signage on site to explain the 'eco' features.

Opinions of Blue-Green Infrastructure

Of the participants, many (89%) agree that the green roof, ditches, pond and wetlands enhance the MSAs; however, only 58% agree that the block paving used for the parking and walkways enhances the site aesthetically. Only 31% of participants agree that the block paving actually adds benefit to the environment; however, 81%

agree that the green roof provides environmental benefit and 89% agree that the ditches, pond and wetland are also environmentally beneficial. Similarly, only 27% of participants agree that the block paving promotes rainwater soak-away (as opposed to 74% appreciating the function of the green roof and 39% understanding the purpose of ditches, pond and wetland as rainwater buffers). This leads to the assumption that favourable public perceptions are more likely to be influenced by nature.

Ultimately, the public expressed a preference for a greener, more aesthetic form of construction; however, they were seemingly not always able to make the connection between attractive green features and sustainable drainage (55% of participants being unable to either agree nor disagree as to whether the ditches, pond and wetland act as buffers to slow rainwater run-off). Only 53% of participants agree that the ditches, pond and wetland can help to control site pollution and as little as 20% are aware that block paving can help to control site pollution as part of a SuDS management train. Indeed, 42% of participants are neither able to agree nor disagree whether the block paving actually promotes rainwater soakaway. Overall, 64% of male participants feel that they are poorly informed about 'Blue-Green' drainage systems, compared with 59% of female participants. As a result, an increase in education about recognising and interpreting sustainable drainage systems is required in order to fuel a public demand for a move away from the traditional mains pipe network. From public opinion, it was apparent that demand for Blue-Green drainage was high (86% of participants want to see these systems used in their neighbourhood and 89% believe they can reduce flood risk, with 72% in agreement that sustainable drainage components promote ecological biodiversity). Participants (75%) want maintenance to be placed into the domain of their Local Authority, despite only 23% believing that sustainable features, such as green roofs, block paving, ditches, ponds and wetlands, are probably hard to maintain.

Discussion

Public perceptions of SuDS are an important indicator of the level of knowledge about these types of system; however, perceptions may not be easy to gauge, especially when SuDS components are often concealed (Sayers, 2012). From our findings it is clear that implementing a shift away from hard 'grey' infrastructure requires the involvement of all stakeholders, including affected public, in developing new practices and behaviours to ensure functionality and sustainability (Everett et al., 2016). The importance of public support for the implementation of any community project is crucial, and understanding public opinion can provide information about what the general population value the most (Sanchez Lopez, 2014).

SuDS can strengthen communities by providing a focus for environmental education and public engagement in environmental protection in a localised area, whilst supporting development resilience to climate change and development pressures (Daly, Jodieri, McCarthy, Pygott, & Wright, 2016). By introducing water to the urban environment, the planning process provides an opportunity to place SuDS within the public arena, addressing the aspirations of the public for a better, cleaner and greener urban environment (Dickie et al., 2010). Ponds and wetlands can be assets to the community, enhancing the quality of life, by providing attractive and tranquil green space within the built environment (Cahill, 2012). Wetlands can attract diverse wildlife species and ponds, with the inclusion of footpaths, benches and picnic tables, can provide attractive community areas to enable social cohesion (Dickie et al., 2010).

Every aspect of SuDS developments requires critical thought and it should be integrated into all types of construction (Illman, 2016). A big push is required by construction professionals and the general public to influence the government to do more to produce SuDS legislation (Corner, 2016). There is currently no one to champion the use of SuDS in the UK, or to raise public awareness of their benefits (Bray, 2016). There is both a lack of effective legislation and a low level of public demand which both need to be urgently addressed (Illman, 2016).

Local planning authorities are not embracing SuDS in developments through a lack of skill, poor policy base, reduced funding and minimal understanding of these systems. This has led to local planning authorities being defeated by developers on appeal (Illman, 2016) and, therefore, the pattern of low delivery on these schemes is maintained, perpetuating poor public awareness (Corner, 2016). There is a great deal of societal malaise and general disinterest in sustainable drainage systems. The public do not appear enthused or interested in changing to a natural solution for managing rainwater simply because they are not making the connection to play space, green areas and bio-diverse wetlands and habitats (Bray, 2016).

Barriers preventing the uptake of SuDS, such as ongoing management and maintenance of public areas, need to be highlighted, worked through and resolved (Illman, 2016). To help gain support for future flood prevention initiatives, it is crucial for local communities to plan for the likelihood of flooding occurrences, and the impact it will have as and when it does happen (Sayers, 2012). A centralised framework must be put in place surrounding the adoption of SuDS by Local Authorities, instead of each local authority pursuing its own agenda without referral to central guidance. Local authorities would generally prefer for the national water companies to adopt and maintain SuDS areas, however, as profit-making organisations there is limited incentive for them to do so (Corner, 2016).

There is minimal concentration on 'amenity', and especially the engagement of local people in embracing and taking responsibility for improving their neighbourhood (Singleton, 2012). This form of surface water management has not been embraced fully by the public, to include the contribution SuDs provide to place-making and the environment (Shaffer, 2016). In areas where the public appeared to be well informed, they also viewed drainage schemes much more positively than in areas where little information had been made available (Apostolaki & Jefferies, 2005). In areas of the USA, such as Orlando, the signage is excellent; however, in the UK there is a reluctance to specify or pay for signage. The client generally does not see signage information as valuable, and quite often feels that SuDS are simply '*foisted upon them by Planning Authorities*' (Bray, 2016).

Conclusions

Blue-Green systems mimic natural drainage processes to reduce the effect on the quality and quantity of run-off from developments and provide amenity and biodiversity benefits. However, embracing these changes and developing the techniques required to construct and maintain SuDS brings with it the opportunity to achieve important environmental and economic benefits through a more effective and sustainable model for the management of storm drainage. Perhaps SuDS should be considered as 'rainfall management' and not as drainage? Because when the term 'drainage' is used, it is considered uninteresting and the public become disengaged.

There is clearly a growing acceptance that we need to develop a more sustainable approach to managing surface water to ensure water quality and reduce reliance on limited sewer capacity, especially with the increase in frequency of severe weather events. However, there is very limited critical thinking, as SuDS are in conflict with conventional drainage. The public must be made to understand that, unlike for sewage, pipes do not have to be the favoured conveyance channel for rainwater.

A SuDS scheme aims to do more than just deal with surface water problems. It is a relatively new and, for some, revolutionary way to seize opportunities with water. Planning requirements should dictate that there has to be SuDS interpretation boards and a budget for P.R. to highlight/publicise SuDS within a development in an effort to increase visibility and boost public demand. An entirely visible process must be engaged as it is critical that people make the connection and understand what SuDS components are doing/how they are linking. Public awareness will help promote the acceptance and use of these systems and encourage a new way of thinking. Public education in the field of stormwater pollution and management must also become an important contributory factor in the public demand for SuDS.

At Gloucester MSAs, it was overwhelmingly apparent that the public wanted to see environmentally friendly forms of construction, and were enthusiastic about the aesthetics of the site. It was also evident that visitors were not consistently forming a link between the look of the MSA and the way it was sustainably managing rainwater, despite the design intention for the landscape to be readily understood. Public education can have a critical role in influencing acceptability of new or innovative practices; therefore, without further education about the benefits of blue-green infrastructure the general public will not be able to demand the sustainable rainwater management systems that underpin the aesthetic forms of construction they enjoy.

Recommendations

- There needs to be greater public awareness and a burden of responsibility on the individual to influence a change in opinion, as well as a corporate responsibility.

- Developers should be required to include sustainable drainage, where practicable, in new developments, built to standards that reduce flood damage and improve water quality. It remains too easy for developers to connect to mains drainage if they argue effectively.
- Local Councils should be given responsibility for maintenance, subject to funding, of vegetative and soft engineered SuDS features meeting the criteria for adoption into the public realm. SuDS schemes generally involve costly ground works or take up large sections of land that could be generating revenue; there is also the perception of a high maintenance liability and overall life cycle costs, which are perceived as a burden rather than a long-term permanent solution. As such there is limited competition for installing and maintaining, meaning that a standard framework should be developed.
- All organisations involved should introduce operational policy to facilitate the establishment of SuDS as the preferred method of surface water drainage for all new developments.
- SuDS planning must be made mandatory and included in every masterplan.
- Future Government policy should seek to increase the promotion of SuDS schemes, especially where new developments will impact on the surrounding area. This should include design and performance standards.

References

Apostolaki, S., & Jefferies, C. (2005). *Social impacts of stormwater management techniques including river management and SuDS*, Final report, SUDS01. Edinburgh: SNIFFER.

Ashley, R., Balmforth, D., Dignan, C., Glerum, J., & Stovin, V. (2012). *Retrofitting to manage surface water*. London: CIRIA Publishing.

Balmforth, D., Digman, C., Kellagher, R. & Butler, D. (2006) Designing for Exceedance in Urban Drainage - Good Practice. CIRIA C635, London.

Booth, C. A., & McLaughlin, A. M. (2017). Contemporary landscapes and buildings of motorway service areas. In S. M. Charlesworth & C. A. Booth (Eds.), *Sustainable surface water management: A handbook for SuDS* (pp. 259–269). Oxford: Wiley-Blackwell.

Bray, R. (2016). *Public perceptions of SuDS*. Interviewed by Mark Gazzard, August 16, 2016.

BWB Consulting. (2016). *Gloucester Gateway MSA* [Online]. Retrieved April 12, 2016 from http://www.bwbconsulting.com/project/N1H1435_gloucester-gateway-msa-m5-j11a-12

Cahill, T. (2012). *Low impact development and sustainable stormwater management*. Hoboken, NJ: Wiley.

Charlesworth, S. E., & Booth, C. A. (2017) Sustainable Surface Water Management: A Handbook for SuDS. Oxford: Wiley–Blackwell.

Carrington, D. (2014, February 1). *The Guardian* [Online]. Retrieved March 14, 2016 from http://www.theguardian.com/uk-news/2014/feb/01/january-uk-wettest-winter-month-250-years

Corner, A. (2016). *Public perceptions of SuDS*. Interviewed by Mark Gazzard, August 12, 2016.

Daly, D., Jodieri, R., McCarthy, S., Pygott, K., & Wright, M. (2016). *Communication and engagement in local flood risk management*. London: CIRIA Publishing.

Defra—Department for Environment, Food and Rural Affairs (2011). *National standards for sustainable drainage systems: Designing, constructing, operating and maintaining drainage for surface runoff* [Online]. Retrieved March 22, 2016 from https://www.gov.uk

Dickie, S., McKay, G., Ions, L., & Shaffer, P. (2010). *CIRIA C687: Planning for SuDS—making it happen*. London: CIRIA.

Everett, G., Lamond, J., Morzillo, A., Chan, F. K. S., & Matsler, A. M. (2016). Sustainable drainage systems: Helping people live with water. In *Proceedings of the ICE—Water Management* 169, 94–104.

Heal, K., Bray, R., Willingale, S., Briers, M., Napier, F., Jefferies, C., & Fogg, P. (2008). *Medium term performance and maintenance of SuDS: A case study of Hopwood Park Motorway Service Area UK* [Online]. Retrieved April 3, 2016 from https://web.sbe.hw.ac.uk/staffprofiles/bdgsa/11th_International_Conference_on_Urban_Drainage_CD/ICUD08/pdfs/251.pdf

Illman, S. (2016). *Public perceptions of SuDS*. Interviewed by Mark Gazzard, August 26, 2016.

Meteorological Office. (2014). *Wettest Winter for England and Wales* [Online]. Retrieved December 15, 2015 from http://www.metoffice.gov.uk

Natural England. (2009). *Natural England's Green infrastructure guidance* [Online]. Retrieved February 12, 2016 from http://publications.naturalengland.org.uk/publication/35033

Pegasus Group. (2010). *Design access statement: Gloucestershire Gateway Motorway Service Area*. Canwell, Sutton Coldfield: Pegasus Planning Group LLP.

Pegasus Group. (2016). *Gloucestershire Gateway MSA* [Online]. Retrieved April 22, 2016 from http://www.pegasuspg.co.uk/index.php/site/project/gloucestershire_gateway_msa

Robert Bray Associates. (2016). *Hopwood Motorway Services* [Online]. Retrieved April 12, 2016 from http://robertbrayassociates.co.uk/projects/hopwood-msa/

Rossmiller, R. (2014). *Stormwater design for sustainable development*. New York, NY: McGraw-Hill Education.

Sanchez Lopez, S. (2014). *Analysing public perception and biodiversity of sustainable urban drainage systems (SuDS)* [Online]. Retrieved May 16, 2016 from http://blogs.nottingham.ac.uk/blue-greencities/2015/03/05/analysing-public-perception-and-biodiversity-of-suds/

Sayers, P. (2012). *Flood risk—Planning, design and management of flood defence infrastructure*. London: ICE Publishing.

Shaffer, P. (2016). *Public perceptions of SuDS*. Interviewed by Mark Gazzard, August 12, 2016.

Singleton, D. (2012). *SuDS in the community: A suitable case for treatment?* [Online]. Retrieved May 9, 2016 from http://www.susdrain.org

Susdrain. (2015a). *Sustainable drainage* [Online]. Retrieved March 16, 2016 from http://www.susdrain.org

Susdrain. (2015b). *An assessment of the social impact of SuDS in the UK* [Online]. Retrieved April 5, 2016 from http://www.susdrain.org

The Design Review Panel. (2013). [Online]. Retrieved April 22, 2016 from http://www.designreviewpanel.co.uk/

The European Union Sustainable Development Strategy. (2006). [Online]. Retrieved March 16, 2016 from https://www.etuc.org/new-eu-sustainable-development-strategy-june-2006

The Flood and Water Management Act. (2010). c.29 [Online]. Retrieved March 19, 2016 from http://www.legislation.gov.uk/ukpga/2010/29/contents

The UK Climate Change Act. (2008). [Online]. Retrieved April 22, 2016 from http://www.legislation.gov.uk/ukpga/2008/27/contents

The University of Nottingham. (2015). *Blue-green cities research project: Delivering and evaluating multiple flood risk benefits in blue-green cities* [Online]. Retrieved April 2, 2016 from http://www.bluegreencities.ac.uk/bluegreencities/index.aspx

Thompson, D. (2014). *Adaptation Committee on climate change, RSPB* [Online]. Retrieved March 14, 2016 from https://www.theccc.org.uk/author/david-thompson/

Water Active. (2015). *ACO helps Gloucester services achieve project sustainability requirements* [Online]. Retrieved April 20, 2016 from http://www.wateractive.co.uk/case_studies

Woods-Ballard, B., Kellagher, R., Martin, P., Jefferies, C., Bray, R., Shaffer, P. (2007a). *Site handbook for the construction of SuDS*. CIRIA C698. London: CIRIA.

Woods-Ballard, B., Kellagher, R., Martin, P., Jefferies, C., Bray, R., & Shaffer, P. (2007b). *The SuDSs manual*. CIRIA C697, London.

Woods-Ballard, B., Wilson, S., Udale-Clarke, H., Illman, S., Ashley, R., & Kellagher, R. (2015). *The SuDS manual*. CIRIA C753, London.

A Decision Support System for Affordable and Sustainable Housing Design and Delivery in Least Developed Countries (LDCs)

John Bruen, John P. Spillane and Su Taylor

Introduction

The right to adequate housing is central within the global legal system and has been recognised from as far back as 1948 in the Universal Declaration of Human Rights to current day recognition as a human right within the United Nations Universal Declaration of Human Rights. The adequacy of housing should not be merely interpreted as just shelter or having a 'roof over one's head'. Housing is a multidimensional and complex area involving many stakeholders from a wide variety of backgrounds. The adequacy of housing needs to reflect this complexity and should be measured against factors such as security of tenure, materials, availability of services, affordability, habitability, cultural adequacy and its contribution to society and sustainable communities so as to ascertain its true value (IASC, 2011).

The need for a sustainable approach to housing has been recognised globally and outlined in a number of globally recognised documents. These strategy and goal documents provide guidance and have promoted and reinforced new approaches to housing provision in LDCs. They include the Global Strategy for Settlement and Shelter and Agenda 21. The need for sustainable approaches to all aspects of development is further recognised in the United Nations Sustainable Development Goals, or SDGs as they are also referred to. Also known as the Global Goals, they are a universal call to end poverty and protect the planet for the enjoyment of all people.

J. Bruen (✉) · S. Taylor
School of Natural and Builit Environment, Queen's University, Belfast, UK
e-mail: john@bruenarchitects.com

J. P. Spillane
University of Limerick, Limerick, Ireland

L. Scott et al. (eds.), *Sustainable Ecological Engineering Design*,
https://doi.org/10.1007/978-3-030-44381-8_9

They consist of 17 interconnected global goals covering a wide range of global issues such as poverty, climate, global inequality and other priorities. Sustainable Goal 11 "Sustainable Cities and Communities" makes specific reference to the built environment, housing, disasters and the importance of preserving culture and identity:

> *By 2030, ensure access for all to adequate, safe and affordable housing and basic services and upgrade slums*
>
> *Support least developed countries, including through financial and technical assistance, in building sustainable and resilient buildings utilizing local materials*
>
> *Strengthen efforts to protect and safeguard the world's cultural and natural heritage*

The SDGs recognise important aspects of sustainable development that are often overlooked in the implementation of housing, for example, culture, natural heritage, use of natural materials and building resilient buildings. However, the SDGs only provide a broad overview of the objectives, targets and goals to be achieved on a macro scale and not a specific approach to achieve them in the specific context of housing provision in individual LDCs to which this study relates. People and how they live are central to global sustainable development and the achievement of the global SDGs. The current and future demands for adequate housing play a central role in achieving the SDGs and fulfilling recognised human rights. However, a number of aspects affect achieving these noble goals and despite its central place in international law over 1 billion of the world's population are not adequately housed (UN-Habitat, 2009).

The aim of this research is to examine the current decision-making process of leading international organisations in the design and delivery of sustainable post-disaster housing in LDCs, and through a case study approach, develop a DSS for use in the development of better sustainable design practices in house construction in LDCs. To achieve this aim, the following objectives will be met:

1. Identify and explore the current challenges/barriers to the design and delivery of sustainable post-disaster housing in LDCs.
2. Identify current approaches to housing delivery in LDCs and post-disaster contexts.
3. Identify the drivers that contribute to sustainable and affordable housing in LDCs and post-disaster contexts.
4. Develop a DSS that will be freely available to NGOs, communities and other relevant stakeholders operating in the field of housing design and delivery in LDCs and post-disaster contexts. The overall objective is that the free dissemination of the decision support system will ultimately aid in enhancing the quality of design and delivery of sustainable post-disaster housing and the communities it serves that are affected by homelessness and poverty.

Literature: Housing Shortages in LDCs

There are various themes and concepts involved in the complex area of housing provision in LDC contexts in order to obtain an overall view of the various aspects for consideration and to inform the research study. The literature explored various aspects to include causes of housing shortage, current approaches to housing in LDCs, current challenges and barriers faced in the provision of housing and also the drivers that can aid effective design and delivery of housing in LDCs.

Population Growth

A number of main factors contribute to the global housing shortage and associated poverty. Population growth is a main contributor to the global housing shortage (UN-Habitat, 2009). World population growth has increased at an unprecedented rate in recent times and is predicted to continue to grow in the foreseeable future (UN, 2017). Global population has grown to almost 7.6 million as of mid-2017 and is projected to continue growing to 9.8 billion by 2050. LDCs are set to contribute disproportionally to the overall global population growth. Population growth in these countries is predicted to nearly double between 2017 and 2050 (UN, 2017). With this upward trajectory in global population growth comes the unavoidable rise in demand of basic human needs such as water, food and shelter.

Disasters

Disasters are a somewhat unknown quantity in terms of when and where they will occur. Historically disasters were considered "acts of god" and nothing could be done to prevent their occurrence (Voogd, 2004). In more recent times disasters were considered "acts of nature", and more recently they are considered an "act of society" (Malalgoda, Amaratunga, & Haigh, 2014). They can be both natural, such as an earthquake and tsunami, and man-made, such as conflict and biological disasters. What is known is that the frequency and destruction caused by disasters increases and LDCs are worst affected. Many LDCs are innately more vulnerable to natural disasters and have experienced disproportionate levels of devastation as a result of disasters, both in the human life, cost and the resulting large numbers of internally displaced populations (Schilderman, 2004). This vulnerability is often caused by common issues which are experienced in many LDCs including lack of preventative actions plans and preparedness, development in areas susceptible to natural disasters, unsettled governments, poor construction standards and techniques, insufficient resources and knowledge in post-disaster recovery and the reconstruction cost relative to GDP in these regions (Toya & Skidmore, 2007). The scale of displacement of populations resulting from disasters is growing. In 2015, natural disasters displaced an estimated 19.2 million people worldwide (IDMC, 2016).

The built environment and construction sector accounts for a significant part of a country's physical assets and economic development contribution. As such, the built environment is a major part of society and it is important to develop appropriate built environments that provide resilience and the ability to adapt to the threats of disasters (Bosher, 2008). However, regardless of preparation the built environment is vulnerable to the effects of disasters and as such disasters have the potential to have a devastating impact on a country and can create significant impacts on social and economic activities (Malalgoda & Amaratunga, 2015).

Disaster Management Cycle

Many organisations involved in works relating to disaster utilise a disaster management approach referred to as the disaster management cycle. White et al. (2004) describes the disaster management cycle as a set of sequential stages that occur during the unfolding of a disaster at which interventions can be undertaken to help lessen or mitigate against the impact of that disaster. This includes not just reactive measures during or after the event but also preventive measures prior to the event. Traditionally there are four stages to the disaster management cycle, namely, (1) preparation, (2) mitigation, (3) response and (4) recovery, which occur at three different periods during the cycle. The three periods include before the event (pre-disaster), impact (disaster) and after the event (post-disaster).

Post-disaster Reconstruction

Post-disaster reconstruction forms only one element of the overall recovery process but a very important one. Sevin and Little (1998) outline that although construction by itself will not eradicate all the impacts of disasters, the construction community and construction has an important role to play in finding and providing appropriate solutions to the continuous threat of disaster. There is also recognition that the construction industry can play a wider role as it assesses, prepares, prevents, responds and recovers from disasters. Post-disaster reconstruction covers many different aspects of the built environment depending on the context and disaster that preceded it. Housing projects often come as first priority in many post-disaster contexts. In LDCs much of the population would have no home insurance as in developed countries and it falls on the government and the international community to endeavour to provide housing for the homeless as a result of the disaster (Hidayat & Egbu, 2010). The process of reconstruction requires considerable medium- and long-term strategic consideration in order to capitalise on an opportunity to improve on the built environment that preceded it.

Sustainable Construction in LDCs

The majority of the world's population currently live in LDCs and future population growth projections anticipate that the majority of growth will also be in LDCs. The need for long-term sustainable approaches in the delivery of construction projects to meet these demands from the construction industry is clear (Du Plessis, 2002). Reffat (2004) states that the concept of sustainability has only recently been introduced into LDCs construction industries and that sustainability and sustainable construction are not yet an essential part of the decision-making process. Banishashemi, Hosseini, Golizadeh, and Sankaran (2017) argue that LDCs put economic development above sustainable requirements with construction demand overshadowing environmental concerns. The demands of sustainable construction in LDCs can differ greatly from that of developed countries. Du Plessis (2002) states that the construction approaches from developed countries, which are highly reliant on technical solutions, adapted by LDCs are inappropriate and must be expanded to address the social and economic pillars of sustainability. Perceived higher costs and underlying socio-cultural factors also contribute to the lower levels of social acceptability of sustainable construction in the main stream affordable housing market (Sibley, Hes, & Martin, 2008). Othman and Ahmed (2013) note that a wide range of challenges exist that affect delivery of sustainable construction projects in LDCs outlining five main headings under which they fall which include human development, technical, managerial, political and the triple bottom line of sustainability (environmental, social and economic).

Decision-Making

Human performance in decision-making has been the topic of research from a number of different perspectives. From a psychological perspective, it is necessary to examine individual decisions in the context of a set of needs, preferences an individual has and values they seek. From a cognitive perspective, the decision-making process must be regarded as a continuous process integrated in the interaction with the environment. From a normative perspective, the analysis of individual decisions is concerned with the logic of decision-making and rationality and the invariant choice it leads to (Kahneman & Tversky, 2000). Different professions refer to decision-making differently, i.e. architecture may refer to decision-making as design. All architects and engineers as designers and project managers, make many decisions on a daily basis in relation to their work. The process of designing reconstruction projects, infrastructure, public space, etc. involves many decisions to be taken on many different levels. On a very simple level a typical design process involves three main stages (Cuff, 1991):

1. Initial concept stage.
2. Design development (problem solving stage).

3. Working drawing/implementation phase.

As every project is unique, these three basic main stages may vary from project to project. However, regardless of what stage a project is at or what context it is located in, design decisions, like all decisions, are based on some rationale or logic as well as being conducted in the context of that particular project (Holm, 2006). Logical decision-making is an important part of all science-based professions within which architects and engineers are deemed to exist, where specialists apply their knowledge in a given area to making informed decisions. Professional decision-making is often seen as being the skilful application of technical knowledge within ethical limitations (Holm, 2006).

Affordability and Sustainability

Affordable housing is an essential concept to LDCs as governments in these countries cannot afford to provide all the required housing Menshawy, Shafik, and Khedr (2016). Affordability and affordable housing is a difficult term to define as it can have different meaning to different people and vary widely from country to country or region to region. In many LDCs, affordability in mainstream housing markets is associated with economic sustainability often with little emphasis on environmental social sustainability (Randolph, Kam, & Graham, 2008). Stone (2006) states that the term "affordable housing" is an unjustified term and that affordability is not a characteristic of housing but rather it is a relationship between housing and people arguing that for some people all housing is affordable regardless of price and for others it is not affordable unless it is free. Stone further argues that "affordable" housing can have meaning and use but only if three essential questions are answered:

1. Affordable to who?
2. On what standard of affordability?
3. For how long?

Incomes in LDCs are lower and as such housing can represent a much higher proportion of a family's income. Other aspects relative to LDCs such as immense population growth with demand out stripping supply, lack of materials, skills and design/construction knowledge, higher levels of natural disasters and fewer utilities often serve to raise the cost of housing relative to the incomes in many contexts. Hayles (2006) states that affordability and sustainability are intrinsically linked, and to be affordable, a house must embrace sustainability principles and be designed and constructed to last. Hayles further states that sustainable affordable housing must meet several goals and be designed so that value to the consumer outweighs its financial cost. In many developed countries this can be largely related to the energy efficiency of the dwelling and the running and maintenance cost over its lifetime. Johnson (2006) further argues that decent affordable housing contributes to the economic, environmental and social health—the sustainability—of communities. McIntosh

(2013) develops this view arguing that any new house, regardless of context, must be within the confines of what is affordable, resilient to future extreme weather events, be culturally appropriate and strengthen community structures.

Research Methodology and Method

The research bridges the domains of architecture and project and design management, in particular the aspect of decision support. A literature review in the relevant areas provided background knowledge of the issues and challenges in this research area and informed the overall identification of gaps in knowledge and the research methodology. A qualitative approach to the research was employed as an appropriate method to ascertain the experiences, attitudes and processes of the various international organisations operating in post-disaster contexts. A multiple qualitative case study approach consisting of 12 individual case studies was undertaken with 9 selected international organisations operating in the area of housing provision in LDCs and post-disaster contexts on a worldwide basis. For environmental and intensity of data collection reasons Sri Lanka was chosen as a country for 10 of the 12 case studies. Sri Lanka was subject to a number recent disasters, e.g. tsunami and recently the ended 30 year civil conflict, and was in the process of rebuilding many parts of the country as a result.

Multiple sources of evidence were utilised, e.g. literature review, 15 semi-structured interviews, documentation and observation of the physical artefact (constructed houses in their context) to collect case study data from the field. Selected approaches were utilised to analyse the data including cognitive mapping, data reduction, logic diagrams and thematic analysis to ascertain the key themes that informed the organisations decision-making and to identify the process followed in undertaking projects in LDC and post-disaster contexts (Figs. 1–4).

Research Results

Research Aim 1 Identify and explore the current challenges/barriers to the design and delivery of sustainable post-disaster housing in LDCs.

The literature identified an extensive list of barriers to the design and delivery of affordable and sustainable housing in LDCs. Many of these barriers were also identified in the field research which further verifies their existence as legitimate barriers in a real-world context and further adds to and confirms the existing knowledge on this topic. On review of the identified barriers, they can be summarised into a number of broad categories some of which relate to the organisations or designers themselves and other wider societal factors which include:

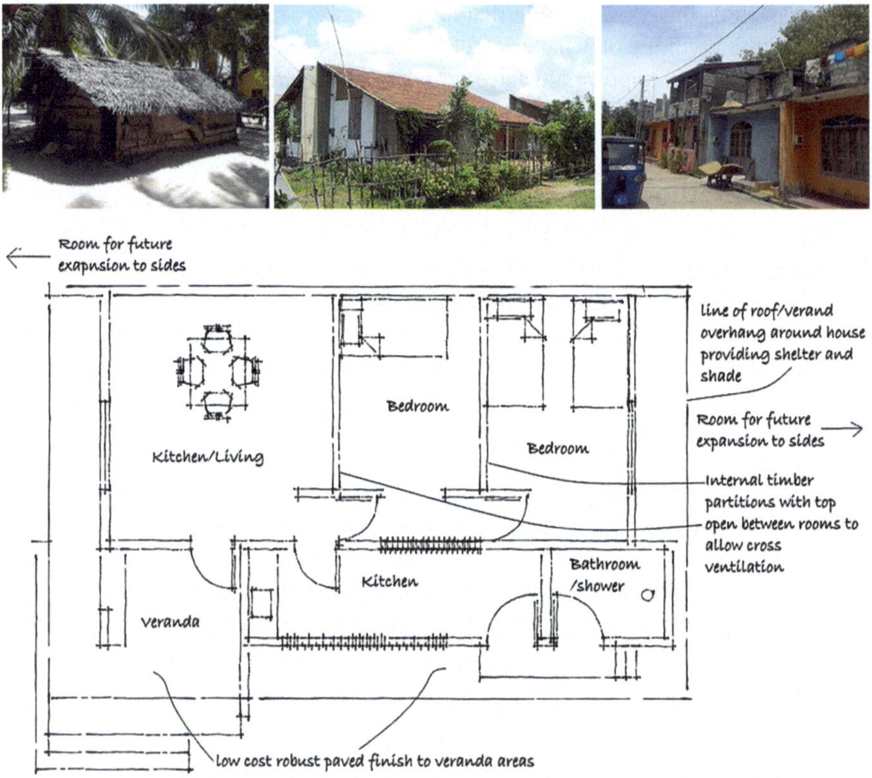

Figs. 1–4 Examples of field research data collection

- Lack of understanding of the true concept of sustainable design and its holistic application in the design and delivery of sustainable housing in LDCs.
- Sustainable design not prioritised in LDCs and post-disaster contexts.
- Governmental and institutional rules and parameters, e.g. design constraints, procurement routes and inappropriate regulations.
- Psychological and sociological issues in relation to sustainable design and its potential output, e.g. materials and approaches that deviate from what is deemed the norm.
- Lack of guidance and guidelines for relevant stakeholders operating in the area.
- Lack of research into this area.
- Poor management in relation to the design and delivery of sustainable housing in terms of both design and project management.

Research Aim 2 Identify current approaches to housing delivery in LDCs and post-disaster contexts.

Top-down strategies from governments and NGOs were identified as a common approach to housing in LDCs and post-disaster contexts. These approaches resulted in little or no consultation or participation with the beneficiaries and were often detrimental to the long-term sustainable performance of the houses and communities. Standardised house design with the use of inappropriate materials and technologies with little reflection of the local culture were identified as common approaches to housing provision. More recent times have seen the greater use of bottom-up macro level self-help approaches. Barriers to the effective implementation of a participation process were identified from the literature. It was observed in both the literature and case studies that in order for participation to be meaningful, the project beneficiaries must be involved from the outset and have genuine control over decision-making. The case studies demonstrate that effective use of a participatory design and delivery process utilising a holistic approach to sustainability can result in appropriate and affordable long-term homes.

Research Aim 3 Identify the drivers that contribute to sustainable and affordable housing in LDCs and post-disaster contexts.

Three main design drivers were identified from the research that contributed to the implementation of sustainable and affordable housing LDCs and post-disaster contexts.

1. Appropriate design and material selection.

   The use of appropriate design and materials were identified as key drivers in the efforts to deliver sustainable and affordable housing LDCs and post-disaster contexts. Localised skills, materials and techniques offered the potential to both deliver appropriate housing and adequately address a holistic approach to sustainability as a result of the spin off effects of this approach. The use of local and traditional materials can generate employment in the community, in the supply of materials and manufacture of building products as well the utilisation and improvement if required, of local knowledge in terms of construction techniques. Practical aspects such as maintenance, future community expansion and employment within the community can all benefit from this approach. Housing typologies should contextually and culturally appropriate of the communities they serve. Case studies highlighted the obstacles to implementing this approach in practice which were predominantly external factors outside of their control, e.g. government or building standards, donors requirements for a particular material or housing typology.
2. Appropriate innovative technology and knowledge transfer specific to developing world contexts.

   Appropriate technology use in conjunction with appropriate design and materials informed by the local conditions and culture can result in long-term sustainable and affordable housing. Western technology imported to a LDC context does not necessarily result in a seamless assimilation into that context. Care should be taken to ensure that the transfer of knowledge and technology by whatever means is firstly well considered on different levels and assessed as to its appropriateness to the context. Achieving this can often require in-depth research into the housing beneficiaries cultural beliefs and the wider cultural context prior to commencing any design work. The research highlighted that this approach was undertaken by a number of organisations and was central to their informed approach with positive outcomes. Small improvements to existing approaches and technology resulted in long-term positive outcomes in a number of instances. However, aspects such as time, cost and acceptance by the community were highlighted as a main obstacle to implementing this approach.

3. Design decision-making assistance tools.

   The literature outlined that limited knowledge and information coupled with the complex relationship between a building and its environment are often the cause of ill informed decisions on long-term sustainable and affordable design and delivery of housing in LDC and post-disaster contexts. Frameworks and structured systematic approaches that enable complex decisions with many competing variables to be made manageable are identified as effective methods to address this. It was highlighted that decision support tools to date were focused on developed country contexts and none were specifically designed for sustainable and affordable housing in LDC or post-disaster contexts. The case study research further confirms this finding. Despite many years' experience on a global basis, none of the organisations had a specific decision support system or framework formulated to aid in the design and delivery process, instead relying on tacit knowledge within the organisation and lessons learned from previous projects.

Research Aim 4 Develop a decision support system that will be freely available to NGOs, communities and other relevant stakeholders operating in the field of housing delivery in LDCs and post-disaster contexts.

A gap was identified for a tool that enabled and assisted architects, project managers and another stakeholders operating in this complex area with their decision-making process from inception to completion. Detailed analysis was undertaken of the case study organisation's approaches to the design and delivery of housing in LDC and post-disaster contexts. Semi-structured interviews with key design decision-makers from each organisation were analysed utilising a combination of cognitive mapping and thematic data reduction to ascertain key themes across all organisations in relation to their decision-making and approaches on projects. Further analysis of logic diagrams constructed for each organisations approach highlighted key stages and aspects considered by the organisation at each stage which enabled a further in-depth understanding of each organisation's overall approach in a graphical format. Analysis of the logic diagrams for each organisation identified key stages common to all organisation as well as additional stages undertaken by some. Combining all research findings a user-friendly decision support system was designed utilising Microsoft Excel as a platform (Fig. 5).

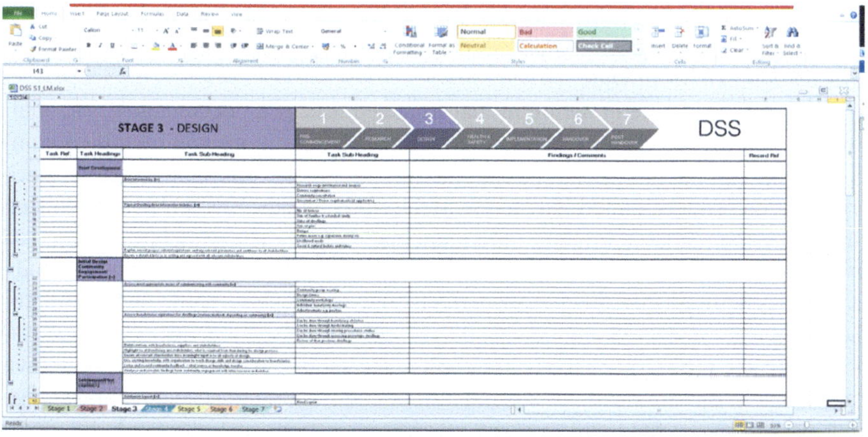

Fig. 5 Screen print of Decision Support System developed from the research

Discussion and Conclusion

The finding and output of this research contribute to both architectural and design management theory and practice. On review of the literature, it was evident that there was a lack of documented knowledge in relation to the decision-making process and management for the design and delivery of sustainable housing in LDC and post-disaster contexts. While the literature has recognised barriers and drivers that organisations face in the design of housing and delivery, the majority of the literature fails to address the processes and stages a typical organisation go through in terms of the design and delivery of housing and the variables that impact their decisions. Through detailed analysis of case study data from a real-world context and the contribution from leading international organisations, the study provides pivotal new information on the various aspects and stages involved in the overall complex process. The study's findings on key themes that influence design decision makers both confirm findings from the literature review and provide additional findings. This combined with the development of a decision support system for use in a real-life context provide a fresh perspective on the overall process and key aspects in decision-making and contribute to the body of knowledge in this area.

Unfortunately many NGOs, INGOs and other organisations often utilise ad hoc and inappropriate or outdated approaches to housing design and delivery in post-disaster contexts that lack meaningful design considerations and engagement with the wider factors that affect the communities they are serving. This is often as a result of organisations not having the requisite expertise or being forced into a role that they are not competent to undertake. This study aims to address these short comings through the detailed analysis of approaches of leading international housing organisations and the development of a decision support system that can be easily disseminated to assist many different stakeholders involved in the field and ultimately benefit the communities served in the long term. Regardless of each organisation's approach, the participating organisations had a high degree of tacit knowledge based on many years' experience working in many LDCs and post-disaster contexts worldwide.

The study is unique in nature in that it gained access to, documented and analysed this knowledge from leading international organisations and identified common themes and approaches for best practice from real-world field research. Each organisation brought something unique to the study and the final proposed decision support system is based on their individual experiences. This was further reflected in the fact that each participating organisation was happy to participate in the study as they recognised that their own methods and approaches, while successful for their organisation, were not guaranteed to be correct or best practice. The organisations were keen to learn from other leading organisations and disseminate their own knowledge in this area with the overall aim of contributing to academic knowledge with data from the field and to assist other organisations and ultimately the communities served.

References

Banishashemi, S., Hosseini, M. R., Golizadeh, H., & Sankaran, S. (2017). Critical success factors (CFFs) for integration of sustainability into construction project management practices in developing countries. *International Journal of Project Management, 35*, 1103–1119.

Bosher, L. (2008). The need for built in resilience. In L. Bosher (Ed.), *Hazards and the built environment—Attaining built-in resilience* (pp. 3–19). Oxon: Routledge.

Cuff, D. (1991). *Architecture: The story of practice*. Cambridge, MA: MIT Press.

Du Plessis, C. (2002). *Agenda 21 for sustainable construction in developing countries: A discussion document*. Pretoria: CSIR Building and Construction Technology.

Hayles, C. S. (2006). An examination of the relationship between sustainability and affordability in residential housing markets. In *12th Pacific Rim Real Estate Conference, Auckland, New Zealand, January 22–25, 2006*.

Hidayat, B., & Egbu, C. (2010). A literature review of the role of project management in post-disaster reconstruction. In C. Egbu (Ed.), *Proceedings of the 26th Annual ARCOM Conference, September 6–8, 2010* (pp. 1269–1278). Association of Researchers in Construction Management: Leeds.

Holm, I. (2006). *Ideas and beliefs in architecture and industrial design*. Doctoral Thesis, Oslo Scholl of Architecture and Design.

IDMC. (2016). *Global report on internal displacement*. Retrieved August 20, 2017 from http://www.internal-displacement.org/assets/publications/2016/2016-global-report-internal-displacement-IDMC.pdf

Inter-Agency Standing Committee. (2011). *IASC operational guidelines on the protection of persons in situations of natural disasters*. The Brookings—Bern Project on Internal Displacement.

Johnson, M. (2006). Decision models for affordable housing and sustainable community development. *Journal of the American Planning Association: The Future(s) of Housing*. Retrieved December 29, 2018 from https://www.researchgate.net/publication/228797899_Decision_Models_for_Affordable_Housing_and_Sustainable_Community_Development

Kahneman, D., & Tversky, A. (2000). *Choice, values, frames*. Cambridge: Cambridge University Press.

Malalgoda, C., & Amaratunga, D. (2015). A disaster resilient built environment in urban cities: The need to empower local governments. *International Journal of Disaster Resilience in the Built Environment, 6*(1), 102–116.

Malalgoda, C., Amaratunga, D., & Haigh, R. (2014). Challenges in creating a disaster resilient built environment. *Procedia Economics and Finance, 18*, 736–744.

McIntosh, J. (2013). The implications of post disaster recovery on affordable housing. In *Approaches to disaster management—Examining the implications of hazards, emergencies and disasters*. ISBN: 978-953-51-1093-4. Retrieved December 29, 2018 from https://www.intechopen.com/books/approaches-to-disaster-management-examining-the-implications-of-hazards-emergencies-and-disasters

Menshawy, A. E., Shafik, S., & Khedr, F. (2016). Affordable housing as a method for informal settlements sustainable upgrading. *Procedia—Social and Behavioural Sciences, 223*, 126–133.

Othman, E., & Ahmed, A. (2013). Challenges of mega construction projects in developing countries. *Organization, Technology and Management in Construction, An International Journal, 5*(1). Retrieved December 28, 2018 from https://www.researchgate.net/publication/243457533_Challenges_of_mega_construction_projects_in_developing_countries

Randolph, B., Kam, M., & Graham, P. (2008). Who can afford sustainable housing. In A. Nelson (Ed.), *Steering sustainability in an urbanizing world*. Aldershot: Ashgate.

Reffat, R. (2004). Sustainable construction in developing countries. In *Proceedings of First Architectural International Conference*. Giza: Cairo University.

Schilderman, T. (2004). Adapting traditional shelter for disaster mitigation and reconstruction: Experiences with community-based approaches. *Building Research & Information, 32*(5), 414–426.

Sevin, E., & Little, R. (1998). Mitigating terrorist hazards. *The Bridge, 28*(3), 156–172.

Sibley, J., Hes, D., & Martin, F. (2008). A triple helix approach: An inter-disciplinary approach to research into sustainability in outer suburban housing estates. In *Proceedings of Methodologies in Housing Research Conference, Stockholm, September 2008*.

Stone, M. E. (2006). What is housing affordability? The case for the residual income approach. *Housing Policy Debate, 17*(1), 151–184. Retrieved October 1, 2017 from www.works.bepress.com/michael_stone/5/

Toya, H., & Skidmore, M. (2007). Economic development and the impacts of natural disasters. *Economics Letters, 94*, 20–25.

United Nations, Department of Economic and Social Affairs, Population Division. (2017). *World population prospects: The 2017 revision, key findings and advance tables*. Working Paper No. ESA/P/WP/248.

United Nations Human Settlements Programme (UN-HABITAT). (2009). *Planning sustainable cities: global report on human settlements 2009*. London: Earthscan.

Voogd, H. (2004). Disaster prevention in urban environments. *European Journal of Spatial Development, 12*, 1–14.

White, P., Pelling, M., Sen, K., Seddon, D., Russel, S., & Few, R. (2004). *Disaster risk reduction: A development concern. A scoping study on the links between disaster risk reduction, poverty and development*. Oversears Development Group, University of East Anglia. Retrieved October 2, 2017 from www.preventionweb.net/files/1070_drrscopingstudy.pdf

Are Construction Professionals Equipped with the Knowledge and Tools to Address the Sustainability Dilemma?

Paul Mundy and Colin A. Booth

Introduction

Assessment of sustainable buildings and the increasing need to frame environmental issues within wider political and social debates (Cole, 2010) has seen the emergence of certification schemes globally (e.g. BREEAM, Green star and Leadership in Energy and Environmental Design (LEED)). The methods of certification provide an objective evaluation of resource use, ecological loadings and indoor environmental quality within a broad culture of performance measurement (Cole, 2010). Kaatz, Root, Bowen, and Hill (2006) agree that these schemes have the ability to impact design and construction practice, to challenge the existing norms and values of designers and constructors whilst validating and promoting a sustainable ethos concurrently.

As with the 'sustainability' and 'sustainable development' discourse, there is a similar debate in relation to construction in terms of 'green building certification' (GBC) and 'sustainable building certification' (SBC) (Newport, Chesnes, & Lindner, 2003). A difference of focus has been identified between the schemes, undertaken in a very narrow context only assessing environmental aspects or they can adopt a broader view to include the social and economic principles as well (Cole, 2010). While SBC covers energy efficiency, it also includes social and environmental aspects (Zuo & Zhao, 2014). The focus of traditional GBC schemes is on environmental aspects (Kaatz et al., 2006) in terms of performance standards of technologies, energy and water savings and physical characteristics in terms of improved internal environments for occupants (Kaatz et al., 2006). Primarily offering mitigation to environmental impacts of the building during construction and operation (Parker, 2012) GBC schemes follow a similar format of credit scoring

P. Mundy (✉) · C. A. Booth
Centre for Architecture and Built Environment Research (CABER), Faculty of Environment and Technology, University of the West of England (UWE), Bristol, UK
e-mail: paul3.mundy@uwe.ac.uk

L. Scott et al. (eds.), *Sustainable Ecological Engineering Design*,
https://doi.org/10.1007/978-3-030-44381-8_10

against categories (Taylor, 2015), with different tools for various types of building (Barlow, 2011).

Mainly categorised as GBC BREEAM Schemes are widely used as a tool for achieving sustainability within the existing UK building stock. An investigation of the importance of these schemes in achieving a sustainable built environment and the impact of these schemes on UK construction and property sectors is provided. Engaging construction professionals to determine their interactions with certification schemes and the wider sustainability agenda to identify knowledge gaps to improve practice. Therefore, this study aims to investigate whether the UK construction professional is equipped with the knowledge and tools, in the form of BREEAM building certification schemes, to aid construction and refurbishment projects to improve the sustainability of the UK construction industry.

Background

To achieve a sustainable built environment it is important to understand how the development itself, along with concurrent impacts on the hinterland, fits within the wider societal sphere (Berardi, 2013). Sustainable construction requires more than current green building certification schemes were conceived to do, requiring a 'systems thinking' approach (Kaatz et al., 2006) that considers the context and location of the development and associated direct and indirect stakeholders. Therefore, sustainability assessments cannot be used to evaluate the quality of building performance, but can be used to transform the context in which the building is developed. Three domains of environmental, social and economic are used to frame sustainability, but it is their points of intersection and understanding the way and extent that the domains affect each other positively or negatively that is considered critical (Cole, 2010). However these less tangible aspects often take a back seat to quantitative carbon reductions prevalent in GBC, as demonstrated through criteria weightings of schemes (Barlow, 2011). On a whole-life basis the main opportunities for reduction in CO_2 emissions is during the use stage, savings can be made in the user operation of the building and their behaviours outside of that. The work of Shove (2002) demonstrates that social practices associated with our lifestyle have become unsustainable through a 'ratcheting up' of unsustainable behaviours. These self-replicating behaviours form new habits and social norms, contributing to the climate change and resource consumption crises through our daily actions. Societal transformation towards sustainable lifestyles involves technological artefacts, but it also requires new user practices, regulations, infrastructures and cultural meanings (Shove, 2010) for lasting change to occur. This view is evident in the work of Williams and Dair (2007), who discuss the link between technological and behavioural sustainability and how the built environment can influence sustainable behaviours, also echoed by Jones, Pykett, and Whitehead (2010) who describe infrastructure layout and architectural characteristics of whole streets affecting the decisions that individuals make. These are the interactions between the domains that Cole (2010) referred to as critical, that are not considered by GBCs.

Social practice theory examines this further stating that people need the material, knowledge and meaning to guide behaviours (Shove, 2010). By offering facility provision, knowledge share and support for more sustainable regimes, sustainable behaviours can develop within a community, sharing sustainable practices further to wider communities. Therefore, the social aspect of sustainable certification schemes should cover psychological well-being and comfort of occupants, accessibility to public facilities and level awareness to wider sustainability issues (Zuo & Zhao, 2014). Kaatz et al. (2006) argue that sustainability assessments should act as educational and emancipating medium that empowers and promotes collaboration among building stakeholders. Consequently, SBC schemes have significant expectations and demands to transform design and construction processes and practices with the construction sector.

There is plenty of research comparing certification schemes (Ameen, Mourshed, & Li, 2015; Zuo & Zhao, 2014); however, the findings of Gil and Duarte (2013) suggest no single tool covers all aspects of sustainability and there is scope to develop new tools or redevelop existing ones. There is no doubt that GBC schemes have contributed to promotion of higher environmental expectations and are directly and indirectly (Cole, 2010) influencing a trend towards sustainable development principles within building design. The current knowledge of the ecological and carrying capacity of the planet is still primitive, so framing the distance to sustainability within these schemes is difficult; however, initiating the discussion is valuable (Cole, 2010) to facilitate progress.

Although BREEAM is not a mandatory requirement of development, it is becoming increasingly required by planning authorities (Parker, 2012), funding agencies or by clients who invest in corporate social responsibility (Barlow, 2011). As with all audit systems BREEAM has limitations, but it does offer verifiable and independent assessment of the performance of building design construction and operation (Barlow, 2011). Therefore, this study attempts to explore the depth of knowledge and understanding of sustainability issues across a range of construction professions and utilises this evidence to reveal if BREEAM is being delivered with due diligence across the industry.

Methodology

The study commences with a literature review to understand the context and the issues within the sustainability dilemma facing the construction industry. Identifying a number of studies that engaged construction professionals and their interactions with BREEAM certification schemes, forming the basis for in-depth semi-structured interviews to be developed to allow examination of sustainability in the built environment within the construction professionals' current reality.

Table 1 Participant information

Participant	Gender	Age	Construction experience	Profession
A	M	34	12	Quantity Surveyor
B	F	33	11	Architect
C	F	23	1	Real Estate Surveyor
D	M	40	18	Project Manager
E	M	55	33	Building Surveyor
F	M	44	22	Energy Manager
G	M	43	21	Client

Construction professionals, who have expressed experience in BREEAM certification and sustainability in the built environment from a variety of disciplines, were selected by responding to invitation to voluntarily undertake an interview to talk about their experiences. A range of ages and length of time in industry are contained within the sample allowing assessment of understanding and engagement of professionals throughout the industry with varying levels of education and industry experience, representative of the industry as a whole (Table 1).

Analysis

For the research to provide deeper understanding of experiences with the UK construction sector, grounded theory is employed allowing themes and theories to be generated through the analysis of the interviews. Due to sample size there may not be enough data generated to develop theories with the rigour required. Therefore, the research adopts template analysis, providing a basis for themes to be identified guiding the research.

The template is generated from the literature review, especially previous surveys undertaken around 10 years ago, that identifies drivers and constraints that were present and whether they are still in effect today forming the basis of the template analysis for use with this research (Table 2).

Results

The debate on the definition and conceptualisation of sustainability (Parkin et al., 2003; Santillo, 2007) has been accused of holding up progress towards achieving sustainable outcomes (Dixon et al., 2008; Pitt et al., 2009; Vallance et al., 2011). The three pillars approach is a common conceptualisation adopted that identifies sustainability is achieved through addressing environmental, economic and social elements (Cole, 2010). Participants were asked how they conceptualise sustainability; responses reflected the literature debate in terms of the diversity of responses.

Table 2 Template for analysis

Themes	Justification for interview questions	References
Conceptualisation of sustainability	The many different definitions and conceptualisations of sustainability, which has been cited as a barrier to progress. Analysis of how the sector views sustainability will identify where focus currently is for the construction professional.	Dixon et al. (2008), Pitt, Tucker, Riley, and Longden (2009), Kaatz et al. (2006), Santillo (2007), and Vallance, Perkins, and Dixon (2011)
Barriers and drivers	Key barriers and drivers for use of BREEAM, identifying these within the interviews will provide deeper understanding of how these are experienced through professional practice.	Dixon et al. (2008), Pitt et al. (2009), Cole (2010), Cinquemani and Prior (2011), and Barlow (2011)
BREEAM in practice	BREEAM is the most common GBC scheme in the UK. Coding interactions with BREEAM in practice will allow identification of construction professional's perceptions of the schemes in use.	Dixon et al. (2008), Pitt et al. (2009), Kaatz et al. (2006), Cinquemani and Prior (2011), and Barlow (2011)
Emergent theories	Some theory will not have been discovered through the literature review. The transcriptions from interviews will be reviewed to identify emergent themes or trends.	

All participants mentioned the environment or environmental aspects, such as reducing CO_2 emissions in one form or another. Although, nearly half of participants only mentioned environmental aspects of CO_2 emission reductions but nothing else. Participant E's response to the question illustrated this point well by saying.

> *Sustainability for me is basically a reduction in carbon footprint generally, and that's in terms of methods of manufacture, emissions involved in movement, vehicular movement of materials to site and also the implementation of those materials in use and how to reduce carbon footprint going forward.*

Sustainability perceived as a reduction in carbon emissions is a common trend throughout all participants; even those who identified deeper understanding understood the baseline of carbon reduction well. However, only discussing this environmental element is not delivering the full scope of sustainability, described as 'traditional sustainability' by participant G who interpreted that '*when asked about sustainability most people assume the baseline is energy conservation*'. Reduction in carbon and energy efficiency is a predominant theme mainly due to media coverage getting the headlines (Ameen et al., 2015) at present, but the focus is also there due to ability for quantification of return on investment from savings (Parker, 2012; Taylor, 2015). Failure to recognise that the sustainability issue goes deeper than just energy efficiency demonstrates a lack of understanding about the core underpinnings in the sustainability discourse and with construction professionals missing this core competency, how can sustainable solutions ever be recognised, let alone realised?

Participant C in contrast responded to the conceptualisation question by drawing on the Brundtland (1987) definition of sustainability, describing the need to design not to hurt future generations, but also saying '*it's not just about the environment, its social and economic aspects as well, and obviously the environmental stuff*'. Participant B also discussed social aspects such as consideration of the design of internal environments having potential for improving health and well-being of occupants and provision of good transport links can allow people economic potential, meaning that only two participants identified aspects of social and economic spheres of sustainability when asked about how they conceptualise sustainability. Participant F referred to the triple bottom line, which represents people, planet and profit, a form of full cost accounting (Elkington, 1998) when discussing their conceptualisation. However, only one bottom line is evident when considering the concept of sustainability:

> *[Sustainability] has to include the triple bottom line, which is this economic appraisal issue. One can take the moral high ground from the point of view of looking at saving energy and saving carbon, that is all well and good but the reality, in commercial business the concept of sustainability is making sure that that business itself is sustainable, economically, for its shareholders who it's accountable too, or the public, if it is a public organisation with public expenditure. Therefore for sustainability to be truly sustainable then it must contain this triple element this third bottom line which is financial sustainability.*

The explanation is focussed in the economic sphere, but solely the economics of the company itself not the wider community or development stakeholders and is truly representative of the capitalist paradigm in which construction in the commercial sector operates. With the commercial sector providing so much investment in the construction sector operating with these intentions, it is difficult for the industry to adopt an alternative approach. Commercial business develops, refurbishes and renovates property regularly, which presents opportunities to improve sustainability within existing buildings, where the construction professional needs to drive the sustainability agenda. Construction professionals need the knowledge to challenge this attitude, to educate their clients into building towards a sustainable built environment.

All participants highlighted cost as a barrier to the uptake of BREEAM schemes but few indicated opportunities offered, such as increased rental and sale prices (Michl et al., 2016), by illustrating these construction professionals may influence client decisions more effectively when giving professional advice. However, as Michl et al. (2016) postulate, increased returns will only be realised in a market that recognises sustainability, the market is moving in that direction but to value sustainability then all stakeholders must have a shared view (Kaatz et al., 2006).

Another key barrier identified was of complexity of BREEAM Schemes and associated cost in terms of time and money in achieving a high rating was shared by most participants. Participant B summarised well by stating: 'If you are not going to score high enough then you won't do it, if it costs too much to achieve you won't do it'. When considering this barrier it is essential to draw upon the previous analysis of conceptualisation responses, it is evident some clients view sustainability in a green building context, focusing on energy efficiency, not acknowledging wider sustainability issues. This narrow view misses sustainability opportunities, such as green roofs and rainwater harvesting, which have lower cost for installation but have less tangible results in terms of return on investment, often taken out during value engineering processes.

Regulatory changes have influenced the uptake of BREEAM schemes in many local authority planning policy documents (Parker, 2012) making a specific BREEAM rating a requirement of planning conditions. Both participants G and F agreed that planning drove this uptake, Participant G comments that Bristol City Council planning policy framework has a 'robust' sustainability policy, when working outside of this area, other councils' requirements differ:

> *In Bath, you could potentially ignore BREEAM, the planning policy doesn't require you to do anything but we feel we give ourselves a better chance of getting a planning decision by being proactive and undertaking a BREEAM assessment... The problem is when developing on behalf of an institution like a pension fund or an investment fund. If they see us spending money when it is not a requirement of a planning application they ask why are we wasting money on this... Then sometimes it is squeezed out of the project, then architects and contractors get wind of it not being a planning condition, they ask why are we spending our time and efforts doing this if it is not a requirement? So it is important to have it as a contractual obligation and a policy at local authority planning or else it tends to get swerved.*

The regulatory driver does have an effect, yet the client can often refuse the 'unnecessary' cost if it is not a requirement. This sentiment is evident amongst construction professionals who question why they have to do additional work if it is not a requirement of planning. This shows a direct conflict between one of the key drivers (legislation in the form of planning) and barriers (cost of complexity) identified by this study. These conflicts are managed in practice, usually to the detriment of sustainability, Participant F has experience of BREEAM as requirement of planning and has identified that sustainability requirements can be 'Swerved'.

> *I think BREEAM can be manipulated in such a way that it supports a planning application and secures your planning application, then that's what the project manager will target at the most cost effective way, not necessarily the best for saving energy... The reality is where biomass boilers were put in to secure BREEAM and planning approval, but those biomass boilers we knew full well, were there to secure the planning approval. We ran them for a couple of months and then we had to bypass them and go back to conventional boilers which was always the unofficial plan but we knew we couldn't get the building through planning if we hadn't done it.*

This identifies that the regulatory driver is used to push through developments with no intention of using the technologies once the BREEAM certification and planning permission has been granted. This behaviour will not help towards achieving a sustainable built environment, making it hard to enforce a voluntary scheme that is a condition of planning based on a score that is interpreted and manipulated in a way that will suit the developers' intentions and not offer the best sustainability outcomes. This demonstrates a reluctance to engage with sustainability, unless it must be incorporated into project. Participant B furthers this by recognising sustainability is viewed as an additional aspect to projects, rather than core to the development, saying: 'some people see it as an add on, like adding on solar panels to tick a box and getting some biodiversity by putting a pile of logs in a project somewhere counts as a point'. Using sustainability as an additional extra and not being core through the design leads to missed opportunities, especially in terms of larger holistic systems thinking approaches (Kaatz et al., 2006). Early sustainability consideration during the design process enables more effective and cost-efficient results (Reed & Gordon, 2000), viewing it as an addition to the project will reduce effectiveness. Participant G indicated frustration and further elaborated on the view of sustainability as an added extra to the role of the construction professional by saying:

> *I wouldn't say it was onerous....If it were any other part of your job like writing a monthly report or doing a planning application, that's onerous but you just get on and do it, because it's your job. But with BREEAM it is seen as an extra... so that's the biggest challenge, to get people to accept it is part of their job and that it is important, not just an add on.*

Construction professional's work is governed by deadlines, with high workloads processing a lot of design and construction information. BREEAM adds more work to the project, therefore more work to undertake and manage. It is why many can view this as an extra, but this attitude needs to change and sustainability needs to be core to the role of the construction professional.

Responsibility is a theme with participants over differing questions. Participant A mentions '*sustainability is to be as responsible as possible*' while Participant B identifies personal responsibility by saying '*Anything we put into that building, I have to be responsible for that*', and recommends sustainability to a client as '*really being more aware and responsible for the building and environment*'. Demonstrating that the responsibility is theirs but also shared with clients and other professionals within the construction industry. Participant E described personal responsibility transcending the professional environment by saying:

> *I think everybody has a duty of care to help us move in the right direction, I personally believe in it, but that goes further, you have to start looking inward in your own house, on a personal level.*

Discussing home life identifies the underlying behavioural sustainability aspect is identified. Through the interview however, the discourse with participant E solely focussed on carbon reduction, so it is apparent that the connection between their role in building design and specification and their ability to support behavioural sustainability through design decisions was not present. The disconnect here could be associated with knowledge and understanding of sustainability, but more so the link between technological and behavioural sustainability (Williams & Dair, 2007) and how the built environment effects behaviour (Jones et al., 2010).

Discussion

Studies agree that a lack of common definition for sustainability is a key barrier to achieving sustainability in construction (Dixon et al., 2008; Pitt et al., 2009). A decade after these studies this issue is still prevalent in this study, leading to missed opportunities and loopholes to avoid good practice in terms of sustainability. Through the interviews, a range of sustainability knowledge was present. The majority of the study not identifying the three spheres of sustainability and the two that did refer to them not discussing them with confidence demonstrates that knowledge of sustainability is not at a level for deployment of effective solutions. To categorise different levels of knowledge or competence within sustainability the four stages of professional competence are used to assign participants a stage of competence. The first category of unconscious incompetence refers to those not aware of what they do not know (Participants A, D, E, F); in this category are those that focussed on CO_2 emission reduction only. This group failed to identify spheres of sustainability but they are not aware that they do not know. Conscious incompetence is demonstrated by one participant (Participant G), where they know that they do not know everything but are still making decisions knowing that they do not know. The third category has the correct intuition of sustainability as a concept and is delivering it in practice where possible; however, some thought is required as it is not second nature to them yet and a lack of confidence is still demonstrated (Participants B and C). The final category is where professionals should have the right intuition and be able to deliver optimum solutions to achieve effective results subconsciously; however, none are in this category which means sustainable solutions are not being delivered.

With the majority of construction professionals unconsciously incompetent, progress is slow. The view of sustainability as solely energy efficiency and CO_2 reductions is hindering progress towards a sustainable built environment. Due to the complexity of sustainability, many factors influence each other causing tipping points; fortunately, solutions are interconnected in their nature. Sustainability research and development operates at a systemic level and requires a systems thinking approach (Kaatz et al., 2006) to realise the real potential that new developments can have for all stakeholders in all spheres of sustainability. Rather than viewing projects in isolation, construction professionals must understand that sustainability is more than energy efficiency, that it includes economic aspects such as providing opportunities of employment and wealth generation, social aspects in terms of providing a healthy cityscape, places for interaction and exercise with comfortable and liveable residential provision.

Conclusions

From the study the following conclusions are derived:

1. BREEAM schemes are being used as a tool to gain planning and regulatory approvals, rather than an opportunity to get the best performance out of the building itself. This is leading to cost making the decisions on the projects rather than best sustainable outcomes; construction professionals are looking for the

lowest cost option to achieve points rather than considering the best design options.
2. The cost element is also prevalent in terms of decision to implement schemes in the first place; if seen to be too costly or complex and time-consuming increasing design costs, clients will not be inclined to do it, causing projects not to adopt sustainability schemes for these reasons alone.
3. Construction professionals view sustainability as energy efficiency and therefore only consider how to reduce energy wastage and emissions. This perception fails to recognise the multifaceted challenge posed to sustainability, leading to missed opportunities for systems thinking design solution.
4. The majority of professionals believe the importance of sustainability and identify with a responsibility. However, it is viewed as an additional aspect to their role, detracting from their workload reducing the inclination to engage deeply in the sustainable design process.

Recommendations

Based on the conclusions the following recommendations are derived:

The research demonstrated the heavily debated conceptualisation of sustainability; this debate was accused of holding up progress a decade ago. Through the interviews it was evident that a clear common conceptualisation is still not present amongst the study sample, if representative of the wider industry, it indicates this barrier is still negatively effecting progress towards solutions, not just with BREEAM schemes but all certification schemes globally. Research into how prevalent this barrier is throughout the industry will allow the formulation of an industry-wide conceptualisation to be developed by an interdisciplinary team of construction professionals, sustainability consultants and social scientists specialising in behaviour change for sector wide adoption at an international level.

It is evident that to achieve sustainability in the built environment a paradigm shift is required, in terms of how sustainability is perceived, but also how designers and clients value design, to design for more than just compliance, to strive for excellence. Sustainability needs to become core to all aspects of the construction professional's role to ensure it is given the attention required to achieve the best networked solutions that tackle the multifaceted problems facing our society. Sustainable must become the standard by which development is permitted; the construction sector can no longer accept the situation and continue with the current 'business as usual' attitude; it must strive collectively and collaboratively to change systemic problems.

References

Ameen, R., Mourshed, M., & Li, H. (2015). A critical review of environmental assessment tools for sustainable urban design. *Environmental Impact Assessment Review, 55*, 110–125.
Barlow, S. (2011). *Guide to BREEAM*. London: RIBA Publishing.

Berardi, U. (2013). Clarifying the new interpretations of the concept of sustainable building. *Sustainable Cities and Society, 8*, 72–78.

Brundtland, G. (1987). Our common future—Call for action. *Environmental Conservation, 14*(4), 291.

Cinquemani, V., & Prior, J. (2011). *Integrating BREEAM throughout the design process: A guide to achieving higher BREEAM and code for sustainable homes ratings (FB 28)*. Abingdon: Taylor & Francis.

Cole, R. (2010). Building environmental assessment methods: Redefining intentions and roles. *Building Research & Information, 33*(5), 455–467.

Dixon, T., Colantonio, A., Shiers, D., Reed, R., Wilkinson, S., & Gallimore, P. (2008). A green profession? A global survey of RICS members and their engagement with the sustainability agenda. *Journal of Property Investment & Finance, 26*(6), 460–481.

Elkington, J. (1998). Accounting for the triple bottom line. *Measuring Business Excellence, 2*(3), 18–22.

Gil, J., & Duarte, J. (2013). Tools for evaluating the sustainability of urban design: A review. *Proceedings of the Institution of Civil Engineers—Urban Design and Planning, 166*(6), 311–325.

Jones, R., Pykett, J., & Whitehead, M. (2010). Governing temptation: Changing behaviour in an age of libertarian paternalism. *Progress in Human Geography, 35*(4), 483–501.

Kaatz, E., Root, D., Bowen, P., & Hill, R. (2006). Advancing key outcomes of sustainability building assessment. *Building Research & Information, 34*(4), 308–320.

Michl, P., Lorenz, D., Lützkendorf, T. & Sayce, S. (2016). Reflecting sustainability in property valuation – a progress report. *Journal of Property Investment & Finance. 34*(6), 552–577. [Accessed 25 February 2018].

Newport, D., Chesnes, T., & Lindner, A. (2003). The "environmental sustainability" problem. *International Journal of Sustainability in Higher Education, 4*(4), 357–363.

Parker, J. (2012). *The value of BREEAM*. Bracknell: BSRIA.

Parkin, S., Sommer, F., & Uren, S. (2003). Sustainable development: Understanding the concept and practical challenge. *Engineering Sustainability, 156*(1), 19–26.

Pitt, M., Tucker, M., Riley, M., & Longden, J. (2009). Towards sustainable construction: Promotion and best practices. *Construction Innovation, 9*(2), 201–224.

Reed, W., & Gordon, E. (2000). Integrated design and building process: What research and methodologies are needed? *Building Research & Information, 28*(5–6), 325–337.

Santillo, D. (2007). Reclaiming the definition of sustainability. *Environmental Science and Pollution Research—International, 14*(1), 60–66.

Shove, E. (2002). Converging conventions of comfort, cleanliness and convenience. *Journal of Consumer Policy, 26*(4), 395–418.

Shove, E. (2010). Beyond the ABC: Climate change policy and theories of social change. *Environment and Planning A, 42*(6), 1273–1285.

Taylor, T. (2015). *Assessing carbon emissions in BREEAM*. Watford: BRE Global.

Vallance, S., Perkins, H., & Dixon, J. (2011). What is social sustainability? A clarification of concepts. *Geoforum, 42*(3), 342–348.

Williams, K., & Dair, C. (2007). A framework of sustainable behaviours that can be enabled through the design of neighbourhood-scale developments. *Sustainable Development, 15*(3), 160–173.

Zuo, J., & Zhao, Z. (2014). Green building research—Current status and future agenda: A review. *Renewable and Sustainable Energy Reviews, 30*, 271–281.

Part II
Sustainability Education

Sustainability in Construction Management Education: A Case Study of Students' Attitudes and Beliefs at Two CM Programmes in Ireland and United States

Lloyd Scott and Blake Wentz

Introduction

By 2030, the world will require 40% more water, 50% more food, 40% more energy, and 40% more timber and fibre (UN Report, 2017). The only way we can meet these demands is by managing our ecosystems smartly and sustainably. For example, by the end of the year 2020, it is estimated that 4.3 billion people will live in cities (UN Report, 2019). This will have a profound effect on urban development, all over the world, and as occupants of earth there is a responsibility to address the sustainability-related challenges with a global view and many different, local perspectives. Planning, designing, managing, and constructing the cities and communities we hope to live will be the responsibility of the students currently in our Universities studying Built Environment programmes. So how will the Architecture, Engineering, Construction and Operations (AECO) professional of the future be equipped to design, manage, construct and maintain such an environment? Will those actors be in a position to make valued decisions around the most appropriate and informed decisions about the fundamental choice that will yield better and more sustainable artefacts? To understand how undergraduate Construction Management students are considering these issues the participants from two University programmes were invited to partake in a sustainability survey.

L. Scott (✉)
Sustainable Construction Research Group, Technological University Dublin, Dublin, Ireland
e-mail: lloyd.scott@tudublin.ie

B. Wentz
Milwaukee School of Engineering, Milwaukee, WI, USA

L. Scott et al. (eds.), *Sustainable Ecological Engineering Design*,
https://doi.org/10.1007/978-3-030-44381-8_11

Defining Sustainable Construction

"Sustainability" is one of the most talked about, but least understood terms related to how the human race exist. It has at the core a clouded existence as differing interpretations compound a tendency to treat the subject matter superficially, as reference to terms like "eco", "green", or "smart" include rhetorical focus. Yet, for those in the public and private sectors, not related to the AECO sector, who do take the issue earnestly, sustainability embraces the environment and its long-term endurance as a matter of concern for all humanity. Just as intrinsic to sustainability are the critical development issues concerning the responsible use of resources within a circular economy and the fostering of social equity through the fair distribution of wealth.

The AECO sector is capable of making a significant contribution to the sustainable objectives, particularly considering the vast amount of material and energy resources required to produce and maintain the built environment, not to mention the sum of emissions and waste generated throughout the entire use-cycle of physical structures. When we consider the global extent of urbanization today while taking into account the pace at which our planet is being further urbanized, it is completely necessary that whatever is built must perform sustainably on all registers—environmentally, economically, and socially.

While there is little or no disagreement about the need for action, the debate continues as to whether it should proceed in small or larger steps: radical or incremental change—that is one of the questions. However, whatever the answer (either systemic overhaul or tweaking the status quo), society does not have the luxury of time to decide how to act. Given that some governments have declared a climate change emergency it is a priority that sustainable construction practice be at the heart of educating the future professionals.

Building a Sustainable Future

Sustainable construction, in line with the stipulations for development outlined in the Brundtland Commission's report "Our Common Future" from back in 1987, aims to meet present-day needs for housing, working environments, and infrastructure without compromising the ability of future generations to meet their own needs for shelter, spaces for work, and service provision (Brundtland, 1987). In meeting these needs now and over time, sustainable construction, in other words, can become more socially supportive in the long run by reducing its overall ecological footprint, while also being innovatively responsive to the ever-increasing demand for built space.

In view of its environmental impact, sustainable construction must involve the design and management of built structures, whether at the scale of buildings, infrastructure, or urban agglomerations; the performance of materials across all scales and throughout their whole use-cycles; and the use of renewable energy resources as well as their attendant technologies in building, operation, and maintenance to reduce global greenhouse gas emissions.

In view of its economic impact, sustainable construction involves the transition from a linear to a circular economy of renewable energy generation, material and waste recycling, water harvesting and preservation, transferable technologies, and the adaptability of structures to changes in use; innovative financing models premised on an economy of means that yields more with less; and the reinvestment of returns back into the common domain for collective benefit.

In view of its social impact, sustainable construction involves adherence to the highest ethical standards in business and industry practices throughout all project phases; the promotion of socially viable living and working environments, including occupational health and safety standards for labour forces and users; and the democratization of all processes pertaining to the production and use of the built environment as a common wealth.

So, what must a curriculum designed to fulfil these objectives include so that sustainable construction professionals on graduation can address the concerns for the aesthetic quality of the built environment, its architecture, its infrastructure, and its urban organization, all attuned to the specificities of local culture as well as global commonalities.

"Target Issues" for Sustainable Construction Education

The AECO sector as an imperative should be defined by its approach to Sustainable Construction which demands commitment to the underlying principles of sustainability, which assert that long-term development of the built environment requires a mutually reinforcing interplay of responsible economic, ecological, social, and aesthetic objectives. Additionally, and in accordance with the Paris Climate Agreement under the UN Framework Convention on Climate Change (adopted in 2015 and ratified in 2016), the Foundation places a premium on the reduction of global greenhouse gas emissions in all construction-related activities throughout the entire use-cycle of built structures.

For this very purpose, Higher Education—in collaboration with its associated universities—should put forth "target issues" as guidelines for sustaining the human made habitat for current and future generations. These "target issues", which need to correlate with the United Nations Sustainable Development Goals (SDGs) as well as the Sustainable Development Strategy of the supporting AEC sector, must serve as criteria for evaluating projects.

Sustainability in built environment education must provide graduates with the capabilities to problem solve, reflect, think systemically and critically, and be able to make timely and insightful decisions (Hegarty, 2008) that will begin to future proof a world constrained by limited natural resources and an ever-increasing population. As sustainability is a value-laden, abstract concept, its interpretation and application within an educational context will result in different educational experiences, learning outcomes, and capabilities (Hegarty, 2008; Holdsworth, Wyborn,

Bekessy, & Thomas, 2008). Langford (2008) refers to construction management as that discipline that has emerged as one that has grasped the sustainability agenda as a profession. Opoku and Ahmed (2013) make the assertion that there is a need for the built environment to embrace sustainability practice through leadership at the appropriate levels in the AECO sector. Given this, sustainable development and related forms of education are contested and open to interpretation. The following presents the widely accepted typologies of education underpinned by a sustainability paradigm drawn from the work of Sterling (2001), a recognized leader in the field of sustainability education and research. Sterling identifies three different typologies, which embody different degrees of understanding, constructing, and translating knowledge:

Education about sustainability, or "learning as maintenance" (p. 60), which does not challenge the current paradigm

Education for sustainability, which is "an adaptive response that equates to second-order learning" (p. 60) based on values and capability

Education as sustainability, or sustainable education, which facilitates third-order learning and change, and a creative and paradigmatic response to sustainability.

Sterling (2011) argues that these three types of sustainability education range from transmissive to transformative in their pedagogical approach and so educators need to consider their approach to creating opportunities to learn for their students. Education about sustainability is based on the transmission of facts or knowledge about sustainability. It involves the transfer of knowledge about sustainability and is usually added to existing subjects that may also present ideas contrary to sustainability. Education for sustainability if designed across a programme of study can adapt into the curriculum content by "building in" sustainability content into modules in what Sterling advocates as "an attempt to teach values and skills perceived to be associated with sustainability…" (2001, p.285). Sterling (2001) in their earlier work makes the argument that the third kind of sustainability education, education as sustainability has the potential to be a "transformative, epistemic education paradigm, which is increasingly able to facilitate a transformative learning experience" (p. 61). So, education as sustainability or sustainable embedded in the curriculum can make a greater impact on those educated.

Sustainability Education in the past has been declared to be critical to the realization of sustainable development. In 2004, the United Nations proclaimed the period 2005–2014 as the Decade of Education for Sustainable Development (UNDESD). This period witnessed heightened global calls and efforts at mainstreaming sustainability education into various institutions and countries. The extent to which this has been achieved is varying. There is a depth of agreement in the need for the graduate professionals of the future to have a sound and well-developed understanding of the current knowledge and best practices that are required within the AEC to embed sustainability as a matter of routine practice (Scott, 2017). We are running out of time!!!!

Research Goal and Objectives

The goal of the study was to compare Built Environment schools programme learning outcomes about essential sustainability knowledge with students' perceptions of requisite knowledge about sustainability to work efficiently in the construction industry. The specific objective was as follows:

1. To determine the perceptions of the students in construction management programmes regarding the required sustainability knowledge necessary to work efficiently in the industry.

Methodology

This research attracts interest for the researchers as there is a view of the need to study and gain a better understanding of students' perceptions and expectations about essential sustainability knowledge to work effectively and efficiently in the construction industry. The methodology applied was determined on the basis of relevance to the focus of this research enquiry but also on the basis of pragmatic positioning, often seen as appropriate for educational settings. This was the case as a different methodological stance would not have allowed the research to be completed within the constraints applicable. According to Creswell (2009) research methodology refers to the systemic approach that a researcher(s) adopts to accomplish the research's aim and objectives. Accordingly, an explorative interpretivist position has been adopted given the nature of the research.

The objectives were achieved by conducting surveys among students enrolled in the Construction Management construction programmes at TU Dublin and Milwaukee School of Engineering. The survey questionnaire for the potential participants was divided into three sub-sections to accomplish the main objectives, which included:

1. Understanding the background of individual respondents
2. Examining the importance of sustainability practices as perceived by the respondents
3. Examining the essential sustainability knowledge required to work efficiently in the industry as perceived by the respondent

The overall research process of the study involved the following steps:

1. Selecting sample students enrolled in the CM programmes
2. Developing the survey instrument
3. Performing cognitive interviews for instrument validation
4. Distributing the survey questionnaire to selected samples
5. Collecting data to examine the industry's expectations and students' perceptions
6. Analysing the collected data

Scope of the Study

The survey questionnaires were restricted in distribution to construction management students taking the related programmes at TU Dublin and Milwaukee School of Engineering located in Ireland and the United States. The authors assumed that the knowledge of an individual was such that they would be in a position to answer the question with a level of knowledge appropriate to their educational experience/ knowledge. While the surveys were sent to students at all levels, responses from students who were within 1–2 years of their graduation were included in the data analyses. This was done to ensure the respondents had well-developed perceptions about the necessary skills required to be successful in the industry. On the other hand, the students at the freshman and sophomore level were relatively new to the programmes and might not have well-developed perceptions about the design and construction industry.

Sample Selection

The population for the study was students in CM at MSOE and TU Dublin, which amount to some 250 students enrolled at both Universities. One hundred and eighteen positive responses to participation in answering the survey were received. The questionnaire responses were collected (35% of the sample of 250). The questionnaires were sent via email with one single reminder where 58 out of the potential 118 students provided their responses. All responses were included in the analyses with a response rate of 53% were included in the data analyses.

Survey Instrument Development

The individual survey questionnaire was composed of three types of questions: (1) close-ended questions with ordered choices, (2) five-point Likert-type scale questions, and (3) open response questions. The survey instrument development process is summarized in Fig. 1 and described in this section.

The authors identified the survey items based on the study's key constructs of interest. Once the first drafts of the survey instruments were developed, a research measurement expert and two academic experts reviewed those in order to ascertain the content validity of the items in terms of relevance, representativeness, and technical quality. Feedback from the subject matter experts was incorporated into the second draft, the pretest version of the survey instruments.

The pretest versions of the survey instrument were next evaluated by two experts from construction education with expertise in sustainability education. Information obtained from the reflective interview sessions were incorporated into final version

Fig. 1 Survey instrument development process

of the survey instruments. Several typographical errors were corrected, and language was revised to ensure clarity of the question, so that the most accurate response might be achieved, and that internal validity would not be an issue.

Distribution of Survey Questionnaires and Collection of Data

The developed and validated survey instruments were encoded using a web survey tool (Google Forms) to facilitate the distribution and collection of the survey questionnaires via internet. After successfully developing the web survey questionnaire, the invitation email along with the survey questionnaire was sent to the study sample. The survey link was open for 2 months to limit the collection period. After 2 weeks of the first invitation to participate in the survey, two wave of reminder emails were sent to motivate the study sample for participating in the survey.

Research Findings

The following results combine the authors' reflection and description of practice with qualitative and quantitative results from the student survey. The results are outlined and discussed based on the elements of sustainability education praxis outlined above, including sustainable construction practice, curriculum development, the inclusiveness of educational praxis, and also the learning experience.

While the majority of respondents are in full-time education, around 7% indicated that they are part-time; this was more obvious among the TU Dublin students where three of the respondents indicated they were part-time (see Fig. 2). There is an opportunity for students at TU Dublin register as part-time, which usually means that the student is in full-time employment but is released to complete their technical education.

The age bracket overall showed that the majority of respondents (78%) were between 18 and 23 years. This would confirm that many students on the CM programmes enter higher education directly after secondary (high) school. No student respondent were older than 35 with just 22% between the ages of 24 and 35. A larger response was confirmed from MSOE, with 55% of the respondents hailing from the US school as can be seen in Figs. 3 and 4.

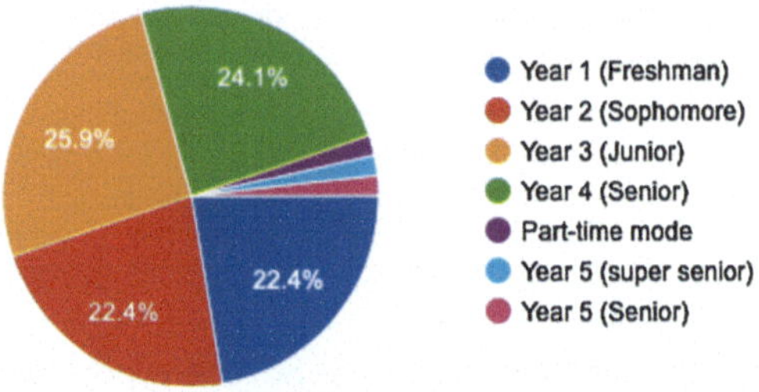

Fig. 2 Year of programme of study

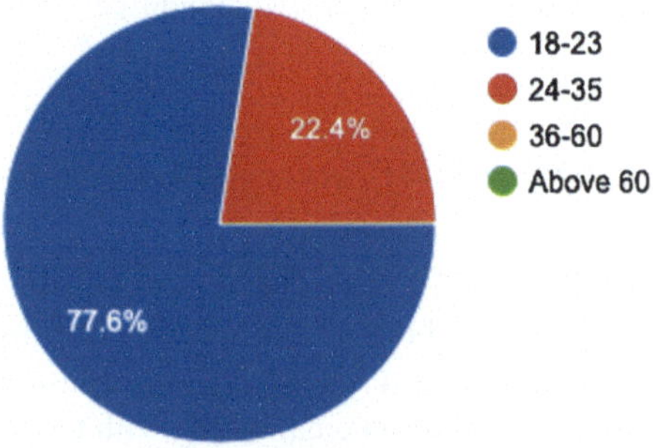

Fig. 3 Age profile of respondents

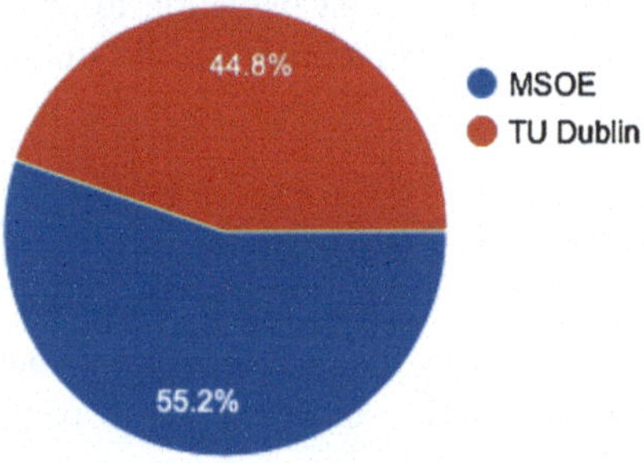

Fig. 4 Programme student respondents are studying

When asked to define sustainability in their own words the 58 survey respondents all offered an answer and opinion. Some really interesting responses were shared but it was quite clear that students have sound knowledge and competence in sustainability in construction. They were clear about the appropriateness of protecting the environment and why this should be the case. Many suggested that society has a responsibility to ensure that people should be responsible and strive to leave the environment in better shape than it ways when they came into it ($n = 10$).

When asked to personally define sustainability the respondents offered varying definitions. While those answers varied in language and length of response, there-was a sound knowledge of what the underlying principles of sustainability are recorded. It would be reasonable to say that all respondents are aware of the need to be sustainable. Some of the more detailed responses are included below:

Avoiding depletion of natural resources to make the earth balanced
Following the triple bottom line (financial, social, and environment needs), as well as working towards cradle to cradle products and ideas
Sustainability is a process to decrease the use of natural resources (halt depletion) and help our environment.
Building things that are environmentally healthy
Building while ensuring the final product is efficient in its energy use while also ensuring that materials used are responsibly sourced and are renewable. While doing all of this it is also important to minimize negative environmental impacts.
Something that can last for an extended period of time with limited maintenance required
Long lasting and environmentally friendly
In simple words, it's the measure in the ability to keep something able to perform the function or task it was made to do.
Having a carbon neutral or carbon negative society, using mostly or completely renewable resources, and having our planets health in mind at every step
It is being conscientious of our impact on the world we are living in, and to mitigate and reduce that impact we have to the best of our abilities
Sustainability is the preservation of resources by employing certain methods to conserve before, during, and/or after a process.
Using resources in a way that does not harm others and allows for continued use of the resource for an infinite amount of time.
Constructing a facility that is both efficient for society, but also constructing it in a way that will not need to be constantly worked on.
Sustainability is the ability of a system to be maintained over a long period of time.
Acting consciously towards preserving the health of the environment and ecology which one will be disrupting.
trying to preserve the environment and trying to use resources efficiently
Trying to maintain the status quo in regard to the environment

Table 1 Response to CO_2 emissions from current building stock question

Yes	40	69%
Not sure	12	21%
No	6	10%

The using of resources which can be renewed and also do not cause harm (or cause limited harm) to the environment appears to have been a common thread in the responses. So it is reasonable to consider that students have an understanding of the term sustainability.

For the question asking about the world's CO_2 emissions created by the construction and usage of buildings, a mixed set of responses arose. There appears to be a solid number of students who are aware of the impact buildings have on the environment as regards CO_2 emissions (Table 1).

There is a need to ensure students have access to factual information and one way of doing this is to make sure students have an opportunity to discuss emerging statistical data from quality resources.

Comments included:

There is no real engagement with sustainable practices
There is no real engagement with sustainable practices
Do you think the drive towards a more sustainable construction sector is having a positive effect on the industry as a whole?
Everyone is taking climate change more seriously
Minimizing carbon footprint and emissions
It is forcing designers and construction contractors to think more about the materials they use and how much waste is generated
Of course, as this protects Mother Earth in the long run and helps slow down global warming
Contractors are more aware of their responsibilities

So, it would appear that students are aware and conscious of the need to address sustainability aspects; they are in the main of the view that personnel in the sector are aware of that need to address the issues. They are also clear that in some cases there is less than a positive engagement in sustainable practices on the job site.

When questioned as to whether a course in sustainable construction should be part of the construction programme, the respondents clearly had a unanimous positive response. In fact, 55 respondents said yes, while the remaining three suggested that it is a maybe included a course. This would strongly suggest, all respondents have a positive orientation to including sustainable construction on the CM programme. As to why sustainability should be included in the curriculum, they believe this as they are of the view that the next generation of AEC professional should have the skills, knowledge, and competences to solve problems with a "*protect the environment mindset*". Some are of the view "*We need the knowledge of this important subject*" and "*Some perhaps don't understand the importance of sustainability or don't know how to build sustainably*".

A very clear message is coming from the future AEC leaders:

> *Yes, as it educates students in how to carry out construction in an environmental way which in turn lowers emission and the construction industries carbon footprint*

To some extent it is already being addressed in both college programmes as some of the respondents have confirmed:

> *It already is at MSOE. I have learned a lot in that course along with the LEED course about sustainability, its affects, and some examples on how to do so.*

In terms of what should be included in a sustainable construction course many of the respondents recommend that students should experience and learn about environmental assessment tools (EAT) such as LEED and or BREEAM. Some were of the view that it should not just be learning about EATs but more about "General sustainability ideas, not just LEED". Just like the academics, there appears to be a mixed opinion as to what should be in the curriculum and how it should be taught but it is clear that all are of the view that there should be exposure to appropriate knowledge in this topic.

When asked as to when students should be exposed to a course in sustainable construction there was a mixed response; however most preferred to have it addressed in either year 2 or 3 (85%) and the reason for this was that students suggested that in year 1 it was too early while in year 4 it was a little late. There was the view that after year 1 most students had a good understanding of where they were heading in the programme.

Three respondents also recommended that addressing sustainability should "permeate through the program". It is heartening to fins such a response as this is an insightful comment. All programme teachers should be aware of where the sustainability aspects fit into their part of the programme. One respondent was of the view that by "*embedding sustainability across the program students will gain a better understanding of its importance in creating a better built environment*". A very mature response and one that would indicate the future is in good hands!

Conclusions and Recommendations

There appears to be an overwhelming view among the respondents that they view knowledge in sustainable construction as important and that programmes should have the most up-to-date information and practice available. There are mixed views as to whether or not it should be provided as a stand-alone element or whether it should be embedded across the programme of study. A sizeable number of respondents believe that students should have the opportunity to study for and get accreditation; some preferred exposure to LEED while others recommended BREEAM. To do this would require programme teams to make decisions as to whether it would be correct to include accreditation as part of the programme offering or as an addition option.

In order to make a difference of such magnitude as indicated by many of the student respondents there is a need for a different type of AEC industry, or, in other words, there is a need to embed sustainable construction practices in all we do. There is an overwhelming call for knowledge to be shared on the topic as identified by the future leaders in construction. Clients and potential clients should be able to expect from all built environment professionals that they have access to and be informed about the world's best sustainable practices and that it be considered "normal" in each AEC situations to aim to apply all known best practice.

The existing research in this area has highlighted the increasing importance of sustainability knowledge. However, the literature in the CM area is still evolving given the lack of clarity over many concepts. There are no existing sustainability education frameworks but there is a need for more to be done to develop this. There is therefore a need for the development of a suitable framework to assist educators align their teaching practices. This research can therefore contribute to addressing the knowledge gap through the development of a systematic evidence-based framework with identification of the elements to be taken into consideration in defining what is needed and recognize what leaners are being exposed in the AEC sector.

This study has engaged with two programmes, one in Ireland and the other in the USA, and so is limited in its scope and generalizability. A recommendation is that a refined version of the study could be used to ascertain the views of a larger cohort of CM students.

A driver of change can often come from students especially if they are viewed as change agents. So, to end with the positive words of one respondent:

> *Yes, as it educates students in how to carry out construction in an environmental way which in turn lowers emission and the construction industries carbon footprint*

References

Brundtland Report. (1987) *Our common future*. World Commission on Environment and Development (WCED).

Creswell, J. W. (2009). *Research design qualitative, quantitative, and mixed methods approaches* (3rd ed.). Thousand Oaks, CA: SAGE Publications.

Hegarty, K. (2008). Shaping the self to sustain the other: Mapping impacts of academic identity in education for sustainability. *Environmental Education Research, 14*(6), 681–692.

Holdsworth, S., Wyborn, C., Bekessy, S., & Thomas, I. (2008). Professional development for education for sustainability. *International Journal of Sustainability in Higher Education, 9*(2), 131–146. https://doi.org/10.1108/14676370810856288

Langford, D. (2008). *Creating a discipline association of researchers in construction management.*

Opoku, A., & Ahmed, V. (2013). Embracing sustainability practices in UK construction organizations: Challenges facing intra-organizational leadership. *Built Environment Project and Asset Management, 4*(1), 90–107.

Scott, L. (2017). *A comparative investigation of Industry, Academics' and Students' Perceptions on Essential Sustainability Knowledge, Competencies and Skills at the International Conference for Sustainable Futures, Bahrain, November 2017.*

Sterling, S. (2001). *Sustainable education—Re-visioning learning and change*. Schumacher Briefing No. 6. Schumacher Society/Green Books, Dartington. ISBN: 1-870098-99-4.

Sterling, S. (2011). Towards anticipative education—Learning by design. In S. Harding (Ed.), *Grow small, think beautiful: Ideas for a sustainable world from Schumacher College*. Edinburgh: Floris Books.

United Nations, Department of Economic and Social Affairs, Population Division. (2017). *World population prospects: The 2017 revision, key findings and advance tables*. ESA/P/WP/248.

United Nations, Department of Economic and Social Affairs, Population Division. (2019). *World population prospects 2019: Volume I: Comprehensive tables.*

Delivering Energy Savings for the Supply Chain Through Building Information Modelling as a Result of the Horizon 2020 Energy BIMcert Project

Barry McAuley, Avril Behan, Paul McCormack, Andrew Hamilton, Eduardo Rebelo, Barry Neilson, Gayle Beckett, António Aguiar Costa, Paulo Carreira, Dijana Likar, Angelina Taneva-Veshoska, Sheryl Lynch, William Hynes, and Toni Borkovic

Introduction

The exploitation and utilisation of energy resources have caused severe ecological and environmental problems, including the production of emissions that contribute to global warming (Enshassi, Ayash, & Mohamed, 2018). The construction industry consumes up to 50% of mineral resources excavated from nature, generates about 33% of CO_2 present in the atmosphere and is responsible for 40% of total global energy through both construction and operational emissions (Ajayi et al., 2016; Zhou & Azar, 2018). This has resulted in the AEC Sector, out of necessity, been

B. McAuley (✉) · A. Behan
School of Multidisciplinary Technologies, Technological University Dublin, Dublin, Ireland

P. McCormack · A. Hamilton · E. Rebelo
Belfast Metropolitan College, Belfast, UK

B. Neilson · G. Beckett
CITB Northern Ireland, Nutts Corner Training Centre, Crumlin, County Antrim, UK

A. A. Costa
CERIS, Department of Civil Engineering, Architecture and Georesources, Instituto Superior Técnico, Universidade Lisboa, Lisbon, Portugal

P. Carreira
iNESC-ID Lisboa, Lisbon, Portugal

D. Likar · A. Taneva-Veshoska
Institute for Research in Environment, Civil Engineering and Energy (IECE), Skopje, Republic of Macedonia

S. Lynch · W. Hynes
Future Analytics Consulting Ltd., Dublin, Ireland

T. Borkovic
Energy Institute Hrvoje Pozar, Zagreb, Croatia

L. Scott et al. (eds.), *Sustainable Ecological Engineering Design*,
https://doi.org/10.1007/978-3-030-44381-8_12

forced to investigate new methods of practice and how best to apply resource-efficient techniques from the extraction of the raw materials to the demolition and disposal of its components.

The realisation that practices now face globalisation, sustainability, and environmental concerns, as well as ever-changing legislation requirements and new skills needed for the information age, has resulted in technologies such as Building Information Modelling (BIM) becoming a key enabler in navigating these concerns (Jaradat, 2014). BIM can be defined as a modelling technology and associated set of processes to produce, communicate and analyse building models (Sacks, Eastman, Lee, & Teicholz, 2018). BIM provides an opportunity for the Architectural, Engineering, Construction, and Operation (AECO) industry stakeholders to evaluate possible solutions and identify potential problems of the final product before the start of actual construction (Badrinath, Chang, & Hsieh, 2016).

However, changing from traditional practices to BIM requires a shift not only in the technology used but also in the way design and construction teams work together (Shelbourn, Macdonald, McCuen, & Lee, 2017). To achieve the associated benefits that are accustomed to BIM, a number of existing challenges to information and communications technology (ICT) utilisation in construction site management must be overcome, which include a lack of knowledge, skills and competence, depth of understanding of decision makers and low ICT literacy (Ozumba & Shakantu, 2017). This BIM movement has also resulted in a clear, direct and automatic impact upon engineering education systems (Jäväjä & Salin, 2014).

To assist in overcoming these barriers, so as to reach EU energy-related targets, a number of funding initiatives have been put in place through Horizon 2020 with a focus on BIM, as a result of it having the potential to rapidly produce energy outputs that enable design teams to analyse and compare the most cost-effective, energy-efficient options. Such an initiative is the Energy BIMcert project, which aims to educate all areas of the supply chain in the use of BIM, to achieve better energy efficiency during the design, construction and ongoing maintenance of an asset.

Energy BIMcert Background

Horizon 2020 is the biggest ever EU Research and Innovation programme, with nearly €80 billion of funding available over 7 years (2014–2020). An initial funding call, as part of this programme, was made available with a focus on supporting innovation through research by way of demonstration of more energy-efficient technologies and solutions. The Energy BIMcert consortium, consisting of industry and academia who are experts in providing BIM solutions, skills and training for the construction industry, backed up by a Technical Advisory Board of stakeholders and external experts, responded to the call. The consortium put forward a proposal to enable the development of a method, materials and micro accreditation for upskilling across the construction supply chain to allow BIM techniques and technologies to be utilised to address energy efficiency requirements.

Energy BIMcert's goal is to develop more efficient and relevant training programme materials that integrate concepts of sustainability and renewables with practical application and integration with technology, as based on real-life industry needs and limitations. The Energy BIMcert consortium consisting of members from Northern Ireland (Belfast Metropolitan College and Construction Industry Training Board (CITB)), Republic of Ireland (Technological University (TU) Dublin and Future Analytics), Portugal (Universidade Lisboa), Macedonia (Institute for Research in Environment, Civil Engineering and Energy (IECE)), and Croatia (Energy Institute Hrvoje Pozar (EIHP)) established five core objectives:

1. To improve the sustainability of the built environment by training its workforce in more efficient and greener ways of designing and constructing through the use of BIM processes, better materials, products and energy sources.
2. To engage with the entire construction sector supply chain via BIM to develop more extensive European links and to encourage a system of peer support across states of varying maturity concerning the delivery of more energy-efficient new and renovated buildings.
3. To encourage greater workforce mobility, continuous upskilling and better employability for all levels of an employee in the construction sector.
4. To create clear pathways of development for individuals and SMEs to upskill from any starting point of knowledge to any required level of the individual, or collaborative expertise, in support of sustainable energy-efficient construction.
5. To develop a pan-European framework for recognition and accreditation of Energy BIMcert's micro accredited learning modules, that will combine to build towards fully standardised skills recognition, linking within existing national and European initiatives and frameworks of accredited courses and awards.

The consortium established a series of work packages which are to be conducted in five stages:

- Stage 1—State of the Art: An open approach to gather state-of-the-art information through direct engagement with project stakeholders across Europe to ensure that the skills gaps identified by SMEs about the implementation of BIM technologies and methods in support of improved energy efficiency in the construction sector are correct.
- Stage 2—Development: Development of the Energy BIMcert platform, which will provide information about the project, share Energy BIMcert outputs, and support stakeholders' communication and collaboration.
- Stage 3—Testing: The rigorous evaluation of the curriculum, the learning materials, and the proposed platform.
- Stage 4—Accreditation: Accreditation of the proposed Energy BIMcert training units and courses.
- Stage 5—Exploitation and Dissemination: The exploitation and dissemination of the project through a broad-ranging outreach campaign.

This paper will focus on Stage 1 with the final four stages outside of scope.

Methodology

The consortium has established a detailed and exhaustive process in determining the training methodologies and associated curriculum. An initial pan European wide survey of the industry was undertaken to ascertain the current level of BIM maturity, knowledge and understanding within Built Environment practitioners and academia and to establish current standards of sustainable design and construction practice. The results from the survey have been cross-referenced with five workshops within the project stakeholders' jurisdictions. The results from the survey and workshops were used to establish the training courses that best-matched industry needs. In parallel, TU Dublin completed a state-of-the-art literature review of the current global status of BIM regarding education and what pedagogical methodologies are applied to deliver these courses. The survey and workshop findings were cross-referenced with the results from the state-of-the-art literature review in the production of a rolling Matrix of Concepts and Methodologies aligned to best practice for knowledge transfer. Both the initial findings for the suggested training courses and Matrix of Concepts and Methodologies have been tested through a series of Reality Check workshops. The outcomes from the Reality Check workshops resulted in the establishment of the final training methodologies, descriptors including learning outcomes, suggested syllabi, and delivery details.

State of the Art

Stage 1 will be broken down into four sections: (1) Energy BIMcert Survey and Workshops, (2) Concepts and Methodologies, (3) Reality Check Workshops and (4) Curriculum and Materials.

Energy BIMcert Survey

A survey was undertaken in all five partner countries by CITB, which was subsequently translated into four languages, resulting in 548 responses. The breakdown of the survey response is illustrated in Fig. 1. To validate the findings, a total of five workshops took place with stakeholders between the 2nd May and 6th June 2018, which were used to gather supplement data to cross-reference with the survey results. The workshops were open to the industry, and representatives of the advisory partners were strongly encouraged to attend. The results found that all respondents recognised that BIM training is required at all levels within their organisations. The survey findings indicate that 57% of respondents have received no formal BIM training, despite 61% using some form of BIM in the workplace. A further 53% of survey respondents indicated they were prepared to implement BIM within the next

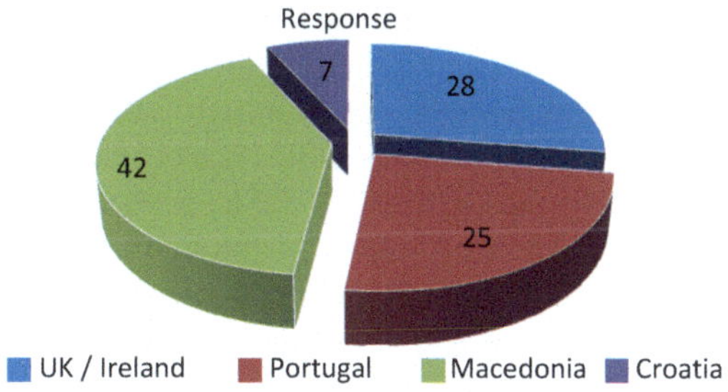

Fig. 1 Response by country

3 years, of those 26% aim to achieve this within the next 12 months. The two main challenges identified within the survey and reiterated at the workshops as most critical were (a) lack of BIM skills (46%) and (b) lack of client awareness of the value of BIM (43%).

Regarding training, 53% of respondents indicated they would prefer a person-led course with practical demonstrations, while the other 47% reported they would prefer online training. Based on the workshop consultations, Energy BIMcert coordinators noticed a discrepancy between industry knowledge and enthusiasm for adopting BIM. The results show that government encouragement for BIM adoption, particularly within the UK and Republic of Ireland, was a strong motivating factor. A lack of Government endorsement within Macedonia and, to a lesser extent, Croatia has resulted in them facing similar concerns to those discussed above.

The results highlighted the necessity to raise awareness of BIM as a sophisticated, sustainable, supportive software, not only for modelling and visualisation tools but furthermore by developing training modules to facilitate the trend. The lack of BIM maturity within some of the partner's jurisdictions' informed the consortium that if potential users were going to understand how BIM is used for energy-related proposes that they would need a fundamental understanding of the core principles, as well as how to access information for review purposes. These findings influenced the consortium in the selection of training courses discussed later in the paper.

BIM Concepts and Methodologies

TU Dublin undertook a state-of-the-art literature review of current BIM teaching initiatives that represent best practice for knowledge transfer. The report identified many potential training methodologies that could be applied to the Energy BIMcert platform. The proposed methods include Problem Project Based Learning (PBL) which is recognised as a capable student-centred pedagogical approach focusing on real-world issues, which allows students to build knowledge and develop critical thinking, creativity, leadership and communication (Badrinath et al., 2016). The PBL approach is fully supported for the software elements of BIM and is widely used in engineering and construction management education to build the ideal scaffold to student learning for sustainable living (Wu & Luo, 2016). The option of narrative videos and in-house video tutorials are also robust learning tools as they enable the student the opportunity to self-learn (Adamu & Thorpe, 2016). The use of narrative videos to teach BIM has become a prevalent theme over the last number of years. Guided Self-Study (GSL) is one of the best ways to facilitate professional development concerning BIM through self-training with continuous peer support and ad hoc internal workshops (Puolitaival & Forsythe, 2016).

Other training methodologies examined include Web-based tutorials which are a combination of media, that allows the user to control, combine and manipulate different types of mediums of communication such as text, graphics, still images and interactive features (Guy & Lownes-Jackson, 2013). Web-based tutorials work in tandem with instructor-led tutoring, which can involve giving step-by-step instructions to students, in a virtual or physical class. Instructor-led tutoring has been the most common BIM tutoring approach in AEC degree programmes (Abdirad & Dossick, 2016). The authors also endorse the use of case studies as tools to enable students to learn and think about real-world challenges and solutions as well as standards and conventions in BIM implementation. Design for Disassembly (DFD) could be applied to any of the discussed methodologies as it involves the process of designing products so that they can be easily, cost-effectively and rapidly taken apart at the end of the product's life so that components can be reused and recycled. Badrinath et al. (2016) explain that through focusing at the issues surrounding sustainable design through an investigation into DFD, one can have a stronger appreciation.

Mastery Learning and Scaffolding Learning approaches were explored as further possible learning methodologies. Mastery learning requires the students to master a more straightforward subject before moving on to the next, more complex one (Shelbourn et al., 2017). The authors also discuss Scaffolding Learning which involves taking steps to reduce the student's degrees of freedom in carrying out some tasks so that the learner can concentrate on the problematic skill they are in the process of acquiring. Other teaching tools and methodologies suggested to the Energy BIMcert consortium included active learning, where both the educator and learners cooperate in formulating and achieving a combined experience whereby the learner has an active role. There is evidence that dynamic learning experiences can enhance the learning of collaborative BIM concepts (Becker, Jaselskis, & McDermott, 2011). The flipped classroom is an instructional strategy and a type of blended learning that reverses the traditional learning environment by introducing the learning material before class, with classroom time then being used to deepen understanding through discussion with peers and problem-solving activities facilitated by teachers. The application of role play can also enable the potential user to gain an understanding of other professional's roles and therefore provide a greater understanding of the importance of data.

The results from this exercise were cross-referenced with the survey and workshop results to ensure that the selected methodology was industry appropriate. Figure 2 provides a matrix of the most suitable concepts and methodologies that best align with the Energy BIMcert survey, workshops, and findings from the

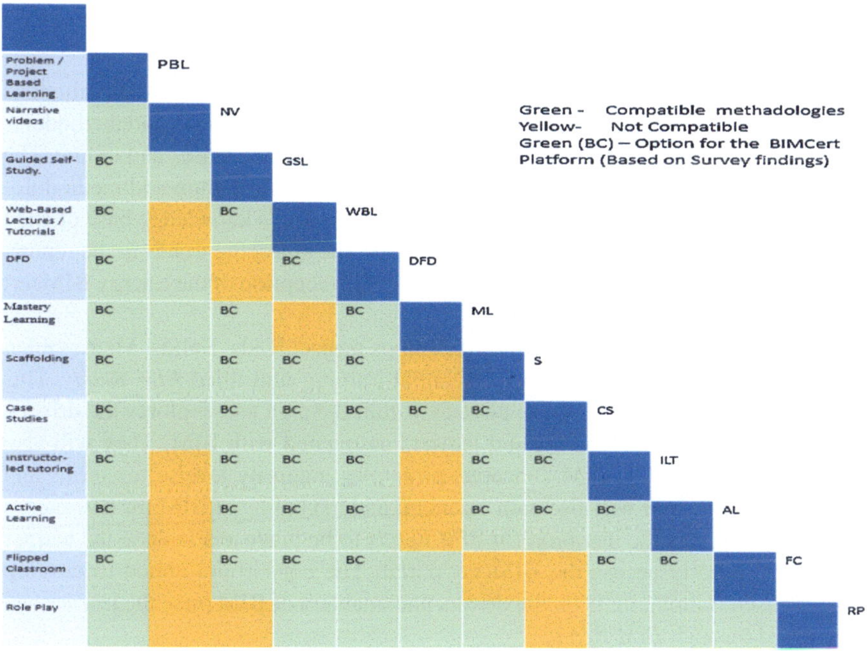

Fig. 2 Matrix of concepts and methodologies

pedagogical and methodologies literature review. Based on this diagram and further cross reference of the literature review the most suitable approach would involve a Scaffolded Learning approach, as users requested a step-by-step method delivered through instructor-led live lecture which would limit the degrees of freedom to ensure a structured approach (Abdirad & Dossick, 2016; Shelbourn et al., 2017). On further review of Fig. 2 and the literature, this would be best complimented through providing an intelligent student centred pedagogical approach focusing on real-world issues (PBL) along with self-training and continuous peer support (GSL). This could be achieved by surrounding the student with a sustainable design through an investigation into DFD (Badrinath et al., 2016; Puolitaival & Forsythe, 2016). However, this may prove difficult due to the nature of the project, and a more blended approach may need to be adopted. This could involve the inclusion of pre-recorded narrative videos to enable the student the opportunity to self-learn, which could be complemented through live lectures to ensure they still have a base to communicate with the lecturer. This would enable a structured pathway for users to move from Scaffolded to Masterly Learning where the user is required to critically understand a topic through a combination of GSL and PBL before moving on to the next area of interest (Adamu & Thorpe, 2016; Shelbourn et al., 2017).

Energy BIMcert Reality Check Workshops

FAC and CITB carried out five trial workshops within the consortiums jurisdictions with key stakeholders to test the suggested curriculum and concepts and methodologies. The workshops were hosted both within the partner city bases and online. The purpose of the workshop was to establish if the current curriculum and methodologies were adequate. The workshop also aimed to develop a knowledge base of BIM awareness on a European level and a rationale for the emerging trends in the reports carried out by CITB and how this would impact the reception of the Energy BIMcert on a national and broader European level.

Before the commencement of the workshops project leads, Belfast Metropolitan College developed the outline for a potential training unit titled *BIM Ready*. This training unit is aimed at complete novices within the area and explores the drivers, enablers, requirements, benefits and barriers associated with BIM. They will also briefly be introduced to BIM concepts, principles, maturity levels, key terms and standards. The learner will also gain a general understanding of BIM for low energy construction. It was the intention for *BIM Ready* to be made available to the learner when they initially access the BIMcert portal. The consortium aimed to develop more focused training units on the drivers and enablers of BIM once the learner had progressed through *BIM Ready*.

In all of the partner countries, BIM was recognised not only as a modelling software but also as a sustainability support tool for energy-efficient and green construction. Throughout the trials, the top five findings were:

1. BIM awareness and skill varies hugely from country to country
2. Some contractors are reluctant to invest in BIM (Need for evidence-based case studies of BIM cost-effectiveness)
3. There is a reluctance to engage with IT as a medium for learning
4. The BIM supply chain is congested (mostly confined to management)
5. Energy BIMcert is a positive and viable enabler/facilitator of BIM upskilling

The favoured learning styles that emerged through the workshops were the blended learning option with self-selection of the micro-sized learning courses. Taking into account the results of the Energy BIMcert Survey and workshops, and given that the platform will be online, the primary methodology suggested of a scaffolded learning environment guided by a series of instructor-led live lectures gained favour. However, a more blended approach involving narrative videos and live lectures which would assist in reducing the requirement of a lecturer and put more focus on the student doing GSL through PBL and DFD before they advance to the next digital badge was also warmly received. The concept of the *BIM Ready* training unit was endorsed, with a recognition that one must gain an understanding of the fundamentals of BIM before they fully appreciate how it can be applied to save energy. Recommendations for the development of further training material in this space were requested, with an active request for courses in working in digital construction, associated workflows, digital skills for construction sites and low energy building construction.

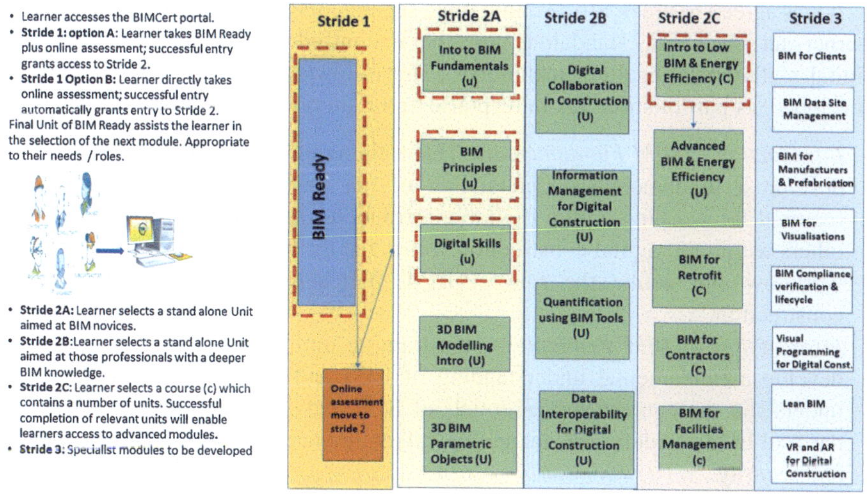

Fig. 3 BIMcert learning pathways

Development of Curriculum and Materials

The development of curriculum and associated materials involved TU Dublin, Belfast Metropolitan College, and IECE working in partnership. While this was ongoing Universidade Lisboa was in contact to ensure that the final platform was suitable to host the potential curriculum. EIHP were also busy continuing to promote the project through established dissemination channels, as well as endorsing the project through the website domain (https://energybimcert.eu). The consortium members decided that the best way forward was to break the development of the curriculum into three strides. The term "stride" was used to avoid confusion with terms such as "level", "stage", and "point", which all have other associated meaning in the BIM Terminology of different EU jurisdictions. Figure 3 identifies the units and courses that best reflect the needs of the industry from the consultation process.

The learner initially accesses the BIMcert portal and will be presented with one of two options. If the learner selects Option A, then they must take the BIM Ready training unit plus online assessment. Successful completion of the assessment grants access to Stride 2. This entry unit is critical to ensuring that all learners have a basic understanding of BIM before they select their next unit within Stride 2. The BIM Ready unit is also vital here as it will serve as a diagnostic tool to assist the leaner in the selection of the next unit. Those who complete BIM Ready will be able to use their assessment marks to automatically pass some of the learning outcomes within the training units offered in Stride 2. This will encourage users to undertake a full unit. Entry Option B is aimed at non-novices and permits the learner to take an online assessment directly.

It was agreed to break Stride 2 into three separate sections. Within Stride 2A, the learner can select many standalone units that will introduce them to BIM principles, digital skills and modelling techniques. The following training units have been developed, as part of stride 2A in response to the survey and workshops findings.

- *Introduction to BIM Fundamentals*: This training unit will enable the learner to develop a fundamental understanding of the information technology skills required for working within digital construction, i.e. understand their role and others within the digital construction sector, list the benefits and value of a BIM workflow and define the role of BIM in achieving improved sustainable construction.
- *Introduction to BIM Principles*: This training unit will enable the learner to develop a fundamental understanding of BIM and associated workflows, i.e. understand the context and essentials of BIM, detail the application and standards of BIM and define the technological requirements for BIM implementation and security.
- *Digital skills*: This training unit will enable the learner to develop a fundamental understanding of the use of digital skills for construction sites, i.e. describe the use of digital skills and devices in construction, demonstrate the use of digital tools to perform a design review and evaluate a BIM model, etc.
- *3D BIM Modelling*: This training unit will enable the learner to develop the fundamental skills for three-dimensional (3D) BIM using industry standard software for their particular profession, i.e. define how BIM may be used within the construction and design industry, create a building model using industry-standard BIM software, implement and manage BIM and develop and publish information using BIM.
- *3D BIM (Parametric) Objects*: This training unit will enable the learner to develop the fundamental skills to create BIM objects using industry standard software, i.e. demonstrate the importance and use of parametric objects in digital construction, generate templates, design and create BIM, export, insert and use BIM objects.

Stride 2B represents units aimed at those more experienced BIM users who wish to advance their knowledge in BIM, e.g. interoperability and collaboration processes. While learning outcomes are developed for these training units, it is not the intention of the Energy BIMcert consortium to develop them any further during this iteration of the Horizon 2020 project. Stride 2C offers the learners the choice of one or more courses, which consists of a series of units. Each unit within a course represents a specific Learning Outcome. This Learning Outcome/Unit will be offered as an individual micro size training option, to ensure that the Energy BIMcert can attract learners who require specific areas of knowledge but do not have the time to complete a standard unit (Stride 2A and 2B) consisting of a series of learning outcomes. After completion of all units associated with the course, the learner will receive a higher award. The learner can take advanced units once they finish the relevant Stride 2C course units, i.e. Advanced BIM & Energy Efficiency. As with Stride 2B, it is not the intention of the consortium to develop all of the courses in this stride. The consortium aims to develop the Introduction to Low Energy Building

Construction course, which will contain some units that will provide an understanding of sustainable and energy efficient design and how to apply them. Stride 3 represents a more discipline-focused stride that represents current specialisations of BIM usage, tools, and concepts. The range of units can be expanded or adjusted in the next stage of the project in response to market needs.

A variety of material from a selection of learning outcomes from the BIM Ready (Stride 1), BIM Fundamentals (Stride 2A), BIM Principles (Stride 2A), Digital Skills (Stride 2A), and Introduction to Low Energy Building Construction Course (Stride 2C) will be developed for testing in Phase 2 of the project. The Energy BIMcert consortium selected these particular learning units because they have the potential to impact significant numbers of construction site workers across Europe rapidly. The selected material can also be delivered through instructor-led live lectures which will enable the opportunity for the lecturer/trainer to engage with the class. It is also the attention to develop material that can be used for GSL to gain an insight into how potential users would interact with this material. Results from the testing of the pilot materials from the second round of partner trial workshops can be found in McAuley et al. (2019).

Conclusion

The paper has discussed how the Energy BIMcert initiative can assist in improving the sustainability of the built environment by training of its workforce in more efficient and greener ways of designing and constructing through the use of BIM processes. The proposed Energy BIMcert platform will provide a direct response to a perceived lack of available micro-sized training units and will cater for all levels of BIM knowledge within the industry. While better energy awareness and sustainable building practices are at the core of the project, the consortium have recognised that a diverse workforce will require a fundamental understanding of BIM principles if they are to fully embrace and engage with the capabilities associated with digital construction. This approach is fully supported by industry who continue to demonstrate an appetite for BIM. The next stage of the project is the piloting and evaluation of material through a series of workshops across the consortiums jurisdictions. These workshops will enable the refinement of the content before being released on the final BIMcert training platform before the final phase of testing.

References

Abdirad, H., & Dossick, C. (2016). BIM curriculum design in architecture, engineering, and construction education: A systematic review. *Journal of Information Technology in Construction, ITcon, 21*, 250.

Adamu, Z. A., and Thorpe, T. (2016) How Universities are Teaching BIM: A Review and Case Study from the UK. *ITcon. 21*, 121–141

Ajayi, S. O., Oyedele, L. O., Kadiri, K. O., Akinade, O. O., Bilal, M., Owolabi, H. A., & Alaka, H. A. (2016). Competency-based measures for designing out construction waste: Task and contextual attributes, Engineering. *Construction and Architectural Management, 23*(4), 464–490.

Badrinath, A. C., Chang, Y. T., & Hsieh, S. H. (2016). A review of tertiary BIM education for advanced engineering communication with visualization. *Visualization in Engineering, 4*(9), 2–17.

Becker, T., Jaselskis, E., & McDermott, C. (2011). Implications of construction industry trends on the educational requirements for future construction professionals. In *International Conference Proceedings of the 47th Associated Schools of Construction, Omaha, Nebraska, United States, April 6–9* (pp. 1–12).

Enshassi, A., Ayash, A., & Mohamed, S. (2018). Factors driving contractors to implement energy management strategies in construction projects. *Journal of Financial Management of Property and Construction, 23*(3), 295–311.

Guy, R., & Lownes-Jackson, M. (2013). Web-based tutorials and traditional face-to-face lectures: A comparative analysis of student performance. *Informing Science and Information Technology, 10*, 241–259.

Jaradat, S. (2014). Educating the next generation of architects for interdisciplinary BIM environments. In *Proceedings of the AAE International Conference on Architectural Education: Common Currency, University of Sheffield, September 3–5* (pp. 127–135).

Jäväjä, P., & Salin, J. (2014). BIM education: Implementing and reviewing "OpeBIM"—BIM for Teachers. In *Proceedings of the Computing in Civil and Building Engineering, Orlando, Florida, June 23–25* (pp. 2151–2158).

McAuley, B., Behan, A., McCormack, P., Hamilton, A., Rebelo, E., Neilson, B., ... Borkovic, T. (2019). Improving the sustainability of the built environment by training its workforce in more efficient and greener ways of designing and constructing through the Horizon 2020 Energy BIMcert Project. In *Proceedings of the CitA BIM Gathering, Galway, September 26–27* (pp. 1–8).

Ozumba, A. O. U., & Shakantu, W. (2017). Exploring challenges to ICT utilisation in construction site management. *Construction Innovation, 18*(3), 1471–4175.

Puolitaival, T., & Forsythe, P. (2016). Practical challenges of BIM education. *Structural Survey, 34*(4/5), 351–366.

Sacks, R., Eastman, E., Lee, G., & Teicholz, P. (2018). *BIM handbook: A guide to building information modeling for owners, designers, engineers, contractors, and facility managers*. Hoboken, NJ: Wiley.

Shelbourn, M., Macdonald, J., McCuen, T., & Lee, S. (2017). Students perceptions of BIM education in the higher education sector a UK and USA perspective. *Industry and Higher Education, 31*(5), 293–304.

Wu, W., & Luo, Y. (2016). Pedagogy and assessment of student learning in BIM and sustainable design and construction. *ITcon, 21*, 218–232.

Zhou, H., & Azar, E. R. (2018). BIM-based energy consumption assessment of the onsite construction of building structural systems. *Built Environment Project and Asset Management, 9*(1), 2–14.

The Application of Virtual Reality to Recreate an Interactive WW1 Camp

Stephen Wilkinson

Introduction

This paper examines the need to preserve our local history, which is part of our "cultural heritage". Many of our historical sites have been lost to the ever-increasing demand for new housing (Historic England, 2017). Much pressure has been placed on many historic places to accommodate badly needed new homes. The research question then is "How can we preserve areas of local history through the application of Virtual Reality (VR)".

The case study presented is of Skipton in North Yorkshire and its contribution to the internment of German prisoners of war at the Raikeswood prisoner of war camp. The paper outlines the contribution to local history of the camp and then goes on to examine the techniques required in order to recreate an online interactive VR model. The aim of the research was to produce an accurate, navigable 3D model of the camp, of both its external and internal features. Finally, initial testing of the model by local school history teachers is presented with conclusion and future work being discussed.

Case Study

The First World War was an event that seems a distant memory, especially to the impact this war had on many communities. Around the country many towns and cities have had various events and exhibitions to commemorate the end of this war.

S. Wilkinson (✉)
Leeds Sustainability Institute, School of the Built Environment, Engineering and Computing, Leeds Beckett University, Leeds, UK
e-mail: s.wilkinson@leedsbeckett.ac.uk

L. Scott et al. (eds.), *Sustainable Ecological Engineering Design*,
https://doi.org/10.1007/978-3-030-44381-8_13

Skipton Council have developed an exhibition to show what had occurred during this time and how it had impacted on their local community. One of the main war time activities in the area was the development of the Raikeswood camp. This camp was initially a number of wooden huts and was primarily used as a training camp by local "Pals" regiments before they were shipped off to the front line in France. Later as the war progressed it was used as a German prisoner of war camp.

However, apart from a few traces, this camp has disappeared under a modern housing estate. What started as a need by local school children to discover more about their local history soon turned into an archaeological dig in parts that hadn't been built over. Further investigation revealed details of the camps location and diaries written by the prisoners. These contained sketches of the camp and cartoons of camp life. Other information discovered were old photographs showing different parts of the camp.

The challenge then of creating a VR camp was to use the above information, maps from 1912 and the topography of Skipton and surrounding area.

Techniques Used in Creating VR Worlds

The main remit of this model was to accurately locate it within the local landscape. The model must have a true representation of the buildings and the function of the main huts must be illustrated, through interaction with the viewer.

Previous research in the application of VR to archaeology, landscapes and buildings is wide and varied and is discussed. For example, Ruffaldi (2008) discuss how 3D VR models may be used to present cultural heritage information in many different ways. A case study of Lucca art (Tuscany, Italy) describes how an art gallery was modelled in 3D graphics and the art collection was placed exactly within the 3D environment and interactivity consisted around information related to each artistic subject.

The impact on the environment of a wind turbine farm on the local community was shown by Jallouli (2008). Their VR model had multi-sensory detail including the noise that the wind turbines would make due to the proximity of the viewer. They were able to assess the impact of the wind farm on the local community through the testing within the VR environment, both in terms of visual and auditory impact.

The issues related to creating natural landscapes for large VR environments is discussed by Belhadj (2007). This paper outlines how landscapes can be generated using fractals, which generates natural terrains using ridges and rivers information. The technique uses a predefined terrain mesh containing ridgelines and a rivers network. This is then modified using an algorithm that displaces these features into the desired natural landscape.

Scheepers (2001) describes how advanced modelling techniques can be used to generate realistic virtual landscapes using satellite geographical information systems (GIS). One such GIS is a Digital Elevation System used within this research enabled an array of elevation points to aid the generation of accurate 3D landscape meshes.

The ability to be able to accurately model and map materials in a 3D VR environment is crucial to the accuracy and realism of the VR world. The need to reduce the number of polygons and vertices of the model is crucial in being able to navigate and interact with the 3D elements. This is regardless of computing power and bandwidth; the efficiency of the model is essential. A number of techniques used to aid the efficiency is shown by Sampaio (2015), who shows how 3D textures, for example, roof tiles and external coatings can be mapped on to a 3D object in order to efficiently describe the model. This research went further, by showing how the modelling of VR facades may be achieved, in order to illustrate how detachment, cracking and fracturing of the facade shows up maintenance problems. A database of facades showing maintenance problems was used and the interactive element illustrated what the anomaly looked like and what the standard procedure for repair was.

The final element of VR is to enhance or augment the feeling of reality by adding extra information, such as sound or text. Amin, Levy, and Boyd (2012) developed an application for mobile phones that recognise the shapes of historical ruins of Arbela in Iraq. The VR element reconstructed the view in 3D to show what it would have looked like before it was destroyed. The sound of ancient Iraq was also played as well as a text window, giving background to each specific historical building being viewed. The project was called Arbela Layers Uncovered (ALU). Other notable augmented VR worlds include fire-fighter training, St Julien & Shaw, 2003; this showed how fire fighters could plot a path through a burning building, in order to safely reach the source safety and extinguish the fire. Realistic fire and smoke was part of the augmented VR world. Other interactions included interactive fire and smoke. The choice of extinguisher decided how quickly the fire was extinguished, or what the result would be if the wrong choice was used. A very useful area of research was carried out by Wolf and Kahlil (2017). They developed a VR system for prototyping interactive architecture. One of the main strengths of this research was the techniques used to navigate and interact with the VR world and the types and combination of technologies available in order to achieve this. Interfaces such as Head Mounted Displays were combined with Screens and Mice to move and grasp objects within the VR buildings.

Research Methodology

From the initial Raikeswood dig by local Skipton school children, as part of their local history lesson into Skiptons part in World War 1. A project group had already been formed by Skipton council in order to document Skipton's part in the First World War. The outcome of this project was to provide an exhibition and a permanent website of this research into local history. The Raikeswood camp project became part of this major project.

From initial project group meetings, an outline of the remit and specification of the VR camp world was made. Research into the development of the environment and the models was also made. The need for these models to exist within a virtual world was also researched. The need to provide interactive information within the virtual world was also investigated. Initial testing was made using local history teachers. This was a small sample, but considered to be heuristic, i.e. using a handful of experts on local history. The conclusions described of these initial findings is also made.

Development of the Virtual camp

The above research how VR environments, models and interactivity may be used in order to recreate historical sites or other VR worlds.

The VR camp being recreated was at its height in 1919 and was nestled to the North West of Skipton. It was just above the wood called Raikeswood. In the distance were the dales, featuring Rough Haw and Sharp Haw, two notable hills in the distance. It was important to achieve the topography for a 360° view of the terrain from within the camp. It was also important to model the detail of the construction of the huts both for the exterior and interior. Within the main huts, items specific to the function of the hut were placed to give a level of authenticity. For example, the infirmary had kidney dishes, jugs and rows of beds. The dormitories had rows of beds, but with storage boxes and tables. The censors office contained tables and chairs, with books and maps.

Each hut had an interior camera, with the exterior being viewed through a drone camera. The exterior had features such as sports grounds, vegetable patches and a view of the plan of the camp as it was laid out in 1919, along with wire fences and lookout posts.

Terrain

From the research previously described, there are a number of ways of developing 3D terrain. The technique used in this case was to use a satellite image of Skipton, showing the different heights of the terrain, described in different shades. White being the highest point, while dark green being a low point. The shades in-between described the terrain in detail. A wire mesh was then displaced using a displacement map. The camp layout was merged into the 1919 map, Fig. 1, as accurately as possible. A separate camp mesh was then shaped to fit the curve of the terrain, according to the position on the terrain, Fig. 2.

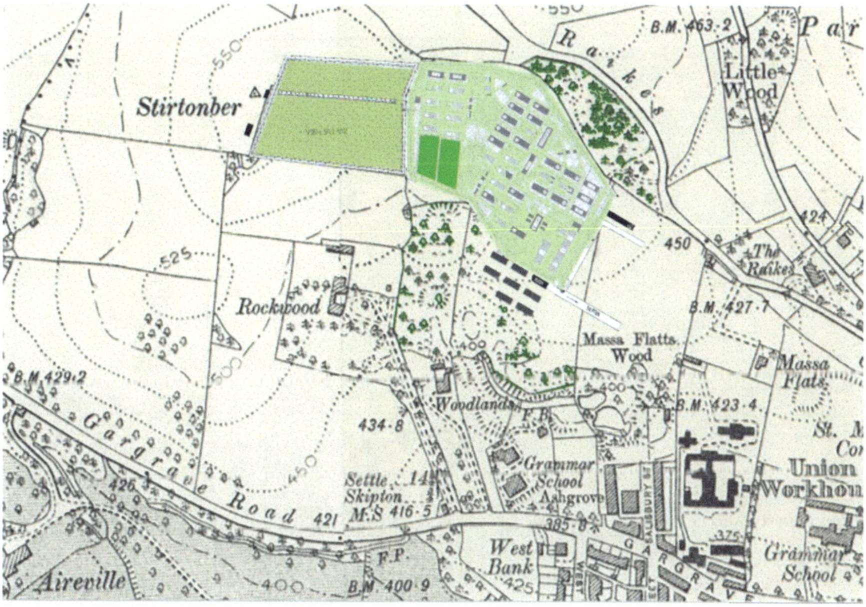

Fig. 1 Camp layout overlaid on to a 1912 map

Fig. 2 Showing the camp terrain made using a displacement map

Huts

The huts were described within 3D geometry, using simple primitives such as cubes. These were then manipulated as a multi sub-object, i.e. extra points were inserted to give more edges to enable extra faces to be created. By moving these points around,

Fig. 3 German POW at the hospital hut circa 1919

the geometry of each hut was achieved. Extra geometry was added, in order to create doors and windows. The impression of construction from wooden planks was produced from the application of normal maps. These are textured maps that bump out the shape of the plank, from the image of the planks on each side of the hut. This material was wrapped around each hut using UV modifier that align to each side of the hut. The aim of each VR hut was to reproduce an authentic representation of internal and external features, using old photographs and sketches from the era of pals and prisoners (Figs. 3 and 4).

Extra Assets

Extra objects or assets in game terminology were created, for example, wheelbarrows for the vegetable garden, fencing and look out posts for the perimeter of the camp, latrine huts, with internal detail and a variety of simple trees for the surrounding woodland.

Internal assets included gramophone record players, books, maps, benches and tables, iron frame beds and plumped pillows and blankets. Other items included kidney bowls, water jugs, trolleys, chests and crates.

Fig. 4 German POW, sketch of the camp circa 1919

Navigation and Interaction

In order to navigate around the camp and to view inside the main huts, a number of virtual cameras were used:

- Drone camera to orbit and zoom around the camp
- Internal camera for the censors office
- Internal camera for the infirmary
- Internal camera for the dormitory
- Internal camera for the standard hut and mess hall

Interaction between cameras and assets was achieved using C# script language. Each camera had a C# script within the object, which allowed the camera to orbit, zoom, pan and pick by using a three button mouse. Each button had a function assigned to it, so holding a particular combination or individual button down and moving the mouse corresponded to a particular function.

The drone camera enabled navigation around the camp and the ability to pick objects of interest, such as the vegetable plot, sports field and each of the main huts (Fig. 5).

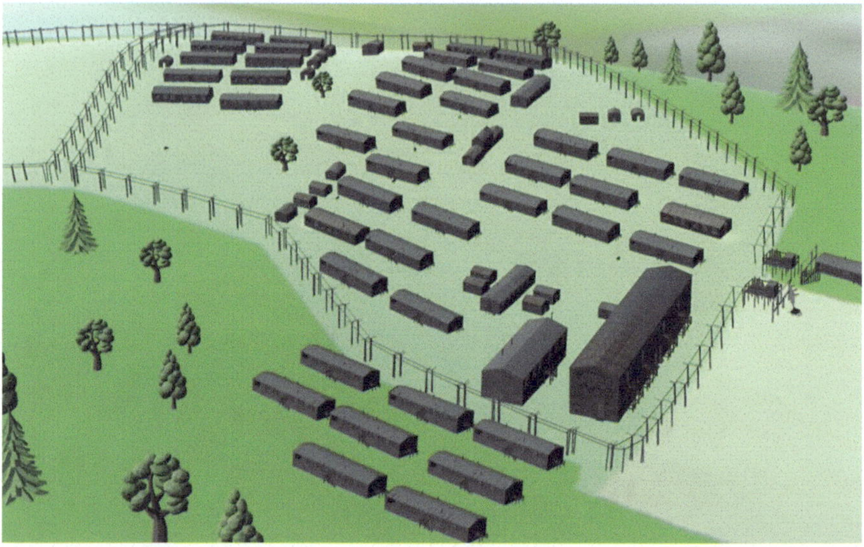

Fig. 5 Drone view of the camp

Fig. 6 Internal view of a hut including interactive window overlay

When a hut was picked, the control was automatically sent to the internal camera. Using the same features as the drone camera it was possible to navigate around the internals of the main huts described previously.

In addition to internal navigation, a simple window appeared that displayed text and an image stating the function of the hust and a little more about the camp and what had happened there. A return button within the image window passed the user back to a drone view and control (Fig. 6).

Exporting Format

A number of formats were used in order to enable different users to participate. For example, mac or PC user had to download the game engine in order to use the VR camp internally on their machine. Different resolutions could be selected in order to achieve this. Later a WebGL version was developed that could be used on the latest web browser that could handle this format. This can be seen at https://skiptonww1camp.co.uk.

Testing

There are a number of ways of testing this VR model. The one chosen was a heuristic model, where it is better to test using 5 experts, rather than 100 non-experts.

To this end the prototype was shown to a number of staff at a secondary school in Skipton. General comments were:

- Overall they loved the model and could see how it would make their history lessons more exciting and dynamic.
- One teacher said she felt it did what traditional textbooks can't do by bringing history alive.
- Another teacher thought that this would have particular appeal to boys in her class who like playing video games and who would otherwise switch off, when it came to learning about local history.
- They all said that it was a great resource of teaching about local history, which is now part of the national curriculum.

Conclusions

The main aim of this research was to produce an accurate, navigable 3D model of the camp, of both its external and internal features. This aim was largely met; however, as with all these types of VR worlds, the level of detail can be increased infinitely. The project team would agree that this is a very good prototype, but more detail and interactions could be included.

One other research question asked of this type of research is "How will recreating the past have an impact in the future". One of the outcomes is to raise awareness and ensure an inquiry into this period, posing further questions, for example:

- Future generations must be told about such important events in our history and how does this effect current thinking?
- What did my local area look like before housing developments covered past sites?
- What was life like in those times and what were conditions like, that people lived in?

- Maps in the hut showed the battle lines at different dates and the territory held by different empires around the world.

In conclusion, this research has opened up another world that is accessible over the internet and gives school children the opportunity to learn more about their local history and hopefully raise more questions then answers as to the why, what and how of the First World War and the UK's place in the modern world.

Future Work

As previously stated, additional detail and interactions could be implemented. More rigorous testing regarding 3D detail, instructions, navigation and interactive information is required. A larger test group of potential users, i.e. school children and college students, using a more analytical testing process should be used.

References

Amin, R. K. M., Levy, R. M., & Boyd, J. E. (2012). Mobile augmented reality for archaeological sites, *PATCH'12, November 2, 2012, Nara, Japan.*

Belhadj, F. (2007, October), Terrain modeling: a constrained fractal model. Proceedings of the 5th international conference on Computer graphics, virtual reality, visualisation and interaction in Africa, Publication: AFRIGRAPH '07, pp. 197–204.

Jallouli, J. (2008, October). VRST '08: Proceedings of the 2008 ACM symposium on Virtual reality software and technology. pp. 257–258. https://doi.org/10.1145/1450579.1450641. Wind turbines' landscape: using virtual reality for the assessment of multi-sensory perception in motion.

Historic England (2017), Research Agenda. Swindon: Historic England

Ruffaldi, E. (2008, September). DIMEA '08: Proceedings of the 3rd international conference on Digital Interactive Media in Entertainment and Arts. pp 113–119. https://doi.org/10.1145/1413634.1413659

Sampaio, A. Z. (2015). Maintenance activity of buildings supported on VR technology. In *VRIC'15, April 8–10, 2015, Laval, France.*

Scheepers, F. (2001). Landscape visualisation. In *AFRIGRAPH, Capetown, South Africa.*

St Julien, T. U., & Shaw, C. D. (2003). *TAPIA'03: Proceedings of the 2003 Conference on Diversity in Computing, Atlanta, Georgia, USA.*

Wolf, K., & Kahlil, R. (2017). Using virtual reality for prototyping interactive architecture. In *MUM 2017, Stuttgart, Germany.*

The Need for the Inclusion of Construction Health and Safety (H&S) in Architectural Education

John Smallwood

Introduction

The South African Construction Regulations (Republic of South Africa, 2014) schedule a range of responsibilities for clients and designers, which were initiated in the 2003 version of the Construction Regulations. However, ironically, 23 years ago the International Labour Office (ILO) (1992) specifically stated that designers should: receive training in H&S; integrate the H&S of construction workers into the design and planning process; not include anything in a design which would necessitate the use of dangerous structural or other procedures or hazardous materials which could be avoided by design modifications or by substitute materials, and take into account the H&S of workers during subsequent maintenance.

Thorpe (2006) in turn states that there is no more important stage in the construction process than that of design as at this stage conceptual ideas are converted into constructable realities. He further states that a variety of considerations need to be balanced simultaneously, *inter alia*, designing for H&S. He highlights that designing for H&S is an integral part of the wider design process and therefore needs to be included in design planning as doing so will result in healthier and safer construction and maintenance of structures and facilities.

Given the aforementioned, prior research conducted by the author (Smallwood, 2002), the Department of Construction Management's focus on the role of all built environment stakeholders in construction H&S, two surveys were conducted. Firstly, among Departments of Architecture and Architectural Technology based at traditional universities and universities of technology respectively to determine, *inter alia*:

J. Smallwood (✉)
Department of Construction Management, Nelson Mandela University,
Port Elizabeth, South Africa
e-mail: john.smallwood@mandela.ac.za

L. Scott et al. (eds.), *Sustainable Ecological Engineering Design*,
https://doi.org/10.1007/978-3-030-44381-8_14

- Whether architectural programmes address construction H&S, and if so, the form in which it is addressed.
- The importance of the inclusion of construction H&S in the tertiary education programmes of built environment disciplines.
- The importance of the inclusion of subject areas relative to construction H&S in an architectural programme.

A second survey was conducted among built environment practitioners to determine the importance of the inclusion of subject areas relative to construction H&S in tertiary built environment programmes.

Review of the Literature

Legislation and Recommendations Pertaining to Designers

Within the context of South Africa, prior to the promulgation of the Construction Regulations in 2003, all designers were required to address H&S, as in terms of Section 10 of the OH&S Act (Republic of South Africa, 1993) designers are allocated the responsibility to ensure that any 'article' is safe and without risks when properly used.

However, the Construction Regulations (Republic of South Africa, 2014) lay down important requirements with respect to clients and designers. Clients are required to, inter alia: prepare a baseline risk assessment (BRA); prepare an H&S specification based on the BRA; provide the designer with the H&S specification; ensure that the designer takes the H&S specification into account during design, and ensure that the designer carries out the duties as scheduled in Regulation 6 'Duties of designers', and include the H&S specification in the tender documents.

The definition of designers in the Construction Regulations includes architects, engineers, interior designers, landscape architects, and quantity surveyors. Construction project managers (CPMs) may also be included depending on the functions they fulfil. Designers are required to, inter alia: ensure that the H&S standards incorporated into the regulations are complied with in the design; take the H&S specification into consideration; include in a report to the client before tender stage all relevant H&S information about the design that may affect the pricing of the work, the geotechnical-science aspects, and the loading that the structure is designed to withstand; inform the client of any known or anticipated dangers or hazards relating to the construction work, and make available all relevant information required for the safe execution of the work upon being designed or when the design is changed; modify the design or make use of substitute materials where the design necessitates the use of dangerous procedures or materials hazardous to H&S, and consider hazards relating to subsequent maintenance of the structure and make provision in the design for that work to be performed to minimise the risk.

Impact of Designers on Construction H&S

Design influences and impacts on construction H&S directly and indirectly. Directly through concept design, selection of structural frame, detailed design, selection of cladding, and specification of materials. Indirectly through the selection of procurement system, conditions of contract, related interventions such as pre-qualification, decision regarding project duration, and selection of contractor (Smallwood, 2008).

Furthermore, Behm (2006) analysed 450 reports of construction workers' deaths and disabling injuries in the USA to determine whether addressing H&S in the project designs could have prevented the incidents. The findings of this research identified that in 151 cases (33.6%), the hazard that contributed to the incident could have been eliminated or reduced if design-for-H&S measures had been implemented.

Obstacles to Designing for Construction H&S

Hecker, Gambatese, and Weinstein (2006) cite the following as obstacles to designing for construction H&S: the narrow specialisation of design and construction practice; limited pre-construction collaboration between the designer and constructor due to the traditional construction procurement system (TCPS); the limited availability of H&S-in-design tools, guidelines, and procedures, and the limited education architects and engineers receive regarding construction H&S.

According to the Construction Industry Development Board (CIDB) (2009), H&S relevant education and training (or lack thereof), at all levels, has a major impact on construction H&S. At the tertiary level, not all construction-related programmes in South Africa include H&S within their curricula.

Health and Safety and the Six Stages of Projects

Within the context of South Africa, it is notable that the South African Council for the Architectural Profession's (SACAP's) Interim Policy on the Identification of Work for the Architectural Profession (Republic of South Africa, 2011) does not mention H&S in any of the six stages of work in terms of providing a standard service: inception; concept and viability; design development; documentation and procurement; construction contract administration, and close out. This is in stark contrast to, inter alia, the identification of work for Construction Managers, and CPMs, which record a range of H&S interventions over the six stages. In the case of CPMs, inter alia, the following: facilitate input from the designer as required by the H&S agent; facilitate and monitor the presentation of the H&S specification by the H&S agent; monitor the preparation of the H&S plan by the PC, and the approval thereof by the H&S agent; monitor the auditing of the PC's H&S performance relative to their H&S plan by the H&S agent, and monitor the production of the H&S file by the PC and the H&S agent.

Research

Research Method and Sample Stratum

The first sample stratum relative to the study 'Inclusion of Construction H&S in Architectural Education' consisted of 12 Departments of Architecture and Architectural Technology based at traditional universities and universities of technology, respectively. The respective heads of departments (HoDs) were surveyed using a self-administered questionnaire delivered per e-mail. Six responses were received, which equates to a response rate of 50%.

The second sample stratum relative to the study 'Tertiary Built Environment Construction Health and Safety (H&S) Education' consisted of a convenience sample, and to a degree, convenience 'snowball' sample, of built environment practitioners. Discipline-specific potential respondents were requested to respond solely relative to their disciplines, or if registered with more than one statutory council, to respond relative to the disciplines concerned. Furthermore, certain respondents indicated that they, for example, studied quantity surveying, but now practised as a construction manager in the form of a managing director of a general contractor. Others practised as both an architect and an interior designer. Construction H&S Agents (CHSAs), Construction H&S Managers (CHSMs), and Construction H&S Advisors such as those employed by the MBAs and SAFCEC were requested to respond relative to all the disciplines.

Research Results 1

66.7% of the responding HoDs stated that their Departments of Architecture addressed/included construction H&S in their architectural programme/curriculum. 33.3% responded in the negative.

In terms of the form construction H&S is addressed, no departments offered a 'Separate subject' construction H&S. 66.7% maintained that it is an 'Issue in design project briefs', and 33.3% responded relative to each of 'Criterion for assessment in design project assessments', 'Component of a subject', 'Module in various subjects', and 'On an ad-hoc basis' (Table 1).

Table 2 indicates the importance of the inclusion of construction H&S in the tertiary education programmes of nine built environment disciplines in terms of percentage responses to a scale of 1 (hardly) to 5 (very), and MSs between 1.00 and

Table 1 Form in which construction H&S is addressed/included in the architectural programme/curriculum offered by responding Departments of Architecture HoDs

Form	Response (%)		
	Un-sure	Yes	No
Separate subject	0.0	0.0	100.0
Issue in design project briefs	0.0	66.7	33.3
Criterion for assessment in design project assessments	0.0	33.3	66.7
Component of a subject	0.0	33.3	66.7
Module in various subjects	0.0	33.3	66.7
On an ad-hoc basis	0.0	33.3	66.7

Table 2 Importance of the inclusion of construction H&S in the tertiary education programmes of nine built environment disciplines according to responding Departments of Architecture HoDs

	Response (%)							
		Hardly				Very		
Discipline	Un-sure	1	2	3	4	5	MS	Rank
Construction Management	0.0	0.0	0.0	0.0	0.0	100.0	5.00	1
Civil Engineering	0.0	0.0	0.0	0.0	25.0	75.0	4.75	2
Electrical Engineering	0.0	0.0	0.0	0.0	50.0	50.0	4.50	3
Mechanical Engineering	0.0	0.0	0.0	25.0	25.0	50.0	4.25	4
Project Management	0.0	0.0	25.0	0.0	0.0	75.0	4.25	5
Landscape Architecture	0.0	0.0	20.0	40.0	20.0	20.0	3.40	6
Architecture	0.0	16.7	16.7	16.7	16.7	33.3	3.33	7
Quantity Surveying	0.0	0.0	25.0	50.0	0.0	25.0	3.25	8
Interior Design	0.0	20.0	20.0	40.0	20.0	0.0	2.60	9

5.00 according to Departments of Architecture HoDs. It is notable that 8/9 (88.9%) of the MSs are >3.00, which indicates that in general the inclusion of construction H&S in the tertiary education programmes of the disciplines can be deemed to be more than important, as opposed to less than important. However, a review of the MSs in terms of ranges provides a more detailed perspective. 5/8 (62.5%) MSs >4.20 < 5.00, which indicates that the importance can be deemed to be between more than important to very important/very important: Construction Management; Civil Engineering; Electrical Engineering; Mechanical Engineering, and Project Management. The disciplines ranked sixth to eighth have MSs >2.60 ≤ 3.40, which indicates the importance can be deemed to be between less than important to important/important: Landscape Architecture; Architecture, and Quantity Surveying. It should be noted that the MS of Landscape Architecture, namely 3.40, is at the upper end of the range. The MS of Interior Design (2.60) falls within the range >1.80 ≤ 2.60 and thus the importance can be deemed to be between hardly important to less than important/less than important. It is important that construction H&S is included in the tertiary education programmes of all nine built environment disciplines.

Table 3 Importance of the inclusion of subject areas relative to construction H&S in an architectural programme according to responding Departments of Architecture HoDs

	Response (%)							
		Hardly				Very		
Subject area	Un-sure	1	2	3	4	5	MS	Rank
Occupational health	0.0	0.0	0.0	16.7	50.0	33.3	4.17	1
Occupational safety	0.0	0.0	0.0	16.7	50.0	33.3	4.17	2
OH&S Act & Regulations	0.0	0.0	0.0	33.3	33.3	33.3	4.00	3
H&S plans	0.0	0.0	0.0	33.3	33.3	33.3	4.00	4
H&S specifications	0.0	0.0	0.0	50.0	16.7	33.3	3.83	5
Role of PMs in construction H&S	0.0	0.0	16.7	16.7	33.3	33.3	3.83	6
Role of construction H&S in project performance	0.0	0.0	16.7	0.0	83.3	0.0	3.67	7
Role of designers in construction H&S	0.0	0.0	0.0	66.7	0.0	33.3	3.67	8
Specifying for construction H&S	0.0	0.0	16.7	33.3	33.3	16.7	3.50	9
Need for construction H&S	16.7	0.0	16.7	16.7	50.0	0.0	3.40	10
Designing for construction H&S	0.0	20.0	0.0	20.0	40.0	20.0	3.40	11
Hazard identification and risk assessment	0.0	16.7	16.7	16.7	16.7	33.3	3.33	12
Designing for construction ergonomics	0.0	16.7	16.7	0.0	66.7	0.0	3.17	13
Role of QSs in construction H&S	0.0	0.0	50.0	0.0	50.0	0.0	3.00	14
Detailing for construction H&S	0.0	16.7	16.7	33.3	33.3	0.0	2.83	15
Role of clients in construction H&S	0.0	0.0	50.0	33.3	16.7	0.0	2.67	16
Environment and construction H&S	0.0	16.7	33.3	16.7	33.3	0.0	2.67	17
Economics of construction H&S	0.0	33.3	33.3	0.0	16.7	16.7	2.50	18
Influence of procurement on construction H&S	16.7	33.3	16.7	16.7	16.7	0.0	2.20	19

Table 3 indicates the importance of the inclusion of nineteen subject areas relative to construction H&S in an architectural programme in terms of percentage responses to a scale of 1 (hardly) to 5 (very), and MSs between 1.00 and 5.00 according to responding Departments of Architecture HoDs. It is notable that 13/19 (68.4%) of the MSs are >3.00, which indicates that in general the inclusion of the subject areas relative to construction H&S in an architectural programme can be deemed to be more than important as opposed to less than important. However, a review of the MSs in terms of ranges provides a more detailed perspective. No MSs $>4.20 \leq 5.00$—between more than important to very important/very important. 9/19 (47.4%) of the MSs $>3.40 \leq 4.20$, which indicates that the importance can be deemed to be between important to more than important/more than important. 'Occupational health' and 'Occupational safety', which are ranked first and second, are important as architects need to be knowledgeable with respect to the related issues. Similarly with respect to third ranked 'OH&S Act & Regulations'. Fifth ranked 'H&S specifications' are of particular importance as in terms of the South African Construction Regulations, clients are required to provide designers with an H&S specification, which is based upon their baseline risk assessment and includes their H&S requirements. Designers are then required to provide the client with a report, which although not stated in the regulations should indicate the residual risk. Clients are also required to provide principal contractors the H&S specification, and to respond in the form of an H&S plan, which is ranked fourth. Therefore, designers should also review such H&S plans to determine the PC and contractors' response to design related construction H&S issues. Sixth ranked 'Role of project managers in construction H&S' is important as project managers are invariably the lead consultant on most large projects in South Africa. Role of construction H&S in project performance, ranked seventh, is important relative to all stakeholders as it constitutes the motivation for their contributing thereto. Eighth ranked role of designers in construction H&S and ninth ranked specifying for construction H&S are important as they empower graduates to contribute positively to construction H&S.

MSs $>2.60 \leq 3.40$ indicate that the importance can be deemed to be less than important to important/important, more so those >3.00. Tenth ranked need for construction H&S and eleventh ranked designing for construction H&S both have MSs of 3.40, which are the upper limit of the range. Stakeholders need to understand and appreciate the need for construction H&S in terms of the motivation for their contributing thereto.

Research Results 2

Table 4 indicates the degree of support for the inclusion of aspects in six tertiary-built environment programmes' construction H&S modules in terms of percentage responses per aspect per discipline, the mean percentage for all disciplines per aspect, and the mean percentage for all aspects per discipline. The total possible responses per discipline (No.) are recorded in the last row, i.e. the number of respondents per discipline.

Table 4 Degree of support for the inclusion of aspects in tertiary built environment programmes' construction H&S modules

Aspect	Response (%)							Rank
	Architecture	Civil engineering	Construction management	Project management	Property development	Quantity surveying	Aspect mean	
Need for construction H&S	100.0	96.2	100.0	100.0	100.0	100.0	99.4	1
Construction Regulations	96.6	100.0	100.0	100.0	100.0	96.0	98.8	2
Role of clients in construction H&S	96.6	92.3	100.0	96.7	100.0	88.0	96.9	3=
Role of construction H&S in project performance	96.6	92.3	100.0	96.7	100.0	92.0	96.9	3=
H&S specifications	96.6	96.2	100.0	96.7	95.5	96.0	96.8	5
Environment and construction H&S	96.6	100.0	100.0	90.0	95.5	92.0	96.4	6
Role of designers in construction H&S	96.6	96.2	96.6	96.7	95.5	88.0	96.3	7=
Economics of construction H&S	93.1	96.2	100.0	96.7	95.5	96.0	96.3	7=
Hazard identification and risk assessment	96.6	96.2	100.0	96.7	90.9	96.0	96.1	9=
Occupational safety	96.6	96.2	100.0	96.7	90.9	92.0	96.1	9=
OH&S Act	93.1	96.2	96.6	96.7	95.5	92.0	95.7	11
Role of project managers in construction H&S	93.1	92.3	100.0	96.7	95.5	92.0	95.6	12
Occupational health	96.6	92.3	100.0	96.7	90.9	92.0	95.4	13
Designer report (to clients)	100.0	96.2	89.7	93.3	95.5	92.0	95.1	14
H&S plans	93.1	92.3	100.0	93.3	95.5	84.0	95.0	15
Role of manufacturers in construction H&S	93.1	92.3	100.0	96.7	90.9	88.0	94.8	16
Influence of procurement on construction H&S	89.7	92.3	96.6	96.7	95.5	88.0	94.5	17

Aspect	Response (%)							Rank
	Architecture	Civil engineering	Construction management	Project management	Property development	Quantity surveying	Aspect mean	
Ergonomics	93.1	92.3	100.0	90.0	90.9	80.0	93.7	18
H&S file	82.8	88.5	96.5	93.3	90.9	76.0	91.4	19
Role of quantity surveyors in construction H&S	72.4	88.5	96.5	86.7	90.9	100.0	88.5	20
Contractor H&S	27.6	84.6	100.0	96.7	27.3	36.0	72.0	21
Managing subcontractor H&S	24.1	80.8	100.0	96.7	27.3	36.0	70.8	22
Other regulations	41.4	69.2	79.3	73.3	27.3	20.0	64.4	23
COID Act	27.6	84.6	93.1	36.7	27.3	24.0	60.9	24
Primary health promotion	17.2	73.1	89.7	40.0	13.6	16.0	54.9	25
Discipline Mean	80.4	91.1	97.4	90.0	80.7	78.1		
Total possible responses per discipline (no.)	29	26	29	30	22	25		

It is notable that in terms of the mean percentage for all disciplines per aspect, 19/25 (76.0%) aspects achieved percentages >90% ≤ 100%, 1/25 (4.0%) aspects achieved percentages >80% ≤ 90%, 2/25 (8.0%) aspects achieved percentages >70% ≤ 80%, 2/25 (8.0%) aspects achieved percentages >60% ≤ 70%, and 1/25 (4.0%) aspects achieved a percentage >50% ≤ 60%. The >90% ≤ 100% range includes 'Need for construction H&S', 'Construction Regulations', 'Role of clients in construction H&S', 'Role of construction H&S in project performance', 'H&S specifications', 'Environment and construction H&S', 'Role of designers in construction H&S', 'Economics of construction H&S', 'Hazard identification and risk assessment', 'Occupational safety', 'OH&S Act', 'Role of project managers in construction H&S', 'Occupational health', 'Designer report (to clients)', 'H&S plans', 'Role of manufacturers in construction H&S', 'Influence of procurement on construction H&S', 'Ergonomics', and 'H&S file'.

Then, in terms of the mean percentage for all aspects per discipline:

- 2/6 (33.3%) disciplines achieved percentages >90% ≤ 100%—Civil Engineering and Construction Management.
- 3/6 (50.0%) disciplines achieved percentages >80% ≤ 90.0%—Architecture, Project Management, and Property Development.
- 1/6 (16.7%) disciplines achieved a percentage >70% ≤ 80%—Quantity Surveying.

In terms of the discipline mean for all 25 aspects, Architecture (80.4%) is ranked fifth after Construction Management (97.4%), Civil Engineering (91.1%), and Project Management (90.0%), and Property development (80.7%). Quantity Surveying (78.1%) was ranked sixth after Architecture.

Discussion

The South African H&S legislative framework is quite clear with respect to the need to address/include construction H&S in tertiary built environment programmes, albeit implicit. The aforementioned is underscored by international research findings, and the implications of such legislation and required practices.

The limited addressing/inclusion of construction H&S in South African tertiary architectural programmes corroborates with the findings of international research.

Study 1's findings are courtesy of architectural academics, as opposed to Study 2's findings, which are courtesy of built environment practitioners. In this context there is a higher level of support for the inclusion of subject areas courtesy of the built environment practitioners, vis-à-vis the architectural academics. Limited international research has been conducted relative to this aspect.

Conclusions

Given the degree of support for the addressing of 19 aspects (Study 1), and 25 aspects (Study 2) in tertiary architectural built environment programmes relative to construction H&S modules, it can be concluded that H&S must be embedded in such programmes, and such embedment must be confirmed. If the aforementioned is not expedited, then tertiary institutions will be marginalising their students and graduates to be.

The limited addressing/inclusion of construction H&S in South African tertiary architectural programmes is likely to be attributable to a lack of knowledge relative to the subject area, and professional associations' councils and the SACAP council not focusing thereon during accreditation visits. The aforementioned is underscored by the limited or non-inclusion of H&S interventions in SACAP's IoW.

Recommendations

Tertiary education architectural programmes should include a module 'designing for construction H&S' as a component of a subject—probably design. The minimum content of such a module would include: H&S legislation and regulations; the role of the various project stakeholders in construction H&S; actions/activities/aspects that impact on construction H&S; the design related aspects that impact on construction H&S, and 'designing for construction H&S' aspects such as design HIRAs, 'design and construction' method statements, the designer report, the H&S specification, and H&S plans. Underpinning knowledge such as the nature of materials and the construction process and its activities is essential. Then, construction H&S should be included among criteria used for evaluating design projects, working drawings, and details. The subject 'History of Architecture' should also address construction H&S. Furthermore, SACAP accreditation reviews of tertiary education landscape architectural programmes should interrogate the extent to which construction H&S is addressed.

Construction H&S should be included in CPD for architects. The South African Institute of Architects (SAIA) and SACAP should develop guidelines/practice notes relative to construction H&S, and SACAP should include construction H&S in their six work stages as documented in their IoW.

References

Behm, M. (2006). *An analysis of construction accidents from a design perspective*. Silver Spring, MD: The Center to Protect Workers' Rights.

Construction Industry Development Board (CIDB). (2009). *Construction health & safety in South Africa status & recommendations*. Pretoria: CIDB.

Hecker, S. F., Gambatese, J. A., & Weinstein, M. (2006). Designing for construction safety in the US: Progress, needs, and future directions. In *Proceedings of the 16th Triennial Congress of the International Ergonomics Association, Maastricht, The Netherlands, July 10–14, 2006*. D:\data\pdfs\art0563.pdf.
International Labour Office (ILO). (1992). *Safety and health in construction*. Geneva: ILO.
Republic of South Africa. (1993). *Government Gazette No. 14918. Occupational Health & Safety Act: No. 85 of 1993, Pretoria.*
Republic of South Africa. (2011). *Government Gazette No. 34645. Architectural Profession Act (44/2000): South African Council for the Architectural Profession's (SACAP's) Interim Policy on the Identification of Work for the Architectural Profession, Pretoria.*
Republic of South Africa. (2014). *No. R. 84 Occupational Health and Safety Act, 1993 Construction Regulations 2014. Government Gazette No. 37305, Pretoria.*
Smallwood, J. J. (2002). The need for the inclusion of construction health and safety (H&S) in architectural education. In S. Rowlinson (Ed.), *Proceedings of the 3rd International Conference of CIB Working Commission W99 Implementation of Safety and Health on Construction Sites, Hong Kong, May 7–10* (pp. 207–212). Hong Kong: The University of Hong Kong.
Smallwood, J. J. (2008). The influence of architectural designers on construction ergonomics. *Ergonomics SA, 20*(1), 40–55.
Thorpe, B. (2006). *Health and safety in construction design*. Aldershot: Gower Publishing Limited.

Who Are the 'Middle Actors' in Sustainable Construction and What Do They Need to Know?

Alice Owen, Kathryn B. Janda, and Kate Simpson

Introduction

The challenge of transforming our built environment, and how we create buildings, to be fit for a sustainable future remains significant. There is a relentless drive, supported by policy and regulation, for more sustainable buildings that use less energy, generate less waste during construction and use, and provide healthy environments for people to live and work in. The pressures are acute for both homes and non-domestic buildings, and for both new build and the existing building stock. New build processes are being driven to change (in the UK) through a 'Transforming Construction' priority in the UK's Industrial Strategy which seeks to improve the quality and consistency of buildings, while also reducing costs and build time, through huge deployment of off-site and module construction. These forms of construction still require knowledge and skills to support sustainable outcomes. However the distribution of *who* has the knowledge and skills and *when* they are deployed must change from the traditional forms of knowledge within individual professional domains developed through vocational training and onsite experience.

Transforming existing buildings, with their histories and multiple functions in terms of shelter and local character, poses a different set of challenges compared to creating new sustainable buildings, a set of challenges that the UK is currently failing to address (CCC, 2019). There are a range of design and technical advances that could be used in existing buildings, from novel space heating, to radical energy

A. Owen (✉)
Sustainability Research Institute School, School of Earth and Environment, University of Leeds, Leeds, UK
e-mail: a.m.owen@leeds.ac.uk

K. B. Janda
Energy Institute, University College London, London, UK

K. Simpson
School of Design Engineering, Imperial College London, London, UK

L. Scott et al. (eds.), *Sustainable Ecological Engineering Design*,
https://doi.org/10.1007/978-3-030-44381-8_15

conservation or whole house 'Energiesprong' approaches. However, implementing these innovations relies on the variety of people involved in working on existing buildings developing and using new skills, working across traditional trade boundaries and reconfiguring construction work processes. We refer here to the 'variety of people working on existing buildings' rather than solely 'the construction industry' because the repair, maintenance and improvement, including renovation, of buildings, particularly homes, is not an entirely professional project. Work on homes can be carried out by home owners with a series of projects over time (Fawcett, 2013) and there is a large element of the homeowners themselves driving towards more or less sustainable construction outcomes (Fawcett & Killip, 2014) as well as a significant DIY element in the delivery of projects. In some cases householders have also equipped themselves to guide builders on technical details for low-energy retrofit (Simpson, 2017; Thomsen & Hauge, 2014). However, to provide some focus to this paper we map only the construction professionals involved in work on existing buildings.

The required direction of travel to more sustainable construction and building performance can no longer be considered 'new' and yet it remains far from the mainstream. To create buildings which are sustainable, we need to consider not only technical and design changes, but how to alter the wider system of construction (Thunshelle et al. 2018). This means we have to understand the relationships that shape construction outcomes, understanding construction as a set of co-evolving actors (Killip, Owen, Morgan, & Topouzi, 2018), operating in networks (Owen, 2015) and as part of a complex supply network, rather than a linear supply chain (Killip, Owen, & Topouzi, 2020).

Questions which guided our data collection and analysis for this paper are:

1. Who shapes construction activity towards more or less sustainable outcomes?
2. What kinds of skills and knowledge do different team members say they need, and are there any additional skills/knowledge gaps not explicitly recognised by those individuals?
3. When in the project cycle do the skills and knowledge that lead to more sustainable outcomes need to be deployed, and by whom?

   These questions are answered in the 'emergent findings' section below, while a fourth question guides the 'discussion' section:
4. Do the answers to questions 1, 2 and 3 help to identify preferred methods and routes for professional development?

To provide some focus in differentiating between individuals and roles in different teams, we use the concept of 'middle actors'.

Who Are the 'Middle Actors' in Construction, and Why Do They Matter?

Social and technological innovations are commonly seen as either being induced from the 'top-down'—e.g., by policymakers—or evolving from the 'bottom-up'—e.g. by citizens. A 'middle-out' perspective focuses instead on agents of change that can promote transition and which are located in the middle, in between the state and its citizens. The Middle-Out Perspective (MOP) developed by Janda and Parag (2013) and Parag and Janda (2014) shows that middle actors can affect change in several different directions: upstream to policymakers, downstream to clients or members and sideways to other middle actors (often by enabling and sharing new professional norms and working practices). By linking the top and bottom, the MOP is both an alternative and complementary to existing 'bottom-up' and 'top-down' efforts to implementing low-carbon innovations and practices in society.

Middle actors are distinct from intermediaries who have been researched in the construction industry (Kivimaa & Martiskainen, 2018) because middle actors have agency, they are not conduits, more or less effective in the transfer of information; middle actors actively shape the decisions and actions taken by others through their own information filtering and interpretation. Middle actors in a construction project are diverse. Between the top-down policy and regulation, including carbon and building regulation, and the bottom-up builder user demands, a range of individuals carrying out diverse roles coax a construction project into fruition.

The MOP is a relatively new and evolving approach to understanding socio-technical energy transitions. It aims to find new ways of approaching existing or wicked problems (Rittel & Webber, 1973). As such, the MOP has attracted a diverse range of research analyses. The initial work of Parag and Janda described how middle actors include (but are not limited to) such groups as building professionals, religious congregations, and commercial property owners (Janda & Parag, 2013; Parag & Janda, 2014). In recent years, these and other authors have further applied and developed the MOP to address a number of other middle actor groups, including providers of housing refurbishment (Janda, Killip, & Fawcett, 2014), heating engineers (Wade, Hitchings, & Shipworth, 2016), community-based organizations (Hamilton, Mayne, Parag, & Bergman, 2014), facilities managers (Goulden & Spence, 2015), social housing providers (Cauvain & Karvonen, 2018), and actors involved in energy storage (Devine-Wright et al., 2017).

Previous research shines a light on how individual professional groups interrelate and how they deploy their skills and knowledge to shape the outcomes of a construction project. Architects seek to interpret standards in a way that allows them to meet client needs and their own professional interests (Fischer & Guy, 2009). Heating engineers use information from others in their professional networks to develop their advice to customers (Wade et al., 2016). Both these groups define areas of legitimacy for their advice, or areas of 'professional jurisdiction' (Abbott, 1988; Wade, Murtagh, & Hitchings, 2018). More broadly, professional groups also work in together in what Abbott (1988) calls a 'system of professions' where different groups claim ownership of particular aspects of socially accepted problems. These then become their 'professional jurisdiction' which they need to defend in the public eye to keep their mastery of the niche, even while they share the work terrain with others. In the construction industry, for example, Janda (1998, 1999) has used Abbott's framework to examine how architects, engineers, and energy efficiency advocates pursue their own professional objectives whilst working on the same project; how a new profession might be needed to enhance residential retrofits (Janda & Killip, 2013); and how architects, engineers and builders excel in different ways in high profile buildings (Janda, 2017).

What Do Middle Actors Need to Know to Deliver Sustainable Construction, and How Do They Learn?

In the UK, the Construction Leadership Council produced a Sustainable Training Guide (2017), which includes recommended learning outcomes for sustainable building, aimed at construction trades, building services engineering trades, managers, supervisors and designers. The construction trades section is divided into: Low energy/low carbon building, Sustainable products, Waste, reuse and recycling, Water and whole build processes. The low energy section (CLC, 2017) includes principles of building performance and associated costs, principles of air-tightness, effective insulation, ventilation and air quality and overheating. Similarly, the Builders Books (ZCH, 2015), the Services Guide (ZCH, 2016a) and the Thermal Bridging Guide (ZCH, 2016b), all developed by the Zero Carbon Hub (ZCH), explain principles such as thermal bridging with illustrated examples of technical solutions for building site use. This work followed the ZCH's work identifying factors contributing to the building performance gap between designed as as-built energy use, which installation detailing was one contributing factor (ZCH, 2014). Previous work with construction teams in Leeds, UK, identified that delivering sustainable construction relies on knowledge of technical detailing for building performance, lean principles of minimising waste and social relations and value involved in construction processes and products (Simpson, Owen, & Chatterton, 2018).

The present UK vocational education training system pursues a top-down management approach, characterised by a task-based system and a lack of self-management for trainees (Clarke, Gleeson, & Winch, 2017). This has major consequences for enabling middle actor agency. In the UK, 73% of construction companies have been found to have no training plan, 81% no training budget and only 19% investing in training (BIS, 2013). Previous studies from across Europe confirmed a lack of energy literacy, plus an increasing demand for multi-skilled actors within areas critical to achieving energy efficiency (ZCH, 2014) but current vocational training is inadequate to this task. For example, for National Vocational Qualifications (NVQ) 2 and 3 in bricklaying, trainees learn about the installation of cavity wall insulation with particular care for junction detailing but there is no mention of air tightness or thermal bridges (Clarke et al., 2017). Similar challenges exist in plastering NVQ Level 2. For groundwork, window installation, loft insulation and labouring no formal vocational training exists. BTEC Pearson released a new specification for courses such as the Level 4 Construction and the Built Environment in 2017 (BTEC, 2017), which does encompass sustainability within units such as Construction Technology and Construction Information, but if students are to gain a full understanding of the issues, their tutors and their sources of information will need to take this integrated view.

Approach

The empirical data used in this paper's exploration of middle actors' skills and knowledge needs were gathered during two projects in Leeds, UK. To allow a flexible and exploratory research approach, research questions were not set at the outset of these projects and we are not presenting the results of a formally designed research inquiry.

The first project acted as a catalyst, gathering interest from local construction stakeholders, including middle actors, with expertise in sustainable building, city wide planning and development or training and skills, from research, public and private sector organisations. The aim was to co-identify the sustainable construction skills required for both retrofit and new-build. The skills needed were identified through a series of telephone conversations with stakeholders ahead of a workshop which gathered insights responding to three questions: what are the benefits of sustainable construction, what are the challenges, and what are the solutions?

The second project further engaged some of those attendees, particularly a local developer of two sites in Leeds and the specialist construction College, also in Leeds. This engagement took place around the completed new office build on one site, which was used as an exploratory case study, and alongside a second office which was at design stage at a second site. Both offices aimed to exceed minimum standards on energy efficiency and the second aimed to become an exemplar of Design for Performance approach for new office buildings, adopted from the Australian NABERS programme. The teams engaged through these developments were initially suggested by the developer with some further suggestions made during each meeting, creating a snowball sample.

The developer has a stated interest in sustainability, which encompasses site accessibility, building use and community support, environmental impact, as well as climate change mitigation and adaptation. The city's further education College specialising in construction training is working in partnership with apprenticeship employers on one of the sites. The initial provision is for 12 Construction Contracting Operation apprentices who are also completing a Level 3 BTEC course in Construction and the Built Environment.

During the second project, informal meetings were held to explore perspectives of the network of middle actors involved in the office developments. To fit the short exploratory project the meetings were unstructured and informal. At the start of each meeting the project was explained, ensuring openness and transparency whilst allowing meeting attendees the opportunity to ask questions or choose to opt out. The meetings focused on sustainable skills required in the sector and gaps the teams had noticed for their peers or others involved in the design and build process. The discussions included aspects of training and development the teams felt were useful and effective, or otherwise.

As the first case study office build was recently completed and the second was at design stage, this allowed teams to reflect on recent or current experiences. The meetings were held within the café situated on the ground floor of the first completed office build providing further authenticity to discussions and allowing particular junction or detail challenges to be pointed out in the structure. This enabled construction practice and site realities to move our previous research beyond theorising and provide reflections with empirical grounding. There is a growing literature using ethnographic methods to understand construction practices (Pink, Tutt, & Dainty, 2013) and while not proceeding as far as autoethnography (Grosse, 2019), this paper recognises the positionality of the researchers as influencers within the construction project system and offers insights from reflective practice within the system, synthesising new primary data with previous experience in order to answer a series of linked questions. In addition, researchers visited training sessions ran by the College, taking place in Portakabins on the site. This allowed conversations with the trainers, apprenticeship providers and the 12 trainees as a group.

Emergent Findings

Middle actors were found to operate as a network. An ecosystem analogy might be useful here as different actors occupied different niches defined by their skills, knowledge and practical influence (Simpson & Owen, 2019). Middle actors identified in these projects were:

- Designers—both architects and structural engineers, who see their purpose as delivering on the architect's vision.
- Designers at the level of delivery—particularly M&E designers who seek to deliver on the architect's vision but also meet other parameters in terms of standards and practice.
- Project Manager—working on behalf of the client or lead contractor and tying together many threads.
- Clerk of Works—also working on behalf of the client, the Clerk of Works lies at the heart of an efficient and effective project once on site. Usually using a background in quantity surveying, the Clerk of Works monitors whether the design drawings and specifications are being implemented correctly. Clerks of Works became less common as a role, often replaced by trade foremen, as cost pressures increased in the construction sector. However, the advent of Building information

Management (BIM) has led to the role re-emerging, albeit with a slightly different profile.
- Tradespeople—construction workers including general builders and specific trades such as electrician, roofer, plasterer, joiner, heating engineer, and glazier.
- Technical advisers—for mechanical and electrical (M&E) services, and for innovative sustainability approaches in building performance, particularly when a building requires accreditation, e.g. BREEAM or WELL or EnergyStar. Advice tends to be framed by the standard that has caught the attention of the client or designer, or has been previously trialled by the advisers. Technical advisers typically work to advise designers and the project manager; they have limited connections to tradespeople or onsite work, but were found to directly liaise with the clerk of works, often on a daily basis during the construction phase. Where the clerk of works needed to clarify a M&E or building performance detail they would photograph the issue and send it to the consultants for advice on an informal basis.

We now turn to the question 'How does middle actors' influence vary?' looking at interest and influence by topic, as a bridge to then considering when they need knowledge.

- Waste management and resource efficiency—driven by regulation and the need for a site waste management plan, but also a visible source of cost and improving resource efficiency in construction and use contributes effectively to a business case for change based on cost and payback. However, reducing waste at source requires thought and care from both designers and the trades who implement design.
- Low energy in use—often considered by designers, in response to regulatory pressures such as compliance with Building Regulations, but realising design intent is dependent on the skills and knowledge of the tradespeople who carry out work. Two areas of skills and knowledge were identified by our project partners within this sphere. First, a range of people across the project team needed to know about the technologies that might enable lower energy use, and second, construction workers, and those evaluating and signing off construction work, needed to understand how individual tasks fitted together into the whole picture of building performance. There was a particular emphasis on the need to understand detailing and the reasons for paying attention to interfaces between materials and different trades' work.
- Low embodied carbon may be considered as a useful benefit by designers, once other performance criteria such as strength and weight have been achieved (Giesekam, Barrett, & Taylor, 2016) but embodied carbon impacts are near invisible to most other construction middle actors.
- Resilience and adaptability—while these terms are most often used in connection with the need to be resilient to the impacts of climate change, particularly flood impacts in the geography of our case study projects, they might also refer to the potential to change building function over time, e.g. allowing flexible

partitioning of space to accommodate changing occupants, or mixing live/work space.
- Health and well-being—health and safety ats work is, with good reason, a high priority for construction sites and all those who work on them. Health and safety in site construction practices is regulated, and carefully monitored with performance scrutinised. However, clients and architects now also consider how the building will contribute to the health and well-being of its users, although there are few explicit connections between this consideration and the onsite practices that deliver the designs.
- Social value—a requirement from all public sector procured construction projects in the UK, social value refers to impacts generated by the project such as employment, skill development and community building. Social value was identified by a range of stakeholders in these construction projects (Simpson et al., 2018) but since it arises both from the building design, and from the project management approach and decisions made about procurement and recruitment, it is the most difficult to grasp of the skills areas identified.

Table 1 summarises what skills and knowledge these actors told us that they needed, i.e. their conscious knowledge needs. The suggestions of when those needs are most acute have been organised by the researchers into stages of the project cycle as defined in the RIBA Workplan (RIBA, 2013).

Discussion

Our mapping shows that the variety of middle actor roles leads to a diverse set of skills and knowledge needs. Talking to construction teams reinforced that both knowing *what* to do, and also knowing *how* to do things are important. In addition, there are a number of softer, process skills that also need to be developed. These blur the boundaries between roles and expertise, requiring construction project team members to be aware of the implications of their decisions on others' work, and also on the eventual building created.

Diverse training needs can only be met through diverse training methods. The UK construction industry's Supply Chain Sustainability School is an example of allowing construction firm employees and project workers to select e-learning modules at a time and location to suit their circumstances. While the School has been able to track activity, i.e. number of e-learning modules completed, it cannot easily track the impact of those modules and this approach, in isolation, leads to a compliance-based approach to knowledge. This is familiar to construction workers (and others) from health and safety knowledge and certification systems. Such an approach may be appropriate for ensuring compliance with, for example, waste management standards. The other approach observed in Leeds is onsite training for apprentices (Simpson & Owen, 2019), which allows trainees to study elements of construction as it takes place in practice. Where contractor teams are

Table 1 Middle actors and their skills/knowledge needs for sustainable construction

Middle actor	Sustainable construction skills/knowledge needs	When in the project cycle is that need manifest? RIBA workplan stages	Preferred learning routes
Designers—architects/ structural engineers	Impact of design decisions—during construction and during building use Efficient building form Material selection Junction detailing Design for re-use Location of plant rooms Building modelling software and links to accurate input variables	Stage 0—Strategic Definition Stage 1—Preparation and Brief Stage 2—Concept Design Stage 3—Developed Design Stage 4—Technical Design Stage 7—post occupancy evaluation	Formal training, CPD accredited Mentoring/sponsorship of innovation and practical learning, informal through project discussions
Designers—M&E	M&E Systems and links to accurate input variables Building modelling software Building Management Systems Building physics and how that translates into materials, fabric and configuration of Communication with on-site project team How users interact with buildings	Stage 3—Developed Design Stage 4—Technical Design Stage 6—Handover and close out Stage 7—post occupancy evaluation	Formal training, CPD accredited, informal through project discussions
Clerks of works	Technical details to be installed Knowledge of all construction contract operations Mechanical and Electrical knowledge (or specialist employed) Relationship between design and sustainable outcomes	Stage 4—Technical Design Stage 5—Construction Stage 6—Handover and Close Out	Formal, on the job—or sponsored by client, informal through project discussions

(continued)

Table 1 (continued)

Middle actor	Sustainable construction skills/knowledge needs	When in the project cycle is that need manifest? RIBA workplan stages	Preferred learning routes
Project Managers	Interaction of decisions and how they affect project outcomes in terms of a range of measures: cost, waste, energy, etc.	Stage 3—Developed Design Stage 4—Technical Design Stage 5—Construction Stage 6—Handover and Close Out	On the job—or sponsored by client. Must be integrated with project manager role and skills, i.e. not a standalone subject, informal through project discussions
Tradespeople	Know what (to do) Know how (to do something) Understanding how their work interacts with others—particularly around detailing and interfaces.	Stage 5—Construction Stage 6—Handover and Close Out	On the job. Training delivered by individuals with experience and credibility Accreditation only beneficial if supported, required and credible, informal through project discussions
Technical advisers—BREEAM	Technical knowledge related to standards and measures Building modelling software Also require process skills to communicate, persuade, encourage, problem solve etc.	Stage 2—Concept Design Stage 3—Developed Design Stage 4—Technical Design Stage 5—Construction Stage 6—Handover and Close Out Stage 7—Post occupancy evaluation	Formal and accredited (to lend value to the work) for technical knowledge. Informal learning for process skills

communicating well with the trainers such training can provide a route to communicating technical details clearly, while ensuring an understanding of the whole building system and also inspiring apprentices. This could be particularly useful for communicating thermal and air-tightness factors, building services systems and minimising waste through on-site operations, among other areas where flexible, responsive deployment of knowledge is needed.

Only the technical advisors on the projects in Leeds recognised that the knowledge gained through previous formal education needed to be continually updated in light of changing information and project needs. They used literature, professional fora, webinars and other means as a way of developing competence in their selected standard. The technical consultants considered it vital that those involved in design have a comprehensive knowledge needs of carbon or energy literacy. However, in translating this knowledge into building performance, other knowledge needs were identified related to the processes which led to sustainable outcomes, e.g. the value and rationale for quality and detailing and, interestingly, how construction might contribute to sustainable outcomes beyond the immediate physical impact of the project, such as generating social value (Simpson et al., 2018).

Working with a specific certification in mind and potentially becoming a benchmark 'base build' was a motivational factor for the developer. However, such standards also make new demands of the knowledge and skills in the construction teams delivering the project. For a small number of middle actors, sustainable construction outcomes is a primary concern. This can be driven by a cocktail of motivations from personal 'green' beliefs and values to commercial value generation or exploiting a niche market opportunity (Killip et al., 2020). For many others, sustainable outcomes are not their main focus. However, there may be other motivating factors which lead to more sustainable outcomes. For example, a focus on accuracy and detailing is important in delivering in practice the intentions and potential of a design. Similarly, sharp attention to avoiding waste and managing effectively within resource constraints may be driven by financial interests but has the effect of reducing waste.

Innovation in methods such as offsite and module construction may be shifting the locus of where skills are deployed, but there are still needs for specialist knowledge. With one construction firm who have moved their entire operation to a purpose-built 'lean' manufacturing unit, they had broken down boundaries between trades and required each member of a multi-trade team to have detailed knowledge of each other's work and skills both so that they could step in and support each other during the whole construction process, but also so that they understood the implications of their decisions and actions on others' areas of responsibility.

Current training focuses narrowly onto design and technical inputs, responding effectively to drivers for accuracy and resource/finance constraint. However, this means that training can easily miss whole building issues such as detailing and quality, and project process issues such as the interaction between different trade activities.

Conclusion

Using the lens of middle actors reveals roles in a construction project beyond the traditional roles of project manager, surveyor and trade. There are many middles in a construction project, as Reindl (2017) found in her study of the retrofit system within Swedish housing companies. Once we have recognised middle actors, and their importance, we must also avoid assuming they are a homogenous group. Beyond the professional roles described here, there are a range of other informal influential roles played by owners and their trusted networks, who will bring their own skills, knowledge and motivation into the conversations about what works needs to be done and how it should be done. Middle actor activities, influencing more or less sustainable construction activities, can arise from being part of a formal role, such as the focus on detailing that a Clerk of Works might provide, or from a personal motivation, such as a technical adviser's commitment to eliminating the performance gap. Either way, seeing these various actors through the lens of the 'middle-out perspective' emphasises that they are not automatons simply performing services demanding of them by others. Instead, they have both agency (the ability to make decisions) and capacity (the ability to enact them). Moreover, they can influence other actors around them, either *up* the supply chain or *down* to their clients, or *sideways* to other actors on the construction team.

The mixture of skills needed for sustainable construction—hard and soft, technical and process—leads to the unsurprising conclusion that training and development routes for sustainable construction need to be more diverse, offering options other than formal college-based learning. Allowing construction team members to develop personalised curricula or learning pathways, which respond to their motivations and project-specific needs, could work well, and different modes of learning such as peer learning, supply chain training, and experiential, site-based learning would all have a place in this kind of personalised approach. However, at present only those roles which have an expectation of continuing professional development in support of certification, or chartered status, are likely to take up such opportunities.

Acknowledgements This project has been sponsored by the Leeds Social Sciences Impact Acceleration Account in association with the ESRC and the University of Leeds EPSRC Impact Acceleration Account. The authors are grateful to CEG and the Leeds College of Building as project partners and to Phil Dyke at CEG and Brian Duffy at LCoB in particular. K. Janda is grateful to the UK Energy Research Centre for initial work on middle actors and the Centre for Energy Epidemiology at University College London for recent developments.

References

Abbott, A. (1988). *The system of professions*. Chicago, IL: University of Chicago Press.

BIS. (2013). *UK construction, an economic analysis for the sector*. London: Department for Business, Innovation and Skills.

BTEC. (2017) *BTEC Higher Nationals Construction and the Built Environment (2017)—RQF Programme Specification Issue 6*. Retrieved April 11, 2019 from https://qualifications.pearson.com/en/qualifications/btec-higher-nationals/construction-and-the-built-environment-2017.html

Cauvain, J., & Karvonen, A. (2018). Social housing providers as unlikely low-carbon innovators. *Energy and Buildings, 177*, 394–401.

Clarke, L., Gleeson, C., & Winch, C. (2017). What kind of expertise is needed for low energy construction? *Construction Management and Economics, 35*(3), 78–89.

CLC (2017). *Construction Leadership Council's Sustainable Training Guide, London*. Available at: https://www.constructionleadershipcouncil.co.uk/wp-content/uploads/2017/05/Sustainable-Building-Training-Guide.pdf.

Committee on Climate Change. (2019). *UK housing: Fit for the future?* London.

Devine-Wright, P., Batel, S., Aas, O., Sovacool, B., Labelle, M. C., & Ruud, A. (2017). A conceptual framework for understanding the social acceptance of energy infrastructure: Insights from energy storage. *Energy Policy, 107*, 27–31.

Fawcett, T. (2013). Exploring the time dimension of low carbon retrofit: Owner-occupied housing. *Building Research and Information, 42*(4), 477–488.

Fawcett, T., & Killip, G. (2014). Anatomy of low carbon retrofits: Evidence from owner-occupied Superhomes. *Building Research and Information, 42*(4), 434–445.

Fischer, J., & Guy, S. (2009). Re-interpreting regulations: Architects as intermediaries for low-carbon buildings. *Urban Studies, 46*, 2577–2594.

Giesekam, J., Barrett, J. R., & Taylor, P. (2016). Construction sector views on low carbon building materials. *Building Research and Information, 44*(4), 423–444.

Goulden, M., & Spence, A. (2015). Caught in the middle: The role of the Facilities Manager in organisational energy use. *Energy Policy, 85*, 280–287.

Grosse, H. (2019). An insider's point of view: Autoethnography in the construction industry. *Construction Management and Economics., 37*, 481–498.

Hamilton, J., Mayne, R., Parag, Y., & Bergman, N. (2014). Scaling up local carbon action: The role of partnerships, networks and policy. *Carbon Management, 5*(4), 463–476.

Janda, K. B. (1998). *Building change: Effects of professional culture and organizational context on energy efficiency adoption in buildings*. PhD dissertation, Energy and Resources Group, University of California at Berkeley, Berkeley.

Janda, K. B. (1999). Re-inscribing design work: Architects, engineers, and efficiency advocates. In *Proceedings of ECEEE Summer Study, June 1–6, 2009, Mandelieu, France* (Vol. 2, pp. 3.11.1–3.11.10). Stockholm: European Council for an Energy-Efficient Economy.

Janda, K. B. (2017). Rafting sustainability in iconic skyscrapers: A system of building professions in transition? *Nordic Journal of Science and Technology Studies, 5*(2), 44–58.

Janda, K. B., & Killip, G. (2013). Building expertise: Renovation as professional innovation. In A. J. Hoffman & R. Henn (Eds.), *Constructing green: The social structures of sustainability* (pp. 35–55). Boston, MA: MIT Press.

Janda, K. B., Killip, G., & Fawcett, T. (2014). Reducing carbon from the 'middle-out': The role of builders in domestic refurbishment. *Buildings, 4*(4), 911–936.

Janda, K. B., & Parag, Y. (2013). A middle-out approach for improving energy performance in buildings. *Building Research and Information, 41*(1), 39–50.

Killip, G., Owen, A., Morgan, E., & Topouzi, M. (2018). A co-evolutionary approach to understanding construction industry innovation in renovation practices for low-carbon outcomes. *The International Journal of Entrepreneurship and Innovation, 19*, 9–20.

Killip, G., Owen, A., & Topouzi, M. (2020). Exploring the practices of UK construction manufacturers and merchants in relation to housing energy retrofit. *Journal of Cleaner Production. 251*, April 2020, https://doi.org/10.1016/j.jclepro.2019.119205

Kivimaa, P., & Martiskainen, M. (2018). Innovation, low energy buildings and intermediaries in Europe: Systematic case study review. *Energy Efficiency, 11*, 31–51.

Owen, A. (2015). Missing the point—The challenge of creating policies and programmes that tap into the motivations of the builders and installers. In *eceee 2015 Summer Study proceedings: First Fuel Now, June 1–6, 2015, Presqu'ile St Giens, France*.

Parag, Y., & Janda, K. B. (2014). More than filler: Middle actors and socio-technical change in the energy system from the "middle-out". *Energy Research and Social Science, 3*, 102–112.

Pink, S., Tutt, D., & Dainty, A. (2013). *Ethnographic research in the construction industry*. London: Routledge.

Reindl, K. (2017). *Implementing energy measures in renovations for multi-family dwellings: Influence and practice of professionals*. Doctoral thesis, Linköping Studies in Arts and Science, Linköping University Electronic Press, Linköping.

RIBA. (2013). *Plan of work*. Available at https://www.ribaplanofwork.com/

Rittel, H. W. J., & Webber, M. M. (1973). Dilemmas in a general theory of planning. *Policy Sciences, 4*, 155–169.

Simpson, K. (2017). *Energy efficiency refurbishment of UK homes: The householders' perspective*. PhD thesis, Loughborough University.

Simpson, K., & Owen, A. (2019). Sustainable construction: Required skills and processes identified by networked industry actors in Leeds, UK. In *ARCOM Conference, Leeds, September 2019*

Simpson, K., Owen, A., & Chatterton, P. (2018). *Equipping construction workers with sustainable building skills: A focus on Leeds*. SRI Briefing Note No. 15.

Thomsen, J., & Hauge, Å. L. (2014). Barriers and drivers for energy efficient upgrade of single-family housing in Norway. In *Sustainable Building Conference, Barcelona, October 28–30*.

Thunshelle, K., Denizou, K., Hauge, A., & Thomsen, J. (2018). *Robust and reliable technology concepts and business models for triggering deep renovation and residential buildings in EU (4RinEU): Protocol for participative deep renovation design and user motivation*. Retrieved April 10, 2019 from http://4rineu.eu/wp-content/uploads/2019/04/4RinEU_WP3_20181031_D3.5_Protocol-for-partecip-deep-renov-design-user-motivation.pdf

Wade, F., Hitchings, R., & Shipworth, M. (2016). Understanding the missing middlemen of domestic heating: Installers as a community of professional practice in the United Kingdom. *Energy Research & Social Science, 19*, 39–47.

Wade, F., Murtagh, N., & Hitchings, R. (2018). Managing professional jurisdiction and domestic energy use. *Building Research and Information, 46*, 42–53.

ZCH. (2014). *Closing the gap between design and as-built performance*. Evidence Review Report. London: Zero Carbon Hub.

ZCH. (2015). *Builder's book: An illustrated guide to building energy efficient homes*. London: Zero Carbon Hub.

ZCH. (2016a). *Services guide: An illustrated guide to building services in new homes*. London: Zero Carbon Hub.

ZCH. (2016b). *Thermal bridging guide: An introductory guide to thermal bridging in homes*. London: Zero Carbon Hub.

Developing a Best Practice Framework for Degree Apprenticeships in Civil Engineering: Initial Findings

Josie Rothera

Introduction

The Institute for Apprenticeships and Technical Education (IFATE) was established by the Enterprise Act 2016 to ensure that standards and qualifications (for apprenticeships) would meet employer demand, sponsored by the Department for Education (DfE) (Powell, 2019). 'Trailblazer' groups, led by a representative group of industry employers, develop an apprenticeship standard for a required role with the corresponding set of knowledge, skills and behaviours (SKB) of the role. Standards provide the details to develop the SKB of that apprentice up to a certain level as set out in the Framework for Higher Education Qualifications (FHEQ) in the UK, which are aligned with the European Qualifications Framework (EQF) and based on the level, draws down different bands of funding from the Apprenticeship Levy. The Degree Apprenticeship qualifications deliver at Level 6 of the FHEQ and EQF.

The Apprenticeship Levy was initiated in April 2017 aligned with the UK Industrial Strategy and requires any employers with a pay bill over £3 million to pay 0.5% into a central fund that would ringfence funding to specially support skills and development training and be topped up by the Department for Education (DfE) at 10% of the maximum funding band. Non-levy organisations, i.e. SMEs, can also access this fund via a 'co-investment' rate, where 90% of the apprenticeship will be paid from the fund, and the other 10% paid for by the company.

In April 2018, a change was introduced to allow levy paying employers in England to transfer up to 10% of their funds to another employer, including their supply chain, and this again increased in April 2019 to 25%, which could be transferred to Apprenticeship Training Agencies (ATAs) (ESFA, 2019). The ESFA at this time also identified contribution to new opportunities and widen participation in apprenticeships, such as the through the trailblazer groups, as a suitable access of funds.

J. Rothera (✉)
Leeds Beckett University, Leeds, UK
e-mail: j.rothera@leedsbeckett.ac.uk

L. Scott et al. (eds.), *Sustainable Ecological Engineering Design*,
https://doi.org/10.1007/978-3-030-44381-8_16

An apprenticeship must last for a minimum of 12 months, combine 'hands-on work' (real work based experience) with the opportunity to train and obtain qualifications (The Apprenticeship Guide, 2019), and have an End Point Assessment (EPA). The qualifications will be delivered and assessed by an Apprenticeship Training Provider (ATP) which could be universities, Higher Education Institutions (HEIs), further education colleges, private training organisations or employer-providers (IFA, 2019). Trailblazer groups are encouraged to work with ATPs and industry professional bodies to contribute the design of the standard. Learning time for the qualifications and further developing skills is based on 20% off-the-job training and is managed by the employer.

The apprenticeship is a tripartite relationship: the apprentice, the employer and the ATP. The parties in this arrangement work closely with one another to ensure a successful outcome, the EPA, for the apprentice. This new way of working to support skills development across a defined period requires careful management of policies, processes and individual commitment.

The Construction Industry

The construction industry currently employs 2.4 million people (7% of the total UK workforce), with an economic output of £113 billion in 2017 (6% of the UK). The *Analysis of the National Infrastructure and Construction Pipeline* presented by the Infrastructure and Projects Authority identifies in the next 10 years, over £600 billion of projected public and private investment will be given, with around £190 million to be spent by 2020/21 (Infrastructure and Project Authority, 2018). The Construction Industry Advisory Board (CITB) in the *Construction Skills Network (CSN) Report* highlighted the significance between this pipeline of work, the forecasted construction output and the current skills levels and argued a case that to meet future demand, the annual recruitment requirement (ARR) for civil engineering professionals will be 1190 (CITB, 2018) (Table 1).

Empowered by the Construction 2025 report (July 2013), and further supported by the new UK Industrial Strategy (HM Government, 2017), specifically the Construction Sector Deal, the Construction Leadership Council (CLC) is responsible for guiding a framework of implementation and learning to support principles as set out in Construction 2025 for the industry. The Skills Workstream will act 'to develop the quality and capacity of skills required to support the Construction

Table 1 Annual Recruitment Requirement (ARR) for professional occupations

	2019	2023	ARR
Other construction professionals and technical staff	208,050	224,480	3260
Surveyors	71,920	77,900	1420
Civil engineers	56,480	60,980	1190
Architects	43,630	48,530	870

sector, including its productivity and modernisation aspirations' (Construction Leadership Council, 2018).

Skills development and relevance is a key area of concern (CIPD, 2016) especially at intermediate technical and professional levels where specialist, job-specific skills are needed (ARCADIS, 2017; UKCES, 2015). Other areas for consideration are gaps between skills the businesses are looking for and the skills that are available (HAYS, 2018), the significance of the levels of employer support of training for new and emerging skills (Farmer, 2016) and the need to reform to the apprenticeship programme in the UK (Leitch, 2006; Richard, 2012).

In the Foraging Futures Report, Universities UK identify the importance of industry and academic collaboration in terms of skills identification: 'If there is a mismatch between the higher level skills people develop through formal education and the higher level skills employers need, this can result in under-utilisation of skills' (Universities UK, 2014).

Degree Apprenticeships in Civil Engineering

The Civil Engineering (66 months) and Civil Engineering (Site Management) (30 months) Degree Apprenticeship standards were approved for delivery in 2017, with the earliest apprenticeships for EPAs around January 2020. The EPA is mirrored upon an Incorporated Engineer (IEng) professional review by the Institution of Civil Engineers (ICE). The Gateway is the point at which an apprentice has achieved: a BSc (Hons) Level 6 Civil Engineering undergraduate degree (accredited by Joint Board of Moderators (JBM)), suitable industry experience, and evidence of continuing professional development for 3 years (CPD). This allows the apprentice to apply to take the EPA.

The level of industry experience required is set out in the standard and is aligned with ICE IEng attributes, governed by the Engineering Council (EngC) who award for 35 professional engineering institutions. The IFA refers to knowledge, skills and behaviours, and the ICE use an established set of nine attributes at knowledge, experience and ability levels. This difference has been reviewed by the ICE mapping the KSBs to the attributes to confirm compatibility across both. This allows the apprentice to engage solely with the ICE's Initial Professional Development (IPD) and CPD systems to satisfy the requirements.

Once the apprentice has applied to take the EPA, confirmation of the EPA date is provided, and 6 weeks before this date, a formal report and extended CV are sent to nominated assessors. On the day of the EPA, the apprentice will deliver a presentation based on the report, answer further questions about the report in an interview, and sit a 2-hour written exam. Once all these components have been passed, the apprentice will have achieved the EPA, and if requested, will become an Incorporated Engineer with the ICE, or have a completed level 6° Apprenticeship.

Criticism

The new Apprenticeship Levy is now 2 years old, and there is concern of the management and flexibility of the scheme (Andrews, 2019; City & Guilds, 2019). The most common issues found were not being able to access the funding in a timely manner leading to a £400 million underspend in 2017/18 (City & Guilds, 2019), the geographical restrictions of the ATPs, and a lack of diversity in the types of apprenticeships being offered.

A recent briefing paper to the House of Commons highlighted over 125,000 fewer apprenticeship starts in England between August 2017 and July 2018, citing the following reasons: complexity of the levy, inflexibility of the levy, requiring non-levy payers to pay 10% of apprenticeship costs and a 20% off-the-job training commitment being high (Powell, 2019).

However, this now in contrast to the warnings that have come from the IFA of significant risk on a budget overspend due to the higher banded standards being more popular to use than others (Burke, 2018). This could mean a £500 million overspend in 18/19 which will increase to £1.5 billion in 19/20. This will adversely influence smaller SMEs that will be looking to access funding for the lower band standards.

More recently in April 2019, ATPs have started to turn away non-Levy paying organisations as the funding pot is running out and to be reserved only for those that are paying into it. This means that SMEs are being prevented from accessing support for skills development.

Methodology

The methodology that would be most relevant to research to draw out participant experience would be derived from a social perspective. The tripartite arrangement meant that consideration should be given to the nature of organisations internally, and how they respond externally, both in terms of employers and ATPs. To understand barriers and success factors from stakeholder's experiences, there needs to be an understanding for the capacity of organisations and ATPs to react to policies regarding the new learning model, and how guiding principles are developed to provide assurance to the process. For apprentice contributions, more analysis into their support network and general knowledge about the process would be required (this is outside of the scope of this paper and will not be addressed here).

There are two main approaches to theory development: inductive and deductive. The deductive approach usually begins with a hypothesis, whilst an inductive approach will usually use research questions to narrow down the scope of the study (Gabriel, 2013). The deductive approach represents a positivist paradigm assuming that 'an objective reality exists that researchers can uncover' (Bergin, 2018), whereas the inductive approach represents a phenomological paradigm. The phenomological paradigm can be further divided into three areas: critical theory, constructivism and realism (Guba & Lincoln, 1994).

Critical theory: Critical theory is focused on behaviours, languages, communication and social construction, to reflect critically on and assess structures and systems in society to understand conditions for change. Used predominantly in the social sciences and humanities, however this is now being considered as applied to the construction industry (McAleenan & McAleenan, 2017; Smiley, Dainty, & Fernie, 2013). To apply this theory would require structures and systems around the phenomena to already be somewhat established, and in the case of degree apprenticeships this is not the case as this approach is new.

Constructivism: Constructivism can be applied from an educational perspective, in terms of both the learner and the pedagogies of that learning, but in the context of a phenomenological paradigm a more relevant emphasis can be placed on the interactions that are experienced within a group or environment, and what phenomena emerge from this. Methodologies to abstract data from this area would include lengthy observations using a large representative sample to attempt to formulate a set of parameters that would better define the phenomena. This approach to the methodology might be appropriate when considering one of the tripartite contributors to the apprenticeship programme.

Realism: Realism can be linked to with an ethnographic approach to research, relating to assumptions that might be made to better define any phenomena that is being investigated. This realism would be a result of a literature review around the topic and any personal experiences of the researcher. Using realism to bracket out the natural attitude of the person based on preconceptions and biases would allow a focus on the phenomological experiences.

Type of Methodology

The author identified the work of Kurt Lewin and the Field Theory as being appropriate to establish a framework in which to set the methodology. The Field Theory states that environments across a person's life become a constellation of mutually independent factors that exist at different points—and can therefore be applied to the tripartite arrangement of the employer, the ATP and the apprentice. The use of this field theory has been recommended to support researchers with an approach to managing change (Burnes & Cooke, 2013), in this case the new degree apprentice ships (Fig. 1).

Phenomological research methods using semi-structured interviews were identified as appropriate to use based on the novelty of the degree apprenticeship scheme and with a desire to understand the experiences of stakeholders engaged with the scheme. To ensure that the contributors could explore their experiences and prevent any unconscious bias stemmed from the author's own experiences, a set of questions were developed that were open ended but formatted in hierarchal levels. Results from discussing experiences would indicate preliminary themes for further investigation that could inform a comprehensive methodology for future research. Purposive sampling was used to identify research participants: Small-Medium

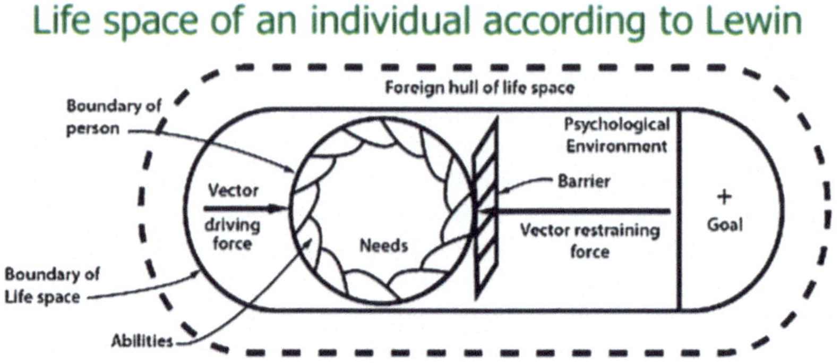

Fig. 1 The Action Research logo is based on Lewinian Field Theory. A person is motivated to meet their individual needs by pursuing various goals. The drawing above illustrates the basic principle and demonstrates the complex interaction of the person and the environment with regard to a particular behaviour

Table 2 A framework to understand levels of influence on a person's experience

	Perceived environment	Typical considerations
Micro	Immediate, direct influence	Work station or space, access to technology, groups
Meso	Guidance around actions	Structure of rules in organisation, support network
Macro	Wider cultural and societal	Influence from professional bodies, government policies

Enterprises (SMEs) that were already engaged at Leeds Beckett by having a registered apprentice learning at the ATP. The SMEs would also fall into the category of non-levy paying. It was assumed that these companies would not have already established and structured organisational management programmes for this type of part-time vocational learning. Five SMEs were identified to take part in the research, of which three completed a semi-structured interview.

A questionnaire was also developed to be used at the outset of summer sessions for apprentices that were aimed at increasing the knowledge and understanding of the scheme. The apprentices had completed their first 9 months of the 30-month Civil Engineering (Site Management) degree apprenticeship. The questionnaire contained six questions based on a Likert scale, with two additional questions.

To allow a reference to the different hierarchal levels of environment, the author considered the social and workplace structure of various influences on an individual. The following were identified (Table 2):

The transcriptions were grouped using a qualitative data software package, NVivo, allowing the apparent learned views of the participants stemming from experience in the built environment, and that which emerged as true experience. Once this data had been grouped, a constructivist approach would provide a subset of the field theory (Burnes & Cooke, 2013) to prioritise the creation of concepts to allow patterns of forces (concepts) helping or hindering the goal of the EPA and to illustrate promising points of intervention in policies, procedures or methods that would allow a best practice framework for managing the degree apprenticeships to emerge for each contributor.

Initial Findings

Primary Themes

Table 3 lists the primary themes that were identified from the transcribed interviews. The themes were then interrogated further to remove any non-contributing text from the node and identify any subsets that would allow a specific focus on an area. An in-built comparison software was used to put the interviews that were now coded against one another as so an indication as to the common themes could emerge.

Early Indicators

Of the six primary themes initially identified, four emerged as dominant as shown in Table 4: the commitment of the participant to the profession, the previous experience of the participant, the personal skills development of the apprentice and the working environment in that order.

The commitment of the participant to the profession was identified as on the macro level theme (Table 3). For all three research participants, their own commitment to the individual and the wider professional community was indicated as likely to increase the probability of a success. All participants related the importance of apprentices to their businesses and of the future of the industry.

The previous experience of the participant provided an indicator around the familiarity of the approach to the End Point Assessment in that it aligns with the Institution of Civil Engineers (ICE) Incorporated Engineer review, and that each of them had either been through that process or one similar.

Table 3 Themes as identified with NVivo software by the author

	Themes
Micro	Previous experience of participant; Working environment
Meso	Personal skills development of the apprentice, Apprentice contributing to business
Macro	Commitment of participant to profession; Awareness of the Degree Apprenticeship

Table 4 Dominant themes emerging from cross-referencing and compounding

	References	Subsets of theme included	Total
Commitment of the participant to the profession	17	Personal support to apprentices, sustainable business, SCE or mentor, own network or support	29
Previous experience of participant	24	ICE system	27
Personal skills development of the apprentice	24	Mentor	27
Working environment	14	Immediate, meso—internal	20

The personal skills development of the apprentice and the working environment were identified at the micro level. For both indicators, managing the apprentice, and sharing with the apprentice knowledge and understanding of civil engineering, whilst empowering them to innovate and problem solve using specialist software, and providing opportunities to put context to their solutions, was important.

Questionnaire

The questionnaire contained six questions that were answered on a Likert scale, with two additional questions. It was given out to all apprentices that attended the first summer session, which was 9 months into the degree apprenticeship, about a third of the way through. Seventeen apprentices responded anonymously to the questionnaire.

The questionnaire showed that 24% of the apprentices were aware of the ICE membership levels, but nearly half of them were unsure about the ICE attributes. The two questions relating specifically to the degree apprenticeship process showed a lack of understanding and confidence in the system from them as individuals and their perceptions of their organisations.

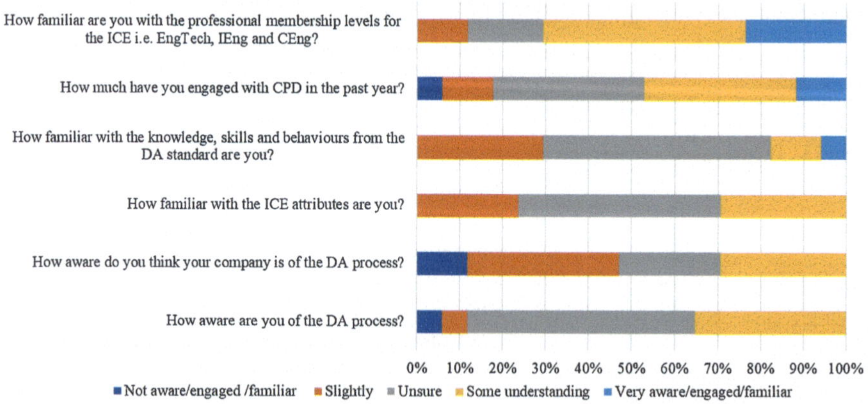

The two other questions that were asked were:

Are you being provided with a structured process that you have been briefed on to support the Degree Apprenticeship?
24% of the respondents indicated that there was a structured process with their employer and that they had been communicated to about it. 41% said this was not the case, and the remaining, over a third, were unsure or did not respond. This question relates to the meso, and somewhat micro, hierarchal levels of inquiry.

Where have you received the most amount of knowledge about the Degree Apprenticeship?
10% of the respondents said that this was from their employer and 60% said that this was from university. Interestingly, 30% said that this was from their own research. This 30% would indicate that they have had access to websites or literature that have helped them inform their choices. This question has links to the macro hierarchal levels of inquiry.

Limitations

Sample Size

Of the five SMEs that were asked to take part in the research, only three did. In phenomological research of this type, a minimum number of four participants would be preferred (Giorgi, 2008). Purposive sampling was used, and SMEs were considered best to approach to allow a more holistic overview from the immediate to the macro environment due to size of business, and roles of those that would be engaged in the interviews. However, this meant that there is no comparison to other organisations and restricted possible areas of improvement or best practice for business. As SMEs are currently not being able to access the Levy, it would further restrict participant numbers. The questionnaire was only taken by half of the potential apprentices at the summer sessions, and only represents the view of Civil Engineering (Site Management) degree apprenticeships. This would help inform specific actions within the 30-month timeframe but needs to be considered alongside further research with the 66-month Civil Engineering degree apprenticeship students.

Time Restrictions

The degree apprenticeship is new, and this initial pilot study has been conducted only 9 months into the programme. To provide robust evidence to the research, follow-up interviews and questionnaires would lend themselves to explore identified themes and subsets in more detail, but this is time bound.

Data Collected

Three transcribed interviews coded and compared is not a significant amount of data to base research on that would eventually lead to a best practice framework. It is expected now that a review of other data sets and how these can be collected will be established, and a schedule of interventions with both SMEs and large organisations will be set out to systematically approach data collection and analysis.

Credibility

The research participants do not form a representative sample of all the tripartite contributors in the scheme, and only offer initial results based on one point of contact. To inform best practice, further work with the participants, the apprentices and the ATPs would need to be carried out. It is not necessarily the number of various participants that take part, as case studies are also important for testing theories (Patton & Appelbaum, 2003). However, even when considering case studies, there would need to be a longer, substantial data gathering period to allow meaningful conclusions to be drawn.

Literature Review on Methodology

A literature review was carried out to identify methodologies in engineering research using a Scopus search, to inform future research based on social approaches (Table 5).

Phenomological

In two reviewed cases (Frimpong & Dansoh, 2013; Gill, White, & Cameron, 2011), purposive sampling (non-probability) was used to obtain 'a sample that can be logically assumed to be representative of the population' (Lavrakas, 2008). Both methodologies used a constructivist approach to data management and analysis (Charmaz, 2005); however one used this to inform the questions for the interviews, whereas the other used this after the interviews to synthesise a composite description of perceptions (Frimpong & Dansoh, 2013). The use of the template analytic technique (Crabtree & Miller, 1999) was a reference tool for an emerging codebook of themes once interviews had taken place in Gill et al. (2011) research. This technique provided a framework for attributes of each theme, extending the initial constructivist approach, and is a more robust method of reviewing the data than 'repeated reading' that was the method that Frimpong and Dansoh (2013) used. In both cases, there is a subjective perspective that is given to the data which could in turn lead to an imbalance in the perceptions presented.

Table 5 Scopus search criteria for literature

Key search word	Filter 1	Filter 2	Filter 3
Phenomological (83)	Limit to Engineering (32)		
Phenomena (1.3M)	+Civil Engineering (32,243)	Limit to Engineering (18,543)	+Skills (112)
Ethnographic (48,840)	+Civil Engineering (390)	Limit to Engineering (101)	
Apprenticeships (6112)	+Civil Engineering (64)	Limit to Engineering (101)	

Phenomena or Phenomenon

The word 'phenomena' identified a significant amount of research to consider. This was narrowed down using filters of '+civil engineering', a limit to the engineering documents, and finally the word 'skills' was included to align any results with the degree apprenticeship field.

In 2014, a Construction Industry Board (CIB) Task Group (TG74) were asked to investigate why the offsite construction market was still relatively small, addressing it as a phenomenon. The research used focus-group discussions which are useful for generating ideas regarding new products and phenomena based on the experts' commonality (Morgan, 1997). Purposive sampling was conducted, and two groups formed. Both importance and time frame of a set of 27 questions generated group scores on the Likert scale. All discussions were transcribed, data coded, then ANOVA used to examine the differences. The results provided statistical data on the topic; however the author contends that this data could have been drawn qualitatively from the transcribed discussions, and the inclusion of statistical data unnecessary. The focus-groups were equal numbers of different experience and competence levels in built environment roles and therefore consideration to be given to how it was monitored that everyone got an equal say, or felt confident to have an equal say, in the eventual collective agreement.

Suresh, Olayinka, Chinyio, and Renukappa (2017) used a mixed-method approach to understand the barriers to adopting knowledge management on construction projects. In-depth interviews were conducted with 25 construction industry experts, and this was followed up by a questionnaire survey of 114 respondents. The data obtained were analysed using thematic analysis and descriptive statistics. The researcher used purposive sampling for its suitability for in-depth qualitative research where the focus is to understand complex social phenomena (Creswell, 2009); however this sample size was not initially set and progressively grew until 'saturation' was reached (Suresh et al., 2017). This 'saturation' was not formulated and therefore it is assumed the researcher was instrumental in deciding this point, thus applying a level of bias to the transcripts. To support the interview process, the researcher reviews four factors that facilitate robust data sets: credibility, dependability, transferability and reproducibility or confirmability (Devault, 2019). Credibility is ensured by prolonged engagement, persistent observation and peer briefing. The outcome of the interviews then informed the questions on the survey that provided quantitative data for analysis.

This is a well-structured methodology with clear focus on what data will be gathered, and how it will be analysed. Even though the research is using a mixed-method ethnographic type approach, it is still investigating phenomena.

Ethnographic

A number of publications (co-authored) by Dainty have identified benefits of using ethnographic research in construction research. Two categories emerge from the researcher's literature: a classic ethnographic approach and a contemporary ethnographic approach. The classical approach stems from long-term engagement with a subject, 'observing behaviours, participating in activities, writing extensive notes, and interviewing and reflecting on one's own role' in the process (Pink, Tutt, Dainty, & Gibb, 2010). This approach when applied to construction is considered time consuming and impractical when considering the project environment as transient in nature and time-bound.

The contemporary approach has therefore emerged to facilitate small-scale studies that allow the researcher to complement the classic interviews, and experiences with data sets taken from broader sources that support the creation of thick descriptions in a reduced time frame. These thick descriptions can 'support trustworthiness in the interpretation and development of themes' (Bornasal, Brown, Perova-Mello, & Beddoes, 2018), and are best supported by a team of researchers with prior experience. This use of contemporary ethnography is supported by examples from an interdisciplinary study concerning migrant workers and communications on UK construction sites (Tutt, Pink, Dainty, & Gibb, 2013).

The contemporary approach has also been used in single case studies which allows the researcher to consider perspectives about organisational rules and procedures to inform potential areas for change (Galea, Loosemore, Powell, & Chappell, 2014), in this case gender equity queries. This single case used data from a wide variety of sources from HR, to policies and communication to contribute to the knowledge, and applied robust and revisable considerations (Lowndes & Wilson, 2003) to gain a better understanding of how entrenched policies and organisational rules currently were.

Rapid ethnography is another alternative to overcome the time-restricted barriers from the classical approach. Similar to the contemporary approach, focus is given to the careful preparation of the research as a key success factor to any outcome and further studies. It needs 'a strong theoretical context and systematic method to be thought out in advance...also requires teamwork, close interaction with informants...new technology, multiple methods...informant sampling and the identification of clearer research questions and propositions in advance' (Isaacs, 2012). Of the case studies that were provided for the cited research none were in the built environment area. The researcher also placed a high value on the use of video recorded data and a team of experienced researchers.

Discussion

The initial findings from the interviews and questionnaires conducted has proved useful experience for methodology and interrogating small amounts of data. However, the volume of data retrieved has been too low for any meaningful discussion, the type of data should be applied to and checked against a set of limiting criteria, and the methodological approach to the entire research project is not at all clear and should consider the stakeholder context. The literature review revealed that investigating phenomena in engineering is common but less so when applied to civil engineering with a social objective. In these civil engineering cases however, the use of ethnographic research rather than phenomological research is prevalent and would be a better application to the research aim.

Ethnographic, Not Phenomological

Ethnographic accounts seek to be both descriptive and interpretive (Gray, 2013), and in the classical approach are a time-consuming method of research. The research aim of this study will be time bound by the journey of the degree apprenticeship and therefore a more contemporary approach would be suitable. The descriptions should have high levels of detail, but in a time bound environment, the careful design by experienced researchers will facilitate data retrieval to aid interpretation. Ethnographic research will consider different types of data, most commonly observation, interviews and documents; however if the rationale for data use is clear on how these will be integrated in the analysis (Ellis, 2006), then data sets can be more variable to complement the complexity or different aspects of the phenomenon (Gray, 2013). This mixed method design should be done by considering the sample population, the required data and the time restrictions of the research.

Defining the Tripartite Stakeholders and Their Context: Sample Population

The tripartite stakeholders involved in the process will have both overlapping and specific objectives in relation to the degree apprenticeship. By identifying these it will allow the researcher to develop an understanding of the data required from each stakeholder to inform the best practice framework. The sample population for the ATP and employers should use profile data such as number of employers, percentage of market operation, previous history of training and typical turnover. In the case of the apprentice themselves, a representative sample would consider previous experience of training schemes, previous and current roles, their support network and possibly salaries. For each stakeholder, the development of 'thick' descriptions to better describe their context would be useful and can 'support trustworthiness in the interpretation and development of themes' (Bornasal et al., 2018).

Gathering the Data

Applying best practice means learning from and through the experience of others (Info Entrepreneurs, 2019). For the ATPs and the employer, a single case study using contemporary ethnography methods could be appropriate to allow a vertical in-depth assessment through policies, documents and management processes, but to inform best practice, a comparison of results would need to be drawn from the sample population. However, to complement this data, horizontal, more general data should be gathered from a larger volume of sources informed by the in-depth studies, and this could be done using questionnaires to achieve quantitative data for analysis.

Applying Time Restrictions

The research is time-bound for the degree apprenticeship delivery, and the contributions made by the individual apprentices. A time dimension should be applied to the design of the methodology that will identify clear periods for data gathering and review.

Conclusions

Degree apprenticeships in civil engineering are a new method of learning and are growing in numbers with employers' keen to use their Apprenticeship Levy contributions to their maximum. In the current absence of any successfully completed apprenticeship, knowledge sharing is based on varied stakeholder experience. This research has the premise to gather this knowledge and create a best practice framework to support each tripartite contributor; however in a time-restricted process a review of the methodology and the data sets that are going to be used should be carried out to provide credibility, dependability, transferability and confirmability to the research. Data taken from a well-designed methodology will act as baseline knowledge that could be then used for future research studies.

References

Andrews, S. (2019). *Is the apprenticeship Levy killing apprenticeships?* [Online]. London. Retrieved June 28, 2019 from https://www.trainingzone.co.uk/deliver/training/is-the-apprenticeship-levy-killing-apprenticeships

ARCADIS. (2017). Arcadis Talent Scale. *The real extent of Britain's construction labour crisis*.

Bergin, T. (2018). *An introduction to data analysis: Quantitative, qualitative and mixed methods*. London: SAGE Publications Ltd.

Bornasal, F., Brown, S., Perova-Mello, N., & Beddoes, K. (2018). Conceptual growth in engineering practice. *Journal of Engineering Education, 107*, 318–348.

Burke, J. (2018). *Levy budget bust: Government agency warns of imminent apprenticeship overspend* [Online]. London. Retrieved June 28, 2019 from https://feweek.co.uk/2018/12/03/levy-budget-bust-government-agency-warns-of-imminent-apprenticeship-over-spend/

Burnes, B., & Cooke, B. (2013). Kurt Lewin's field theory: A review and re-evaluation. *International Journal of Management Reviews, 15*, 408–425.

Charmaz, K. (2005). *Grounded theory in the 21st century*. Thousand Oaks, CA: SAGE.

CIPD. (2016). *Over-skilled and underused. Investigating the untapped potential of UK skills* [Online]. London: Chartered Institute of Personnel and Development. Retrieved May 8, 2019 from https://www.cipd.co.uk/Images/over-skilled-and-underused-investigating-the-untapped-potential-of-uk-skills_tcm18-48001.pdf

CITB. (2018). Construction Skills Network: 2019–2023.

City & Guilds. (2019). *The languishing Levy pot: Accessing unspent apprenticeship funds* [Online]. City & Guilds. Retrieved June 28, 2019 from https://www.cityandguilds.com/news/march-2019/the-languishing-levy-pot-accessing-unspent-apprenticeship-funds

Construction Leadership Council. (2018). *Skills workstream: 2018 Strategy and action plan*. London: BEIS.

Crabtree, B. F., & Miller, W. L. (1999). *Doing qualitative research*. London: SAGE Publications.

Creswell, J. (2009). *Research design: Qualitative, quantitative, and mixed methods approaches*. Thousand Oaks, CA: SAGE.

Devault, G. (2019). *Establishing trustworthiness in qualitative research* [Online]. The Balance Small Business. Retrieved July 2, 2019 from https://www.thebalancesmb.com/establishing-trustworthiness-in-qualitative-research-2297042

Ellis, J. (2006). All inclusive benchmarking. *Journal of Nursing Management, 14*, 377–383.

ESFA. (2019). *Apprenticeship funding: How it works* [Online]. London. Retrieved June 28, 2019 from https://www.gov.uk/government/publications/apprenticeship-levy-how-it-will-work/apprenticeship-levy-how-it-will-work

Farmer, M. (2016). *Modernise or die—Farmer review*.

Frimpong, S., & Dansoh, A. (2013). A phenomenological study of private homebuilders' perceptions on the employment of professional services C3. In *Proceedings of the 29th Annual Association of Researchers in Construction Management Conference, ARCOM 2013* (pp. 425–433).

Gabriel, D. (2013). *Inductive and deductive approaches to research* [Online]. Retrieved June 19, 2019 from https://deborahgabriel.com/2013/03/17/inductive-and-deductive-approaches-to-research/

Galea, N., Loosemore, M., Powell, A., & Chappell, L. (2014). Gender equity in construction professions: A new institutionalist perspective (pp. 1111–1119).

Gill, L., White, L., & Cameron, I. D. (2011). Service co-creation in community-based aged healthcare. *Managing Service Quality: An International Journal, 21*, 152–177.

Giorgi, A. (2008). Difficulties encountered in the application of the phenomenological method in the social sciences. *Indo-Pacific Journal of Phenomenology, 8*, 1–9.

Gray, D. (2013). Chapter 17: Ethnography and participant observation.

Guba, E. G., & Lincoln, Y. S. (1994). Competing paradigms in qualitative research. In N. K. Senzin & Y. S. Lincoln (Eds.), *Handbook of qualitative research*. Thousand Oaks, CA: SAGE.

HAYS. (2018). *The Hays global skills index.*
HM Government. (2017). *Industrial strategy building a Britain fit for the future* [Online]. London: Crown copyright. Retrieved May 8, 2019 from https://assets.publishing.service.gov.uk/government/uploads/system/uploads/attachment_data/file/664563/industrial-strategy-white-paper-web-ready-version.pdf
IFA. (2019). *Forming a Trailblazer Group* [Online]. London. Retrieved July 1, 2019 from https://www.instituteforapprenticeships.org/developing-new-apprenticeships/forming-a-trailblazer-group/
INFO ENTREPRENEURS. (2019). *Best practice* [Online]. Retrieved July 7, 2019 from https://www.infoentrepreneurs.org/en/guides/best-practice/
Infrastructure and Project Authority. (2018). *Analysis of the National infrastructure and construction pipeline*. IPA.
Isaacs, E. (2012). The value of rapid ethnography.
Lavrakas, P. J. (2008). *Encyclopedia of survey research methods* [Online]. Thousand Oaks, CA: SAGE Publications. Retrieved June 12, 2019 from https://methods.sagepub.com/reference/encyclopedia-of-survey-research-methods
Leitch, S. (2006). *Prosperity for all in the global economy—World class skills*. London: HM Treasury.
Lowndes, V., & Wilson, D. (2003). Balancing revisability and robustness? A new institutionalist perspective on local government modernization. *Public Administration, 81*, 275–298.
McAleenan, C., & McAleenan, P. (2017). Critical theory: Understanding the impact language has on workers' safety and health.
Morgan, D. L. (1997). *Focus groups as qualitative research* (2nd ed.). Thousand Oaks, CA: SAGE Publications.
Patton, E. & Appelbaum, S. (2003). The case for case studies in management research. *Management Research News, 26*(5), 60–71.
Pink, S., Tutt, D., Dainty, A., & Gibb, A. (2010). Ethnographic methodologies for construction research: Knowing, practice and interventions. *Building Research and Information, 38*, 647–659.
Powell, A. (2019). *Apprenticeships and skills policy in England*. London House of Commons Library.
Richard, D. (2012). *The Richard review of apprenticeships*. London: Department for Business, Innovation and Skills.
Smiley, J.-P., Dainty, A., & Fernie, S. 2013. Analysing the discourses surrounding change and reform in the construction sector: The need for a marriage between critical theory and a socio-historical cultural perspective.
Suresh, S., Olayinka, R., Chinyio, E., & Renukappa, S. (2017). Impact of knowledge management on construction projects. *Proceedings of Institution of Civil Engineers: Management, Procurement and Law, 170*, 27–43.
The Apprenticeship Guide. (2019). *What is an apprenticeship?* [Online]. London. Retrieved July 1, 2019 from http://www.apprenticeshipguide.co.uk/what-is-an-apprenticeship-2/
Tutt, D., Pink, S., Dainty, A. R. J., & Gibb, A. (2013). Building networks to work: An ethnographic study of informal routes into the UK construction industry and pathways for migrant up-skilling. *Construction Management and Economics, 31*, 1025–1037.
UKCES. (2015). *Employers skills survey 2015*. UKCES.
Universities UK. (2014). *Forging futures: Building higher-level skills through university and employer collaboration.*

User Satisfaction of a Green Star-Rated Literary Museum in South Africa

Kenneth M. Rampou, Sharon Dent, Gerrit J. Crafford, and Katharina Crafford

Introduction

In green buildings, '….those occupants, who are engaged with the building's "green" identity, are more likely to tolerate any environmental discomfort' than they do with conventional buildings (Brown & Cole, 2009; Leaman & Bordass, 2007; Monfared & Sharples, 2011). Occupants experiencing poor overall indoor environment quality (IEQ) while having high expectations of the green building are likely to be dissatisfied with the building. On the other hand, occupants with pro-environmental attitudes are closely associated with stronger tolerance, also known as the forgiveness factor observed in green buildings (Leaman, Stevenson, & Bordass, 2010). These occupants will generally demonstrate a higher level of satisfaction (Deuble & De Dear, 2012). In a recent study, the age of a building was found to be a significant factor in the occupants' satisfaction. Perceived satisfaction levels of occupants were higher directly after occupying a green building owing to the honeymoon effect (Liu, Wang, Lin, Hong, & Zhu, 2018).

Post-occupancy evaluation (POE) is the process of evaluating the buildings after they have been occupied for some time (Preiser, Rabinowitz, & White, 1988). POE information is beneficial for the control of the internal environment of the building to enhance user comfort, well-being and satisfaction (Fieldson & Sodagar, 2017). POE is a well-researched topic globally. The majority of post-occupancy evaluations relate satisfaction with IEQ, while others only address one aspect of IEQ, namely thermal comfort (Gou, Lau, & Shen, 2012; Zuo et al., 2017). Of those POE

K. M. Rampou · S. Dent · G. J. Crafford (✉)
Department of Quantity Surveying, School of the Built Environment,
Nelson Mandela University (North Campus), Port Elizabeth, South Africa
e-mail: gcrafford@mandela.ac.za

K. Crafford
Department of Construction Management, School of the Built Environment,
Nelson Mandela University (North Campus), Port Elizabeth, South Africa

L. Scott et al. (eds.), *Sustainable Ecological Engineering Design*,
https://doi.org/10.1007/978-3-030-44381-8_17

studies addressing comfort and user satisfaction in green buildings, the user satisfaction aspect's theoretical foundation has not been well established.

Nature regulates the human comfort of hygiene, vision, thermal and audio aspects. Once indoors, the building envelope modifies these comfort factors and combines them to become IEQ (Adan & Bluyssen, 2009). Occupants who are comfortable with environmental stressors of thermal, lighting, indoor air quality and audio-IEQ are not necessarily satisfied with the building's spatial aspects of design and needs in relation to their productivity and health. To ensure satisfaction, there is a way to design more humane buildings (Schneider, Gruman, & Coutts, 2012). This process is known as social design, and it is a participatory approach which involves people in the planning and management of their spaces (Schneider et al., 2012). It is therefore crucial for post-occupancy evaluations to take the human aspects of health and well-being (productivity, comfort and satisfaction) into consideration.

In previous related studies by Baird (2010) and Gou, Prasad, and Lau (2013), comfort is clearly articulated through IEQ. The green building assessment ratings, for instance, GBCSA (2016) also use IEQ as a rating in relation to one of the social requirements of sustainability. User satisfaction in the above two studies is, however, only determined through the satisfaction index, which implies a model construct of correlation and causation (Wright, 1921). This study attempts to advance satisfaction in the context of a model and fill this gap in research. This will enable the researcher to connect the data sets with confidence, as data without a model is mere noise (Anderson, 2008). Ignoring satisfaction as one of the factors of POE will lead to a proliferation of POE models constructed on correlation and causation, thus constraining academic and industry development.

The objectives of this study are therefore:

- To determine the extent to which users are satisfied with the overall performance of the National English Literary Museum
- To determine the extent to which users are comfortable within the environment of the National English Literary Museum
- To determine to what extent users tolerate the National English Literary Museum environment

Literature Review

Benefits of Post-occupancy Evaluations

POE has several benefits in the short-, medium- and long-term horizons (Baird & Oosterhof, 2010; Parkinson, Reid, McKerrow, & Wright, 2017; Preiser et al., 1988). In the short term, POE information is beneficial for the control of the internal environment of the building to enhance user comfort, well-being and satisfaction (Fieldson & Sodagar, 2017). In the medium to long term, project owners benefit from the knowledge that the occupier needs are satisfied. The lessons learned can provide a feed-forward impetus to improve the performance of future buildings through informed decision-making at the design stage. POE information is further used for fine-tuning a building's operations to enhance its performance (Fieldson & Sodagar, 2017).

POEs of buildings provide invaluable insights into the environmental performance and user behaviour of buildings. The broader aspects of performance measured by a POE can address issues such as space utilisation, management, environmental impact and cost during the operating stage of the building (Li, Froese, & Brager, 2018). POE can also be used as a learning loop to feed forward the lessons learnt to better inform the decision-making process at the design stage (Fieldson & Sodagar, 2017).

Post-occupancy Evaluation User Satisfaction Indicators

User satisfaction is a concept with a variety of definitions owing to its contextual application and subjective nature (Giese & Cote, 2000). In the built environment, user satisfaction can be defined as to how close a completed and operating building is to matching the user's expectations and needs (Applegate, 1993; Leaman et al., 2010). User satisfaction is important because occupants must spend significant parts of their lives working, residing, or relaxing in the setting (Schneider et al., 2012).

Satisfaction with a building is dependent on many parameters, some of which are productivity, health and well-being, contentment with features and facilities (Newsham et al., 2009). In addition, several other factors creating conflict in research are the purpose of measurement, the timing, the building age, the locality, the macro and microclimates of the building, and the vacancy rate (Gou et al., 2013; Kim & De Dear, 2012; Parkinson et al., 2017). Moreover, human satisfaction levels change with time (Sinha, Gupta, & Kumar, 2013). This list is non-exhaustive.

In an attempt to reduce the conflict of satisfaction parameters, Kim and De Dear (2012) adapted Kano's model of satisfaction into the built environment. This comprehensive study on approximately 43,200 survey samples of mixed office buildings used the Center for the Built Environment (CBE) POE database from University of California to identify IEQ factors affecting the satisfaction of the users.

In the adaptation of Kano's model, Kim and De Dear (2012) used the four standard dimensions of IEQ, namely thermal comfort, air quality, lighting and acoustic quality, and expanded the list to include office layout, office furnishings, cleanliness and maintenance, and overall satisfaction. The study identified a nonlinear relationship between IEQ factors and overall satisfaction. Fieldson and Sodagar (2017) used the comfort variables of air temperature, humidity, ventilation and air movement, air quality, daylight, artificial lighting and overall noise to determine user satisfaction. However, this study is conducted on a single building, is not generalisable and does not make use of a benchmark database for comparison.

Numerous studies have made use of the BUS survey benchmarks to determine the levels of indoor environmental satisfaction on a wide range of buildings

(Frontczak et al., 2012; Goh & Sun, 2016; Paevere, Brown, Leaman, Luther, & Adams, 2008). These studies compared the relationships between occupant satisfaction and IEQ parameters. Other studies relative to this study comprehensively assessed the occupants' comfort in comparison with IEQ and the overall satisfaction of the occupants with the building (Baird, 2010; Gou et al., 2013).

In mediating these divergent approaches in research, the discussion on user satisfaction adopts the approach outlined in Baird (2010) of grouping user satisfaction indicators and operational factors in facilitating POE as they are interrelated. Despite this best attempt to state POE satisfaction parameters as discrete, a building functions as a unit devoid of segregated parameters. These indicators are therefore bound to overlap:

Satisfaction indicators:

- Design
- Needs
- Productivity
- Health

Operational factors:

- Space in the building (Occupation density)
- Image to visitors
- Availability of meeting rooms
- Suitability of storage arrangements
- Facilities meet work requirements
- Space at desk
- Effect of building on behaviour
- Cleaning

Post-occupancy Evaluation Comfort Indicators: Indoor Environmental Quality (IEQ)

The relationship between the indoor environment quality, health and comfort of building occupants is complex and still not fully understood (USGBC, 2018). This complexity arises from human factors such as heritage, individual expectations, local customs and norms, and occupants' activities. The building factors are its design, construction, operation and the control the occupants can exert on their environment (Adan & Bluyssen, 2009).

Human comfort is a general feeling expressed in terms of well-being and involves personal health. The measures of human health and well-being use models to combine IEQ parameters into a single index to predict the comfort of users with their environment (MacNaughton et al., 2017; Pei, Lin, Liu, & Zhu, 2015).

For the purpose of this research, the building comfort index parameters, IEQ and POE comfort indicators refer to the following:

- Thermal comfort
- Visual comfort
- Indoor air quality (IAQ)
- Noise levels

Post-occupancy Evaluation Tolerance Indicator

Tolerance (Deuble & De Dear, 2012) is also known as the forgiveness factor. The results of a POE survey on green buildings in Australia revealed that occupants with pro-environmental attitudes were closely associated with the stronger forgiveness factor observed in green buildings. Occupants with pro-environmental behaviour tend to '…reduce their use of some resource, to pollute less, or to recycle more' (Ewoldsen & Beverly, 2013).

Occupants with pro-environmental behaviours also indicate higher overall satisfaction with green buildings and tend to overlook shortcomings related to performance (Deuble & De Dear, 2012; Kurisu, 2016; Leaman & Bordass, 2007).

Another dimension to the forgiveness factor warrants consideration, namely that of time. The age of a building is a significant factor in the occupants' satisfaction. In a recent study by Liu et al. (2018), perceived satisfaction levels of occupants were higher directly after occupying a green building owing to the honeymoon effect. This phenomenon is attributable to occupants being impressed with the features of the new building, which is more glamorous compared to an older or more conventional building.

The honeymoon effect is prevalent immediately post-occupancy and up to a period of 12 months. This reinforces the recommendation for conducting POEs after this period has elapsed (Leaman et al., 2010; Liu et al., 2018; Preiser et al., 1988), together with the defects liability period and the building fine-tuning having taken effect.

The forgiveness of green buildings can be nurtured in occupants through re-education and demonstrating the impact of buildings on 'global climate change' (Deuble & De Dear, 2012, p. 226).

The Building: The National English Literary Museum (NELM)

The National English Literary Museum (NELM) is a cultural facility housing a comprehensive collection of South African English literature resources comprising manuscripts, photographs, posters, theatre programmes, published works and reference resources. It is the first green museum in South Africa. The building is a double-storey, concrete-framed structure with brick infill panels and extensive concrete coffered slab roofs. Its footprint is 4001 m^2 with a total building area of 5004 m^2 on a site area of 14,800 m^2. At the time of the study, the building occupancy period was 24 months, and the beneficial occupation had taken place on 1 July 2016 (Fig. 1).

The study building is located in a region bordering three climatic conditions, namely the temperate coastal, the sub-tropical coastal and the temperate interior (South African Bureau of Standards, 2011).

The Client's Need and Design Brief

The client's need was a prominent, contemporary and functional building providing sufficient accommodation for the safe storage and seamless retrieval of valuable literary artefacts and adequate space for research, learning, performances, events and exhibitions (NELM, 2017). The broader key objectives were a limited environmental impact, cultural and educational significance, information organisation and design, social upliftment and community acceptance.

Design Approach of NELM Building

In articulating the client's objectives, the design team adopted a sustainable approach through the GBCSA, and the building served as a pilot for the development of a rating tool for public and education buildings. The GBCSA uses rating tools as a standard measurement for green buildings to promote integrated designs, raise awareness of green building benefits, recognise stewardship of the environment and reduce its impact of development. The study building was awarded a Green Star five-star rating on completion (GBCSA, 2018).

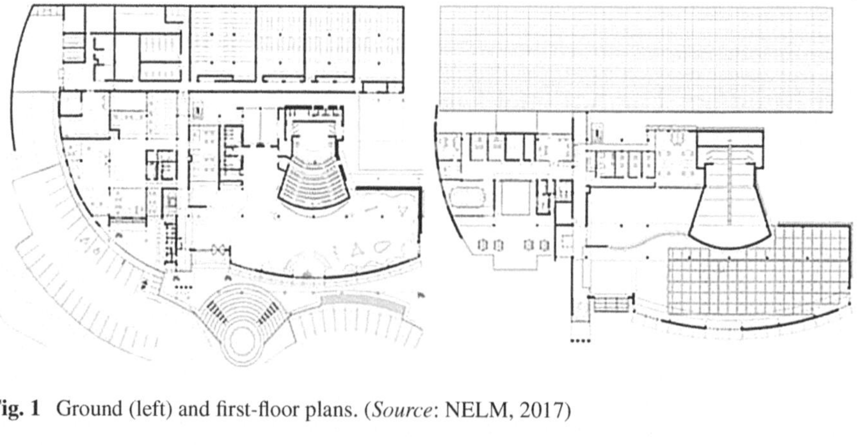

Fig. 1 Ground (left) and first-floor plans. (*Source*: NELM, 2017)

Methodology

A positivist stance was the underlying philosophy adopted for this study as it established a generalisation amongst NELM building occupants. In line with positivists who assume a deductive approach and use quantitative methods to justify observed relationships between the constituent parts of reality, this study adopted a deductive approach. A cross-sectional case study survey research design was used to identify characteristics of a population by asking a number of questions of individuals in that population related to an issue. In this case, the BUS questionnaire was employed to determine the occupants' satisfaction, comfort and forgiveness factors.

BUS Survey Questionnaire

The licence prohibited the publication of the details of the BUS survey questionnaire. The one-year user licence endorsement was in May 2018. The questionnaire consists of 11 sections (Leaman, 2011, p. 4) and includes the following information:

- Background information about age, gender, time in the building, time at a desk, time at visual display unit (VDU), workgroup size, window seats and other basic information about the sample and the respondents.
- Ratings and feedback for design, needs, image, cleaning, storage, meeting facilities.
- Response times for key variables, perceived productivity and health, thermal comfort, ventilation, lighting (including glare), noise (including interruptions).
- Other workplace performance variables, including perceived control.

The BUS questionnaire utilised a seven-(7)-point Likert scale, the mid-scale values for comfort being neither comfortable/uncomfortable and the mid-scale values for satisfaction factors being neither satisfactory/unsatisfactory. The means were interpreted as follows:

Satisfaction	Comfort
>6.14 ≤ 7.00 (Very satisfactory)	>6.14 ≤ 7.00 (Very satisfactory)
>5.29 ≤ 6.14 (Satisfactory)	>5.29 ≤ 6.14 (Satisfactory)
>4.43 ≤ 5.29 (Slightly satisfactory)	>4.43 < 5.29 (Slightly satisfactory)
>3.57 ≤ 4.43 (Neither satisfied/dissatisfied)	>3.57 ≤ 4.43 (Neither satisfied/dissatisfied)
>2.71 ≤ 3.57 (Slightly unsatisfactory)	>2.71 ≤ 3.57 (Slightly unsatisfactory)
>1.86 ≤ 2.71 (Unsatisfactory)	>1.86 ≤ 2.71 (Unsatisfactory)
≥1.00 ≤ 1.86 (Very unsatisfactory)	≥1.00 ≤ 1.86 (Very unsatisfactory)

The researcher used the hard copy version of the BUS survey questionnaire instead of the Internet for data control and enrichment. The on-site dissemination of the questionnaires with open-ended sections was in a museum staff meeting. The collection of populated questionnaires took place 2 days later.

Population

The occupant population for the building under study was small and comprised 25 occupants. Sampling included the entire population of the building (*N* = <100). A response rate achievement of 92% resulted from 23 responses.

Data Analysis

The data were analysed using descriptive statistics, comparative analysis and percentiles for Objectives 1 and 2. A mean score was calculated for each factor in the user satisfaction indicators and comfort indicators. Subsequently, the comparison of the NELM building's means to the relative means of factors on the BUS database took place. Since the BUS benchmarks are intellectual property, it restricted the publication of the numerical database benchmarks. However, in the discussion of the results, the authors did indicate whether the NELM building's means were lower than, similar to, or higher than the benchmark means.

Additionally, Objective 1 required the calculation of the satisfaction index, which was calculated by averaging the scores for the specific satisfaction variables. Objective 2 required the calculation of the comfort index, which was calculated by averaging the scores for the particular comfort variables.

Objective 3 required the calculation of a forgiveness factor. The forgiveness factor formulation was derived by dividing individual building mean scores for the variable 'Comfort overall' by the average of scores for the specific comfort variables ('Temperature in summer', 'Temperature in winter', 'Air in summer', 'Air in winter, 'Lighting', and 'Noise'). Higher values indicate that occupants are more tolerant or 'forgiving' of the conditions.

Results

The focus of this section is on the analysis of results in Table 1, indicating the satisfaction factors' scores compared to BUS benchmarks.

Table 1 Satisfaction factors' scores compared to BUS benchmarks

			Comparison to benchmark		
Satisfaction factors	Mean (scale 1–7)	Mean interpretation	Worse	Similar	Better
Design	6.04	Satisfactory			●
Health	4.50	Slightly satisfactory			●
Needs	5.81	Satisfactory			●
Satisfaction factor	*%Increase/decrease*		*Worse*	*Similar*	*Better*
Productivity	16.82				●

Objective 1 To determine the extent to which users are satisfied with the overall performance of the green building.

Satisfaction Factors

Table 1 depicts the satisfaction factors pertaining to the study.

The satisfaction parameter of the overall design of the building (6.04) significantly outperformed the perceived respondents' health (4.50). The satisfaction with design supports the participatory approach of social design of Schneider et al. (2012) which in this case was implemented through the integrative design processes of the GBCSA (2018).

The articulation of the client's needs for the NELM building commenced with the needs assessment converted into a norms document to define the spatial requirements. The project planning meetings served as a platform to conceptualise and convert the client's requirements into a design. As a result, the occupants are satisfied that their needs were met. The 'Needs' result agrees with the literature, which states that green buildings perform better in meeting the client's needs because of the integrative approach to design. The process of signing off the client's requirements at each project milestone ensures proper inclusion of the client's needs and desires (GBCSA, 2018; Sant'Anna, Dos Santos, Vianna, & Romero, 2018).

Early longitudinal studies on green buildings employing the case study method reveal green buildings increase worker productivity and reduce absenteeism (Romm & Browning, 1998). POEs consistently reveal similar results, indicating a good office environment can facilitate an increase in productivity by up to 20% through an improvement in lighting, heating and cooling (Al Horr et al., 2016). The NELM building occupants indicate a perceived increase in productivity by almost 17% (16.82%), which is in line with the literature.

The NELM building scored better than the BUS benchmarks in all the factors.

Operational Factors

Table 2 depicts the operational factors pertaining to the study.

The operational factors of a building relate to the overall management of the facilities to provide a conducive environment to occupants. These factors link the activities of the occupants and satisfaction with the physical environment (Preiser et al., 1988) and affect how well the physical arrangement and equipment within the building satisfy the needs of the occupants. In terms of the eight operational aspects, the occupants range from being very satisfied to satisfied. The NELM building also scored better than the BUS benchmark in all the factors.

Table 2 Operational factors' scores compared to BUS benchmarks

Operational factors	Mean (scale 1–7)	Mean interpretation	Comparison to benchmark		
			Worse	Similar	Better
Image	6.68	Very satisfactory			●
Safety	6.54	Very satisfactory			●
Meeting	6.31	Very satisfactory			
Storage	6.22	Very satisfactory			●
Work requirements	6.18	Very satisfactory			●
Space—building	5.95	Satisfactory			●
Space—desk	5.59	Satisfactory			●
Effect—behaviour	5.41	Satisfactory			●

Table 3 Comfort factors' scores compared to BUS benchmarks

Comfort factors	Mean (scale 1–7)	Mean interpretation	Comparison to benchmark		
			Worse	Similar	Better
Comfort overall	5.72	Satisfactory			●
Air overall—winter	5.54	Satisfactory			●
Temp—winter overall	5.54	Satisfactory			●
Air overall—summer	5.50	Satisfactory			●
Light—overall	5.50	Satisfactory			●
Temp—summer overall	5.36	Satisfactory			●
Noise—overall	4.95	Slightly satisfactory			●

Satisfaction Index

The study building's satisfaction index derived from satisfaction factors of design, health, needs and productivity scored 0.71, which indicates that the occupants are slightly satisfied with the NELM building.

Objective 2 To determine the extent to which users are comfortable within the environment of the green building.

Comfort Factors

Table 3 depicts the comfort factors pertaining to the study.

The relationship between the indoor environment quality, health and comfort of building occupants is complex and still not fully understood (Choi, Loftness, & Aziz, 2011; Heinzerling, Schiavon, Webster, & Arens, 2013; USGBC, 2018). This complexity arises from human factors such as culture, heritage, individual expectations, local customs and norms, and occupants' activities. However, the occupants of the NELM building perceived all the comfort factors as satisfactory, except for overall noise, which was slightly satisfactory. The noise factor partially supports the assertion by Leder, Newsham, Veitch, Mancini, and Charles (2016) that acoustic performance in green buildings is low in that the source of noise in the study building as reinforced by comments is mainly attributed to the dialogue during break times, trolley movement and some operational factors. Daylight glare remains the primary source of visual discomfort (Hirning, Isoardi, & Cowling, 2014; Pei et al., 2015) compared to electric lighting.

Control Factors

Table 4 depicts the control factors pertaining to the study.

The respondents indicated some control for cooling, heating and ventilation. The ventilation mean (3.59) was the lowest of the three. There was little control of lighting and almost no control of noise and this supports the observations by Leaman and Bordass (2007) and Gou et al. (2013) that where occupants lack control of conditions exceeding the perceived levels of comfort, it becomes a source of dissatisfaction.

Comfort Index

The study building's comfort index derived from comfort factors of temperature in summer, the temperature in winter, the air in summer, the air in winter, lighting, and noise which scored a mean of 1.02 (Slightly comfortable). The comfort index score (1.02) is above the mid-scale point and indicates the building on average performs better than the benchmark buildings.

Objective 3 To determine to what extent users tolerate the green building's environment.

Figure 2 depicts the forgiveness factor pertaining to the study.

The study building achieved a forgiveness factor score of 1.05. The typical value of the forgiveness factor ranges from 0.80 to 1.20, with a value exceeding one (1)

Table 4 Control factors' scores compared to BUS benchmarks

Control factors	Mean (scale 1–7)	Mean interpretation	Comparison to benchmark		
			Worse	Similar	Better
Control—cooling	4.31	Some control			●
Control—heating	4.31	Some control			●
Control—ventilation	3.59	Some control			●
Control—lighting	3.27	Little control	●		
Control—noise	2.59	No control		●	

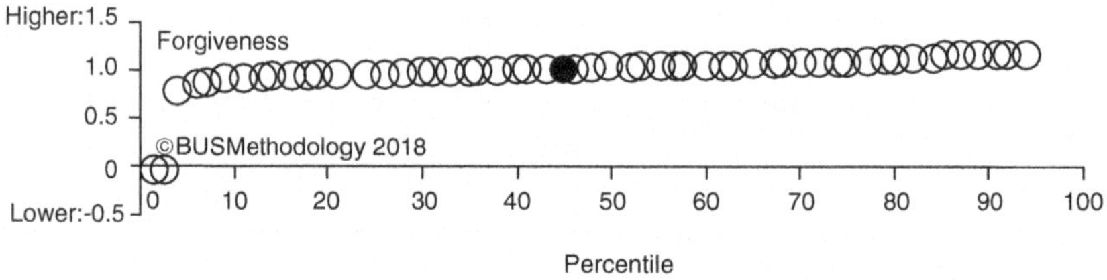

Fig. 2 Forgiveness factor

indicating the occupants may be more tolerant of the environmental conditions of the building (Baird, 2010; Leaman & Bordass, 2007).

Conclusion and Recommendations

The study reviewed the existing POE protocol to arrive at a conceptual framework in the context of academic research. The selected POE in terms of the approach was a hybrid of indicative-diagnostic, possessed high academic suitability and was low in cost. The BUS methodology was deemed the most suitable for meeting these criteria.

The findings indicate the perception of occupants with building design rated highest in terms of satisfaction, which is attributable to the integrative design process of the GBCSA and some elements of social design. Based on the comments of the users in relation to health, the POE revealed the air conditioner as the element of the building with the majority of concerns.

This information is beneficial to the facilities managers to enable occupants to exercise more control over the thermal conditions of the building. In terms of future designs and HVAC control strategies, it would be beneficial to install systems that learn and mimic the desired conditions of the occupants.

The occupants' perception of the overall comfort of the building rated the highest whereas comfort with the overall noise scored the lowest. Lack of control of noise where environmental conditions of the building exceed the perceived levels of comfort was identifiable as a source of dissatisfaction. In benchmarking the building with similar buildings, the study building performed better in terms of all aspects of comfort.

This information is beneficial to the building owners and managers in introducing noise reduction strategies in traffic areas such as floor padding for noise attenuation. Benefits could arise from a noise awareness campaign and the installation of a visible or audible decibel warning system as additional strategies. The summary finding indicates the users of the building are slightly comfortable with the environment of the green building.

The study building occupants demonstrated more tolerance with the building's environment. Tolerance is higher after the occupation of building owing to the honeymoon effect, which is characterised by an average occupancy period of 12 months. The occupancy of the study building was 24 months at the time of the POE. The tolerance of the building environment by occupants may be attributable to pro-environmental behaviours.

Facilities managers are encouraged to use POE for improving a building's performance over its life cycle as well as for troubleshooting and finding solutions to aspects of building performance affecting the behaviour of occupants.

In this fourth industrial revolution, the amount of information gathered through a POE is enormous and can serve as business intelligence to various users and institutions. It is how this information is interpreted and utilised that a competitive advantage can be created.

References

Adan, O. C. G., & Bluyssen, P. M. (2009). *The indoor environment handbook: How to make buildings healthy and comfortable*. London: Earthscan.

Al Horr, Y., Arif, M., Kaushik, A. K., Mazroei, A., Katafygiotou, M., & Elsarrag, E. (2016). Occupant productivity and office indoor environment quality: A review of the literature. *Building and Environment, 105*, 369–389.

Anderson, C. (2008). The end of theory: The data deluge makes the scientific method obsolete. *Wired Magazine*. Retrieved from https://www.wired.com/2008/06/pb-theory/

Applegate, R. (1993). Models of user satisfaction: Understanding false positives. *RQ, 32*(4), 525–539.

Baird, G. (2010). *Sustainable buildings in practice: What the users think*. Oxon: Routledge.

Baird, G., & Oosterhof, H. (2010, February). Users' perceptions of health and productivity in sustainable buildings. *Environment Design Guide* (pp. 1–10).

Brown, Z., & Cole, R. J. (2009). Influence of occupants' knowledge on comfort expectations and behaviour. *Building Research and Information, 37*(3), 227–247.

Choi, J.-H., Loftness, V., & Aziz, A. (2011). Post-occupancy evaluation of 20 office buildings as basis for future IEQ standards and guidelines. *Energy and Buildings, 46*, 167–175.

Deuble, M. P., & De Dear, R. J. (2012). Green occupants for green buildings: The missing link? *Building and Environment, 56*, 21–27.

Ewoldsen, D. R., & Beverly, R. (2013). Applying social psychology to the media. *Journal of Chemical Information and Modelling, 53*(9), 1689–1699.

Fieldson, R., & Sodagar, B. (2017). Understanding user satisfaction evaluation in low occupancy sustainable workplaces. *Sustainability, 9*(10), 1720.

Frontczak, M., Schiavon, S., Goins, J., Arens, E., Zhang, H., & Wargocki, P. (2012). Quantitative relationships between occupant satisfaction and satisfaction aspects of indoor environmental quality and building design. *Indoor Air Journal, 22*(2), 119–131.

Giese, J. L., & Cote, J. A. (2000). Defining consumer satisfaction. *Academy of Marketing Science Review, 1*, 1–24.

Goh, B. H., & Sun, Y. (2016). The development of life-cycle costing for buildings. *Building Research and Information, 44*(3), 319–333.

Gou, Z., Lau, S. Y. L., & Shen, J. (2012). Indoor environmental satisfaction in two LEED offices and its implications in green interior design. *Indoor Built Environment, 21*, 503–514.

Gou, Z., Prasad, D., & Lau, S. S. Y. (2013). Are green buildings more satisfactory and comfortable? *Habitat International, 39*, 156–161.

Green Building Council South Africa (GBCSA). (2016). *Initiative credit claimed: GBCSA's ruling INN-1 innovation strategies*. Cape Town: GBCSA.

Green Building Council South Africa (GBCSA). (2018). *Leading the transformation of the South African property industry to environmental sustainability*. Cape Town: GBCSA. Retrieved May 14, 2018 from https://gbcsa.org.za/#

Heinzerling, D., Schiavon, S., Webster, T., & Arens, E. (2013). Indoor environmental quality assessment models: A literature review and a proposed weighting and classification scheme. *Building and Environment, 70*, 210–222.

Hirning, M., Isoardi, G., & Cowling, I. (2014). Discomfort glare in open plan green buildings. *Energy and Buildings, 70*, 427–440.

Kim, J., & De Dear, R. (2012). Nonlinear relationships between individual IEQ factors and overall workspace satisfaction. *Building and Environment, 49*, 33–40.

Kurisu, K. (2016). *Pro-environmental behaviors*. Tokyo: Springer.

Leaman, A. (2011). *BUS occupant survey method: Details for licensees*. BUS Methodology.

Leaman, A., & Bordass, B. (2007). Are users more tolerant of 'green' buildings? *Building Research and Information, 35*(6), 662–673.

Leaman, A., Stevenson, F., & Bordass, B. (2010). Building evaluation: Practice and principles. *Building Research and Information, 38*(5), 564–577.

Leder, S., Newsham, G. R., Veitch, J. A., Mancini, S., & Charles, K. E. (2016). Effects of office environment on employee satisfaction: A new analysis. *Building Research and Information, 44*(1), 34–50.

Li, P., Froese, T. M., & Brager, G. (2018). Post-occupancy evaluation: State-of-the-art analysis and state-of-the-practice review. *Building and Environment, 133*, 187–202.

Liu, Y., Wang, Z., Lin, B., Hong, J., & Zhu, Y. (2018). Occupant satisfaction in three-star-certified office buildings based on comparative study using LEED and BREEAM. *Building and Environment, 123*, 1–10.

MacNaughton, P., Satish, U., Guillermo, J., Laurent, C., Flanigan, S., Vallarino, J., … Allen, J. G. (2017). The impact of working in a green certified building on cognitive function and health. *Building and Environment, 114*, 174–186.

Monfared, I., & Sharples, S. (2011). Occupants' perceptions and expectations of a green office building: A longitudinal case study. *Architectural Science Review, 54*(4), 344–355.

National English Literary Museum (NELM). (2017). *The first green museum in South Africa*. Grahamstown: NELM.

Newsham, G., Brand, J., Donnelly, C., Veitch, J., Aries, M., & Charles, K. (2009). Linking indoor environment conditions to job satisfaction: A field study. *Building Research and Information, 37*(2), 129–147.

Paevere, P., Brown, S., Leaman, A., Luther, M., & Adams, R. (2008). Indoor environment quality and occupant productivity in the CH2 building. In *SB08: Proceedings of the 2008 International Scientific Committee World Sustainable Building Conference, Melbourne, September 21–25* (pp. 222–229). Balnarring, VIC: ASN Events.

Parkinson, A. T., Reid, R., McKerrow, H., & Wright, D. (2017). Evaluating positivist theories of occupant satisfaction: A statistical analysis. *Research and Information Building, 46*(4), 430–443.

Pei, Z., Lin, B., Liu, Y., & Zhu, Y. (2015). Comparative study on the indoor environment quality of green office buildings in China with a long-term field measurement and investigation. *Building and Environment, 84*, 80–88.

Preiser, W. F. E., Rabinowitz, H. Z., & White, E. T. (1988). *Post-occupancy evaluation*. New York, NY: Van Nostrand Reinhold.

Romm, J. J., & Browning, W. D. (1998). *Greening the building and the bottom line: Increasing productivity through energy-efficient design*. Boulder, CO: Rocky Mountain Institute.

Sant'Anna, D. O., Dos Santos, P. H., Vianna, N. S., & Romero, M. A. (2018). Indoor environmental quality perception and users' satisfaction of conventional and green buildings in Brazil. *Sustainable Cities and Society, 43*, 95–110.

Schneider, F., Gruman, J., & Coutts, L. (2012). *Applied social psychology: Understanding and addressing social and practical problems* (2nd ed.). Thousand Oaks, CA: SAGE Publications.

Sinha, A., Gupta, R., & Kutnar, A. (2013). Sustainable development and green buildings. *Drvna Industrija, 64*(1), 45–53.

South African Bureau of Standards (SABS). (2011). *SANS 10400-XA: 2011: South African National Standard—The application of the National Building Regulations*. Pretoria: SABS.

United States Green Building Council (USGBC). (2018). *LEED v4: Building operations and maintenance guide*. USGBC. Retrieved May 14, 2018 from https://www.usgbc.org/guide/om

Wright, S. (1921). Correlation and causation. *Agricultural Research, 20*(7), 557–585.

Zuo, J., Pullen, S., Rameezdeen, R., Bennetts, H., Wang, Y., Mao, G., … Duan, H. (2017). Green building evaluation from a life-cycle perspective in Australia: A critical review. *Renewable and Sustainable Energy Reviews, 70*, 358–368.

Addressing Accreditation Criteria Related to Sustainable Construction in the United States: A Case Study

Richard Burt

Introduction

The majority of construction management programs in the United States are accredited by either the Accreditation Board for Engineering Technology (ABET) or the American Council for Construction Education (ACCE). ABET accredits construction management programs through its Applied and Natural Science Accreditation Commission. In general, construction management schools housed in engineering colleges are accredited by ABET and those housed in architecture colleges are accredited by ACCE. The largest construction management programs based on student enrollment are all accredited by ACCE.

Prior to July 2015, the standards and criteria for accreditation for bachelor's degree construction programs set by the ACCE were prescriptive in nature requiring construction programs to demonstrate the inclusion of a required number of credit hours of prescribed subject matter and topical content. There was no requirement in these standards to include subject matter specifically to sustainable construction issues. These standards did allow enough flexibility for environmental sciences to be used to satisfy core science requirements and environmental issues were identified as a parameter to be addressed while addressing project planning (Tinker & Burt, 2004).

R. Burt (✉)
McWhorter School of Building Science, Auburn University, Auburn, AL, USA
e-mail: rab0011@auburn.edu

L. Scott et al. (eds.), *Sustainable Ecological Engineering Design*,
https://doi.org/10.1007/978-3-030-44381-8_18

Development of the Student Learning Outcome Based ACCE Standards

In 2011, a small task force, chaired by the author, was charged by the ACCE Standards Committee to develop a set of standards based on student learning outcomes. During the 2011/2012 academic year a series of workshops was conducted with construction industry participants to gather data in order to develop a set of learning outcomes that would define the capabilities of students graduating from a 4-year ACCE accredited program. During the three workshops held in Atlanta, Dallas, and Phoenix, the following statements related to sustainability were identified by the attendees:

- *Sustainability: how projects are evaluated; what are the variables? Life cycle of decisions*
- *Understanding of sustainability and how it relates to construction*
- *Life Cycle of Buildings*
- *LEED & Sustainability*
- *Green Building Practices*
- *Green Building*
- *LEED/Environmental-Considerations*
- *Apply Sustainable Construction Practices*
- *Analyze life cycle costs*
- *Understand Recall Green Construction Practices*

The results of the industry workshops were presented to the Standards Committee at the ACCE annual meeting in July 2012. Following analysis of the results a draft set of learning outcomes was developed. In defining the learning outcomes, the task force used action verbs to determine the cognitive level consistent with Bloom's taxonomy. The draft set of learning outcomes were then sent to ACCE member schools for review. The draft learning outcomes were evaluated by two distinct groups: members of the school's industry advisory council and academic faculty. In evaluating the learning outcomes respondents were asked to pay particular attention to action verbs used in each learning outcome as it defines the cognitive level of learning expected. Respondents were asked to identify if the cognitive level of learning should be at a higher or lower level. A total of 312 surveys were received, of which 128 identified themselves as faculty and 178 identified as industry advisory council members.

Respondents were asked to rate across a 5-point Likert scale how strongly they agreed with individual student learning outcomes being included in the new standards and to make any comments related to the proposed outcome. The draft student learning outcome related to sustainability was: *Upon graduation students should be able to "Understand sustainable principles for a variety of construction activities."* This outcome was numbered 29 of 31. The draft outcomes were ranked based on their rating average. The following responses were obtained for this draft student learning outcome (Fig. 1).

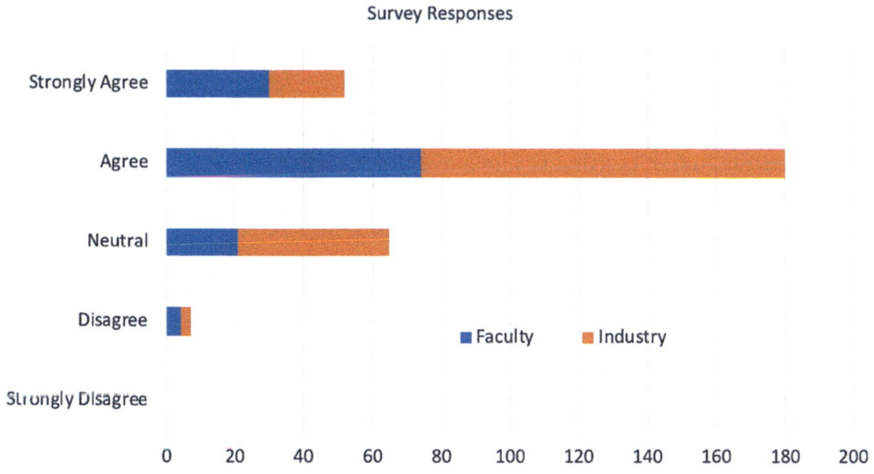

Fig. 1 Survey responses for student learning outcome 29: *Understand sustainable principles for a variety of construction activities*

Table 1 Additional comments from faculty and industry survey

Additional faculty comments	Additional industry comments
LEED use	This technology is changing and must be up to date. Research projects and problem-solving activity are best
This is an entirely overused word (sustainable). I would prefer to see "efficiency," "reduction in waste generation," or some such term. Architects in particular overuse this word	LEED accreditation would be nice, though should not be a requirement. Understanding the principles is important
Change to analyze	Prefer graduates to be LEED accredited in commercial work to the fullest extent that they can be
Explain not understand	This is becoming more and more important; plus taking sustainability to include the general environment of the community around the project
I think this should be apply	

In addition, the following comments were received (Table 1):

The above results ranked this outcome 25th out of 31 outcomes, with little difference between the ranking of faculty and industry respondents. Following analysis of these results by the task force, a recommendation was made to the Standards Committee to include as the 18th of 20 outcomes, the following student learning outcome related to sustainability: *Understand the basic principles of sustainable construction.* The revised standards have been in effect since July 2015.

Building a Sustainable Construction Curriculum Around a Student Learning Outcome Based Accreditation Standard

The incorporating of sustainable construction subject matter and topical content into the curriculum of construction management schools had been a subject of discussion in the United States since the early 2000s following significant growth in adoption of the LEED Green Building program (Nobe & Dunbar, 2004; Tinker & Burt, 2004). Similar discussions were also happening in the United Kingdom (Cotgrave & Alkhaddar, 2006). These discussions among the academic community led to some construction management programs incorporating sustainable construction content into their undergraduate curriculum. The McWhorter School of Building Science first incorporated sustainable construction subject matter and topical content into its Building Science (BSCI) curriculum as part of a major curriculum redesign in 2010.

Teaching Sustainable Construction Under the Previous ACCE Standards

Prior to the publication of the revised ACCE standards in 2015 the undergraduate degree in Building Science taught sustainable construction topics during a one credit-hour sophomore level course called *Introduction to Sustainable Construction*. All students that graduated from this program were required to take this course. The course met once a week for 1 h over a 15-week semester and the following topical content was typically covered:

- Definitions; Objectives, what is Sustainable Construction? Introduction to LEED
- Sustainable Site Work; SWPPP & Water Resource Protection
- Conserving Water Consumption
- Minimizing nonrenewable energy consumption
- Building Commissioning
- Sustainable Construction Materials & Products
- Construction Waste Management
- Indoor environmental quality
- Green Building Requirements: LEED Revisited
- Green Building Examples

As part of the school's annual assessment measures, graduating seniors are asked to complete an exit survey. One of the questions asked is to rate the value of the courses they have taken on a 5-point Likert scale. The Likert scale had one representing *not valuable at all*, through to five representing *highly valuable*. Data began to be collected for the *Introduction to Sustainable Construction* course at the end of the fall 2013 semester and continued through fall 2018. A total of 16 semesters worth of data was collected. The average response rate for each semester's graduating cohort ranged from a low of 2.10 to a high of 3.86, with an average across all 16 semesters of 3.07 in evaluating the *introduction to Sustainable Construction* course. In a somewhat simplified comparison this equates to a range of between *seldom valuable* and *valuable,* with an average of *somewhat valuable*. The comparison with all other BSCI courses the responses ranged from a low of 3.83 to a high of 4.46, with an average across all 16 semesters of 4.08. This is a significant difference and led the faculty to review the way sustainable construction was delivered within the curriculum.

Adapting to a Student Learning Outcomes Based Accreditation Standard

Following the publication of the new ACCE Document 103: Standards and Criteria for Accreditation of Bachelor's Degree Construction Education Programs, faculty at the McWhorter School of Building Science at Auburn University began a curriculum review process to adapt its BSCI curriculum to the new standards. This process began with an analysis of each individual student learning outcome to identify the knowledge and skills each outcome sought to cover and the appropriate cognitive level that needed to be achieved. For the student learning outcome—*Understand the basic principles of sustainable construction,* this began with identifying what the faculty believed were the principles of sustainable construction.

After a reviewing the literature, together with the requirements of the LEED green building rating system and consultation with BSCI faculty with expertise in the area of sustainable construction, the following basic principles of sustainable construction were identified:

- Consideration of environmental impacts of construction activities on the site and its surroundings
- Minimize nonrenewable energy consumption use
- Use environmentally preferable products in the construction process
- Reduce construction waste
- Protect water resources and conserve water consumption
- Enhance indoor environmental quality
- Operate buildings using practices that are cognizant of their environmental impacts

Based on these principles the faculty identified draft course learning outcomes relating to this outcome. In developing these course learning outcomes faculty were attentive to the appropriate cognitive level and used action verbs in the framing of these outcomes, consistent with the cognitive level of understand such as explain and describe. These were reviewed and revised by industry stakeholders at two curriculum development workshops in Atlanta, GA and Birmingham, AL. Finally, faculty with expertise in the subject matter refined these to develop the final course learning outcomes. The course learning outcomes included in the new curriculum were:

1. Identify green building requirements relating to construction
2. Explain how sustainable construction practices minimize nonrenewable energy consumption use
3. Explain how the use of certain types construction materials and products can minimize environmental impact
4. Describe how construction companies reduce construction waste
5. Demonstrate how sustainable construction practices protect water resources and conserve water consumption
6. Describe how sustainable construction practices enhance indoor environmental quality
7. Describe how buildings are commissioned

Once these course learning outcomes were finalized, the next decision taken was how and where these course learning outcomes should be taught and ultimately assessed. As mentioned previously student's perception of the value of the *Introduction to Sustainable Construction* course was significantly less than other BSCI courses. The faculty felt that this was mainly due to the timing of when the course was offered and because the course was only worth one credit hour when the majority of the other BSCI classes were worth between three and four credit hours. The BSCI curriculum model is arranged such that students usually took this course in the first or second year of a 4-year program. The majority of the students took this course before they had taken courses in materials and methods, mechanical, electrical, and plumbing systems, and construction management topics. Many of the topics discussed in the *Introduction to Sustainable Construction* required them to have some basic exposure to and understanding of some of the topics taught in these classes. For example, it is very difficult for students to understand the commissioning process if they have little or no knowledge of mechanical, electrical, and plumbing systems. After a proposal to teach this class later in the curriculum was discussed and rejected, it was decided to integrate the course learning outcomes and associated topics across the revised curriculum.

Courses are taught in BSCI in each of the 4 years of the program. Due to the requirements to teach the university core curriculum, only four BSCI classes are taught in the first 2 years (freshman and sophomore) of the program. It was therefore decided to spread approximately 13 h of instruction across seven different BSCI classes, these are set out in Table 2.

Assessment of Student Learning Outcomes

One of major changes set out in ACCE's Document 103B: Standards and Criteria for Accreditation of Bachelor's Degree Construction Education Programs is a requirement for programs to demonstrate how students have achieved all of the 20

Table 2 Mapping course learning outcomes to BSCI classes

Course learning outcome	Class	Hours
Identify green building requirements relating to construction	BSCI 1100—Introduction to Construction	2
Explain how sustainable construction practices minimize nonrenewable energy consumption use	BSCI 4700—Mechanical Systems in Buildings	2
Explain how the use of certain types construction materials and products can minimize environmental impact	BSCI 2300—Construction Materials and Methods	2
Describe how construction companies reduce construction waste	BSCI 4350—Construction Project Analysis	2
	BSCI 4360—Construction Field Lab	1
Demonstrate how sustainable construction practices protect water resources and conserve water consumption	BSCI 4700—Mechanical Systems in Buildings	2
Describe how sustainable construction practices enhance indoor environmental quality	BSCI 4700—Mechanical Systems in Buildings	1
Describe how buildings are commissioned	BSCI 4700—Mechanical Systems in Buildings	1
	BSCI 4750—Electrical Systems in Buildings	1
Total		14

ACCE Student Learning Outcomes. The main requirement is for programs to use a direct assessment method that can evaluate a student's learning through their academic performance. There is also a requirement for programs to use the results of this assessment as part of a Quality Improvement Plan (American Council for Construction Education, 2018).

All students in the BSCI program are required to take BSCI 4990 *Thesis* in their final year. This had been a requirement of the program dating back to the 1950s. Although titled *Thesis,* the requirements for the class are more like a traditional capstone class. Students are required to create an individual project demonstrating mastery of curriculum content through the application of skills/knowledge to a theoretical construction company and a project. Students are required to put together a written thesis and make an oral defense of their work. As the teaching of sustainable construction topics was spread across multiple BSCI courses, it was decided that the assessment of the student learning outcome related to sustainable construction would be conducted in BSCI 4990 *Thesis.*

As a requirement of BSCI 4990 *Thesis* students are required to obtain a complete set of construction documents for a building where the cost of project is between \$1,000,000 and \$3,000,000 and the building floor area between 9000 and 12,000 SF. These documents are used by the students to respond to a series of prescribed tasks such as creating an estimate and schedule. The requirements related to sustainable construction are set out in Table 3 below. This portion of the thesis accounts for 5% of their final grade.

Table 3 BSCI 4990 thesis requirements related to sustainable construction

You are to conduct an assessment of your project building to demonstrate that you understand how the design and construction of your building reflects the basic principles of sustainable construction. Set out below are specific tasks to complete that relate to principles of sustainable construction. You are to answer these as they relate specifically to your building
1. The USGBC through its LEED certification program has different certification programs for different construction projects
• Select the current certification program that would be applicable to your construction project and locate and reference at least two resources or tools from the USGBC website that set out the requirements of the specific certification program
2. During construction, a contractor should consider the environmental impacts of construction activities on the site and its surroundings. A LEED Pre-requisite on any LEED certified project is to reduce pollution from construction activities by controlling soil erosion, waterway sedimentation, and airborne dust
• Review your project documents to identify any national, state or local requirements that control site erosion and sedimentation
• Identify six specific measures incorporated into your project that reduce pollution from construction activities. Describe how each measure helps to reduce pollution using illustrations obtained from project documentation or other sources
3. Sustainable construction projects seek to minimize nonrenewable energy consumption, protect water resources and conserve water consumption. This is accomplished through good design and operating the building using sound environmental practices. A LEED Pre-requisite on any LEED certified project is to provide fundamental commissioning and verification to support the design, construction, and eventual operation of a project that meets the owner's project requirements for energy, water, indoor environmental quality, and durability
• Review your project documents to identify and summarize any specific requirements related to commissioning and verification
• Review the document New Construction Building Commissioning Best Practice by the Building Commissioning Association. Identify and describe the contractor's commissioning responsibilities during the construction phase for a project such as yours
• Identify six pieces of commissioned equipment from your project that might be included in the construction checklist and describe the specific commissioning process for at least one piece of equipment
4. Sustainable construction projects seek to use environmentally preferable products in the construction process. The LEED certification process seeks to minimize the embodied energy and other impacts associated with the extraction, processing, transport, maintenance, and disposal of building materials and gives credit for using construction products that provide building product disclosure and optimization
• Choose three construction products used in your project that you believe are environmentally preferable
• Locate the product manufacturers website and use the information available to explain how these products seek to minimize the embodied energy and other impacts associated with the extraction, processing, transport, maintenance, and disposal of building materials
5. Another sustainable construction principle is to reduce construction and demolition waste disposed of in landfills and incineration facilities by recovering, reusing, and recycling materials
• Identify three material streams used in your project where waste materials could be diverted from landfill or incineration

(continued)

Table 3 (continued)

• Describe with specific reference to your project how the three waste materials streams will be collected during the construction phase and processed locally after they leave the site. Your description should include a site utilization plan highlighting key features specific to waste management (i.e., dumpsters, salvage material lay down)
6. The quality of the indoor environment is essential in sustainable construction projects. LEED certification seeks to promote the well-being of construction workers and building occupants by minimizing indoor air quality problems associated with construction and renovation. This is achieved by developing and implementing an indoor air quality (IAQ) management plan for the construction and preoccupancy phases of the building
• Identify the requirements and procedures and describe how you would protect the air distribution system (for example, ductwork) during construction
• Give an example specific to your project and describe how you would protect absorptive materials stored on-site and from moisture damage

In order to obtain data to evaluate student performance, a grading rubric is used to record individual student performance. The grading rubric measures student performance across five separate criteria and performance is evaluated across a 5-point grading scale. The grading rubric used is set out in Fig. 2 below.

Data is collected from the grading rubrics every semester this class is offered which is during the spring, summer, and fall semesters. Data has been collected continuously since the fall of 2015. The average performance score for each of the five assessment criteria is reported and incorporated into the annual Quality Improvement Report. The results are shared with both the faculty and the school's Industry Advisory Council members. The results compiled from the grading rubrics are set out in Fig. 3 below.

Conclusion

This paper has presented one US construction's schools' journey to revise its curriculum following significant and fundamental changes to the national accreditation standards. The shift from a prescriptive-based standard to a student learning outcomes-based standard required the program to review not only what was being taught in the curriculum but also how students were assessed on what they had learned. Based on previous shortcomings in the previous curriculum model, a decision was made to integrate sustainable construction topics across the revised curriculum and evaluate the students learning in their final semester during their capstone experience. Results obtained from assessment tools conducted since the fall of 2015 suggest students are consistently meeting the performance criteria. These results reflect mainly students that took the *Introduction to Sustainable Construction* class. Students that began learning about sustainable construction through the courses set out in Table 2, began being assessed in the spring of 2019

Sustainable Costruction Rubric

Criteria	Key Metric	Grading Scale 5	4	3	2	1	0	Student Score
LEED Certification Program Selection	Correct Program Selected							
Environmental impacts of construction activities on the site (5 Points)	Identify requirements that control site erosion and sedimentation.	All project specific reqs identified	Most project specific reqs identified	Some project specific reqs identified	Reqs identified are not project specific	Some generic non project specific reqs identified	No reqs identified	
	Identify and describe 6 specific measures	6 project specific measures identified and fully described and illustrated	Less than 6 project specific measures identified and/or descriptions lacking detail	Less than 4 project specific measures identified and/or descriptions lacking detail	Measures identified and described but are not project specific	Measures identified but little attempt to describe them	No measures identified	
Fundamental commissioning and verification (5 Points)	Identify and summarize project commissioning & Verification	Documents reviewed and all reqs identified					No evidence of project docs review	
	Identify and describe contractors responsibilities	Identification and description consistent with best practice	All responsibilities identified and some description	All responsibilities identified	Some responsibilities identified but little description	Some responsibilities listed	No responsibilities identified	
	Identify 6 pieces of equipment and describe one process in detail	6 pieces of equipment identified and process well described	6 pieces of equipment identified and process somewhat described	6 pieces of equipment identified	4 pieces of equipment identified	2 pieces of equipment identified	No equipment identified	
Environmentally preferable products (5 Points)	Choose 3 environmentally preferable products	3 products correctly identified					No products identified	
	Explain how products are environmentally preferable	Website information used to explain all ways the 3 products reduce environmental impact	Website information used to explain some of the ways ways the 3 products reduce environmental impact	Website information used to explain some of the ways 2 products reduce environmental impact	Non-product information used to explain some of the ways 3 products reduce environmental impact	Non-product information used to explain some of the ways 2 products reduce environmental impact	No explanation	
Reduce construction waste (5 Points)	Identify 3 material streams	3 material streams identified					No material streams identified	
	Describe how materials collected and processed	Collection & processing of material streams for all 3 are described & are project specific	Collection & processing of material streams for all 3 are described but are not project specific	Collection & processing of material streams for 2 are described & are project specific	Collection& processing of material streams for 2 are described but are not project specific	Only 1 Collection & processing of material stream described	Collection and processing not described	
	Site utilization plan for CWM	Site utilization plan shows project specific locations of dumpsters, salvage material lay down etc. for all 3 streams	Site utilization plan shows project specific locations of dumpsters, salvage material lay down etc. for 2 streams	Site utilization plan shows project specific locations of dumpsters, salvage material lay down etc. for 1 stream	General information about locations of dumpsters, salvage material lay down etc. given but not project specific		No site utilization plan	
Indoor air quality (IAQ) management plan (5 Points)	Identify requirements and procedures and describe protecting the air distribution system	Requirements, and procedures are project specific and description is consistent with industry best practice		Requirements, and procedures are not project specific and description is not consistent with industry best practice			No Requirements, and procedures identified	
	Example of protecting absorptive material	Example is project specific and description is consistent with industry best practice		Example is not project specific and/or description is not consistent with industry best practice			No example given	
Total Score	(Sum of all points above)	0					**This score to overall rubric**	
% of total Points	(Total Score/25)	0.00%						

Fig. 2 Sustainability grading rubric

and future results will be analyzed to identify any significant changes. There is no doubt that changes to the ACCE accreditation standards introduced in 2015 have had a significant affect on the teaching of sustainable construction topics in the United States. From a position in the early 2000s when there was no formal requirement to incorporate to sustainable construction subject matter, there is now a

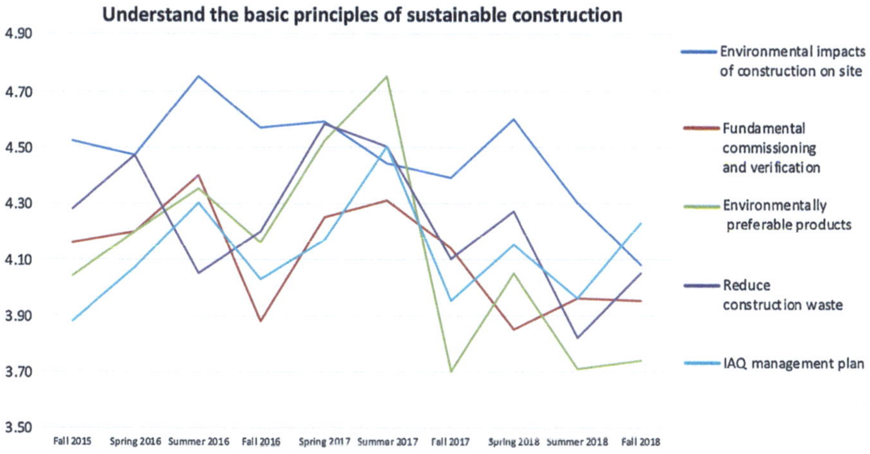

Fig. 3 Results from sustainability grading rubric

requirement not only to demonstrate that this topic is being taught but also to document that students have an understanding of the *basic principles of sustainable construction.*

References

American Council for Construction Education. (2018). *Document 103B: Standards and criteria for accreditation of bachelor's degree construction education programs.* Retrieved from https://www.acce-hq.org/images/uploads/Doc_103B_Final_Updated_0727182.pdf.

Cotgrave, A., & Alkhaddar, R. (2006). Greening the curricula within construction programmes. *Journal for Education in the Built Environment, 1*(1), 3–29. https://doi.org/10.11120/jebe.2006.01010003

Nobe, M. C., & Dunbar, B. (2004, April). Sustainable development trends in construction. In *Proceedings of ASC Proceeding of 40th Annual Conference.*

Tinker, A., & Burt, R. (2004). "Greening" the construction curriculum. *International Journal of Construction Education, VIX*(2), 26–33.

Part III
Project Management in Sustainability

Stakeholder Management: Proposal for Research—Do Successful Project Managers Employ 'Interest-Based Negotiation' to Create Successful Project Outcomes?

John Heathcote, Colin Butlin, and Hadi Kazemi

Introduction

Simplistic generalisations (APM, 2006) have proposed that stakeholders to the project are merely 'satisfied', which leads to the likelihood that the original business case (or value proposition) is compromised in order to meet others' objectives, objectives that may well be partisan. Around 10 years ago, Murtoaro and Kujala (2007) claimed that the subject of negotiation had not been studied very systematically in the project context, and based on the findings from a literature review, not much has changed. On a more positive note however, academic literature on project management and general negotiation are both extensive, therefore enabling this research to study each of these subjects independently with the aim of gaining a greater understanding of negotiations in the context of project management. In terms of project management literature, current thinking on this subject is shifting from a 'hard systems' model that emphasises on the planning and control dimensions, to a 'soft systems' model that includes relationship and resource management (Pollack, 2007) or more collectively worded, stakeholder management. Corresponding with this shift, project management success is defined as 'the appreciation by the various interested parties of the project outcomes', with the interested parties being 'people or groups who are interested in the performance and/or success of the project, or who are constrained by the project' (APM, 2006; Karlsen, 2008; Morris, 2013) and, therefore, it is proposed that project success is best judged by the project stakeholders (Turner & Zolin, 2012). Given that project stakeholders have the ability to impact on the success of a project however, or simply interpret its

J. Heathcote (✉) · C. Butlin · H. Kazemi
Leeds Sustainability Institute, School of the Built Environment and Engineering, Leeds Beckett University, Leeds, UK
e-mail: j.heathcote@leedsbeckett.ac.uk

L. Scott et al. (eds.), *Sustainable Ecological Engineering Design*,
https://doi.org/10.1007/978-3-030-44381-8_19

success differently to other stakeholders (Turner, 2009), negotiation is considered a key component of project management (Morris, 2013) although the lack of research in this specific area may suggest otherwise.

In terms of the 'general' negotiation literature, negotiation parties typically adopt one of two major approaches, either positional/distributive or interest-based/integrative and while an integrative approach is generally recommended (Fisher, Ury, & Patton, 1991; Thompson, 2011), the literature also promotes the advantages of a distributive approach for some negotiation situations (Baarveld, Smit, & Dewulf, 2015), such as those concerned with the division of a single resource whereby only one issue is under negotiation. As established from the project management literature however, projects involve many stakeholders, all of whom have different interests at different stages of the project (Turner, 2009), and as such, this would suggest that distributive negotiations may be an ineffective strategy for project management. Further supporting this point, it has been identified that stakeholders complicate project negotiations by the problem of their relative power (Mitchell et al., 1997) which is further compounded by the fact that their power can change during the project life cycle (Aaltonen and Kujala (2010) and, consequently, multiple negotiations are necessary throughout the project to avoid stakeholders from using their power to block the process (Van Rijswick & Salet, 2012). As highlighted across the literature, powerful stakeholders can use a variety of tactics to reduce awareness and concern about their own actions while raising the alarm about others and consequently selecting the right negotiation strategy (Pruitt & Rubin, 1986) is a critical part of the project management process. Due to the complexities and variability of the power held by stakeholders, distributive or bargaining approaches to negotiation would appear unsuitable for negotiations by project managers with stakeholders. The basic idea of stakeholder management, largely agreed by contemporary writers on stakeholder management (Forester, 2009; Baker et al., 2012; Liu & Wang, 2014), is that they should be *engaged*, and so an interest based/integrative negotiation is the only negotiation method that would be suitable and Achterkamp and Vos (2008), Thompson (2011), Morris (2013) and Susskind, Mnookin, Rozdeiczer, and Fuller (2005) appear to support this for projects, with Murtoaro, Kujala, and Artto (2005) taking this point a step further by suggesting that a project may be conceived as a single continuum of recurring negotiations with multiple parties, at different times, and with varying concerns.

As Fisher et al.'s (1991) interest based/integrative approach to negotiation (one they refer to as: 'Principled negotiation') is best described, and it offers a process for interest-based negotiation than can be reduced to a set of principle components, which the presence of those principles in the negotiation with stakeholders on projects can be tested in a closed question survey.

Research Review

Work has been conducted to scope out the content of 'stakeholder management' literature. Suggestions on how to 'manage' stakeholders in the project's environment are limited to generalisations, though 'stakeholder engagement' is an emerging theme. Separately to the work on stakeholder management is the idea of 'interest-based negotiation'(IBN), and this approach offers an approach to assist project managers and teams in meaningful engagement that offers a way to both maintain the project's outcome objectives accommodate stakeholders' interests and to do so without compromise of the project's raison d'etre. To bridge this gap in project management process and to investigate whether negotiation played the important role literature was purporting, the researchers were led to hypothesise that:

Null-hypothesis H0:	IBN principles **cannot** be correlated with successful stakeholder negotiation outcomes in projects
Alternative hypothesis H1:	IBN principles **can** be correlated with successful stakeholder negotiation outcomes in projects

Research Methodology

Hypothesis development is an important step in the logical positivist approach to the application of the scientific method (Polit & Beck, 2012). In the case of this proposed study the work on the secondary data is prolific enough to allow for the hypothesis to be generated. Asking about the potential for the use of interest-based negotiation allows for an established set of principles that can be utilised to test the theory. Furthermore in the course of the development of the study, the researchers were able to uncover significant anecdotal evidence from several case studies that had been recalled by practitioners. Consequently, the researchers were able to use this as an interpreted observation (Polit & Beck, 2012) that in itself could form a start point from which to hypothesise.

The proposed research method would adopt a deductive approach, and set a set of closed questions, to which Likert scale responses would be requested. Deductive closed question surveys can be employed to test a hypothesis (Easterby-Smith, Thorpe, & Lowe, 1995; Emery & Cooper, 1991; Polit & Beck, 2012). Adopting a positivist position and applying quantitative research methods, this research tests the hypothesis through a comparative study that uses a questionnaire to compare successful and unsuccessful stakeholder outcomes when using and not using IBN principles during project stakeholder negotiations. Choosing how to quantify these variables, however, is of the utmost importance (Polit & Beck, 2012) and therefore requires careful design. It is suggested that the key to a comparative design is its ability to allow the distinguishing characters of two or more cases to act as a springboard for theoretical reflections about contrasting findings (Bryman & Bell, 2003), and thus this research selected a questionnaire to compare contrasting 'outcomes' when IBN principles are and are not applied.

After a series of screening questions to verify that the respondent sample's experience included projects likely to include some stakeholder interest complexity, respondents would be asked firstly to '*Thinking of a time when you have had an* ***successful*** *stakeholder outcome…*'. This prepositioning of the respondents would then be followed up by series of questions that did not mention 'interest-based negotiations' (IBN) but identified conditions and approaches that could be directly related to the component processes of 'interest-based negotiation'. In the proposal this set of questions relate to $n = 5$ questions, questions 1–5 set out and justified in Table 1. Following this the respondent would again be prepositioned with the statement: '*Thinking of a time when you have had an* ***unsuccessful*** *stakeholder outcome…*'. Questions 6–10 questions would repeat questions 1–5.

Table 1 identifies the principles of interest-based negotiation that guided the construction of the survey questions.

Table 1 Comparison table for IBN principles and survey questions

IBN principles (Fisher et al., 1991)	Research questions	IBN outcome
Separate the people from the problem: Disentangle the people problems from the substantive problems and work on each separately	*Question 1* The discussions focused on tackling a joint problem, not each other	*Question 5*: The meeting/negotiation concluded with a win/win solution with both parties satisfied
Focus on interests, not positions: Identify and make explicit the needs or interests that the people want satisfied from the negotiation	*Question 2* The interests of both parties were identified and understood	
Generate a variety of options for mutual gain: Before deciding upon a specific solution, invent a variety of alternatives that advance shared interests and creatively reconcile differing interests	*Question 3* New/alternative solutions were identified through joint discussion	
Insist that the result be based on some objective standard: Where interests conflict, make a decision based on some fair standard independent of the naked will of either side	*Question 4* A mutually agreed 'criteria' was used to reach a solution	

Discussion and Conclusions

If the hypothesis can be supported and 'interest-based negotiation' approaches support better stakeholder outcomes, then we would expect a positive correlation between IBN responses and successful stakeholder outcomes, and the converse for unsuccessful stakeholder outcomes.

The expected result allows for a negative correlation with unsuccessful stakeholder outcomes, so that the absence of the key principles of IBN and the results for these are accommodated in the design of the study.

A positive correlation between successful stakeholder outcomes and the principles of IBN (and specifically which of the identified four principles), would provide specific guidance for project managers around how to approach the 'engagement' of interest-based (or integrative) negotiation. It would bridge the current gap between stakeholder management processes that exist in normative project management. While interest-based negotiation attracts theoretical support from contemporary writers on project management (and anecdotal and case study support from other fields), this study if it finds a positive correlation would be able to provide some deductive/quantitative support were the study able to achieve statistical significance. For this reason, it is proposed that a statistical significance test (such as the Mann-Whitney U test or similar) be applied to establish a 'P' value, which will show the probability of whether the results are down to chance along, or whether they are reliable. Were the results to show an unexpected correlation, for example, for the unsuccessful stakeholder outcomes, or a negative correlation, then these results will potentially offer findings that will permit new lines of inquiry.

References

Aaltonen, K., & Kujala, J. (2010). A project lifecycle perspective on stakeholder influence strategies in global projects. *Scandinavian Journal of Management, 26*, 381–397.

Achterkamp, M. C., & Vos, J. F. J. (2008). Investigating the use of the stakeholder notion in project management literature, a meta-analysis. *International Journal of Project Management, 26*, 749–757.

Association for Project Management. (2006). *APM body of knowledge* (5th ed.). Princes Risborough: Association for Project Management.

Baarveld, M., Smit, M., & Dewulf, G. (2015). Negotiation processes in urban redevelopment projects: Dealing with conflicts by balancing integrative and distributive approaches. *Planning Theory & Practice, 16*(3), 363–384.

Bryman, A., & Bell, E. (2003). *Business research methods* (Vol. 539). Oxford: Oxford University Press.

Easterby-Smith, M., Thorpe, R., & Lowe, A. (1995). *Management research: An introduction*. London: Sage Publications.

Emery, C. W., & Cooper, D. R. (1991). *Business research methods*. Burr Ridge, IL: Irwin.

Fisher, R., Ury, W., & Patton, B. (1991). *Getting to yes: Negotiating agreement without giving in* (2nd ed.). New York: Penguin Books.

Forester, J. (2009). Dealing with Differences: Dramas of Mediating Public Disputesby Oxford University Press.

Karlsen, J. T. (2008). Forming relationships with stakeholders in engineering projects. *European Journal of Industrial Engineering, 2*, 35–49.

Liu, S., & Wang, L. (2014). Understanding the impact of risks on performance in internal and outsourced information technology projects: The role of strategic importance. *International Journal of Project Management, 32*(8), 1494–1510. Elsevier Ltd.

Mitchell et al. (1997). The Conspicuous lCorporationBusiness, Public Policy, and Representative Democracy University of Michigan Press.

Morris, P. W. (2013). *Reconstructing project management*. Hoboken: Wiley.

Murtoaro, J., & Kujala, J. (2007). Project negotiation analysis. *International Journal of Project Management, 25*, 722–733.

Murtoaro, J., Kujala, J., & Artto, K. (2005). Negotiations in project sales and delivery process: An application of negotiation analysis. Espoo, Helsinki University of Technology, Laboratory of Industrial Engineering and Management, Report 2005/3.

Polit, D.F., and Beck, C.T. (2012), Nursing research: Generating and assessing evidence for nursing practice 9. Reference and Research Book News, Vol. 26(3).Philadelphia, PA: Lippincott Williams and Wilkins

Pollack, J. (2007). The changing paradigms of project management. *International Journal of Project Management, 25*, 266–274.

Pruitt and Rubin (1986). Pruitt, D. G. and Rubin, J. Z. 1986. Social conflict: Escalation, stalemate, settlement, New York: Random House.

Susskind, L., Mnookin, R. H., Rozdeiczer, L., & Fuller, B. (2005). What we have learned about teaching multi-party negotiation. *Negotiation Journal, 21*(3), 395–408.

Thompson, L. (2011). *The mind and heart of the negotiator*. Upper Side River, NJ: Pearson Education.

Turner, R, (2009). A Model of project complexity: distinguishing dimensions of complexity from severity. In International Research Network of Project Management Conference (IRNOP), 2009-10-11-2009-10-13.

Turner, R., & Zolin, R. (2012). Forecasting success on large projects: Developing reliable scales to predict multiple perspectives by multiple stakeholders over multiple time frames. *Project Management Journal, 43*(5), 87–99.

Van Rijswick, M., & Salet, W. (2012). Enabling the contextualization of legal rules in responsive strategies to climate change. *Ecology and Society, 17*(2), 18.

Effective Management of Hazardous Asbestos Waste Within a Confined Water Main Pipeline Construction Project: Multiple Case Study Review

Paul Collins and John P. Spillane

Introduction

As cities around the globe become more established, it is foreseeable that there will be an increase in urban development. One of the key sectors to facilitate this development is that of the construction industry. With the construction industry becoming more and more time, quality and cost centric, working in such confined site locations brings with it increased managerial difficulties. This is particularly the case where sites contain hazardous waste, which must be extracted, handled and removed from site, safely. For the purposes of this study, the confined construction site in question are that of pipeline construction, which are conducting decommissioning works on asbestos pipes, where they are typically installing replacement concrete or ductile iron pipes. Asbestos is a hazardous material that was widely used across the construction industry, mainly due to its insulation and fire resistance properties, along with being lightweight. However, such usage was banned when the dangers of the material were discovered (Li et al., 2014).

Due to the increased importance of health and safety on construction sites, it is imperative that such hazardous waste be removed and disposed of safely. To achieve this, particularly where spatial constraints exist, such as those that occur on inner city, confined urban construction sites, can be challenging, primarily due to the lack of storage and handling space available. To further such concerns, it is estimated that materials account for 45–60% of the on-site cost (Spillane & Oyedele, 2017); thus, effective measures to assist the management and mitigation of the risks that handling hazardous waste poses to those in and around site must be considered.

P. Collins · J. P. Spillane (✉)
Construction Management and Engineering, School of Engineering, University of Limerick, Limerick, Ireland
e-mail: john.spillane@ul.ie

L. Scott et al. (eds.), *Sustainable Ecological Engineering Design*,
https://doi.org/10.1007/978-3-030-44381-8_20

Furthermore, Begum et al. (2006) state that, based on the average breakdown of waste around the world, 28% is generated from the construction and industrial industry, which amplifies the need for further research on waste management in the construction industry and its current unsustainable approach to material and waste minimisation.

Based on this premise, the aim of this paper is to investigate effective functions, if any, of increasing the health and safety of workers and reducing costs, on confined pipeline construction sites, which are removing hazardous asbestos waste. The results will enable site management to acknowledge and address the concerns of managing the safe handling, extraction and removal of hazardous asbestos waste, within a confined water main pipeline construction project.

Management of Materials Within a Confined, Hazardous Waste Construction Site: A Thematic Literature Review

To garner further insight into the topic and highlight resultant themes, a literature review was carried out. This also aided in identifying the gap in knowledge on confined pipeline construction sites dealing with asbestos removal. This research investigated several approaches which have been implemented into construction sites previously, to reduce costs and increase health and safety. This includes Materials Management, Space Scheduling, Critical Space Analysis and Supply Chain Management, all of which were considered by Spillane and Oyedele (2017), in a study focusing on confined site construction, but not that of hazardous waste removal. Based on this premise, these four themes were used as a base on which to investigate and develop the literature reviewed and, subsequently, the discussion of the results from the case studies identified.

Materials Management

Site materials management can be defined as the allocation of delivery, storage, and handling, spaces and resources for supporting the labour force and minimising inefficiencies (Thomas, Riley, & Messner, 2005). Spillane, Oyedele, Von Meding, and Konanahalli (2010) identify factors, such as a lack of storage space, over-crowding, lack of room for the effective handling of materials, and the difficulty in transporting materials within the site, as the main causes contributing to poor production on confined construction sites. While Thomas et al. (2005) believe materials on construction sites should be portioned into three areas: semi-permanent storage, staging areas and workface storage; however, this fails to consider sites with hazardous waste.

To achieve a satisfactory materials management system, Patel and Vyas (2011) outline that the functions of the materials management model need to be categorised into two functions: primary and secondary functions. Patel and Vyas (2011) identify primary functions as materials requirement planning, purchasing and inventory planning and control, while also mentioning secondary functions such as standardization and simplification and forecasting and planning.

Thomas et al. (2005) highlight that the approach used in managing the materials on site at the beginning of a project will not necessarily be of benefit in the middle or latter stages of the project. Navon and Berkovich (2006) note that the construction industry invests only 0.15% of the production costs into materials management and control; further evidence to suggest that more efficient methods of materials management need to be implemented.

Space Scheduling

Space Scheduling is the allocating of workspace to resources associated with activities in a schedule as it changes over time (Tommelein & Zouein, 1993). In this context, the physical construction works take up the majority of the construction site; space scheduling was an essential asset to implement on site. Winch and North (2006) believe that any space scheduling system should be used as a decision support tool, not a decision-making tool, and also merge with existing applications used by the management team on-site, along with being user friendly. Choo and Tommelein (1999) focus on the application of lean construction to space scheduling, stating the importance of considering the flow of the conversion process in construction projects over the sequence of activities which represent the conversion process. Tommelein and Zouein (1993) focus on the benefits of the MovePlan model over other solutions, which uses a two-dimensional representation, to allocate space. Such applications may have merit on sites dealing with the disposal of hazardous waste.

Zouein and Tommelein (2001) state that site layout planning and the scheduling of activities need to be decided at the planning stage, to avoid losses in time and costs, although this fails to take confined construction sites into consideration, where space is at a minimum. Choo and Tommelein (1999) identify the WorkMovePlan, which combines the Last Planner methodology implemented in the WorkPlan software with Space Scheduling. The WorkMovePlan has the potential to increase efficiency on confined pipeline construction projects, by allowing a project manager to specify site spaces on a day-to-day basis, for labour, equipment and materials (Choo & Tommelein, 1999).

Critical Space Analysis

This approach applies methods of analysis into the project planning in construction, which allows a project manager to not only plan works on site each week but also allocate space for these works to take place. North and Winch (2002) define critical space analysis as the examination and optimisation of the spatial loads placed on a construction project's available space, by its scheduled tasks. Critical space analysis is presented as a possible solution to space scheduling of materials and waste on construction sites, especially those of a confined nature. Critical space analysis is a decision support tool for the management team on a construction site, to address space scheduling on site.

Winch and North (2006) note that the average worker needs 19 m^2 of space to conduct his/her job effectively, and production will be halved if this is cut down to 10.4 m^2 (resulting in costs doubling in wages alone) and also note that maximum production occurs at 30.2 m^2 which is rarely available on a confined urban construction site. Winch and North (2006) express the benefits of the VIRCON System (Virtual Construction Site), which is capable of processing data files from AutoCAD and incorporate this with data from Microsoft™ Project, to improve the strategic planning on site. There is of course the possibility that the VIRCON System would be applicable on a confined pipeline construction site dealing with hazardous waste, though there is little evidence to suggest that it was taken into consideration.

Dawood and Mallasi (2006) identify a software tool known as PECASO (patterns execution and critical analysis of site-space organisation), which was developed with the methodology of critical space analysis to 'assist site managers in the assignment and identification of workspace conflicts'. The methodology of critical space analysis applied in PECASO focuses on dynamic visual planning, identification of activity execution workspace, critical space analysis and quantification and varying execution pattern and production rate (Dawood & Mallasi, 2006).

Although research on critical space analysis is proven to increase production on confined construction sites, there is little regard for those sites which are dealing with hazardous waste in the form of asbestos.

Supply Chain Management

Supply chain management can be defined as 'the network of organisations that are involved, through upstream and downstream linkages, in the different processes and activities that produce value in the form of products and services in the hands of the ultimate customer' (Vrijhoef & Koskela, 2000). Fernie and Thorpe (2007) furthers this in stating that it is often viewed from a perspective that seeks to strategically manage and strongly position a firm within markets. However, Vrijhoef and Koskela (2000) note that construction supply chain can be defined as a temporary supply chain producing one-off construction projects. Spillane et al. (2011) provide context; if increased logistical management is needed, supply chain management is essential in the overall material management process, including aspects such as location of the site entrance to minimise delivery implications. Vrijhoef and Koskela (2000) consolidate that construction sites may be one of the following four specific roles:

- The impact of the supply chain on site activities
- The impact on the supply chain itself
- The impact on transferring activities from the site to earlier stages on the supply chain
- The impact on the integrated management and improvement of the supply chain, and the site production which are not exclusive and can be used together

Ala-Risku and Kärkkäinen (2004) focus on the negative impact that poor material delivery procedures can have on a construction project's supply chain and identify possible solutions to combat this. This includes providing the task planner (project manager) with reliable information on material availability and to guarantee material availability for project tasks, without building up unnecessary inventory.

More recently, Cheng and Kumar (2015) identify a Building Information Modelling (BIM) based framework model which incorporates 'Material Logistics Planning' that promotes tidy construction sites and efficient project delivery. This BIM-based technology uses the 'Revit Application Programming Interface' that extracts geometric and material information from BIM models and integrates this information with the schedule for formulating a dynamic construction site layout model (Cheng & Kumar, 2015).

Research Method

To carry out this study, a qualitative analysis approach was used. This research includes a literature review, from where three case studies of confined pipeline construction projects where the removal of hazardous asbestos waste was being undertaken.

A qualitative approach was chosen to provide industry insight on managing confined pipeline construction projects, where hazardous asbestos waste is being decommissioned and removed. Therefore, a qualitative analysis was employed, where an extensive review of literature and interviews from case studies can all be compiled to clarify if Materials Management, Space Scheduling, Supply Chain Management, etc. can be implemented to assist in the management of the extraction and removal of hazardous waste from site. The case studies incorporated nine interviews (three from each case study), with construction managers, site agents, civil engineers and foremen, who overseen the pipeline construction and asbestos removal process on each of the respective confined sites studied.

Case Studies

Each of the three case studies selected is a confined pipeline construction sites, which incorporate the removal of hazardous asbestos pipes and installation of replacement concrete and ductile iron pipes. These construction sites are heavy civil engineering works which involve pipeline construction, flood relief and water treatment plants. Each of the three case studies is in the south of Ireland. Selective sampling was applied, where sites were chosen, where the removal of hazardous waste pipeline construction was incorporated, and that the projects were being undertaken in a confined site environment. Initially, six sites were identified, from where three were identified as best meeting the criteria identified, a confined construction site pipeline project, dealing with the removal of hazardous asbestos waste. The three case studies are analysed based on the information gathered from the interviews carried out with each of the project management teams. The interviews will be used to establish the most frequent problems in the management of confined pipeline construction sites which are removing hazardous asbestos waste. The interview transcripts are then analysed and thematically coded, using the themes identified from the literature as a basis.

Case Studies: Qualitative Results and Analysis

Case Study 1 consists of a drainage scheme in a coastal town in County Cork in the South of Ireland. The project is designed to protect a highly populated area from fluvial flooding from several rivers and also tidal flooding from the Celtic Sea. The project also includes the installation of storm sewer pipework, which includes the removal of asbestos piping, three pumping stations and stabilisation works on seven bridge abutments. The main contractor has a vast amount of experience in this area, having delivered similar projects nationwide. On-site interviews with the Construction Manager and two Civil Engineers were carried out, lasting an average of 20–30 min each.

Case Study 2 consists of a project which includes the installation of fourteen underground pumping stations, which are all of caisson construction, over 30 km of new sewers around the adjacent harbour, and the repair and replacement of existing sewers. Interviews were carried out with the Construction Manager, Site Agent and a Civil Engineer, all of which took place on-site. The interviewees had an average of 15 years' experience working on similar projects.

Case Study 3 was a water main and sewer upgrade project in County Cork. This project involved the replacement of 10 km of water mains and upgrading over 5 km of sewer network, which included the removal of asbestos piping. This project was located in highly populated areas; therefore, all of the works were undertaken in a confined site environment. Interviews took place with the Construction Manager, Civil Engineer and Foreman, with interviews lasting an average of 20–30 min.

Using the themes identified in the literature, each of the subsequent interviews was thematically analysed and coded, providing an overview of the findings from each interview and associated case studies, related back to these themes and authors (Table 1).

Discussion

Subsequent to the above, each of the four themes is discussed, reverting back to the literature in addition to comments made by the interviewees, in the identification of the factors within each theme.

Theme 1: Materials Management

On review of the case study interviews, factors resonating from Materials Management were prolific. Construction Managers, Civil Engineers, Site Agents and Foremen repeatedly referred to how they used materials management practices to maintain high production levels, while also considering the welfare and safety of workers and reducing overall costs on-site. The main factors which appeared over the course of the qualitative analysis include: to support the labour force in minimising inefficiencies, and use three separate material storage areas, such as semi-permanent storage, staging areas and workface storage, to avoid collisions.

In all nine interviews, the participants noted the importance of supporting the labour force to minimise inefficiencies as one of the key components in managing a confined pipeline construction project. Separate materials areas for equipment and supplies were recognised with high importance by 7 of 9 interviewees. However, each participant acknowledged that there was a lack of resources available to support the management team when attempting to address the need for separate materials areas.

Several interviewees noted that materials management on a confined pipeline construction project which is removing asbestos piping is one of the most difficult environments a site engineer/manager can face. This was due to the majority of the site being open excavation, to remove hazardous waste and install the new pipeline. A management method that two interviewees specified had improved the overall production levels on site was placing materials on stillage's which could be easily transported to and from site.

Table 1 Themes, factors, authors and case study results

Theme	Factor	Authors	Case study 1			Case study 2			Case study 3		
			Int #1	Int #2	Int #3	Int #1	Int #2	Int #3	Int #1	Int #2	Int #3
Materials management	Support the labour force and minimise inefficiencies	Spillane et al. (2010)	X	X	X	X	X	X	X	X	X
	Sling every pipe	Patel and Vyas (2011)						X			
	Use three separate material storage areas	Thomas et al. (2005)	X	X	X	X		X		X	
	Keep materials on stillages	Navon and Berkovich (2006)				X	X				
Space scheduling	Decision support tool for project Managers	Tommelein and Zouein (1993)					X				
	The MovePlan Model used	Winch and North (2006)	X	X	X	X	X	X	X	X	
	Segregate space on-site	Choo and Tommelein (1999)			X	X	X			X	
	Moveable containers						X				
	Implemented at the planning Stage	Zouein and Tommelein (2001)	X	X				X	X	X	X
Critical space analysis	Allocate space for works to take place on-site		X	X	X		X		X	X	
	Installing platforms for extra space	Dawood and Mallasi (2006)	X								
	Stacking welfare facilities	North and Winch (2002)							X		
	Possible support tool solution	Winch and North (2006)		X	X		X	X	X	X	
Supply chain management	Regulate supplies to reduce inventories	Fernie and Thorpe (2007)	X		X	X	X	X	X	X	X
	Base number of pipes on-site by length of site and number used daily	Spillane et al. (2011)					X		X		
	Pallet hazardous waste and remove in bulk	Ala-Risku and Kärkkäinen (2004)								X	
	Remove hazardous waste whole			X				X			X
	Finish work as it progresses	Cheng and Kumar (2015)						X			
	Increase transparency and alignment of the supply chain	Vrijhoef and Koskela (2000)	X		X		X	X	X	X	X

Theme 2: Space Scheduling

Space scheduling is a decision support tool which can be used to allocate space on-site for works to take place. The factors which were identified from the interviews included:

- Implementing methods at the planning stage to reduce losses in time and profit.
- Allocating space on a two-dimensional software to increase visualization of construction within the management team.
- Segregate space on-site (Hazardous Waste, New Materials, etc.)

In six of the interviews, the planning stage was identified as being critical to achieving project success in terms of quality, cost and time. Although there has been little research on space scheduling over the last decade, most of the research conducted has not become obsolete. Zouein and Tommelein (2001) mentioned the importance of outlining the sites layout at the planning stage to schedule where works are to take place throughout the construction phase.

Four interviewees referenced the importance of segregating space on-site for materials and hazardous waste, to avoid collisions within the supply chain, a point echoed by Choo and Tommelein (1999) as one of the primary benefits of the WorkMovePlan software, which combines a Last Planner methodology with Space Scheduling.

In terms of segregating space on-site it should also be acknowledged that Tommelein and Zouein's (1993) work highlights the MovePlan Model, which can implement a two-dimensional plan. This is used to allocate space on-site during the planning stage, which has the potential to increase the efficiency of these confined pipeline construction sites which are removing asbestos piping.

Theme 4: Supply Chain Management

On analysis, Supply Chain Management was referred to on numerous occasions in each interview. Supply Chain Management, which was originally applied on the manufacturing industry, aims at regulating supplies to drastically reduce inventories to effectively regulate supplier's interaction with the production line (Vrijhoef & Koskela, 2000). In 8 out of the 9 interviews, the management team agreed that regulating supplies to reduce inventories is one of the key elements to a successful confined pipeline construction project. The main factors referenced during the interviews included:

- Base number of pipes on-site by length of site and number amount used daily
- Placing hazardous waste on pallets and removing in bulk
- Remove hazardous waste whole
- Finish work as it progresses (manholes, connections, tar, testing, etc.)
- Increase transparency and alignment of the supply chain.

With the construction supply chain accounting for a large quantity of waste on-site, more effective management is needed (Vrijhoef & Koskela, 2000). The factors noted above have the potential to realign the supply chain, to maximise quality and minimise costs associated. Five of the nine interviewees noted the importance of informing delivery drivers of set down areas and delivery locations in advance. This aided in avoiding the confined site from being overrun with materials, waste and plant, thus demonstrating the need to, and resultant benefits of, increased transparency and alignment of the supply chain (Vrijhoef & Koskela, 2000). This could also be implemented by identifying, in the planning stage, the optimal location for the site entrance, to minimise delivery implications (Spillane et al., 2011).

Similarly, Ala-Risku and Kärkkäinen (2004) illustrate the negative impact that poor material delivery procedures can have on a construction project's supply chain and proposes possible solutions. These include providing the project planner with reliable information on material availability and to guarantee material availability, without building up unnecessary inventory.

Conclusions and Recommendations

The responsibility of governing bodies to consistently rehabilitate and replace the underground water main and sewage services within urban centres will continue to be an ever-growing concern worldwide. This has led to the emergence of the confined pipeline construction site, which forms the basis of this research, to analyse and investigate possible methods of effective management of materials, in, around and off site, thus increasing the overall efficiency of these sites in terms of production levels, health and safety and on-site costs.

After analysing numerous journals and articles on confined site construction, four key themes emerged and were then used as the basis for the subsequent case study interviews: Materials Management, Space Scheduling, Critical Space Analysis and Supply Chain Management. During the literature review, it was evident that a gap in research conducted on confined construction sites, focusing on decommissioning and removing hazardous waste, was required. To gauge the actuality of the difficulties in managing confined construction sites where hazardous waste needs to be removed, three asbestos pipeline projects were identified. In total, nine interviews were undertaken, to ascertain if the four themes above could be applied in this instance. Following the analysis carried out over the three case studies, the qualitative results revealed Materials Management and Supply Chain Management as the themes which were most frequently applied by construction managers, site agents, civil engineers and foremen to increase production levels, reduce costs and increase the overall health and safety of the workers on site. Throughout the case study analysis, Space Scheduling was consistently acknowledged with high importance by all members of the management team, despite many being unfamiliar with the terms. Segregating space on-site into areas, such as hazardous waste and new materials, was prioritised by four interviewees as being the key component to a successful pipeline construction project.

This was further supported whereby experienced professionals who expressed the apparent absence of technology available to assist the management team, in the designing of site layout at the planning phase of such projects. This was of particular focus due to the fact that such project boundaries are fluid and dissimilar to other projects, and are not static in nature. This, therefore, leads to an increased necessity for further site layout planning, due to the dynamic nature that such pipeline projects face. This subsequently makes the implementation of these principles all the more difficult, but yet important, particularly in the context of handling and removing hazardous asbestos waste on confined pipeline construction projects.

As a result, it is recommended that members of the management team over confined pipeline construction projects which are decommissioning and removing hazardous asbestos waste should consider implementing methods of Materials Management, Space Scheduling, Critical Space Analysis and Supply Chain Management. The above-mentioned themes, formed on an extensive literature review and qualitative analysis, have provided evidence to suggest an increase in efficiency by means of increasing production, increasing the health and safety of workers and the general public, and reducing costs on these projects.

References

Ala-Risku, T., & Kärkkäinen, M. (2004). Material delivery problems in construction projects: A possible solution. *International Journal of Production Economics, 104*(1), 19–29.

Begum, R. A., Siwar, C., Pereira, J. J., & Jaafar, A. H. (2006). A benefit–cost analysis on the economic feasibility of construction waste minimisation: The case of Malaysia. *Resources, Conservation and Recycling, 48*(1), 86–98.

Cheng, C. P., & Kumar, S. (2015). *A BIM-based framework for material logistics planning*. 23rd International Group for Lean Construction Conference (IGLC 2015), held at Perth, Australia, 28–31 July.

Choo, H. J., & Tommelein, I. D. (1999). *Space scheduling using flow analysis*. 7th International Group for Lean Construction Conference (IGLC 1999), held at University of California, Berkeley, California, USA, 26–28 July.

Dawood, N., & Mallasi, Z. (2006). Construction workspace planning: Assignment and analysis utilizing 4D visualization technologies. *Computer-Aided Civil and Infrastructure Engineering, 21*(7), 498–513.

Fernie, S., & Thorpe, A. (2007). Exploring change in construction: supply chain management. *Engineering Construction and Architectural Management, 14*(4), 319–333.

Li, J., Dong, Q., Yu, K., & Liu, L. (2014). Asbestos and asbestos waste management in the Asian-Pacific region: Trends, challenges and solutions. *Journal of Cleaner Production, 81*, 218–226.

Navon, R., & Berkovich, O. (2006). An automated model for materials management and control. *Construction Management and Economics, 24*(6), 635–646.

North, S., & Winch, G. M. (2002). *VIRCON: A proposal for critical space analysis in construction planning*. 4th European Conference on Product and Process Modelling in the Building and Related Industries (ECPPM). pp. 359–364.

Patel, K. V., & Vyas, C. M. (2011). *Construction materials management on project sites*. National conference on recent trends in engineering & technology, held at Anand, Gujarat, India, 13–14 May, pp. 1–5.

Spillane, J. P., & Oyedele, L. O. (2017). Effective material logistics in urban construction sites: a structural equation model. *Construction Innovation, 17*(4), 406–428.

Spillane, J. P., Oyedele, L. O., von Meding, J., Konanahalli, A., Babatunde, J. E., & Iyabo, T. K. (2011). Challenges of UK/Irish contractors regarding material management and logistics in confined site construction. *International Journal of Construction Supply Chain Management, 1*(1), 25–42.

Spillane, J. P., Oyedele, L. O., Von Meding, J. K., & Konanahalli, A. (2010). *Material management in confined construction sites*. The Construction, Building and Real Estate Research Conference of the Royal Institution of Chartered Surveyors (RICS), held at Dauphine Université, Paris, 2–3 September.

Thomas, H. R., Riley, D. R., & Messner, J. I. (2005). Fundamental principles of site material management. *Journal of Construction Engineering and Management, 131*(7), 808–815.

Tommelein, I. D., & Zouein, P. P. (1993). Space scheduling for construction progress planning and control. *Automation and Robotics in Construction, 1993*, 415–422.

Vrijhoef, R., & Koskela, L. (2000). The four roles of supply chain management in construction. *European Journal of Purchasing and Supply Management, 6*(3), 169–178.

Winch, G. M., & North, S. (2006). Critical space analysis. *Journal of Construction Engineering and Management, 132*(5), 473–481.

Zouein, P. P., & Tommelein, I. D. (2001). Improvement algorithm for limited space scheduling. *Journal of Construction Engineering and Management, 127*(2), 116–124.

Construction Programme Failure and the Impact on Waste, Resource Efficiency, and Natural Capital in Construction Project Environments

Neil Pickavance, Andrew Ross, and Damian Fearon

Introduction

Delayed and disrupted construction projects negatively affect the natural environment. This can be via excessive material waste, labour resource inefficiency, and misuse of natural capital (e.g. water and fuels). In addition, increased CO_2 emissions result from delayed projects, particularly from vehicle deliveries arriving over a longer-than planned time period. Further, projects that exceed their completion date can suffer cost over-runs, supply chain disruption, contractual disputes, quality and re-work issues, workforce discord leading to churn and instability, and occupational uncertainty to the end-user. These characteristics sustain the causation of poor sustainability from delayed and disrupted construction projects. With Critical Path Method, the de facto project planning and scheduling technique in construction, despite the availability of alternative project management systems such as Lean Construction, Last Planner® System, Agile Project Management, and PRINCE2, this research aims to investigate the actual application of CPM at a sample construction organisation in the UK. Questions focus on aspects of construction planning and programme: key challenges, development, co-ordination, integration, procedures, methods, techniques, software, training, development, and process frameworks.

Emitting 13 million tonnes of CO_2 per annum the UK construction industry could lower that figure by reducing project delays, disruption, late completion, and handover of built assets. On the whole, complex construction projects in the UK are likely to be finished more than 6 months late (CIOB, 2008) with circa 50% of UK construction projects delivered late. Despite improvements in supply chain management, collaborative working practices, early contractor involvement, partnering, non-adversarial approaches, and adoption of digital technology, there appears

N. Pickavance (✉) · A. Ross · D. Fearon
Liverpool John Moores University, Liverpool, UK
e-mail: n.m.pickavance@2017.ljmu.ac.uk

L. Scott et al. (eds.), *Sustainable Ecological Engineering Design*,
https://doi.org/10.1007/978-3-030-44381-8_21

limited success in arresting the trend of late project completion. With the industry characterised by excessive sub-letting, work package fragmentation, transient self-employed labour, and poor trade-to-trade integration, this research assesses approaches, techniques, and behaviours in project planning and scheduling in the UK construction industry.

Construction project planning and scheduling research, and construction sustainability research, generally sit apart in literature. This paper adopts a trans-disciplinary view to link, investigate, and form an understanding of how improved project planning and scheduling can improve surety of project completion from both a commercial and contractual perspective and also in terms of better project sustainability. Taking a qualitative approach to the problem the paper seeks to understand, via semi-structured interviews, the techniques, methods, actions, approaches, behaviours, and decision-making around project planning and scheduling within project teams on a major UK construction project. Figures 1, 2, 3, and 4 indicate various examples of site conditions encountered on construction projects suffering with programme delays and disruption.

Literature: Construction Planning and Programme Management

In construction the Critical Path Method of project planning and scheduling remains the most commonly used technique in project environments (Ammar, 2012), despite newer techniques such as Lean Construction and Last Planner® System being available. This could possibly be due to familiarity and historical popularity of CPM by contractors and almost universal specification of CPM by construction clients. In construction dispute resolution CPM is commonly used to demonstrate effects of delay, disruption, and time entitlement (Livengood, 2016). However, despite widespread use, CPM can be criticised for its effectiveness as a planning tool (Tommelein,

Fig. 1 Site conditions on delayed UK construction projects (2014)

Fig. 2 Site conditions on delayed UK construction projects (2014)

Fig. 3 Site conditions on delayed UK construction projects (2014)

Riley, & Howell, 1999) and is deemed too rigid, deterministic, and a need to be too predictive in environments that requires flexibility and continual adjustments in the project work schedule.

Historically, considerable research into the Critical Path Method of project control has studied the mathematical, computational, and algorithmic characteristics of network analysis (Aquilano & Smith, 1980; Kelley, 1961). An extensive body of knowledge from the late 1950s to the present day exists that views CPM as a logical system of process flow control. With CPM's origins as a network analysis method in industrial and production environments the transfer of CPM into a construction environment as a process control tool occurred in the early 1960s: '*A non-computer approach to the critical path method for construction*' (Fondahl, 1962).

A neural network approach to construction planning was assessed in '*Key determinants for construction schedule performance*' by Kog, Chua, Loh, and Jaselskis (1999), again indicating the scientific and quantitative approach. In contrast, in Laufer and Tucker (1987) their focus turned to whether construction project

Fig. 4 Site conditions on delayed UK construction projects (2018)

planning was doing its job, a query aimed at the integrity of those undertaking the construction planning role. Winch and Kelsey (2005) also directly questioned the planner role with '*What do construction project planners do?*' (Gibson, Wang, Cho, & Pappas, 2006) turned attention towards a lessons-learned study of the pre-construction planning process with '*What is preproject planning, anyway?*'.

New lean project management methodologies (transferred from the automotive, manufacturing, and aerospace industries) offer alternative project planning management approaches to traditional CPM technique (Ballard, 2000; Daniel, Pasquire, & Dickens, 2014). Generally categorised as Collaborative Planning these comprise techniques and systems such as Lean Construction and Last Planner® System. Other related methods offered to construction that have emerged in the last 10–20 years are Agile Project Management, AgiLean Project Management, Six Sigma, and Lean Six Sigma. These systems are characterised by greater collaboration and inclusivity throughout the project management life cycle, particularly of the site work-crews, and are seen as systems adopting social trust compared to the rigid, deterministic, and predictive, technical-rational nature of CPM planning.

Construction texts that address project planning and scheduling tend to do so in an approach that is limited to explanations and diagrams of work schedules, task bars, logic linkages, constraints, activity relationships, and resources. An example can be seen in Lester (2006). While HM Government construction industry reports have a history of demanding improved project and programme surety (Egan, 1998; Latham, 1994), they, and publications such as British Standards Institution (2010), Association of Project Management (2012), and Chartered Institute of Building (2016), provide no guidance on the application of CPM project planning and scheduling.

Research Methodology and Method

This research design adopts a qualitative study for interpretation of data, discovering concepts and relationships (Corbin & Strauss, 2008). The collation of data is by interview and the research position is interpretivist in nature, with the qualitative strategy emphasising social reality research over quantification of data (Bryman, 2015).

The data collection was undertaken at a construction project (£300.0 M) in the UK Midlands area that had entered its contract phase. The main contractor is a large construction organisation who operate predominantly on the UK mainland. In addition, two sub-contract companies engaged on the same project were included in the study, one operating in the steel erection sector and the other operating in the mechanical and electrical building services sector.

Interviewees were selected as a cross-section sample of construction management personnel. Four staff from the main contractor, and one member of staff from the sub-contractors were interviewed. Each participant was either a project manager or project planner with between 15 and 32 years' experience in construction. The mean of the sample was 26.2 years.

The questions posed were open and exploratory to elicit rich data. The eleven questions span common topics in project planning and scheduling within a practitioner environment. The interviews were semi-structured on a personal (face-to-face) basis with the five key topics:

Programme Key Challenges

1. What do you perceive are the main challenges for contractors planning with Critical Path Method?

Programme Development, Co-ordination, and Integration

2. How much do attitudes and cultures within a company influence the successful development and co-ordination of an integrated construction programme?
3. The development of construction programmes require many iterative passes around the supply chain. How well do you feel this process is managed?
4. How do you feel the programme development and co-ordination process could be improved?

Programme Procedures, Methods, and Techniques

5. What are the methods, techniques and protocols that construction organisations use to develop, co-ordinate and integrate critical path programmes?
6. Who has the responsibility for undertaking the project planning role at your organisation?
7. What processes do you follow to establish and integrate the various sub-contract trade sections of the construction programme?

Planning Software, Training, and Development

8. What programme scheduling software is used at your organisation?

9. What planning and programming training, both formal and informal, is provided to those responsible for undertaking the planning role?
10. How is the planning role nurtured and supported to allow stability, consistency and longevity within the organisation?

Programme Process Framework

11. How do you feel a process similar to the RIBA's Plan of Work would be of use for developing a co-ordinated and integrated programme?

The interview data has been transcribed and an open-coding analytical process undertaken. NVivo 11 qualitative analysis computer software package was used to code at the sentence level to identify some initial categories and concepts. This provided an exploratory investigation into the subject matter. Open coding allowed the text to be opened up and expose the thoughts, ideas, and meanings contained therein, to uncover, name, and develop concepts (Corbin & Strauss, 2008).

Results

Question 1 related to the perceived challenges for contracting organisations planning with Critical Path Method. The emerging key concepts from the open-coded categories from the transcript data are: (1) fundamental understanding of programmes, (2) resourcing of programmes (plant, materials, and labour), and (3) stages and trade sections of programme. Table 1 provides example statements from the interview data, main category per interviewee, and emerging concepts.

Question 2 related to the effects that attitudes and cultures within companies have on the programme development. The emerging key concepts from the open-coded categories from the transcript data are: (1) operational cultures at companies, (2) shared understanding of programme, (3) trust required in programmes, and (4) team attitudes to programme development. Table 2 provides example statements from the interview data, main category per interviewee, and emerging concepts.

Question 3 related to the process of integrating individual supply chain trade programmes as part of the overall programme development. The emerging key concepts from the open-coded categories from the transcript data are: (1) the requirement for a central programme direction for the disparate trade programmes and (2) ownership of those individual programme contributions. Table 3 provides example statements from the interview data, main category per interviewee, and emerging concepts.

Question 4 related to a view of how current programme development and co-ordination processes could be improved. The emerging key concepts from the open-coded categories from the transcript data are: (1) use of programme development frameworks and (2) a supply chain framework to provide process consistency. Table 4 provides example statements from the interview data, main category per interviewee, and emerging concepts.

Table 1 Data examples, main categories, and emerging key concepts from responses to question 1

<table>
<tr><th colspan="3">Programme key challenges</th></tr>
<tr><td rowspan="4">1. What do you perceive are the main challenges for contractors planning with Critical Path Method?</td><td colspan="2">Data (examples)</td></tr>
<tr><td colspan="2">→ People understanding CPM and buying into it
→ Understanding CPM usage and outputs
→ Understanding the critical 'spine' through the CPM programme
→ Early supply chain involvement so they understand programme aspirations
→ Supply chain nominating their correct person so they understand it
→ Communicating the critical 'spine' to the supply chain
→ Better identification of critical and non-critical activities
→ Resourcing to suit the master programme constraints
→ Identifying the correct commissioning period.
→ Preserving the commissioning period [from being compressed by delayed installation]</td></tr>
<tr><td>Main category per interviewee (via open coding)</td><td>Emerging key concepts</td></tr>
<tr><td>→ Understanding CPM and Benefits
→ Supply Chain Understanding
→ Understanding Critical Items
→ Understanding and Focusing on the Critical Path
→ Plant Resource to Support Programme
→ Programme—Commissioning Integration</td><td>→ Understanding of Programmes
→ Resourcing of Programmes
→ Stages of Programmes</td></tr>
</table>

Question 5 related to what programme methods that the contracting organisations currently use to co-ordinate and integrate Critical Path Method programmes. The emerging key concepts from the open-coded categories from the transcript data are: (1) programme development being generally supply chain led and (2) programme integration being a relatively informal process. Table 5 provides example statements from the interview data, main category per interviewee, and emerging concepts.

Question 6 related to the responsibility of undertaking the project planning role. The emerging key concepts from the open-coded categories from the transcript data are: (1) shared ownership of the role and (2) evidence of diffused responsibility for undertaking the planning role. Table 6 provides example statements from the interview data, main category per interviewee, and emerging concepts.

Table 2 Data examples, main categories, and emerging key concepts from responses to question 2

Programme development, co-ordination, and integration		
2. How much do attitudes and cultures within a company influence the successful development and co-ordination of an integrated construction programme?	Data (examples)	
	→ Culture is a main influencer of successful programme development → Construction has a poorer planning culture than petrochemical or nuclear sectors → Cultures affect knowledge and understanding of programmes → The right attitude and culture is critical for programme development → Culture affects dissemination of supply chain strategy to their own leads → Company cultures critically influence the tender programme calibre → Poor cultures can allow programme provision to just be a 'tick-in-a-box' → Fragmented construction supply chains cause fragmented project cultures → Cultures have significant influence on approaches to programme development	
	Main category per interviewee (via open coding)	Emerging key concepts
	→ Company Culture and CPM → Supply Chain Understanding → Programme Reliability → Collaborative Attitude to Programme → Company Culture and CPM → Supply Chain Understanding	→ Operational Cultures → Shared Understanding → Trust in Programmes → Team Attitudes

Question 7 related to the process currently used to specifically integrate the sub-contract trade programmes. The emerging key concepts from the open-coded categories from the transcript data are: (1) limited processes seem to exist for this crucial process in master programme development and (2) the lack of a formal framework appeared to support the limited processes currently used. Table 7 provides example statements from the interview data, main category per interviewee, and emerging concepts.

Question 8 related to the programme scheduling software used by the contracting organisations. The emerging key concepts from the open-coded categories from the transcript data are: (1) a standardised adoption of the main three planning software packages was evidenced. Table 8 provides example statements from the interview data, main category per interviewee, and emerging concepts.

Table 3 Data examples, main categories, and emerging key concepts from responses to question 3

Programme development, co-ordination, and integration		
3. The development of construction programmes require many iterative passes around the supply chain. How well do you feel this process is managed?	Data (examples)	
	→ The iterative development process not managed at all; it's hit-and-miss → Sub-contractor buy-in comes after they have an order, it's too late → Iterative programme development not done at tender stage → Common to get programme advice from a sub-contractor but then go with another → Iterative development is managed well if you afford the time and it's early enough → We are getting better [at programme development] but it requires commitment to it → It is managed well [relating to taking sub-contractor advice] → The process can be a tick-in-a-box [just to say it has been completed]. → Poorly managed, compared to workshops I've seen in the past → Visibility and transparency of the iterative programme development process is poor	
	Main category per interviewee (via open coding)	Emerging key concepts
	→ Programme—Evolution → Supply Chain—Understanding → Supply Chain—Early Involvement → Supply Chain—Sub-Letting → Team Collaboration → Programme Development – Unstructured	→ Programme Direction → Programme Ownership

Question 9 related to the existence of training in planning and programming techniques for the project staff at the organisations. The emerging key concepts from the open-coded categories from the transcript data are: (1) a semi-formal approach to the provision of planning training appeared common, (2) an underlying apathy towards formal and structured planning training, and (3) didn't seem to be an awareness of a central responsibility towards provisioning that training; the feeling was it was somebody else's problem. Table 9 provides example statements from the interview data, main category per interviewee, and emerging concepts.

Table 4 Data examples, main categories, and emerging key concepts from responses to question 4

Programme development, co-ordination, and integration		
4. How do you feel the programme development and co-ordination process could be improved?	Data (examples)	
	→ Need to move from tender through to construction with the same sub-contractors → Once the critical 'spine' is established use programme co-ordination workshops → Supply chain need to understand what really makes an activity work → Full process understanding: design details, procurement, resourcing → Engage with supply chain, get heavily involved, make them understand programme → Get the outline of the programme, the critical path, major elements of work → Using resource schedules and histograms → Front-end engagement with the supply chain is invaluable → Early involvement [supply chain]. → Forming a clear, defined project programme strategy	
	Main category per interviewee (via open coding)	Emerging key concepts
	→ Supply Chain—Consistency → Supply Chain—Consistency → Supply Chain—Consistency → Supply Chain—Consistency → Supply Chain—Early Involvement → Supply Chain—Early Involvement	→ Programme Framework → Supply Chain Framework

Question 10 related to evidence and examples of how the planning role specifically was nurtured and sustained at the organisations. The emerging key concepts from the open-coded categories from the transcript data are: (1) appeared a low investment by companies in the planning role and (2) what little there was appeared to be semi-formal such as by just having a planner, and ensuring there was always a planning presence, and replacing if they left. Table 10 provides example statements from the interview data, main category per interviewee, and emerging concepts.

Question 11 related to the development and implementation of a formal process framework for integrating the various sub-contractor trade programmes, similar to how the RIBA Plan of Work supports staged design development. The emerging key concepts from the open-coded categories from the transcript data are: (1) need for strategic programme collaboration between project teams and (2) the adding of a formal process to pre-construction (pre-planning) stage may be contingent on only

Table 5 Data examples, main categories, and emerging key concepts from responses to question 5

Programme procedures, methods, and techniques		
5. What are the methods, techniques and protocols that construction organisations use to develop, co-ordinate and integrate critical path programmes?	Data (examples)	
	→ Programme creation by main contractor for supply chain to follow → Supply chain involvement to familiarise them with the project → Understanding your supply chain and control of the supply chain → Stability of the supply chain → Supply chain advice → Internal team input and co-ordination → Programme certainty at sub-contract pre-order stage → Steer them [the supply chain] as best we can → Programme creation by sub-contractor for their supply chain to follow → Early [programme] workshops need to happen	
	Main category per interviewee (via open coding)	Emerging key concepts
	→ Programme—Evolution → Supply Chain—Programme Resourcing → Programme—Development (Structured) → Programme—Commissioning Integration → Team Collaboration → Supply Chain—Engagement	→ Supply Chain Led → Informal Processes

being permissible by the client. Table 11 provides example statements from the interview data, main category per interviewee, and emerging concepts.

Discussion

Previous literature on project planning and scheduling provides insights quite varied to the investigation undertaken here, with focus on schedule logic, the role of the Planner, and lean and agile project management methodologies. A discussion of the current findings is provided here.

Challenges of traditional CPM planning appear to be fundamentals such as basic understanding of CPM, the correct use of resourced programmes, and establishing

Table 6 Data examples, main categories, and emerging key concepts from responses to question 6

Programme procedures, methods, and techniques		
6. Who has the responsibility for undertaking the project planning role at your organisation?	Data (examples)	
	→ Initially the project planner; needs to be full team plan for buy-in → Project manager has overall responsibility → Planner evolves programme but no back-up [from the team] → Planner leads but it's a team effort as we can only deliver as a team → Planner controls programme, but input and development is the team → Planner; ultimate responsibility is the project manager and director → Dedicated planner, with project manager and team development → Project manager [at sub-contractor] → Planner, or project manager on smaller projects	
	Main category per interviewee (via open coding)	Emerging key concepts
	→ Programme—Creation → Planning—Responsibility → Planning—Responsibility → Planning—Responsibility → Planning Role—Project Manager → Planning—Responsibility	→ Shared Ownership → Diffused Responsibility

the various stages of the programme. Poor grasp of these basics appears across the sample and makes it difficult to determine whether CPM is at fault, the users of CPM, or other reasons.

Cultures, shared understanding, trust in programmes, and team attitudes influenced the success of integrating programmes. Trade sub-programmes, crucial to the master programme, require direction and ownership. Improvements to programme development and co-ordination require consistency and early team involvement via improved collaborative processes.

Current programme techniques to develop, co-ordinate, and integrate CPM programmes appear supply chain led and adopt informal processes. The responsibility is shared across the planner and project manager, sometimes the team; however this diffused responsibility can cause ownership issues. Integration of trade programmes lacks a framework to adhere to.

Electronic scheduling software is predominant at the organisations, providing a standardisation to the CPM programmes. Wider training in planning and programming skills is lacking, with a semi-formal training structure. Apathy and low interest

Table 7 Data examples, main categories, and emerging key concepts from responses to question 7

Programme procedures, methods, and techniques		
7. What processes do you follow to establish and integrate the various sub-contract trade sections of the construction programme?	Data (examples)	
	→ Workshops and analysis of each programme iteration → Team familiarisation and collaboration → Develop a programme critical 'spine' of the four or five main trades → Aligning the supply chain with a [our] pre-set construction programme → Iterative passes of the [developing] programme around the supply chain → Supply chain experience → Programme creation by sub-contractor for their supply chain to follow → Communication and meetings to understand programme dependencies → Programme creation by main contractor for supply chain to follow → [Work] scope and interfaces identification then programme integration	
	Main category per interviewee (via open coding)	Emerging key concepts
	→ Team Collaboration → Team Collaboration → Communication with Supply Chain → Programme—Site Co-ordination → Programme—Integration Framework → Programme—Development (Unstructured)	→ Limited Processes → Lack of Framework

towards training was observed and the nurturing of the planning role appeared to have low investment and was semi-formal.

The suggestion of a process flow framework for CPM programme development, co-ordination, and integration appeared a requisite in contracting and was well understood. Its use as a collaboration tool was extolled; however it was deemed contingent on clients agreeing to something which may require more front-end time for programme development.

Conclusion

Project planning and scheduling, via CPM, operated inconsistently on the sample project. There appears lack of structured and authoritative implementation, apathy towards CPM, issues with training, supporting, and nurturing the planning role,

Table 8 Data examples, main categories, and emerging key concepts from responses to question 8

Planning software, training, and development		
8. What programme scheduling software is used at your organisation?	Data (examples)	
	→ Currently ASTA Powerproject → Used Microsoft Project and Oracle Primavera → Used Project Commander → Microsoft Excel for short-term programming in a simple format → Microsoft Project and ASTA Powerproject → Oracle Primavera on power station work → ASTA Powerproject, work in Microsoft Project but transfer to ASTA → ASTA. Other variants but that's standard protocol here	
	Main category per interviewee (via open coding)	Emerging key concepts
	→ Planning Software—ASTA Powerproject → Planning Software—ASTA Powerproject → Planning Software—Microsoft Project → Planning Software—Oracle Primavera → Planning Software—ASTA Powerproject → Planning Software—ASTA Powerproject	→ Standardisation

programme integration, and understanding team responsibilities. The sample project may be representative of contracting organisations.

The pursuit of sustainability in construction often looks at material technology, off-site manufacture, and lean methods. This paper concludes that serious problems appear to exist with the fundamental application of construction project planning and scheduling. Construction sustainability executives need to acknowledge programme management as a key driver.

Further investigation is necessary to study CPM planning within construction project environments in the UK. Study across pre-construction and construction stages may give a greater understanding of how CPM is adopted and applied, as it appears (here) to have a poor framework of application.

Table 9 Data examples, main categories, and emerging key concepts from responses to question 9

Planning software, training, and development		
9. What planning and programming training, both formal and informal, is provided to those responsible for undertaking the planning role?	Data (examples)	
	→ No general planning course, mainly experience from site → Software training from ASTA and Oracle → Generally self-taught, books, literature on planning → Nothing formal, mainly informal → Our graduates spend time in the planning department for experience → [Graduate] training is structured; formal and measureable → There isn't a structure [planning training pathway] → No formal training in terms of planning → We use internal people with site experience [as planners] who understand sequence → Not exactly sure; all staff can attend courses	
	Main category per interviewee (via open coding)	Emerging key concepts
	→ Planning Training—Lack of Structure → Planning Training—Structured Placement → Planning Training—Lack of Structure → Planning Training—In-House → Planning Training—College → Planning Training—Lack of Structure	→ Semi-Formal → Apathy and Low Interest → Someone Else's Problem

Table 10 Data examples, main categories, and emerging key concepts from responses to question 10

Planning software, training, and development		
10. How is the planning role nurtured and supported to allow stability, consistency and longevity within the organisation?	Data (examples)	
	→ Generally, most contractors don't keep [planning] records → Programme is nurtured by keeping same key staff through a project → People think all you're doing [as a planner] is just colouring bars in → People see planning as a serious role, they see you stressed → It is changing now…project managers realise they get benefit → The business recognises it's an important discipline → There is a department lead…who is highly valued by the business → If a planner leaves they are replaced with another → I don't believe they [planners] are [supported], they seem on their own → I don't think we [managers] get close enough [to the planner]	
	Main category per interviewee (via open coding)	Emerging key concepts
	→ Planning—Focus Via Techniques → Planning—Support of Role → Planner Allocation → Programme—Site Co-ordination → Programme—Managing Impact of Change → Planning—Support of Role	→ Low Investment → Semi-Formal

Table 11 Data examples, main categories, and emerging key concepts from responses to question 11

<table>
<tr><th colspan="3">Programme process framework</th></tr>
<tr><td rowspan="4">11. How do you feel a process similar to the RIBA's Plan of Work would be of use for developing a co-ordinated and integrated programme?</td><td colspan="2">Data (examples)</td></tr>
<tr><td colspan="2">→ Especially [needed] for integrating sub-contractor programmes
→ Very useful, a set structure [for] risk, output rate, schedule uncertainty
→ Would provide consistency to programmes over current ad hoc [development]
→ Supply chain would understand it [programme development] better
→ Would we have time for that [process]? It would be a challenge
→ It would be hard [to do]; only if the client bought-in would it be a success
→ If we needed more time on pre-construction [to do this] clients wouldn't be interested
→ Clients would need to buy-in, but they would [benefit from] receiving a 'guarantee'
→ Work could be properly planned, a side-effect being a lot safer site as well
→ Stages [for programme development] might work, gets the whole team together
→ [Would provide] better integration and more in-depth programme development</td></tr>
<tr><td>Main category per interviewee (via open coding)</td><td>Emerging key concepts</td></tr>
<tr><td>→ Programme—Development (Structured)
→ Team Understanding
→ Programme—Integration Framework—Client Buy In
→ Programme—Integration Framework
→ Programme—Strategy
→ Team Input to Programme</td><td>→ Collaborative Strategy
→ Contingent on Client</td></tr>
</table>

References

Ammar, M. A. (2012). LOB and CPMP integrated method for scheduling repetitive projects. *Journal of Construction Engineering and Management, 139*(1), 44–50.

Association of Project Management (Ed.). (2012). *APM body of knowledge* (6th ed.). Princes Risborough, Buckinghamshire: Association of Project Management.

Aquilano, N. J., & Smith, D. E. (1980). A formal set of algorithms for project scheduling with critical path scheduling/material requirements planning. *Journal of Operations Management, 1*(2), 57–67.

Ballard, H.G., 2000. *The last planner system of production control.* Doctoral Dissertation, University of Birmingham.

British Standards Institution. (2010). *BS6079-1:2010: Project management. Principles and guidelines for the management of projects.* Retrieved April 20, 2018, from https://www.shop.bsigroup.com/.

Bryman, A. (2015). *Social research methods.* Oxford: Oxford University Press.

Chartered Institute of Building (CIOB). (2008). *A research on managing the risk of delayed completion in the 21st century.* Ascot: CIOB.

CIOB. (2016). *Code of practice for programme management in the built environment.* Chichester: Wiley Blackwell.

Corbin, J., & Strauss, A. (2008). *Basics of qualitative research: Techniques and procedures for developing grounded theory.* Los Angeles, CA: Sage.

Daniel, E.I., Pasquire, C., & Dickens, G., (2014, September). *Social perspective of planning in construction: The UK experience.* In Proceedings of the 30th Annual Association of Researchers in Construction (ARCOM) Conference (pp. 1–3).

Egan, J. (1998). *Rethinking construction.* London: Department of Environment, Transport and the Region.

Fondahl, J. W. (1962). *A non-computer approach to the critical path method for the construction industry.* Stanford, CA: Department of Civil Engineering, Stanford University.

Gibson, G. E., Jr., Wang, Y. R., Cho, C. S., & Pappas, M. P. (2006). What is preproject planning, anyway? *Journal of Management in Engineering, 22*(1), 35–42.

Kelley, J. E., Jr. (1961). Critical-path planning and scheduling: Mathematical basis. *Operations Research, 9*(3), 296–320.

Kog, Y. C., Chua, D. K. H., Loh, P. K., & Jaselskis, E. J. (1999). Key determinants for construction schedule performance. *International Journal of Project Management, 17*(6), 351–359.

Latham, M. (1994). *Constructing the team.* London: HMSO.

Laufer, A., & Tucker, R. L. (1987). Is construction project planning really doing its job? A critical examination of focus, role and process. *Construction Management and Economics, 5*(3), 243–266.

Lester, A. (2006). *Project management, planning and control: managing engineering, construction and manufacturing projects to PMI, APM and BSI standards.* Amsterdam: Elsevier.

Livengood, J.C. (2016). *Use of project schedules and the critical path method in claims.* In Construction contract claims, changes, and dispute resolution (pp. 175–200).

Tommelein, I. D., Riley, D. R., & Howell, G. A. (1999). Parade game: Impact of work flow variability on trade performance. *Journal of Construction Engineering and Management, 125*(5), 304–310.

Winch, G. M., & Kelsey, J. (2005). What do construction project planners do? *International Journal of Project Management, 23*(2), 141–149.

An Investigation into the Gap Between Programme Management Theory and Practice

John Heathcote and Ghizlane Ben Baha

Introduction

One of the approaches entrenched in the conventional "control and monitoring" aspects of the project management theory seems to be the confinement of project success to the scope, budget and time parameters of the iron triangle with little regard to benefits realisation and value creation, or what Midler (1995) termed "projectification". Maylor, Brady, Cooke-Davies, and Hodgson (2006) and Chih and Zwikael (2015) argue that this "projectification" or the over focus on the iron triangle as measure of success has created an output focused mentality that limits projects effectiveness and ultimately leads to failure. Additionally, based on their research on the effects of "projectification" Maylor et al. (2006) state that there is *"apparent dichotomy between the very mangerialist ideal of project as a means of control, and the claimed benefits, flexibility, and the less bureaucratic structures that accompany projects"* (Maylor et al., 2006, p. 1). Therefore, it has been concluded that the current scope of project management must be broadened beyond the simple delivery of outputs, and that the emphasis must be placed on the generation of benefits and value creation in alignment with business strategy as the true measure of success (Maylor et al., 2006; Winter, Smith, Cooke-Davies, & Cicmil, 2006). In order to achieve this, Maylor et al. (2006) suggest a shift towards the "programmafication" of the project management theory, identifying "programme management" as necessary benefits realisation tool and considering it *"more holistic approach to effecting fundamental and transformational change in organisations than projectification"* (Maylor et al., 2006, p. 671).

J. Heathcote (✉) · G. Ben Baha
Leeds Sustainability Institute, School of the Built Environment and Engineering, Leeds Beckett University, Leeds, UK
e-mail: j.heathcote@leedsbeckett.ac.uk

L. Scott et al. (eds.), *Sustainable Ecological Engineering Design*,
https://doi.org/10.1007/978-3-030-44381-8_22

The original intention behind programme management as being the creation of greater coordination between a group of projects that use similar resources with the aim of maximising benefits and value is broadly accepted in literature (Lycett, Rassau, & Danson, 2004; Maylor et al., 2006). However, the debate regarding what Programme Management is remains confined between the argument that is just a scaled-up version of project management, and the argument that whilst programme management borrows and extends elements of project management, the two serve completely different purposes (Walenta, 2015). Regardless, the insinuation behind both of these arguments it seems to be that the two concepts of project and programme do in fact overlap. The implications of this however present two problems. The first problem is that with the transition towards "programmafication" intermediately comes with more complex, magnified project challenges. This conclusion can be corroborated through research performed by which finds that only 16% of large programmes are successful whilst 53% of them underperform and 31% ultimately get cancelled due to similar issues that cause project failure. Lycett et al. (2004) agreed with this, further arguing that even though programme management is perceived by various authors as a solution to project failure and a way of bridging the gap between project delivery and organisational strategy as well as the tensions that arise from it, this approach perhaps exacerbates these tensions. The second issue here comes from the fact that although project management theory is more developed and has an established body of knowledge, some authors still claim that it is not fully functioning and describe it as fragmented whilst some go even further to claim that it is obsolete making the theoretical foundations of its relatively new extension "programme management", even less well-established.

Building on what has been discussed thus far, as well as Meskendahl's (2010) research which found that 66% of organisational strategy is not successfully realised, it appears that much like project failure, programme failure may be caused by a disconnect between the current programme management theory and its applications in practice.

In this respect, this research aimed to investigate the alleged gap between programme management theory and practice, through a qualitative inquiry using a focus group of programme and project management experts. The conclusion as to if programmes in practice underexploit the theory of programme management and in which ways it does so has been drawn through a comparison between a critical examination of the current theoretical framework and a comprehensive analysis of findings gathered from the experiences of the focus group participants.

According to Gardiner (2005), this identity crisis is due to the overwhelming inconsistencies regarding programme management definitions, goals, and practices within the available literature, which he associates with the novelty of programme management as a disciple. Ferns (1991) instead correlates this confusion to the fact that programme management often being *"used interchangeably with project management"* (Ferns, 1991, p. 148). Shehu and Akintoye (2009) speculate that a gap exists between what is presented as theory and what is being practiced.

As a pioneer of the Programme management theory, Reiss first described it as the *"coordinated management of a portfolio of projects which call upon the same resources"* (Reiss, 1996, p. 1). Nonetheless, in a more recent publication and in collaboration with several other authors he succinctly described programme management as *"the orchestration of organisational change"* (Reiss, Chapman, Leigh, Pyne, & Rayner, 2006, p. 8). Ferns (1991) further expands on this notion by suggesting that programme management might be a mechanism for coordinating projects in a way that allows otherwise unrealisable benefits to be extracted. He elaborates further by stating that it is the *"the coordinated support, planning, prioritisation and monitoring of projects to meet changing business needs"* (Ferns, 1991, p. 149). Ferns's (1991) definition is substantiated by that of the Project Management Institute who propose that a programme is essentially "*a group of related projects, managed in a coordinated way to obtain benefits and control not available from managing them individually*" (PMI, 2016, p. 6). Despite agreeing with the latter, Pellegrinelli (1997), who has written extensively about programme management, further adds that it is an approach by which projects are utilised, framed, and directed towards achieving strategic goals and engender organisational change. It can be surmised that by exploiting shared resources to enable the prioritisation and adjudication between sometimes competing projects, programme management bridges the gap between organisational strategy and the delivery of projects (Van Buuren, Buijs, & Teisman, 2010). Besides the previously mentioned definitions, further ones include: "*The integration and management of a group of related projects with the intent of achieving benefits that would not be realised if they were managed independently*" (Lycett et al., 2004); "*The process of coordinating the management, support and setting of priorities on individual projects, to deliver additional benefits and to meet changing business needs*" "*The coordination management of a series of interconnected projects and other non-project work, for the delivery of a specific package of benefits*" (Maylor et al., 2006); and "*Enterprise Program Management is the structures ad processes creating tight linkages between organisational strategy and the totality of its projects and related change activity*".

Authors seem to be offering different perspectives on a common theme, which is putting the delivery of beneficial change outcomes and value at the centre of programme management theory. Therefore, it may be argued that organising and coordinating a group of projects is not the goal behind the programme management approach, but rather a tool it utilises to determine the benefits of each one of those projects and on the basis of which it prioritises them to better serve the strategic vision of an organisation.

As an area of academic interest, the underpinning theory of programme management is still relatively new. This is evident in the inconsistency of the research conducted surrounding it thus far. Therefore, it is crucial to reference that the examination of the programme management theory literature including some of its most comprehensive works such as Lycett et al. (2004), Pellegrinelli (2010), Reiss et al. (2006), and Maylor et al. (2006) has revealed that it has little theoretical foundation and or lack scientific evidence supporting findings. This review of the theory of programme management led to the development of a problem statement: *Programmes in practice appear to under exploit the theory of project management.* This was in turn developed into a Research Aim: *Examine the application of programme management theory in contemporary practices and determining the gaps between the two.*

Research Review and Methodology

Understanding the "lived experience" might provide a useful comparison to help identify the challenges of programme management in praxis. It is evident that this research is exploratory in nature and therefore lends itself to the interpretivist philosophy (Symon & Cassell, 1998). Following this established philosophical position and given that research objectives aspire to investigate actual problems faced within the practical applications of programme management theory. Punch (1998) urges that this goal requires general-guiding, open-ended questions as well as a free-flowing, loosely structured design that does not allow for data to be pre-structured and prescribed; therefore for this purpose, a qualitative research design was found to be more suitable than a quantitative one as it encompasses a wider range of information that cannot be exclusively presented in numerical form. Consequently, and in order to sufficiently support the direction this study aimed to take, this research used a qualitative data approach over a cross-sectional time horizon using a focus group method (although the experiences of the participants are longitudinal) in order to then allow for theory to be inducted after the data analysis process. Silvester (1994) showed how responses from open question unstructured interviews or focus groups could be analysed using "attributional coding". This was the method used to make sense of the data and involved researcher interpretation and was used in conjunction with thematic coding (Symon & Cassell, 1998; Polit & Beck, 2012). The study design of a focus group and to achieve an inductive output of new theory; theory which can be treat more quantitatively in a future follow up study. Whilst providing rich insights and illustrative context, this sought of qualitative data includes a small sample and is subject to research bias in interpretation; however these limitations are thought to still provide useful research outputs providing the limitations are understood (Symon & Cassell, 1998; Polit & Beck, 2012), and any claims to truth are suitable caveated. The findings are presented in Table 1.

Table 1 Provides a summary of the focus group findings

	Thematic code	Quotes	Attributions
1	Strategic importance of projects	*"As a programme manager myself, I want everyone involved in the programme to be value creators not just project expeditors".* ***PMO1*** *"Changing human behaviour, and concentrating on projects we actually need, so it is the interaction that brings complexity between all stakeholders. Interests change, and there can be a lot of prospects change but continuous change makes it complex".* ***PMO5*** *"We find that if we do not invest we might not exist anymore, and if we do invest in the wrong thing, we might not exist within 2 years, so it is about making sure where the value is and establishing the benefits upfront".* ***PMO1*** *"Programme managers should listen to the project managers and more importantly be seen to be doing something about the issues that they are raising especially if it is to do with strategy, because the best information for you as a programme manager at the strategic level comes from the project manager level. And then use their examples to convince the people at the top…. we must use their tangible examples to show the things that are not being done, allow the project managers to actively contribute".* ***PMO1*** *"So, do you have to be a strategic thinker to deliver a programme? it is an inevitability that if you're good at delivering projects you start to think about that as part of the challenge, and you start to see the strategic connexion".* ***Facilitator***	Debating what "value" is Facilitated programme mgt Stakeholder negotiation The need for innovation Risk perception
2	Stakeholder negotiation	*"Project managers need to be challenging the end-users on their vastly over-egged briefs…we are not using an interest-based negotiation approach yet…the estates now have for the first time ever said that we're going to reduce carbon emissions and energy, we're going to have a sustainable estate that is the right size and works hard... usually they're not used to getting this sort of narrative into the negotiation, it's just tell me what you want and then we'll write it down and give it to an architect to design".* ***PMO1*** *"The thing about that issue is that these contracts have slipped into supply chain and I think part of why we have done that is because once you start to think about your programme you start to think about the coordination opportunities and I would say that a major opportunity around that is the way we operate procurement. Egan talked about bringing the team together in the supply chain, it seems to me to be a precursor to bringing those suppliers and getting them involved in what you're doing".* ***Facilitator*** *"Actually, it became very clear to me since being on this course that connectivity and emotional intelligence that they develop over time and then using that to influence, which is something you can't learn in a text book. It is too much focus on process and methodology so not enough soft skills like stakeholder engagement".* ***PMO5*** *"We set up frameworks across the programme so people are not just contracting, however we have to go through the proper procurement process which sets you up automatically to be adversarial, but we go through that just as a mean to an end to get frameworks on board so then we are working with what we now call "strategic partners", where it might be a firm of architects, in the past we did use them for one job and then they're off, but if you got the longevity of the framework and they know they're going to be working on the next project and the ones after that, these issues that you're talking about disappear or get easier to manage, to the point where the design team can say that this is not a building that you actually need, just put a big flag up and that'll do the marketing, because they know they're not going to get a building project out of it but it means that that money will be spent on a building project that we actually do need".* ***PMO1***	Debating what "value" is External stakeholders as the market Contracting issues Aligned contract suppliers with value-adding strategies Contracts management

	Thematic code	Quotes	Attributions
3	Governance needs to support strategic decision-making on projects	*"The estates department is traditionally seen in poor relation within any organisation, but actually they're one of the most strategic change providers in any institution. However, the culture of the organisation views the department the same as it was 20 years ago, and basically want them to just do what people at the top say". **PMO1*** *"It has become very political. And you start to think, where am I on this diagram?" **PMO3*** *"I have noticed during my experience as PMO in the government sector primarily is that the people at the top level do not really know what is going on at the ground level, so us as project delivery and programme managers kind of know where the complexity is and where the issues are inside the organisations, however I am not convinced that some of those issues are actually going to the top … there is a disconnect in the governance process". **PMO6*** *"When it comes to the governance structure around the project, the temptation is: someone wants something so we do it, regardless if it aligns with the organisational strategy or the department's strategy, in some ways it is even more fundamental than that, the university finally has a realistic strategy that the faculty strategy needs to tie into, so the programme in the governance system is where every capital request should be able to deliver the strategy". **PMO1*** *"We were really struggling due to the fact that we were not managed at a programme level, we were operating under a portfolio of many individual projects and we needed much more coordination into a programme…the board of directors offered our partners lots of jobs in order to reduce problems, and while that worked for a couple of weeks, we ended up 200 workers that are sitting down in the street because we are unable to get them enough jobs, because we couldn't plan that far ahead. So, now we are struggling to scale the projects to a level that would achieve any real benefit because it just hasn't been planned out and taught through on a programme level". **PMO4*** *"This especially happens in big organisations where there is too much focus given to where people are on the hierarchal structure. I would say that programme and project management need not to concern itself with the hierarchal structure and just be absolutely focused on the goal". **PMO1***	Governance not appropriate/inflexible Motivational theories Hierarchical issues
4	Lack of strategic focus	*"…they're very much stuck in the traditional way of customer service, speed, delivery and maintenance, there has never been any standard specifications across the department of estates". **PMO1*** *"There isn't much coordination in my department… to give you an example our programme is completely reactive, it is to do with repair and maintenance… however on our push to reduce a leakage problem, we have completely taken over all the resources from the other projects to go and work on the leak issues… so, we do not have the work basket to be doing all the work in the road, so why are dragging everyone off another project? Just to be sat in the vans doing nothing". **PMO4*** *"When someone wants something, so we do it, regardless of its value and benefit and if it aligns with the organisational strategy". **PMO1*** *"In the public sector the government is not going to go bust, they have plenty of capital to spend, so they bring in external contractors who are going to tell them what they want to hear, and strategy can get lost in that". **PMO6***	Service support culture Need for standardisation Coordination issues Engineering paradigm Seeing projects as being just about delivering "kit" or "product" rather than strategic benefit/ value

	Thematic code	Quotes	Attributions
5	Value issues	*"…it's about building rapport and understanding where each of us are coming from, often I find that if I can get involved earlier and say look you can have your new facility, but it's going to be smaller than you think, but it's going to be much more higher quality than what you are used to, if we get off on the right foot then we get where we need to be much faster. But I just need the rest of the department to do the same because they do not".* ***PMO1*** (discussing where value lies and when to invest) *"it is interesting that you draw a very narrow window of a 3 years sensitivity, or the do-nothing option versus do the wrong thing. The margins of getting this right seem to be extremely narrow in terms of time, the same thing applies to projections of numbers of students and the finances that come out of the back of all those. So, the dimensionality of the business case can be tricky".* ***PMO3***	Debating what "value" is Difficulty of identifying worthwhile value-adding projects
6	Need to innovate	*"Our strategic view is that there is a national shortage in 18–22-year olds, which is our demographic, therefore we need to diversify our income stream, we need to innovate".* ***PMO1*** *"Starting to think about your programme will lead you to start seeing more and more coordination opportunities".* ***PMO3*** *"…there is a need to invest and actually there is a really significant risk if you do not invest, and that particular bit is probably quite misunderstood or underplayed. If you are a business that's not investing, you're going out of business sometime in the near future".* ***Facilitator***	Need for proactive strategies
7	Blame culture/ frustration	*"I think a major part of it is to do with the building of trust and credibility as well as trying to have a voice that will be heard. I have a colleague who is very focused on gathering evidence, because actually nobody can refute evidence and from what gather pretty much all successful project managers do the same thing which is to follow the facts".* ***PMO3*** *"Ultimately when you've done all of that then it basically comes down to a choice, if they're still not willing to listen to facts of how this is how we can save this money, this how we can do this project better, as an individual you have a simple decision to make: can I stand this? Can I live like this in my professional career? Or should I be moving on somewhere else? In my experience when you do gather your evidence and you have built your credibility you do get somewhere it just takes time".* ***PMO1***	Inappropriate governance Frustration/lack of emotional intelligence
8	Risky projects	*"The speeding up and slowing down of projects is the difficult bit because if you are slowing down and you are in a project team, the programme starts to be affected if you are not linked into that overall vision. But more than that from a finance director point of view: that is just pure flexible gold dust, because you're in a situation where you can say, we can move quickly, or we can move less rapidly and reduce the risk if we need to spread it out. So that kind of strategic flexibility is at least the promise of programme management but to get to that you must get to that overlap structure".* ***Facilitator*** *"… construction like any other project is an investment, so the client wants to do more with less capital, therefore capital is a big risk so therefore the financial risk to the organisation is significant enough that just the 10% if the projects in the programme has the capacity to bankrupt the organisations".* ***Facilitator***	Dealing with strategic risk

	Thematic code	Quotes	Attributions
9	Honest reporting	*"This observation and talking to people about it has lead me to the realisation that as an organisation rose, this function in the middle becomes so vital because it transacts all the other information, narratives develop, and those narratives do not necessarily serve the best interest of the beneficiaries or those who are commissioning the programmes and project, and as you have previously said, all kinds of misleading measures are created to support the false narratives, when in reality it is just miss-information upon miss-information". **PMO2***	Culture Hierarchy
10	Programmes and projects not well understood	*"I started off as project manager …but for me, it kind of evolved, no one said now you are a programme manager, the stuff I did on my undergraduate degree was about learning fractional networks and how to set up a project management methodology, there was no handbook on how to evolve to a programme mgt role. You just see that this project is suffering from this problem and that project is as well so let's start thinking about this more broadly.... So, I think you've got to identify that earlier then really try to set up the programme mgt systems the same as you would a project structure, and it took me a long while to realise that, and then start thinking more strategically, what do we do about procurement? What do we do about contract mgt? Across a programme, because there is no handbook on how to do this". **PMO1***	Ambiguity surrounding projects and programme and their management
11	Information not going up the organisation	*"I am not convinced that some of those issues are actually going to the top because in the government sector, the government is not going to go bust so it is not commercially tied or dependant in that sense, they have plenty of budget they can spend on technology and things like that, so it is more about political pressures and stuff like that so it is all about what information they get at the top, and I think some of the challenges we happen because the real message is kind of lost in the middle there on its way to the top, so the people wonder why most programmes fail or are not delivered according to time, budget and quality which brings up the issue of accountability". **PMO6***	Communication Hierarchy Misrepresentation Honest reporting
12	Benefits tracking	(In response to an inquest about benefits tracking over a time frame and setting a maximum and minimum criterion in order to help understand the range of business case potential inside a specific programme) *"no, not at all, and I have worked at two similar establishments now and I know that the academic strategy is always lagging behind, so all we have done is provide a framework as to what we can do with the information that was given to us, and how to deliver it". **PMO1***	Benefits tracking failure

Findings summary, Authors, 2018

Focus Group Findings

Discussion and Conclusions

The qualitative data analysis and thematic coding revealed thirteen overarching themes and patterns across the focus group discussion. The addressed themes appear to largely be related to strategic alignment issues, challenges connected to deciding and debating what "value" is and where it lies, as well as the challenges resulting from inflexible governance systems, hierarchal structures, and the blame culture which the latter accompanies. Following the identification of themes were interpretations and attributions made by the researching in an attempt to speculate and link these issues to possible root causes.

The Strategic importance of projects and **lack of strategic focus** were dominant themes in the focus group discussion. In relation to these themes, it is critically important to mention that before the dialogue officially commenced and without being prompted by a question, the key respondent in the focus group began by contextualising the programme he was heading, elaborating on the sector, explaining the structure of the industry and the strategy the organisation is perusing, the importance of competitor investment and then the investment challenges for his organisation was facing, before fully diving into the discussion about programmes. This observation insinuates that from this respondent's point of view, programmes were about strategy delivery and the key purpose of being able to align the projects within the programme to the organisational strategy, which is in accordance with programme management theory and how it should work in practice. Although other participants agree that this is how programmes should operate, there was a general consensus amongst them that this is not the case within their organisations. In fact, the experts expressed that their programmes are completely failing to achieve that and speculate that their organisations might be measuring the wrong things. In this respect, this lack of strategic focus issue could be termed as a portfolio effect where data about projects is available; however there is not enough of an alignment towards the strategy. Regardless, the experts all agreed that across their varying industries, the intention is to have alignment with strategy, however, that is not how it is their projects and programmes are working. The participants further highlighted this disconnect from strategy through talking about the problems they are facing in regard to the selection and prioritisation of projects inside programmes they have worked on or are currently working on. Examples given included the misuse of resources resulting from infective or non-existent coordination where different departments are working in isolation rather than sharing resourcing to fix projects issues and the repercussions resulting from that. Other issues discussed involved what these participants find to be a problem with placing the focus on costlier projects at the expense of smaller ones and the effects that they have experienced as a result. Interpretations here attributed this particular problem to the "engineering paradigm" and the "service support culture" where projects are not treated as means for outcomes and benefits generation but rather as being just about delivering "kit", "product", and "service" and that being the full extent of their involvement in strategy delivery. As well as this, risk perception is another attribution here where organisations associate risk to cost rather than to benefits which can lead to a distortion in the strategic vision.

Challenges regarding selection, prioritisation, and coordination of projects within programmes can also directly tie to **the debate about what value is and where it lies** which was another prominent theme in the findings. There were many expressed concerns across the focus group regarding the difficulty of identifying worthwhile, value-adding projects. Interpretations of the discussion here speculate that this "value" debate might be due to a lack of understanding of the organisation's strategy which can lead to people being unable to select the appropriate projects that will serve it and deliver its objectives. Moreover, "value" and "benefits" might be undermined due to another identified theme which is the **risky nature of projects** and human risk perception issues as well as "strategic misrepresentation", where the projects are presented as more beneficial than they are, leading to a fear of investing in the wrong projects as elaborated on in the quotes mentioned above. The focus group participant also attested that they have experienced **benefits tracking failure** within their programmes. This theme could be found to originate from to the earlier challenges of the strategic misrepresentation, the lack of consensus surrounding value, and the disconnect from organisational strategy where it becomes difficult to set objective measurable benefits criteria in the midst of this strategy distortion. The key respondent elaborated on the importance of benefits tracking stating that when objective measures of benefits are applied throughout the different levels of the programme, fluctuating performances can be easily identified and compared so that issues can be rapidly rectified, and benchmarks of performance can be set for each area of the programme. He argued that through this, a culture of knowledge transfer and good practice can become embedded in the organisation where project performance is constantly monitored, compared and learnt from and benefits are regularly tracked.

The themes and the issues discussed thus far may be rooted in another theme extracted from the data which is that **Programmes and projects are not well understood.** Even the most seasoned respondent in the focus group agreed that there is an ambiguity surrounding these theories of project and programme management and went on to say that evolving from project manager to programme was not aided by any framework or handbook that directs one on how to do so, but rather by the development of their strategic thinking gained through years of experience in this field. This ambiguity can be corroborated by and directly attributed to what the literature review uncovered as being the novelty of these disciplines which results in the theories being not very well established and lacking a certain level of sophistication that might otherwise help eradicate a majority of the issues people are facing in practice.

Stakeholder management and negotiation issues was another prevalent theme in the conversation. The majority of the experts identified the adversarial nature of contracts to be a major point of contention within programmes and all expressed a need for more collaboration, alliances, and partnering in order to surpass contracting, sub-contracting and supply chain problems. This issue can be attributed to number of causes including the debate about value where stakeholders are not in agreement about worthiness of certain projects, contracting issues where separate parties are only concerned with protecting their interests, but mainly to a lack of finding common ground and not dealing with stakeholder management as an ongoing interest-based negotiation where long-term partnerships can be established allowing all parties to benefit and work towards a unified goal. Some of the respondents who actually perform this collaborative work within their supply chains revealed that this can lead to valuable projects being readily identified and selected, thus allowing for contract rates to be re-considered and for more focus to be placed on the strategy rather than on dealing with stakeholder management problems.

The need for governance to support strategic decision-making on projects is a theme that took a considerable amount of time in this discussion and was revealed to have multiple facets including challenges pertaining to honest reporting, information not getting through to the top, as well frustrations linked to blame culture. Project managers unanimously agreed in the focus group that Programmes are heavily influenced by hierarchical structures, although governance is necessary as it plays a big role when it comes to project selection, prioritisation, and authorisation processes; this is often impacted by bureaucracy and politics where they feel that the best interests of the organisation are not always put as a priority. The participants voiced their frustrations regarding the inflexibility of their governance systems which they viewed to not always be in support of strategic decision-making. It is believed that there is a predominant culture of expecting the people at project level to just do as they are told and ensure outputs delivery which they think reduces their roles to project expeditors rather than value creators and considerably limits their abilities to innovate. The key respondent agreed and expressed that throughout his long experience in this field, he found that programmes are frequently in a state of fragmented chaos due to bureaucracy and the political pressures associated with governance; he recommends that in order to move forward both programme and project management must become hieratically blind. Some respondents believe that this governance inflexibility makes it difficult to carry out the task of reporting especially as they feel that false narratives can develop based on incorrect measures that have already been put in place, whilst some blame the lack of communication between people at the top and people at project level for these issues. The matters discussed thus far appears to be in direct conflict with what the literature presents as the role and the main goal of governance bodies, where it was argued that not only should governance be in charge of prioritisation, coordination, and authorisation within the programme but also provide culture of learning, transparency, and accountability. This disconnect between project managers and the people at the top of the governance structure was interpreted to have a direct connection to motivational theories. It could be speculated through this interpretation that a flattening of the hierarchy, inclusion, having a bottom-up approach to strategy execution rather than top-down and allowing people at project level to have a degree of autonomy and contribution towards value creation and strategy development might be a solution to this issue.

The **need for innovation** is another identified theme. Nearly all the respondents voiced that there is a necessity for more proactive strategies that create exiting coordination opportunities and allow for greater value and benefits to be attained, and view that most of their organisation devote little chances to that and in some cases do not concern themselves about the long term. This lack of innovation experienced could be attributed to risk perception where people find it safer not to upset the status quo and risk losing investment, as well as to strategies that are not well developed or understood as mentioned previously. This theme could also be tied to the inflexible governance structures where barriers to innovation can be present leading people to merely do as they are told.

The experts' testimonies uncovered a variety of issues and challenges which upon preliminary exploration and examination appear to corroborate the hypothesis that there is indeed a gap between programme management theory and practice or that perhaps programmes in practice underexploit this theory.

Based on this, it could be argued that not only does the current theory of programme management not offer specialised frameworks on how to accomplish strategy delivery through coordination prioritisation and strategic alignment, but rather it just describes and speculates on its possible effects and advantages, it also completely fails to deal with the residual effects of the engineering paradigm and further entirely disregards the effects of risk perception amongst other human cognitive behaviours over key elements of managing programmes such as the strategic decision-making involved in deciding which projects hold the most value and benefits realisation.

Although the concept of "value" is not heavily mentioned in the APM's (2012) body of knowledge, it was a rather large topic of discussion in the primary data results. The experts elaborated that when attempting to create a programme and then prioritising and coordinating the projects inside it, issues concerning what stakeholders perceive value to be and where it lies, arose. They argued that the subjectivity attached to "value" makes it increasingly difficult to identify worthwhile, value-adding projects which in turn results in organisations failing to establish and attain strategic benefits. They further speculate that the debate about value might be a result of strategies that are either not well formed or not well communicated to the people working on the projects.

It has been argued that the selection of projects by value and the realisation of benefits heavily rely on the careful choice of objective measures placed on the benefit (Lumsden & Heathcote, 2011) and that the management of benefits within a programme should facilitate for strategy alignment by positioning the projects undertaken by the organisation as building blocks of strategy and vehicles for its implementation (Lycett et al., 2004). Nonetheless, failure to track benefits and challenges in recognising strategic risk were issues identified in the primary data and attributed to the debate about value and the disconnect from strategy. The experts explained that it is especially difficult to set measurable benefits criteria when value is undermined and when projects are presented as more beneficial than they actually are, leading to a fear of investing in the wrong projects. Whilst the literature mentioned the importance of setting key performance indicators to measure the beneficial outcomes attained by the programme against the strategy criteria (Kerzner, 2011; Letavec, Rollins, & Altwies, 2008) and recommends "programme maturity" (Reiss et al., 2006), it fails to put into consideration the issues of "strategic misrepresentation" and risk perception as per the respondents' experiences, it also offers no tools on how to do it, instead the majority of publications recommend the re-purposing of project management methodologies related to monitoring and control or propose models that have been proven to contain weaknesses and errors (Jugdev & Thomas, 2002).

Also related to the theme of "value" was the issue of innovation. The primary data revealed that most people involved in programme and project management express a need for more proactive strategies that can create more coordination opportunities. The experts believed that project proposals frequently disregard the idea of delivering sustainable benefits through the achievement of sufficient innovation and explain that this often results in selecting projects that offer no benefits as elaborated on in the NAO's (2016) study. This issue was attributed to the current programme processes being more inclined towards "less risky" solutions that can replicate *tried and tested* results in hopes of meeting the time and budget targets of programmes. Although the focus group discussion placed great emphasis on the importance of innovation, the available literature does not make mention of this topic. This is another gap in the literature which the focus group testimonies suggest needs to be addressed.

Governance, more specifically its role in supporting strategic decision-making, was found to be a key element of programme management in both the literature and the focus group discussion and is believed to have a crucial impact on the success or failure of strategy delivery (Muller, 2009; Reiss et al., 2006; Sanderson, 2012). Amongst the many roles of governance, the literature identifies the prioritisation of projects as one of the most important ones. However, this seemed to be a problematic area for the respondents who find that the reviewing of the priorities is an assumption in the theory of programme management that seems less readily achieved in practice. The reasons of why this is the case were attributed to bureaucracy and hierarchical structure issues where challenges including honest reporting, information not getting through to the top, and the incapability to innovate appeared to be the results of inflexible governance approaches and blame culture with the key respondent stating that "*"when it comes to the governance structure around the project, the temptation is: someone wants something so we do it, regardless if it aligns with the organisational strategy or the department's strategy"*. In this respect, the participants recommend that in order to move forward, both programme and project management must become hieratically blind, they further urge that in order to ensure accuracy, selection and prioritisation processes must involve the beneficiaries and the project managers who are working on these projects, not only to encourage strategic decision-making but to also make room for innovative ideas to emerge. Furthermore, they speculate that having a neutral, flexible, and subjectivity free governance process, where projects are selected, prioritised, and reviewed strictly on the basis of the benefit they bring and no other possible ulterior motives from the people involved in the organisation, will protect against objectivity and prevent the conflicts resulting from bureaucracy to distract from the bigger picture of greater value creation and benefits realisation.

The previous point appears to fall in line with Sanderson's (2012) opinion that the reason programmes are currently underperforming might be connected to the fact that current governance mechanisms are underdeveloped and thus are not conducted in a facilitative manner and are actually contributing in the failure of strategy delivery.

Connected to the challenges regarding coordination and prioritisation was the topic of visualisation; although not discussed extensively, the focus group participants pointed to this as being another missing element in the theory of programme management. Despite literature describing the advantages of prioritisation and coordination (Reiss et al., 2006), it does not propose any tools that can help the people overseeing this prioritisation process do so. However, in the focus group it was argued based on personal experiences in practice that infographics regardless of how simplistic they may be, have significant effects on spotting strategic links and dependencies, facilitating the removal of complexity and providing more clarity that allows for beneficial opportunities to emerge. It is believed that all of these key elements can ease the governance' prioritisation and coordination.

Stakeholder management and negotiation was another prevalent topic of discussion that has direct ties to the previously mentioned challenges of determining "value", governance, and strategic alignment. This was found by the experts to be difficult to apply in practice and attributed this difficulty to the disagreements regarding the value of projects which they believe is made worse by the adversarial nature of contracts, resulting in problems with supply chains and strategy execution delays. Although the literature mentions the importance of stakeholder engagement in programmes, with Reiss et al. (2006) specifically urging that the governance system should act a vehicle for reporting to all stakeholders, it can be noticed that perhaps the full extent of this stakeholder involvement is not thoroughly explored in literature.

Partnering, establishing long-term alliances and dealing with stakeholder management as an ongoing interest-based negotiation where common goals and interests are outlined, and benefits are collaboratively reached, were seen to be core components of stakeholder management that can positively influence strategy formulation and execution. The experts also believed that there are financial benefits associated with implementing these components, explaining that when collaborative efforts were made within their organisation's supply chains, it was noticed that valuable projects were being readily identified and selected which allowed for contract rates to be re-considered thanks to the element of trust introduced by "partnering". Experts also testified that this also led to more focus being placed on the strategy and minimised stakeholder management problems and conflicts. It may be argued that the theory of programme management must explore the concept of stakeholder management further by investigating the efficacy of interest-based negotiation.

Whilst the current theory of programme management places a great deal of importance on detailed integrated planning, it does not pay a lot of attention to the human behavioural aspect of managing projects. The experts consistently brought up the importance of emotional intelligence, negotiation skills, and dealing with risk perception as ways to deal with the issues and challenges discussed thus far. PM05 stated that *"Changing human behaviour and concentrating on projects we actually need, so it is the interaction that brings complexity between all stakeholders"*, implying that the human behavioural facet has a role in regard to programme management complexity. In this respect, it may be suggested that in its focus on achieving a high level of detail and control, the current theory of programme management undermines this important, unavoidable aspect.

The results are interpreted allowing for the hypotheses: *H1 "The systems nature of projects and programmes means that the two should not be treat as separate disciplines, but rather interconnected and overlapping systems"; H2 "That without further development, that addresses the control emphasis of some programmes, that programmes may serve to limit the creation of value by projects"; H3 "Economics behavioural theory should be employed to mitigate the partisan and subjective decision making programme governance might be vulnerable to"; H4 "Group decision making can be improved through the use of visualisation/infographics aimed at providing a holistic programme perspective".*

References

Chih, Y., & Zwikael, O. (2015). Project benefit management: A conceptual framework of target benefit formulation. *International Project Management Journal., 33*(2), 352–362.

Ferns, D. (1991). Developments in programme management. *International Journal of Project Management, 9*(3), 148–156.

Gardiner, P. D. (2005). *Project management: A strategic planning approach* (1st ed.). London: Palgrave Macmillan.

Jugdev, K., & Thomas, J. (2002). Project management maturity models: the silver bullets of competitive advantage? *Project Management Journal, 33*(4), 4–14.

Kerzner, H. (2011). *Project management metrics, KPIs, and dashboards: A guide to measuring and monitoring project performance*. New York: International Institute for Learning.

Letavec, C., Rollins, S. C., & Altwies, D. C. (2008). *Project Management Professional (PgMP): A certification study guide with best practices for maximising business results*. Fort Lauderdale, FL: J Ross Publishing.

Lumsden, A. J. & Heathcote, J. (2011). Programme Management: The Potential of Programme Management to improve the delivery value of projects, Online Journal of Project Management, Reviews, No. 4, 1st Quarter 2011.

Lycett, M., Rassau, A., & Danson, J. (2004). Programme management: A critical review. *International Journal of Project Management, 22*, 289–299.

Maylor, H., Brady, T., Cooke-Davies, T., & Hodgson, D. (2006). From projectification to programmification. *International Project Management Journal., 24*(8), 663–674.

Meskendahl, S. (2010). The influence of business strategy on project portfolio management and its success: a conceptual framework. *International Journal of Project Management, 28*, 807–817.

Muller, R. (2009). *Project governance*. Surrey, England: Gower Publishing Limited.

National Audit Office. (2016). *Delivering major projects in government: A briefing for the committee of public accounts*. National Audit Office. Retrieved June 25, 2018, from https://www.nao.org.uk/wp-content/uploads/2016/01/Delivering-major-projects-in-government-a-briefing-for-the-Committee-of-Public-Accounts-Summary.pdf.

Pellegrinelli, S. (1997). Programme management: organising project based change. *International Journal of Project Management, 15*(3), 141–149.

Polit, D. F., & Beck, C. T. (2012). *Nursing research: Generating and assessing evidence for nursing practice* (9th ed.). Wolters Kluwer: Lippincott Williams & Wilkins.

Project Management Institute. (2016). *Governance of portfolios, programs, and projects: A practice guide*. Newtown Square, PA: PMI.

Punch, K. (1998). *Introduction to social research: Quantitative and qualitative approaches*. London: Sage.

Reiss, G. (1996). *Programme management demystified: Managing multiple projects successfully*. London, New York: E & F Spon.

Reiss, G., Chapman, J., Leigh, G., Pyne, A., & Rayner, P. (2006). *Gower's handbook of programme management*. Hampshire: Gower Publishing Limited.

Sanderson, J. (2012). Risk, uncertainty and governance in megaprojects: A critical discussion of alternative explanations. *International Journal of Project Management, 30*(4), 432–443.

Shehu, Z., & Akintoye, A. (2009). Construction programme management theory and practice: Contextual and pragmatic approach. *International Journal of Project Management, 27*(7), 703–716. Retrieved May 28, 2018, from www.sciencedirect.com

Silvester, J. (1994). Attributional coding. In C. Cassell & G. Symon (Eds.), *Qualitative methods in organisational research: A practical guide*. London: Sage.

Symon, G., & Cassell, C. (Eds.). (1998). *Qualitative methods and analysis in organizational research*. London: SAGE Publications.

Van Buuren, A., Buijs, J., & Teisman, G. (2010). Program management and the creative art of coopetition: dealing with potential tensions and synergies between spatial development projects. *International Journal of Project Management, 28*, 672–682. Print.

Walenta, T. (2015). *Projects & programs are two different animals, don't underestimate the gap. 9th World Congress International Project Management Association (IPMA) 2015. IPMA WC 2015. 28–30 September–1 October 2015*. Panama: Westin Playa Bonita.

Winter, M., Smith, C., Cooke-Davies, T., & Cicmil, S. (2006). The importance of 'process' in rethinking project management: the story of a UK government funded research network. *International Project Management Journal, 24*(8), 650–662.

Obstacles of Sustainable Construction Project Management in South Africa Construction Industry

Nokulunga Mashwama, Didi Thwala, and Clinton Aigbavboa

Introduction

The construction industry is an engine for economic growth of any country; hence investment in infrastructure would promote the smooth flow of goods and services (Mashwama, Aigbavboa, & Thwala, 2016). Construction industry plays a huge role in influencing the competitive advantage of enterprise within the economy and it also attracts foreign investors to invest in it (Forbes & Ahmed, 2011). It is the responsibility of the government and construction stakeholders to ensure that the construction industry keeps improving to keep up with new ways of building, and be able to adapt to new methods, systems and strategies of construction industry (Forbes & Ahmed, 2011; Kubba, 2010). In addition, green building is increasingly being acknowledged in the construction property sector market as a response to the growing demand for environmentally friendly buildings (Chang, 2011). However, there are various barriers to sustainable building; mostly relating to poor government involvement and pioneering it, green building standards are perceived difficult to be understood and people's perceptions about green building being expensive (Marco & James, 2016). According to Häkkinen and Belloni (2011), the barriers of sustainable construction in developing countries are lack of motivation, lack of economic incentives and weak legislation enforcement. Sustainable construction is a green building model applied throughout the design, construction and maintenance

N. Mashwama (✉) · D. Thwala · C. Aigbavboa
Department of Construction Management and Quantity Surveying, University of Johannesburg, Johannesburg, South Africa

SARChI in Sustainable Construction Management and Leadership in the Built Environment, Faculty of Engineering and the Built Environment, University of Johannesburg, Johannesburg, South Africa
e-mail: nokulungam@uj.ac.za; didibhukut@uj.ac.za; caigbavboa@uj.ac.za

L. Scott et al. (eds.), *Sustainable Ecological Engineering Design*,
https://doi.org/10.1007/978-3-030-44381-8_23

period or life cycle of the building, while protecting the environment (Claassen, 2017; Griskevicuis, Tybur, & Van den Bergh, 2010). Sustainable construction is using sustainable construction principles/practices from inception to completion of a construction project while lessens the negative effect on the environment (GBCSA, 2017; Roper & Beard, 2006). Most greenhouse emissions of about 41% come from building structures (Ampofo-Anti, Dumani, & Van Wyk, 2015; Claassen, 2017).

Sustainable Construction Project Management

Ecology and Environmental Friendliness

Conservation of natural environment is one of the major drivers of sustainable project management (Chevallier, 2015; Claassen, 2017). Moreover, South Africa has an obligation to reduce its carbon emission by putting in place environmental sustainability strategies and start practicing them (Windapo, 2014). The growth of sustainable project management is directly related to the advance of environmental consciousness (Sánchez, 2015; Tharp & Silvius, 2013). Moreover, if environmental consciousness gains motivation, sustainable project management simultaneously experiences huge growth and development (Chevallier, 2015; Robichaud & Anantatmula, 2011). It is rather a tedious task to protect the natural environment; however, various endeavours have explored constructive outcomes of the reception of green thinking in building design and construction (Hoffman & Henn, 2008).

Reduced Operational Cost of Buildings

The main economic benefit associated with green buildings is reduced operation cost (Gowri, 2004). Green buildings are designed and built in a manner at which there is minimal usage of electricity to power them and there is minimal usage of water, as green buildings have built-in water recycling mechanisms (Kubba, 2010). Sustainable project management is motivated by financial gain through the need to diminish the cost of building activities or operational costs after construction to yield higher returns from the investments (Windapo, 2014). Giving rise to the concept of life cycle costing which is concerned with improving an incentive for money in the possession of corporeal properties by considering the overall cost factors identifying with the property amid its operational life cycle (Woodward, 1997). Moreover, energy and water proficient buildings have possessed the capacity to lessen their running cost expenses notably, with utility being reduced by 50% than that of conventional buildings (Forbes & Ahmed, 2011).

Social Responsibility and Status

Stakeholders are required to fulfil commercial social responsibility; hence people are driven to sustainable project management for status and noticeable conservation (Tharp & Silvius, 2013). Furthermore, status expands the desire for green buildings, and thus an increased demand for sustainable project management, regardless of the fact that going green is perceived to be expensive than the conventional way of doing things (Silvius & Schipper, 2012). Moreover, quickly rising media consideration, straightforwardness and responsibility on ecological care and conservation have framed a social trend to significantly amplify green construction (Lu, Cui, & Le, 2013; Zhang, Shen, & Wu, 2011). Green building is becoming more of a fashionable trend and therefore inducing the growth of sustainable project management (Mazmanian & Kraft, 2009).

Hindrances to Sustainable Project Management

Green Buildings Are Expensive

The main barrier restricting the growth and development of green building is the perception that green building are expensive (Zhang et al., 2011). The green building councils worldwide have shown that although green buildings are initially expensive to construct, the economic benefits in the long term are impeccable (GBCSA, 2017). The expenditures related to the construction of green buildings is regularly no more prominent that the expenditures of constructing a conventional building and the investment can be often acknowledged in the cost of energy and electrification and water over the long term (Forbes & Ahmed, 2011; Lambeck & Eschemuller, 2009). Furthermore, sustainable project management has the capability of bringing down construction costs as well as the amplified risk of delays in construction (Sánchez, 2015).

Lack of Expertise or Experience

Green building projects differ from traditional construction projects as they require the utilization of special materials and building practices to accomplish sustainability (Lu et al., 2013). The concept of green building is accepted by a few South African construction companies, due to the lack of awareness, skill and expertise of sustainable project management (Ampofo-Anti et al., 2015). Additionally, a green project design procedure is intended to be a multiple effort from a numerous stakeholders to achieve all-inclusive outline arrangements that boost building performance as well as yield a profitable investment (Forbes & Ahmed, 2011, p. 284). Sustainable project management incorporates an extensive variety of experts to build up a mutual vision of the project, with the goal that the broad green building regulation can be consolidated in the design from the very beginning (Marco & James, 2016; Sánchez, 2015).

The Complexity of Green Building Rating Systems

Architects, Engineers, Project managers and construction labourers are confronted with a test to cater for the demand of new facilities, which should limit the negative effect on the natural environment (AlSanad, 2015). The building should yield economic gain on investments and giving ecological advantages to the investor (Kubba, 2010). The green rating systems have been set up to decrease and encourage the construction of sustainable building which are friendly to the environment (GBCSA, 2017; Gowri, 2004). The reason behind the commissioning of green building rating systems is to educate the design teams and the construction team of what is broadly acknowledged in terms of green buildings, as well as the objectives that are proposed to be achieved (GBCSA, 2017).

Resistance to Change

The crucial component in sustainability is change in the form of transformation towards additional sustainable practises (Silvius & Schipper, 2012). Furthermore, the progress required for sustainability will thus be the implied arrangement of qualities that experts possess and that impact or directs conduct. Moreover, Hoffman and Henn (2008) contend that ecological advancement and green thinking in the built and engineering industry will keep on stalling if the substantial societal and mental boundaries that endure are not confronted.

Claassen (2017) state that the development of green building is not hampered by an absence of advancements and assessment strategies, yet rather is liable to different authoritative and procedural challenges caused by the reception of new techniques or technologies. Resistance towards new advances is due to the fact that they require process changes that include unforeseen dangers and expenses (Claassen, 2017). The GBCSA (2017) reiterates by saying that sustainable headship requires one to take courage as it requires doing thing in a non-conventional manner. Furthermore, requires altering the way of thinking through ambition and bravery to eradicate set limitations.

Research Methodology

Research Area, Approach and Design

The research study was based in the Mpumalanga Province where new commercial property development projects are emerging. The commercial property sector plays a fundamental role in the economic development of the province. This sector comprises both the public and private sector. The public sector performs an administrative, financier and guiding role in the commercial property market and the private sector becomes the primary implementing agent. For the purpose of the study we focused on the public sector who employs contractors and professionals from private sector to undertake the construction of primary and supporting structure. Quantitative approach method was adopted to investigate the hindrance of sustainable project management in the construction industry of South Africa. 80 Questionnaires were distributed and 65 were brought back which were all valid and usable. A well-structured questionnaire was distributed to different professional companies in Mpumalanga Province, such as civil engineers, project managers, directors, quantity surveyors and construction managers. The questionnaires were sent via e-mails, some were delivered to the known construction firms by the researcher. The secondary study was conducted from reliable scholarly sources such as articles, journals, books, publications, websites and site experience on the field.

Statistical Package for the Social Science (SPSS)

The quantitative data collected was analysed with Statistical Package for the Social Science (SPSS), a computer programme which is used for analysing data concerned with social phenomena. The software was used to generate various statistical, including descriptive statistic, which provides a basic summary of all variables in the data (SPSS, 2004). The benefits of using SPSS is that it allows for scoring and analysing quantitative data at speed and it can also be used to perform multivariate analysis. SPSS also helps to present the data in a logical format (SPSS, 2004), thereby reducing time spent on calculating scores. However, accuracy in results is highly dependent on inputs, hence the need to accurately capture data from the questionnaire.

Computation of the Mean Item Score (MIS)

Five-point Likert scale was adopted for the study which gave a wider range of possible scores and increase statistical analyses that are available to the researcher. The five point scales were transformed to mean item score abbreviated as MIS.

The computation of MIS was calculated from the total of all weighted responses and then relating it to the total responses on a particular aspect. The MIS was adopted to rank the factors from highest to lowest and expressed and calculated for each item as follows:

$$\text{MIS} = \frac{1n_1 + 2n_2 + 3n_3 + 4n_4 + 5n_5}{\Sigma N}$$

where
n_1 = Number of respondents for strongly disagree
n_2 = Number of respondents for disagree
n_3 = Number of respondents for neutral
n_4 = Number of respondents for agree
n_5 = Number of respondents for strongly agree
N = Total number of respondents

Findings

Promoters of Sustainable Project Management

Respondent were asked based on their experience as to which factor promotes the use of sustainable project management in construction projects. Public and market demand for green buildings was ranked first with (MIS = 4.56; STD = 0.985); educating and training of stakeholders and clients on the future benefits of green buildings was ranked second with (MIS = 4.45; STD = 1.148); heavier taxes and penalties on unsustainable construction practices was ranked third with (MIS = 4.33; STD = 0.823); project team to create a culture for the adoption of green building practices was ranked fourth with (MIS = 4.31; STD = 0.972); interest free lending schemes provided by government to overcome financial barriers was ranked fifth with (MIS = 4.29; STD = 1.117); upgrading courses to supplement knowledge regarding sustainable construction was ranked sixth with (MIS = 4.20; STD = 0.991); bonuses for staff in obtaining green mark award for their building with (MIS = 4.01; STD = 0.878); global warming scare was ranked eighth with (MIS = 3.98; STD = 0.962); subsidy or incentives from government for green building projects was ranked nine with (MIS = 3.88; STD = 1.147); green buildings help increase health of occupants was ranked ten with (MIS = 3.75; STD = 1.020); and simple green building practices for the transition from conventional to green practices was ranked eleven with (MIS = 3.65; STD = 1.138) (Table 1).

Table 1 Promoters of sustainable project management

Promoters of sustainable project management	Mean	STD	*R*
Public and market demand for green buildings	4.56	0.985	1
Educating and training stakeholders, clients on the future benefits	4.45	1.148	2
Heavier taxes and penalties on unsustainable construction practices	4.33	0.823	3
Project team to create a culture for the adoption of green building practices	4.31	0.972	4
Interest free lending schemes provided by government to overcome financial barriers	4.29	1.117	5
Upgrading courses to supplement knowledge regarding sustainable construction	4.20	0.991	6
Bonuses for staff for obtaining green mark award for their building	4.01	0.878	7
The global warming scare	3.98	0.962	8
Subsidy or incentives from government for green building projects	3.88	1.147	9
Green building help increase health of occupants	3.75	1.020	10
Simple green building practices framework to adopt in the transition from conventional to green practices	3.65	1.138	11

Obstacles of Sustainable Project Management

Respondents were asked based on their experience as to which factor is an obstacle of sustainable project management in construction projects. Resistance to change from conventional to green practices by organization employees and lack of awareness of green building methods and technologies were ranked first with (MIS = 3.99; STD = 1.089 and 1.123, respectively); absence of dependable exact cost data/information was ranked second with (MIS = 3.88; STD = 0.981); limited government involvement and the complexity of codes and regulations on green building and sustainable construction were ranked third with (MIS = 3.75; STD = 0.876 and 1.058, respectively); high costs of green building materials was ranked fourth with (MIS = 3.68; STD = 1.189); attract a higher risk of delays in construction was ranked fifth with (MIS = 3.61; STD = 0.889); lack of experience and perception that green buildings are expensive was ranked sixth with (MIS = 3.59; STD = 0.875 and 0.989, respectively); lack of expertise or training and lack of governmental support were ranked seventh with (MIS = 3.47; STD = 1.189 and 0.986, respectively); lack of interest from clients was ranked eighth with (MIS = 3.33; STD = 0.857); lack of knowledge regarding green building principles and limited access to green building material supplier were ranked ninth with (MIS = 3.05; STD = 0.978 and 0.969); limited financial incentives was ranked tenth with (MIS = 2.92; STD = 1.109); lack of communication and interest among project team members and absence of general familiarity with the benefits of green building was ranked eleventh (MIS = 2.87; STD = 1.160 and 1.111); and lack of management and time to implement green construction practices was ranked twelfth with (MIS = 2.65; STD = 0.907) (Table 2).

Table 2 Obstacles hindering SPM

Obstacles hindering the adoption of sustainable project management	Mean	STD	*R*
Resistance to change from conventional to green practices by organization employees	3.99	1.089	1
Lack of awareness of green building methods and technologies	3.99	1.123	1
Absence of dependable exact cost data/information	3.88	0.981	2
Limited government involvement	3.75	0.876	3
The complexity of codes and regulations on green building and sustainable construction	3.75	1.058	3
High costs of green building materials	3.68	1.189	4
Attract a higher risk of delays in construction	3.61	0.889	5
Lack of experience	3.59	0.875	6
Perception that green buildings are expensive	3.59	0.989	6
Lack of expertise or training	3.47	1.189	7
Lack of governmental support	3.47	0.986	7
Lack of interest from clients	3.33	0.857	8
Lack of knowledge regarding green building principles	3.05	0.978	9
Limited access to green building material supplier	3.05	0.969	9
Limited financial incentives	2.92	1.109	10
Lack of communication and interest among project team members	2.87	1.160	11
Absence of general familiarity with the benefits of green building	2.87	1.111	11
Lack of management and time to implement green construction practices	2.65	0.907	12

Conclusion

Resistance to change from conventional to green practices by organization employees; lack of awareness of green building methods and technologies; absence of dependable exact cost data/information; limited government involvement; the complexity of codes and regulations on green building and sustainable construction; high costs of green building materials were the obstacles of sustainable construction project management. Public and market demand for green buildings; educating and training stakeholders, clients on the future benefits; heavier taxes and penalties on unsustainable construction practices; project team to create a culture for the adoption of green building practices were the promoters or influencers of sustainable project management. From the findings the government of South Africa needs to play a huge role implementing sustainable construction project management in public buildings, which will benefit the government and the end users and reduce the nation's carbon emission emanating from construction sector. Sustainable buildings are initially expensive to construct, the economic benefits in the long term are impeccable. To effectively ignite rapid growth of sustainable project management, the government ought to finance development costs associated with the construction of green buildings.

References

AlSanad, S. (2015). Awareness, drivers, actions, and barriers of sustainable construction in Kuwait. *Procedia Engineering, 118*, 969–983.

Ampofo-Anti, N.L., Dumani, N., & Van Wyk, L. (2015). *Potential for reducing greenhouse gas emissions in the South African Construction sector.* DBSA Green Fund Research and Policy Development to advance a Green Economy in South Africa: Research Report.

Chang, N. (2011). *Analysis for sustainable engineering: Theory and application. Journal of green manufacturing and systems engineering.* McGraw-Hill Professional, Access Engineering.

Chevallier, R. (2015). *The importance of protecting the globe's most important ecological assets.* Retrieved from https://www.saiia.org/research/the-importance-of-protecting-the-globes-most-important-ecological-assets/.

Claassen, S. (2017). *Mitigation of GHG emissions now an obligation for South African companies.* Retrieved from http://crown.co.za/latest-news/electricity-control-latest-news/5517-mitigation-of-ghg-emissions-now-an-obligation-for-south-african-companies.

Forbes, L. H., & Ahmed, S. M. (2011). *Modern construction: Lean project delivery and integrated practices.* New York: CRC Press.

Gowri, K. (2004). Green building rating systems: An overview. *ASHRAE Journal, 46*(11), 56–58. 60.

Green building Council South Africa. (2017). About us. Retrieved from https://gbcsa.org.za/certify/green-star-sa/

Griskevicuis, V., Tybur, J., & Van den Bergh, B. (2010). Going green to be seen: Status, reputation, and conspicuous conservation. *Journal of Personality and Social Psychology, 93*(3), 392–404.

Häkkinen, T., & Belloni, K. (2011). Barriers and drivers for sustainable building. *Journal of Building Research & Information, 39*(3), 239–255.

Hoffman, A. J., & Henn, R. (2008). Overcoming the social and psychological barriers to green building. *Sage Organization and Environment, 21*(4), 390–419. Retrieved from http://oae.sagepub.com

Kubba, S. (2010). *Green construction project management and cost oversight.* Oxford: Butterworth-Heinemann

Lambeck, R. & Eschemuller, J. (2009). *Green construction.* Chapter in Urban construction project management. McGraw Hill Construction Series, Access Engineering.

Lu, Y., Cui, Q., & Le, Y. (2013). Turning green to gold in the construction industry: Fable or fact. *Journal of Construction Engineering and Management, 139*(8), 1026–1036.

Marco, Z. & James, C.A. (2016). *The perceived barriers to the construction of green conference proceedings of the 9th CIDB postgraduate conference held in Cape Town.* Conducted by the Construction Industry Development Board: Cape Town, South Africa.

Mashwama, N. X., Aigbavboa, C., & Thwala, D. (2016). Investigation of construction stakeholders, perception on the effects and cost of construction dispute in Swaziland. *Procedia Engineering, 164*, 196–205.

Mazmanian, D. A., & Kraft, M. E. (2009). *Toward sustainable communities: Transition and transformation in environmental policy.* Cambride MA, England: The MIT Press.

Robichaud, L., & Anantatmula, V. (2011). Greening project management practices for sustainable construction. *Journal of Management in Engineering, 21*(1), 48–57.

Roper, K. O., & Beard, J. L. (2006). Justifying sustainable buildings-championing green operations. *Journal of Corporate Real Estate, 8*(2), 91–103.

Sánchez, M. A. (2015). Integrating sustainability issues into project management. *Journal of Cleaner Production, 96*(2015), 319–330.

Silvius, G., Schipper, R., & Nedeski, S. (2012). Sustainability in project management: Reality bites. *Journal of Procedia—Social and Behavioral Sciences, 2*(2), 1–11.

SPSS. (2004). *A guide to the processing, analyzing and reporting of (research) data.* Netherlands: Wolters-Noordhoff BV.

Tharp, J., & Silvius, G. (Eds.). (2013). *Sustainability integration for effective project management*. Hershey, PA: Business Science Reference.

Windapo, A. O. (2014). Examination of Green Building Drivers in the South African Construction industry: Economics versus Ecology. *Journal of Sustainability, 6*(9), 6088–6106.

Woodward, D. G. (1997). Life cycle costing theory: Information acquisition and application.*International. Journal of Project Management, 15*(6), 335–344.

Zhang, X., Shen, L., & Wu, Y. (2011). Green strategy for gaining competitive advantage in housing development: A China study. *Journal of Cleaner Production, 19*(2), 157–167.

Part IV
Energy and Energy Efficiencies

Long-Term Durability of Solar Photovoltaic Modules

Chibuisi Chinasaokwu Okorieimoh, Brian Norton, and Michael Conlon

Introduction

Solar photovoltaic cells convert solar energy into electrical energy through the photovoltaic effect. Solar energy can reduce emissions of carbon dioxide (CO_2) associated with the generation from fossil fuels as the only CO_2 emissions are those embodied in their manufacture (Norton, 1999). The electricity generated by solar PV is more environmentally friendly as it is carbon-emission free at the point of generation when compared to fossil fuel generation. Solar PV panels experience long-term performance degradation resulting in lower like-per-like efficiency and performance ratios when compared with their initial performance.

Reducing rates of PV module degradation aim to maintain efficiency of solar PV systems (Li, 2016). As manufacturers usually guarantee the life span of PV modules for more than 20 years (Li, 2016), it is therefore necessary to track and mitigate the degradation of PV modules over this period. Both during and beyond this period knowing degradation behaviour is essential for operation, maintenance and repair.

Distinguishing Transient Performance Changes from Longer-Term Degradation

PV module output varies with solar irradiance and module temperature. It is also affected by shading, rain and dust (Dunlop & Halton, 2006; Tiwari, Mishra, & Solanki, 2011). All these variations are transient on a variety of timescales and/or reversible. Degradation refers to loss of output due to physical degradation or

C. C. Okorieimoh (✉) · B. Norton · M. Conlon
Dublin Energy Lab, School of Electrical and Electronic Engineering, Technological University Dublin, Dublin 8, Ireland
e-mail: d18125175@mydit.ie

L. Scott et al. (eds.), *Sustainable Ecological Engineering Design*,
https://doi.org/10.1007/978-3-030-44381-8_24

damage to the PV cell, the effects are not reversible. It refers to effects that will ultimately require the replacement of a PV cell for the system to return to its initial performance. Transient effect caused by increase in PV cell ambient temperature can lead to reductions in output and efficiencies (Okorieimoh et al., 2019). Degradation is measured by changes mean efficiency and/or performance ratio over the long-term as illustrated indicatively in Fig. 1. It can also be observed in perturbation caused by cell failure in the current-voltage (*I*–*V*) curves for an array.

This paper discusses the long-term durability of solar photovoltaic modules with particular emphasis on:

1. How to achieve long-term durability of solar photovoltaic modules
2. What affects the durability of solar photovoltaic modules
3. What are the remedies to solar photovoltaic (PV) degradation
4. How the durability and reliability of PV modules can be improved

Individual module degradation can be attributed to intrinsic property changes in the PV materials caused by external effects such as (a) potential induced degradation (PID) (Pingel et al., 2010) and (b) light induced degradation (LID) (Sopori et al., 2012).

The outdoor operation of cells as part of a module in an array means mechanisms external to solar cell such as corrosion in interconnections and solder bonds play a significant role in performance degradation (Li, 2016). This makes it important to determine the degradation rates under outdoor operational conditions rather than indoor testing of isolated modules. Li (2016) classified the major difficulties in evaluating degradation rates of PV modules from real operational data into:

1. Large fluctuations of the operational data due to uncontrollable external parameters such as weather conditions like solar radiation, rain, cloud movement, wind velocity and ambient temperature together with unexpected changes of factors external to PV systems such as unexpected shading, inverter problems and control failures.
2. Systematic 'degradation' in the measurement of PV module operational performance caused by control sensor drifting with time as a result of electronic ageing of components such as the drifting of irradiance sensors. The energy output of a PV system depends on weather conditions (Osterwald, Anderberg, Rummel, & Ottoson, 2002), (Tiwari et al., 2011), (Li, Prawiradiraja, & Battul, 2013).

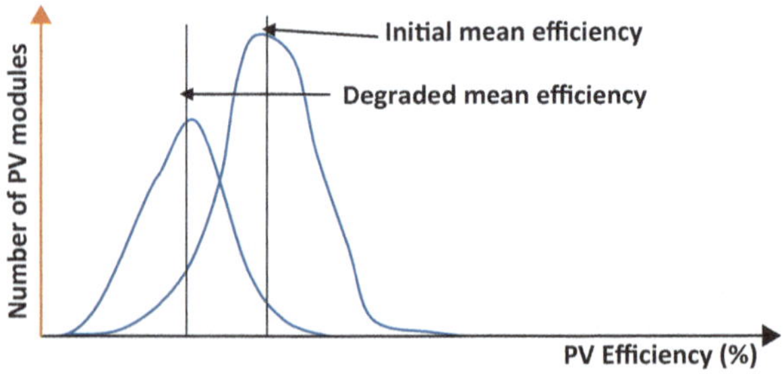

Fig. 1 Degradation on solar PV system

According to Osterwald et al. (2002), the degradation rate of silicon PV modules is around −0.7% per year of maximum power rating.

Degradation Rates

Jia, Thomas, Armin, and Timothy (2014) investigated performance degradation of the following types of PV modules: monocrystalline silicon (m-Si) (such as glass-back sheet with frame and glass-glass without frame), heterojunction crystalline silicon, monocrystalline silicon back-contact, multi-crystalline silicon, double-junction "micromorph" silicon, single-junction/double-junction amorphous silicon (a Si), cadmium telluride (CdTe), and copper indium gallium diselenide (CIGS) as seen in Fig. 2, 3 years outdoor monitoring data showed the degradation paths of each module.

Statistical decomposition techniques were used to draw out paths for the performance ratio (PR), short-circuit current (I_{SC}), open-circuit voltage (V_{OC}), and fill factor (FF). Degradation rates for the monocrystalline silicon (m-Si) modules were found to be equal to or less than −0.8% per year as a result of the decrease in I_{SC}. Multi-crystalline silicon modules exhibited a higher degradation rate of −1.0% per year. The a-Si, micromorph silicon and CdTe modules showed a degradation rate of about −2% per year. The CIGS module showed a degradation rate of −6% per year higher than CdTe.

The study of annual degradation rates of recent crystalline silicon photovoltaic modules were carried out by Tetsuyuki and Atsushi (2017). Six crystalline silicon PV modules connected to an electric power grid were analysed. Three indicators were used for the annual degradation rates of the different crystalline silicon

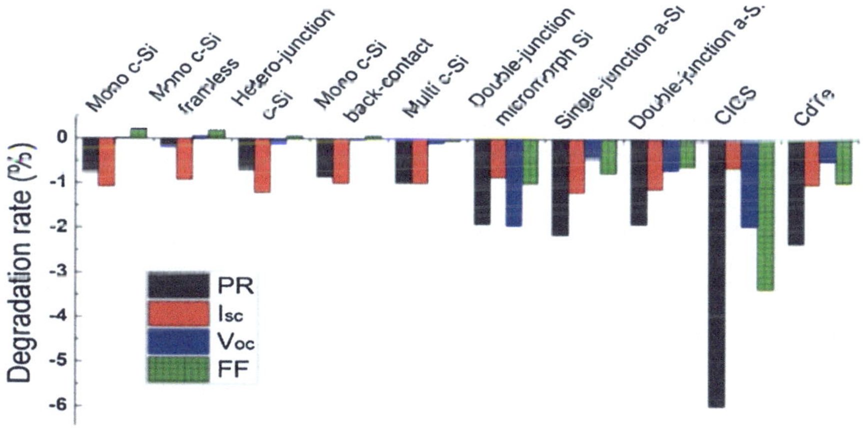

Fig. 2 Mean annual degradation rates (%) of performance ratios (PRs) and the I–V curve components: short-circuit current (I_{SC}), open-circuit voltage (V_{OC}), and fill factor (FF) for ten module types (Jia et al., 2014)

PV: energy yield, performance ratio and indoor power. The performance of the module was evaluated from electricity output measurements taken over 3 years. The following trends were found in the three indicators; energy yield: 0.0, −0.4% per year, 0.0, 0.1% per year, 1.5% per year and 0.5% per year, performance ratio: 0.0, −0.4% per year, −0.1% per year, 0.0, 1.4% per year and 0.5% per year and indoor power: 0.1% per year, −0.3% per year, 0.2% per year, 0.0, 0.7% per year and 0.6% per year were similar. The performance of the newly installed PV modules were found to decrease by over 2% as a result of initial light-induced degradation (LID) after installation (Tetsuyuki & Atsushi, 2017).

The power output of an outdoor PV module has been shown to reduce as a result of thermal cycling causing crack formation between solders and metals (Nochang, Jaeseong, & Changwoon, 2014). Dunlop and Halton (2006) studied degradation of PV modules in outdoor conditions for 22 years. They monitored the electrical power outputs of monocrystalline silicon, polycrystalline silicon and amorphous silicon modules. They found 8–12% decrease of maximum power output of the PV modules (P_{max}) after 20 years outdoor exposure. Their research showed that about 80% of the reduction was due to corrosion and the remaining 20% was attributed to dust accumulating on the PV modules.

An experimental study of degradation modes and their effects on photovoltaic module was conducted after 12 years of field operation (Saadsaoud, Ahmeda, Er, & Rouabah, 2017). Their investigation found that degradation led to annual reductions in output power ranging between 2.08% and 5.2%. Short circuit current (I_{sc}) reduced by between 2.75% and 2.84% annually. The open-circuit voltage (V_{oc}) was found to be the least affected, with annual reductions ranging between 0.01% and 4.25%.

Degradation Influences

The existence of only one highly degraded PV module in a PV system reduces daily output from (Takatoshi, Tomoya, Tadashi, & Yoshitaka, 2018):

1. 19.8 kWh to 18.7 kWh during sunny days
2. 11.3 kWh to 10.8 kWh during partly cloudy sunny days
3. 5.5 kWh to 5.3 kWh during cloudy days

Pramod, Tiwari, Sastry, Birinchi, and Vikrant (2016) investigated degradation of mono-crystalline photovoltaic modules after 22 years of outdoor exposure. They studied 90 mono-crystalline silicon PV modules installed on the rooftop of the National Institute of Solar Energy (NISE) near New Delhi, India, after 22 years of outdoor operation. They carried out visual inspection, thermal imaging, current-voltage characteristic curve analysis and insulation resistance measurement and in addition calculation of the degradation rate. The mean power reduction rate of 90 PV modules over the period of 22 years was found to be about 1.9% per year at a peak rate of power reduced by 4.1% per year and the minimum rate of power reduction was 0.3% per year. The result of electrical resistance of insulation measurements of 90 PV

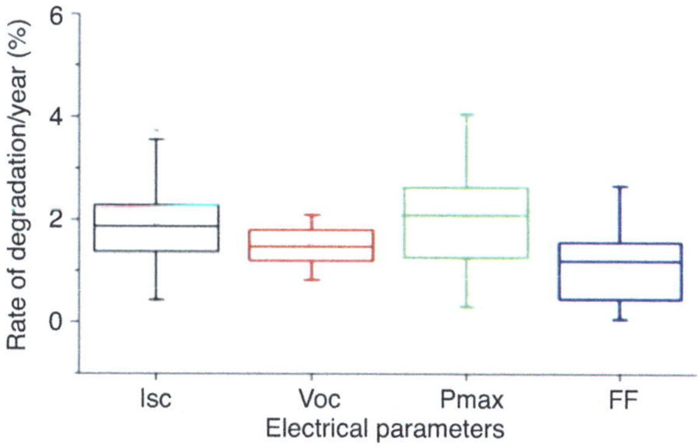

Fig. 3 Degradation rates for different electrical parameters (Pramod et al., 2016), where I_{SC} short-circuit current; V_{OC} open-circuit current; P_{max} maximum power and FF fill factor

modules (both in dry and wet conditions) showed that only 2 PV modules showed insulation electrical resistance of less than 400 MΩ in dry conditions. Analysis of electrical parameters shown in Fig. 3 indicated that there was degradation of short circuit current, from 0.4% to 3.7% per year with a mean value of 1.8% per year. The open circuit voltage ranged from 0.8% to 2.1% per year with a mean value of 1.4% per year and fill factor ranged from 0.7% to 2.6% per year with a mean value of 1% per year. The maximum power P_{max} reduction rate ranges from 0.3% to 4.1% per year with a mean value of 1.9%/year. The reduced power output was mainly due to the degradation in short circuit current.

Diagnosis of Degradation Mechanisms

The reliability and degradation of solar PV modules was investigated by David, Muyiwa, Gabriel, and Isaac (2017) as part of a case study of polycrystalline modules in Ghana. Fourteen polycrystalline modules were installed on the concrete roof in a hot humid environment. They were evaluated after continuous outdoor exposure for 19 years. They used a visual inspection checklist to document the physical state of the modules. The PV modules were also evaluated by current-voltage (*I*–*V*) characterization and thermal imaging. Their results showed that the modules were found to be in good physical state with the exception of some bubbles developing on the front side. There was insignificant corrosion found at the edge of the cells. The performance change of the PV modules over the exposure duration was: nominal power, 21% to 35%; short circuit current, 5.8% to 11.7%; open circuit voltage, 3.6% to 5.6% and 11.9% to 25.7% for fill factor, respectively.

Zhengpeng, Timothy, and Armin (2011) carried out a study on PV module durability under high voltage biased damp hot and humid conditions. They made use of ten photovoltaic module technologies which comprisal (a) five thin-film technologies and (b) five silicon wafer based technologies. The PV modules were subjected to accelerated ageing tests in a climate dark chamber under temperature conditions of 85 °C and relative humidity of 85% and electrical bias for a period of 650 h

(27 1/12 days). They applied a bias voltage of ±1000 V DC between the active circuit of each module and the module frame. Their results showed biased stressing conditions in damp heat could significantly degrade the electrical performance and cause several defects including delamination, glass surface deterioration, frame corrosion, and metal grid discoloration, depending on module type and bias polarity.

PV modules can be damaged by weather, temperature variations, soiling effect and ultraviolet exposure. Typical faults are summarized in Table 1. Performance monitoring of PV systems aims are to maintain the power output from PV systems, thus increasing economic viability (Parveen & Saurabh, 2019). To evaluate the degradation of PV modules, Parveen and Saurabh (2019) suggested a clustering-based technique with different arrangements. They estimated the performance ratio (PR) of the PV modules without physical inspection on-site, making the suggested model useful for real-time estimation of PR. This may, in turn, lead to stronger forecasting of PV array power output. Their model calculated the degradation in output solar power for amorphous silicon (a-Si), polycrystalline silicon (p-Si), and silicon hetero-junction with an intrinsic thin layer (Si-HIT) over 3 years. The degradation rate for a-Si was lowest at 0.85% per year, and was highest for Si-HIT technology at between 0.95% and 2.03% per year. Their results showed good agreement with the standard procedure used for performance evaluation in a similar earlier study, but as data was taken from a range of other studies, furthers corrections for environmental factors may be necessary. Therefore, the suggested model has an advantage over other methods that real-time estimation is possible as it does not require physical inspection and imaging.

To create a PV panel simulation which is effective under changed environmental conditions, Murari et al. (2017) developed a model using MATLAB of an equivalent circuit. This allowed them to perform joint simulation of a PV device with power electronics interfaces. Such simulations allow optimization of the design of solar arrays and power systems.

Table 1 Typical faults associated with PV modules

Manufacturing defects	• Hot spots (bad soldering) (Kuitche, Pan, & Tamizhmani, 2014); • Micro-cracks "snail trail" (Köntges et al., 2008); • Contamination (discolouration) (Köntges et al., 2008)
Installation faults (PVTRIN, 2011)	• Incorrect design of the PV system; • Low inverter and module ventilation; • Loose or very tight cables; • Sensors placed badly; • Lack of lightening protection; • Actions that lead to corrosion
Degradation	• Connection issues with solder bonds (Li, 2016); • Sensor drifting and packaging of materials (Li, 2016); • Delamination (Zhengpeng et al., 2011); • Micro-cracks (Köntges et al., 2008)
Catastrophic failure	• Fire outbreak; • Failing of tree branches

Spagnolo, Del Vecchio, Makary, Papalillo, and Martocchia (2012), Krenzinger and De Andrade (2007), Buerhop, Schlegel, Niess, and Vodermayer (2011) King, Kratochvil, Quintana, and McMahon (2000) and Ancuta and Cepisca (2011) have stated that observations using imaging methods, such as infrared (IR) thermography, are valuable tools for inspecting PV-plants. This method is efficiently relevant to inaccessible roof mounted PV systems as well as to extended field plants because it is fast, reliable, contact free, non-destructive and involves measurements during operating conditions but requires no clouds and no wind. Many infrared (IR)-based analysing methods are used to investigate PV modules (Breitenstein, Rakotoniaina, & Al Rifai, 2003). Another method for image failure of PV modules is electroluminescence (Johnston, Call, Phan, & Ahrenkiel, 2009). Köntges et al. (2008), for instance, employed this method in order to analyse the influence of micro-cracks in PV modules on power loss.

Buerhop-Lutz and Scheuerpflug (2015) inspected PV-plants using an aerial, drone-mounted infrared thermography system. They carried out their measurement using an unpiloted drone (Multikopter), a lightweight infrared (IR)-camera PI 450 (Optris), a visible camera GoPro and equipment for navigation. They presented frequently detected failure modes of installed PV modules by focusing on crystalline modules from residential and industrial roofs as well as from solar parks in the field.

Concluding Observations

When a PV module maintains acceptable efficiency for a long duration:

- More energy are produced over a module life.
- The installation is more economically viable as the same initial cost provides for a longer period of energy generation.
- Manufacturers and/or installers have lower outlays to satisfy insurance of performance guarantees.
- Inspection and maintenance costs are lower.
- Enhanced reliability provides more inherent resilience to electricity generation.
- The expected warranty of 20–25 years is assured.
- There is less environmental impact as the rate of PV module disposal is reduced.

When installed PV system costs were high, electricity (market interventions such as feed-in tariffs) were made. They encourage take-up by providing favourable revenue for electricity sold to the grid. They also paid-back the cost of installation, usually in less than 20 years (McCormack & Norton, 2013). As PV generated electricity prices reach grid-parity, the need for such market interventions is obviated. The drivers underlying PV generation are also changing; there is greater interest in PV electricity self-consumption and in security of electrical supply (Castillo-Cagigal et al., 2011). Both the latter require assured long-term PV performance; thus understanding factors determining durability and reliability of PV systems will become more important in initial design. This is also likely to lead to innovations in diagnostics in system operation.

References

Ancuta, F. & Cepisca, C. (2011). *Fault analysis possibilities for PV panels*. Proceedings of the 3rd International Youth Conference on Energetics (IYCE), pp. 1–5.

Breitenstein, O., Rakotoniaina, J. P., & Al Rifai, M. H. (2003). Quantitative evaluation of shunts in solar cells by lock-in thermography. *Progress in Photovoltaics: Research and Applications, 11*, 515–526. https://doi.org/10.1002/pip.520

Buerhop, C., Schlegel, D., Niess, M., & Vodermayer, C. (2011). *Quality control of PV-modules in the field using infrared-thermography*. Proceedings of the 26th PVSEC, Hamburg, Germany, pp. 3894-3897. https://doi.org/10.4229/26thEUPVSEC2011-5AO.5.2.

Buerhop-Lutz, C. L., & Scheuerpflug, H. (2015). Inspecting PV-plants using aerial, drone-mounted infrared thermography system. *Proceedings of the 3rd South African Solar Energy Conference, 3*, 422–429. http://hdl.handle.net/2263/49538

Castillo-Cagigal, M., Caamano-Martin, E., Matallanas, E., Masa-Bote, D., Gutierrez, A., Monasterio-Huelin, F., & Jimenez-Leube, J. (2011). PV self-consumption optimisation with storage and active DSM for the residential sector. *Journal of Solar Energy, 85*(9), 2338–2348.

David, A. Q., Muyiwa, A. S., Gabriel, T., & Isaac, A. E. (2017). Reliability and degradation of solar PV modules-case study of 19-year-old polycrystalline modules in Ghana. *Technologies, 5*, 22. https://doi.org/10.3390/technologies5020022

Dunlop, E. D., & Halton, D. (2006). The performance of crystalline silicon photovoltaic solar modules after 22 years of continuous outdoor exposure. *Progress in Photovoltaics, 14*(1), 53–64. https://doi.org/10.1002/pip.627

Jia, Y. Y., Thomas, R., Armin, G. A., & Timothy, M. W. (2014). Performance degradation of various PV module technologies in tropical Singapore. *IEEE Journal of Photovoltaics, 4*(5), 1288–1294. https://doi.org/10.1109/JPHOTOV.2014.2338051

Johnston, S.W., Call, N.J., Phan, B., & Ahrenkiel, R.K. (2009). *Applications of imaging techniques for solar cell characterization*. Proceedings of the 34th IEEE, Photovoltaic Specialists Conference (PVSC), pp. 000276-000281. doi:https://doi.org/10.1109/PVSC.2009.5411681

King, D. L., Kratochvil, J. A., Quintana, M. A., & McMahon, T. J. (2000). Applications for infrared imaging equipment in photovoltaic cell, module, and system testing. *Proceedings of the Twenty-Eighth IEEE Photovoltaic Specialists Conference, 2000*, 1487–1490. https://doi.org/10.1109/PVSC.2000.916175

Köntges, M., Kunze, I., Naumann, V., Richter, S., Hagendorf, C., Berghold, J., & Roericht, M. (2008). Snail tracks, worm marks and cell cracks. Photovoltaic power systems programme. *IEA International Energy Agency, 7–8*, 36–40.

Krenzinger, A., & De Andrade, A. C. (2007). Accurate outdoor glass thermographic thermometry applied to solar energy devices. *Journal of Solar Energy, 81*, 1025–1034.

Kuitche, J., Pan, R., & Tamizhmani, G. (2014). Investigation of dominant figures modes for field-aged c-Si modules in desert climatic conditions. *IEEE Journal of Photovoltaics, 4*, 814–826.

Li, X., Prawiradiraja, T. P., & Battul, D. (2013). The role of humidity in energy output of solar panels in coastal regions. *GSTF Journal of Engineering Technology, 2*, 72–76.

Li, X. Y. (2016). Degradation analysis of photovoltaic modules based on operational data: effects of seasonal pattern and sensor drifting. *IOP Conference Series: Earth and Environmental Science, 40*, 012063. https://doi.org/10.1088/1755-1315/40/1/012063

McCormack, S. J., & Norton, B. (2013). In E. Michalena & J. M. Hills (Eds.), *The shadows cast by inadequate energy governance: Why more sun does not necessarily mean more photo-*

voltaic electricity, in renewable energy governance: Challenges and insecurities. Heidelberg: Springer-Verlag.

Murari, L. A., Pradip, K. S., Soumya, D., Biplab, S., Anagh, G., Arvind, P., & Biswas, R. (2017). An improved approach to design a photovoltaic panel. *Indonesian Journal of Electrical Engineering and Computer Science, 5*(3), 515–520. https://doi.org/10.11591/ijeecs.v5.i3.pp515-520

Nochang, P., Jaeseong, J., & Changwoon, H. (2014). Estimation of the degradation rate of multi-crystalline silicon photovoltaic module under thermal cycling stress. *Microelectronics Reliability, 54*, 1562. https://doi.org/10.1016/j.microrel.2014.03.021

Norton, B. (1999). Renewable electricity what is the true cost? *Power Engineering Journal, 13*, 6–12.

Okorieimoh, C. C., Norton, B., & Conlon, M. (2019). *Effect of transient performance changes on photovoltaic modules output*. 10th Annual Graduate Research School Symposium, Technological University Dublin, Ireland. https://doi.org/10.13140/RG.2.2.28516.94083.

Osterwald, C.R., Anderberg, A., Rummel, S., & Ottoson, L. (2002). *Degradation analysis of weathered crystalline-silicon PV modules*. Proceedings of the 29th IEEE Photovoltaic Specialist Conference. pp. 1392-1395. Retrieved from http://www.osti.gov/bridge

Parveen, B., & Saurabh, B. (2019). Clustering-based computation of degradation rate for photovoltaic systems. *Journal of Renewable and Sustainable Energy, 11*(1), 014701. https://doi.org/10.1063/1.5042688

Pingel, S., Frank, O., Winkler, M., Daryan, S., Geipel, T., Hoehne, H., & Berghold, J. (2010). *Potential induced degradation of solar cells and panels* Proceedings of the 35th IEEE Photovoltaic Specialist Conference. pp. 2817-2822.

Pramod, R., Tiwari, G. N., Sastry, O. S., Birinchi, B., & Vikrant, S. (2016). Degradation of mono-crystalline photovoltaic modules after 22 years of outdoor exposure in the composite climate of India. *Journal of Solar Energy, 135*, 786–795. https://doi.org/10.1016/j.solener.2016.06.047

PVTRIN. (2011). WP2_D2.6_catalogue of common failures and improper practices on PV installations and maintenance_ver 1, EPIA, created June 2011.

Saadsaoud, M., Ahmeda, A. H., Er, Z., & Rouabah, Z. (2017). Experimental study of degradation modes and their effects on reliability of photovoltaic modules after 12 years of field operation in the steppe region, Special issue of the 3rd International Conference on Computational and Experimental Science and Engineering (ICCESEN 2016). *Journal of Acta Physica Polonica A, 132*, 3–11. https://doi.org/10.12693/APhysPolA.132.930

Sopori, B., Basnyat, P., Shet, S., Mehta, V., Binns, J., & Appel, J. (2012). *Understanding light induced degradation of c-Si solar cells*. Proceeding of the IEEE Photovoltaic Specialist Conference. pp. 5200-54200.

Spagnolo, G.S., Del Vecchio, P., Makary, G., Papalillo, D., & Martocchia, A. (2012). *A review of IR thermography applied to PV systems*. Proceedings of the Environment and Electrical Engineering (EEEIC). 11th International Conference, pp. 879-884.

Takatoshi, H., Tomoya, N., Tadashi, T., & Yoshitaka, I. (2018). Influence of degradation in units of PV modules on electric power output of PV system. *Journal of International Council on Electrical Engineering., 8*(1), 118–126. https://doi.org/10.1080/22348972.2018.1477095

Tetsuyuki, I., & Atsushi, M. (2017). Annual degradation rates of recent crystalline silicon photovoltaic modules. *Progress in Photovoltaics: Research and Applications, 25*, 953–967. https://doi.org/10.1002/pip.2903

Tiwari, G. N., Mishra, R. K., & Solanki, S. C. (2011). Photovoltaic modules and their applications: A review on thermal modelling. *Applied Energy, 88*, 2287–2304.

Zhengpeng, X., Timothy, M. W., & Armin, G. A. (2011). PV module durability testing under high voltage biased damp heat conditions. *Energy Procedia, 8*, 384–389. Retrieved from www.sciencedirect.com

Impact of Public Charging Infrastructure on the Adoption of Electric Vehicles in London

Shaherah Jordan, Darryl Newport, Stephanie Sandland, and Paula Vandergert

Introduction

As part of the London Mayor's drive to make London a zero-emission city by 2050, the London Mayor has pledged to have 150 rapid charging points installed in the capital by the end of 2018. Understanding how public charging infrastructure impacts consumer behaviour and ultimately the adoption of plug-in electric vehicles (PEVs) in London is important due to the high cost of installation.

Public charging forms part of the built environment, and the built environment has been found to influence changes in travel behaviour across studies on different continents after accounting for attitude-induced self-selection (Milakis, Efthymiou, & Antoniou, 2017). Behaviours researched across the studies covered modes such as driving, walking and cycling; and related to behaviours like a switch in transport mode such as driving to walking, or the reduction of the use of transport which considerably contributes to high emissions such as driving (Milakis et al., 2017).

As charging time, range anxiety and purchase price are the most commonly cited barriers to PEV adoption (Carley et al., 2013; Egbue & Long, 2012; Hidrue et al., 2011), addressing the role of public charging infrastructure in the equation could be advantageous in the efforts to encourage uptake. According to Carly et al. (2013, p. 45) 'range anxiety could be addressed by increasing the number and visibility of public charging stations'.

Although self-efficacy is key to behaviour change, 'capability, opportunity and motivation must exist for any behaviour to occur' (Michie, Atkins, & West, 2014, p. 59) and an adequate public charging network addresses capability and opportunity where the operation of an electric vehicle is concerned. While there has been some discussion regarding the importance of access to residential charging as a factor for PEV adoption in the UK, analysis of London PEV registration data shows

S. Jordan (✉) · D. Newport · S. Sandland · P. Vandergert
Sustainability Research Institute, University of East London, London, UK
e-mail: s.jordan@uel.ac.uk; d.j.newport@uel.ac.uk

L. Scott et al. (eds.), *Sustainable Ecological Engineering Design*,
https://doi.org/10.1007/978-3-030-44381-8_25

that prevalence of public charging points could be a greater factor in the adoption of PEVs. Furthermore, as residential parking is less prevalent in urban areas, understanding how and where the public charging infrastructure should be installed is important in ensuring that it allays range anxiety.

Literature Review

The current body of work regarding charging type and PEV adoption is arguably divided with some studies indicating that access to residential charging is the key factor in PEV adoption and other studies indicating that public charging infrastructure is the key factor.

Stated preference studies (Hackbarth & Madlener, 2013; Hidrue et al., 2011) indicate that having a place to charge a PEV increased the likelihood of respondents to consider purchasing a PEV. In addition to the limitations associated with stated preference, much of the research exploring access to charging as a driver of PEV uptake was conducted in the United States where at least half of the homes have access to residential charging.

In their study amongst new car buying households (Axsen & Kurani, 2012), more than three quarters of the respondents who expressed an interest in making their next vehicle an PEV had the ability to install a charge point.

In addition to academic literature, a report from the London Environmental Committee (2018) states that residents with off street parking tended to be the early adopters of PEVs due to the accessibility of charging.

The literature focused on public charging infrastructure points towards it being a positive factor, however, the conditions under which it has the most impact varies, as do the potential reasons why.

Charge infrastructure has been shown to reduce the impact of range anxiety (Neubauer & Wood, 2014, p. 20) and it has been argued that it will be crucial to the early stages of PEV diffusion due to its practical and psychological function (Bakker, 2011).

A study by the Tokyo Electric Power Company exploring the driving and charging habits of PEV drivers before and after the installation of a charging point found that there was little use of the public charging point but that the distance drivers travelled increased considerably after installation and they returned with less stored battery power than before (Bakker, 2011).

The real potential for impact on PEV growth may be found in the combination of access to both public and residential charging, with public charging seeing as the supplement to residential charging (Lin & Greene, 2011).

However, it has also been posited that awareness of public chargers was not deemed to be 'the most significant determinant of PEV interest' (Bailey, Miele, & Axsen, 2015, p. 9). One of the limitations flagged in the study by Bailey was that 'consumer interest, perception and technology acceptance were not tested'.

Finally Bonges and Lusk (2016) looked into various factors associated with public charging points such as how the design of parking spaces and chargers for PEVs could limit access to the charging point. The social side of public charging points such as etiquette, best practice and the current limits to the enforcement of best practice was also explored in Bonges' study.

Due to the disparity in literature pertaining to charging, the relationship between PEV registrations and housing type was explored statistically, with the results indicating that access to residential charging may not be as strong of a factor as previously believed.

Using data from the Department for Transport (DfT) and Nomis, a correlation was sought between the number of PEV registrations per borough and the number of detached and semi-detached houses per borough. Housing stock was chosen as the variable as detached and semi-detached properties are the property types most likely to have the potential for off street parking and ultimately home charging installation. Furthermore, analysing the data in this way was reflective of how the data was analysed in the study by Axsen and Kurani (2012).

Using this approach, the correlation between detached and semi-detached houses was found to be weak (0.131 and 0.009, respectively). In fact, the strongest correlation was found between flats and PEV registrations in 2011/2012 (0.529) when PEV registrations were first being recorded. This correlation was statistically significant.

The results of this analysis led to questions as to whether using the same methodology would illustrate a similar trend with public charging points.

This paper seeks to understand the impact of public charging infrastructure on PEV adoption in London using existing data. For the purpose of this study, public charging refers to charging points that are installed in public.

Hypotheses

The total number of public charging points per borough is a factor in PEV adoption.

There is no difference between the type of charging point (speed) and the adoption of PEVs per borough.

Charging points in combination with housing type have a positive effect on PEV registrations.

Research Method

Secondary analysis of existing data was selected as a research method as the data was readily available and could be analysed quickly. Given the nature of the research question, the analysis of existing data was able to more quickly answer the research question than undertaking primary research involving respondents. Furthermore, previous studies that suggest that public charging is a factor in PEV adoption are based on stated preference choice models (Achtnicht, Buhler, & Hermeling, 2012; Batley, Toner, & Knight, 2004; Hackbarth & Madlener, 2013).

Statistical analysis was conducted on (1) the number of public charging points per borough, (2) percentages of housing stock type per borough and the (3) number of registered PEVs per borough.

Charging point data for Q2 of 2018 was sourced from Zap Map who hold the data on the charging points in the UK—both publicly and privately funded. The data was segmented into rapid chargers (43 kw), fast chargers (7–22 kw) and slow chargers (3 kw) and the total number of connectors.

Housing stock data was sourced from the Office of National Statistics and was segmented into five categories: detached, semi-detached, combined detached and semi-detached, terraced and flats. Detached and semi-detached properties were combined as a segment because percentages of each type of housing stock were relatively low across London.

The data on PEV registrations per borough in Q2 2018 was sourced from the Department for Transport.

Correlation and chi tests were performed on the charging point and PEV registration data to understand if there was a relationship between the number of public charging points available per borough and the number of PEV registrations per borough, and how strong the correlation was.

The same tests were performed on the housing type and PEV registration data, after which a pcor test was performed, where the PEV registration was x, housing type was y and the number of charging points was z. Each of the different types of charging points were analysed in this way.

After this a correlation test was run to ascertain the strength of the correlation between connector types and PEV registrations per borough.

The main limitation to this approach is the sample size. There were 2.665 m cars on London roads in Q1 of 2018 of which 14,466 were plug-in cars, LGVs and quadricycles (VEH0131, 2018). The ideal sample size for this type of analysis would be 19,800 based on a confidence level of 95% and a confidence interval of 4.

In addition to this, this data contains details of cars registered to car clubs in London. In many instances cars in car clubs would be registered to one address in a borough which may slightly skew the data, particularly as the sample size is already small.

Lastly, although the car registration data was available at borough level, a more accurate correlation may have been able to be drawn had the data been available at ward level.

Results

Hypothesis: The total number of public charging points per borough is a factor in PEV adoption.

A positive high degree of correlation was found between the number of registered PEVs per borough and the total number of connectors (0.530).

Hypothesis: There is no difference between the type of charging point (speed) and the adoption of PEVs per borough.

When testing the correlation between PEV registrations per borough and the different types of connectors (slow, fast and rapid), the strength of the correlation between rapid chargers was the strongest 0.601. The strength of the correlations between slow chargers and fast chargers was relatively similar with correlations of 0.466 and 0.437, respectively.

Hypothesis: Charging points in combination with housing type have a positive effect on PEV registrations.

The introduction of housing type as a variable had only a slight impact on the correlation between PEV registrations and total connectors per borough.

There was a positive high degree of correlation between total connectors and PEV registrations per borough (0.530). This level of correlation remains the same when detached properties (0.576), semi-detached properties (0.551) a combination of detached and semi-detached properties (0.560) and flats (0.529) were introduced as factors.

This drops slightly to a moderate degree of correlation when terraced properties (0.490) are introduced as a factor.

Discussion

These results are promising in as much that they support the case for investment into public charging infrastructure; however, factors such as location are thought to be critical in the efforts to increase PEV adoption (Azadfar et al). Commercial locations and areas where people linger, such as restaurants, hotels, shopping malls, churches and entertainment venues, are considered more promising locations for installation of EVSE (Bakker, 2011, p. 50).

The strong correlation between rapid chargers and PEVs is another positive result; however, due to the potential for accelerated battery degradation with overuse of rapid chargers, it would be advantageous to educate drivers accordingly. Furthermore, although access to rapid chargers is likely to encourage longer journeys and ultimately appeal to the public to adopt PEVs more rapidly (Azadfar, Sreeram, & Harries, 2015), usage patterns remain very low (Bakker, 2011, p. 56).

Industry insight indicates that fuelling behaviour differs from internal combustible engine (ICE) to PEV driver in that PEV drivers are more opportunistic in the way that they charge, and are more likely to charge because of convenience as opposed to battery depletion (Azadfar et al., 2015).

Conclusion

Although the role of the built environment on transport has been explored from various angles and looks at various modes, PEV ownership is at a unique cross-section of greener travel, technology adoption and a change in fuelling behaviour as opposed to travel behaviour. This makes identifying the factors for behaviour change more challenging.

Built environment characteristics related to perceived outdoor spaciousness such as off-street parking have been found to influence automobile ownership after accounting for attitudes (Cao, Mokhtarian, & Handy, 2007). Therefore, it is plausible that the presence of an PEV charger could influence the type of vehicle someone chooses to purchase.

This study only looks at whether the number of public charging points is a factor in PEV adoption and does not explore the location, visibility, price or aesthetics of the charging points. However, given the level of investment required to roll out public charging infrastructure, this study indicates that more research needs to be done to determine the factors that make the installation of public charging more impactful in the efforts to drive PEV growth.

Yet still, evidence from this analysis shows that there is a significant enough correlation between the number of charging points per borough and the number of PEV registrations per borough to put more effort into the public charging infrastructure in a holistic way involving, quantity, location, design and visibility. Ultimately, access to charging adds to the capability that potential PEV drivers need in order to change their vehicle. This study supports existing research that identifies public charging infrastructure as an adequate response to the real and perceived barriers to PEV adoption.

Acknowledgments Special thank you to Zap Map for providing the charging point data for this study.

References

Achtnicht, M., Buhler, G., & Hermeling, C. (2012). The impact of fuel availability on demand for alternative-fuel vehicles. *Transportation Research Part D: Transport and Environment, 17*(3), 262–269. https://doi.org/10.1016/j.trd.2011.12.005

Axsen, J., & Kurani, K. (2012). Who can recharge a plug-in electric vehicle? *Transportation Research Part D: Transport and Environment, 17*, 349–353. Retrieved from https://pdfs.semanticscholar.org/928a/cd77c7e0eb3b49f6878ee16b7f4c1a17282d.pdf

Azadfar, E., Sreeram, V., & Harries, D. (2015). The investigation of the major factors influencing plug-in electric vehicle driving patterns and charging behaviour. *Renewable and Sustainable Energy Reviews, 42*, 1065–1076. https://doi.org/10.1016/j.rser.2014.10.058

Bailey, J., Miele, A., & Axsen, J. (2015). Is awareness of public charging associated with consumer interest in plug-in electric vehicles? *Transportation Research Part D: Transport and Environment, 36*, 1–9. https://doi.org/10.1016/j.trd.2015.02.001

Bakker, J. (2011). Contesting range anxiety: The role of electric vehicle charging infrastructure in the transportation transition. *Eindhoven University of Technology, 18(May)*.

Batley, R. P., Toner, J. P., & Knight, M. J. (2004). A mixed logit model of U.K. household demand for alternative-fuel vehicles. *International Journal of Transport Economics, 31*(1), 55–77.

Bonges, H. A., & Lusk, A. C. (2016). Addressing electric vehicle (EV) sales and range anxiety through parking layout, policy and regulation. *Transportation Research Part A: Policy and Practice, 83*, 63–73. https://doi.org/10.1016/j.tra.2015.09.011

Cao, X., Mokhtarian, P. L., & Handy, S. L. (2007). Cross-sectional and quasi-panel explorations of the connection between the built environment and auto ownership. *Environment and Planning A, 39*(4), 830–847. https://doi.org/10.1068/a37437

Carley, S., et al. (2013). Intent to purchase a plug-in electric vehicle: A survey of early impressions in large US cites. *Transportation Research Part D: Transport and Environment, 18*(1), 39–45. https://doi.org/10.1016/j.trd.2012.09.007

Egbue, O., & Long, S. (2012). Barriers to widespread adoption of electric vehicles: An analysis of consumer attitudes and perceptions. *Energy Policy, 48*(2012), 717–729. https://doi.org/10.1016/j.enpol.2012.06.009

Environmental Committee. (2018). *Electric Vehicles*. London. Retrieved from https://www.london.gov.uk/moderngov/documents/s67614/xx EVs Appendix - scoping document.pdf

Hackbarth, A., & Madlener, R. (2013). Consumer preferences for alternative fuel vehicles: A discrete choice analysis. *Transportation Research Part D: Transport and Environment, 25*, 5–17. https://doi.org/10.1016/j.trd.2013.07.002

Hidrue, M. K., et al. (2011). Willingness to pay for electric vehicles and their attributes. *Resource and Energy Economics, 33*(3), 686–705. https://doi.org/10.1016/j.reseneeco.2011.02.002

Lin, Z., & Greene, D. (2011). Promoting the market for plug-in hybrid and battery electric vehicles. *Transportation Research Record: Journal of the Transportation Research Board, 2252*, 49–56. https://doi.org/10.3141/2252-07

Michie, S., Atkins, L., & West, R. (2014). *The behaviour change wheel. A guide to designing interventions*. London: Silverback Publishing.

Milakis, D., Efthymiou, D., & Antoniou, C. (2017). Built environment, travel attitudes and travel behaviour: Quasi-longitudinal analysis of links in the case of Greeks relocating from US to Greece. *Sustainability (Switzerland), 9*(10), 1–17. https://doi.org/10.3390/su9101774

Neubauer, J., & Wood, E. (2014). The impact of range anxiety and home, workplace, and public charging infrastructure on simulated battery electric vehicle lifetime utility. *Journal of Power Sources, 257*, 12–20. https://doi.org/10.1016/j.jpowsour.2014.01.075

Table VEH0131. (2018). Plug-in cars and vans licensed at the end of quarter by location of registered keeper: United Kingdom in Ultra Low Emission Vehicles, Department for Transport, 2018, Retrieved February 10, 2019, from https://www.gov.uk/government/statistical-data-sets/all-vehicles-veh01.

Performance Evaluation Based Claims Process for Insuring Energy Performance of New Dwellings

Rajat Gupta, Matt Gregg, and Agnese Salvati

Introduction

There are two key converging housing-related issues in the UK: housing supply shortage and a legal commitment to reduce carbon emissions. First, there is an under-supply of housing in England. This is seen through overcrowding, acute affordability issues, etc. In response, the 2015 Government set out an ambition to deliver one million net additions to the housing stock by the end of 2020 (Wilson, Barton, & Smith, 2017). Simultaneously, the UK is legally committed to achieve net zero carbon emissions by 2050 (BEIS, 2019). According to the UK government's Department for Business, Energy and Industrial Strategy's (BEIS) *Clean Growth Strategy* (2018), the UK has outperformed the target emissions reductions of the first carbon budget (2008–2012) by 1%; however, this success to date is primarily attributed to significant improvement in the power and waste sectors. To continue progress the other sectors, e.g. housing, will need to meet its share of reductions.

These motives drive the need for new dwellings to be built with high standards of insulation, air-tight with managed ventilation, high-efficiency heating systems and renewables. In turn, regulatory bodies and the building sector need/have needed to verify their ability to deliver new homes built to specific standards (e.g. (previously) Code for Sustainable Homes (CSH), Passivhaus). Building performance evaluation (BPE)[1] is an effective process to evaluate the effectiveness of design and

[1] "Building performance evaluation is the process of systematically comparing the actual performance of buildings, places and systems to explicitly documented criteria for their expected performance. It is based on the post-occupancy evaluation (POE) process model developed by Preiser, Rabinowitz, and White (1988)" (Preiser & Vischer, 2006) p. 7

R. Gupta (✉) · M. Gregg · A. Salvati
Low Carbon Building Research Group, Oxford Institute for Sustainable Development, School of Architecture, Oxford Brookes University, Oxford, UK
e-mail: rgupta@brookes.ac.uk

L. Scott et al. (eds.), *Sustainable Ecological Engineering Design*,
https://doi.org/10.1007/978-3-030-44381-8_26

construction to meet the above objectives. BPE findings allow for the improved delivery of dwellings and materials that match as-designed expectations, performance verification and improvement of the installation of systems and materials, to identify the performance gap between as-designed and in-use, and to make improvements on homes immediately evaluated, thereby reducing the performance gap.

Performance Gap in Housing

Common objectives for housing projects and consequently their evaluations are to achieve energy, comfort or health and well-being targets often defined by a standard. Housing providers and individual homebuyers pay for not only a new dwelling but also a specific performance whether they are aware of this or not. Recent studies (Monahan & Gemmell, 2011; Thompson & Bootland, 2011) demonstrated in-use energy use up to three-five times more than design predictions. This energy performance gap (EPG) between the predicted energy performance of a building (domestic or non-domestic) and its measured performance has been highlighted by several studies (Bordass & Leaman, 2005; Gaze, 2014a, 2014b; Gill, Tierney, Pegg, & Allan, 2010; Gupta & Kapsali, 2014; Lowe, Wingfield, Bell, & Bell, 2007; Stevenson & Leaman, 2010; Stevenson & Rijal, 2010; Williamson, Soebarto, & Radford, 2010; Wingfield, Bell, Miles-Shenton, & Seavers, 2011). Clearly national policy targets for carbon reduction cannot be met without understanding, quantifying and minimising this performance gap. For more information on this also refer to Zero Carbon Hub's (2014) report *Closing the gap between design and as-built performance.*

Corresponding with the findings of Zero Carbon Hub, BPE studies that also evaluated the in-use energy performance of new dwellings (Gaze, 2014a; Wingfield et al., 2011; Wingfield, Bell, Miles-Shenton, South, & Lowe, 2011) indicate the reasons for the performance gap can generally be attributed to discrepancies that arise across the building process, from the design and modelling tools used to design the building, through build-ability, materials and build quality (*as-designed* and *as-built*), systems integration and commissioning but also handover and operation, as well as the understanding, comfort and behaviour of the occupants. In fact, occupant behavioural patterns (which can be positive, e.g. wearing warmer clothes in addition to heating during the winter, or negative, e.g. leaving windows open when the heating is switched on) have been found to impact energy consumption by a factor of 2–3 in physically identical homes (Gram-Hanssen, 2010; Steemers & Yun, 2009).

I-Life Project: Development of Insurance Product and Claims Process

The research presented in this paper is from a 33-month project (October 2015–June 2018) funded by Innovate UK aimed at developing a new commercial insurance product designed to underpin a whole-life warranty for the energy performance of new housing. The project aimed to develop an integrated approach to building performance assurance, supported by specification and long-term building testing, socio-technical monitoring protocols and a building information modelling execution plan to underpin the warranty. The I-Life insurance framework would be set out in three phases: (1) Sell the insurance to developers or house purchasers; (2) Construction phase: I-Life construction checklists are developed and used to evaluate process and (3) Insured phase.

In theory, if deficiencies in the fabric or services of an I-Life insured dwelling are found to cause excessive energy consumption, the insurance would cover these deficiencies. Conversely, I-life would not insure excess energy consumption resulting from climatic or occupancy factors. The need to parse out the difference between these factors necessitates BPE. BPE helps to identify and reduce the gap between 'as designed' and 'in use' performance through a systematic collection and analysis of qualitative and quantitative information related to fabric performance, energy performance and environmental conditions. Design and site review are used to check, verify and establish the reason for the EPG. Reviewed documents include design drawings, details, specifications, calculations, air permeability test, commissioning tests and documents, third party site review to confirm as-built conforms to as-designed. Following the BPE, the builder is notified of any construction issues and given opportunity to justify action or remedy. The main insurable variables are U-value, linear thermal bridges, thermal mass parameter, air permeability, services plant and system efficiency.

Methodology

This paper guides the reader through the process as if a claim request were raised. Figure 1 shows the process of evaluating the validity of the claim and the use of BPE to determine the cause. Because excess energy consumption resulting from occupancy factors is not insured, the differentiation between physical (fabric or systems) and occupancy factors of excessive energy use is important.

BPE Methods

First, energy bills were collected to establish the EPG and theoretical validity of a claim. To determine the likely cause of greater gas and/or electricity consumption, the following are performed/evaluated: (a) Fabric testing: thermal imaging and

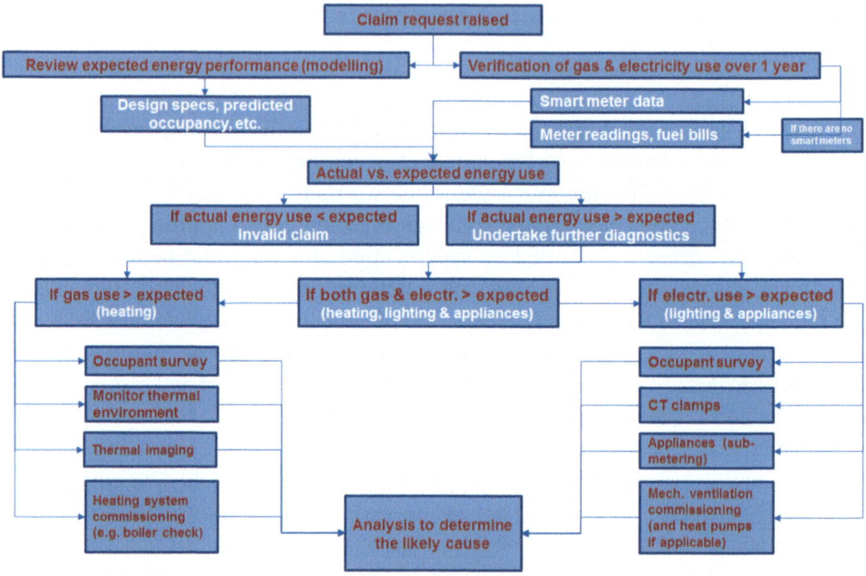

Fig. 1 Methodology flow chart: monitoring and testing protocol in the event of a claim

air-permeability document review; (b) Energy assessment: gas (30 min) and electricity (5 min) remote monitoring for 12 months (1 Jul 2017–30 Jun 2018); (c) Domestic Energy Audit and Reporting Method (DomEARM) to disaggregate regulated from unregulated loads through end use audits of all lighting and appliances; (d) Environmental/occupant behaviour monitoring: indoor and outdoor temperature and relative humidity (RH) (5 min), CO_2 concentration (5 min), radiator temperatures (30 min), window/door opening state change; and (e) Occupant characteristics and behaviour: interviews, occupant surveys and activity logging sheets.

Case Study Flats

The dwellings used as case studies for the pilot monitoring are four apartments of a recent housing development built in Southeast England, completed and occupied in 2017. The complex is composed of a mix of housing typologies—townhouses, cottages, apartments and penthouses—which surround a central courtyard. The case study flats differ in size, orientation and location within the four storey buildings (Table 1). All the flats are rated B in their Standard Assessment Procedure (SAP) documents, with primary energy varying from 75.7 to 85.2 kWh/m^2/year.

Table 2 shows the occupancy of the flats. The tenant in flat N31 worked from home, using many office/electric appliances (monitors, desktop computer, laptops, amplifiers, etc.) and was at home most of the time during both weekdays and weekends. In flats N15 there were two tenants plus one guest staying regularly on

Table 1 Case study form details

No.	Total floor area (TFA) (m^2)	Glazed orientation	Ratio of ext. surface area to TFA	Ratio of windows to TFA	SAP Energy efficiency rating	SAP Primary energy (kWh/m^2/year)
N31	67	NW—SE	1.40	0.18	83 (B)	85.2
N15	72	NW	0.39	0.10	84 (B)	75.7
N33	86	NW—NE—SE	1.48	0.19	83 (B)	84.1
N06	95	SE	1.43	0.15	84 (B)	78.8

Table 2 Case study occupancy details

No.	Beds/floor	No. of occupants	Avg. occupancy/SAP assumption	Age range	Occupancy type
N31	2 bed/4th (top)	1	1.0/2.2	46–55	Working from home
N15	2 bed/2nd	2 (often 3)	2.5/2.3	16–25 and 26–35	1 works at home 1–3 days per week
N33	3 bed/4th (top)	1	1.0/2.5	over 65	Most of the time at home
N06	3 bed/ground	1	1.5/2.7	56–65	Works at home/out on weekends

weekends. The tenant of flat N33 spent most of the time at home, but mostly in the living room, which he used to cook, eat, relax and socialise. The other three rooms were used as master bedroom, study room and guest room. In the study room and guest room the heating was often off because they were rarely used. The tenant of N06 lived alone but frequently had guests staying overnight (normally 1–3 days per week).

The flats have all been designed with the same performance standards, higher than building regulation requirements at the time (Table 3). External walls are composed of one layer of concrete blocks and one inner layer of insulation (mineral wool) and plaster. The external surface is either plastered or with brick claddings and windows are double glazed.

Limitations of the Study

- Primary energy in SAP includes energy for space heating, water heating, lighting and pumps/fans, but does not consider energy for cooking and appliances. (Note that in the flats all cooking and other appliances consume only electricity.) For this reason, the present study will only focus on these aspects of energy use in the dwellings for the purpose of evaluating the reasons behind the EPG.
- This empirical study is based on a large amount of high-frequency energy and environmental data. However, the proportion of energy attributed to domestic hot

Table 3 Case study design details

Fabric U-values/air permeability	Designed	UK building regs. Part L Limiting parameters
Windows U-value (Wm^2/K)	1.4	2.00
Wall U-value (Wm^2/K)	0.16	0.30
Roof U-value (Wm^2/K)	0.10	0.20
Floor U-Value (Wm^2/K)	0.16	0.25
Air permeability $m^3/(h\ m^2)$ at 50 Pa	4	10

water (DHW) use had to be estimated due to the unavailability of sub-metering data. The percentage of gas use for DHW was estimated correlating the monthly gas consumption with the number of heating degree days (HDD) and considering the intercept of the resulting trend line as the estimated share of gas use for DHW. The estimations were confirmed by information gathered from occupants through surveys and interviews. The estimations are deemed accurate enough for the purpose of the study.

- Building fabric performance tests beyond air permeability and thermal imaging were not performed, e.g. fabric U-values and thermal bridging.
- At the time infrared thermography was performed the weather conditions were not ideal due to limited cloud cover, allowing some incident solar radiation on the building fabric. For this reason, the survey was restricted to interiors only. Consequently, thermal imagining excluded important issues with the building fabric performance.

Results

The climate over the monitored heating season was rather in line with the average of the region (Met Office, 2018). Temperatures were slightly higher than average in January, but lower than average in February and March. Overall, it can be said that the monitoring study is representative of a typical heating season for the region.

Energy Use Over The Monitoring Period: 1 July 2017–30 June 2018

Annual energy use for the monitored period is shown in Table 4. The table also shows SAP modelled projection for the flats, gas and HDD correlation, and the EPG for the flats. In comparison with the typical domestic values reported by the Office of gas and electricity markets (Ofgem) (2019), the flats performed better than

Table 4 Energy use and performance gap

No.	Total In-use kWh (1 Jul 17–30 Jun 18)			SAP (modelled) kWh			Energy performance gap			Monthly gas/HDD (Coef. Of Deter.) (Oct 17–April 18)
	Gas	Elect[a]	Tot[a]	Gas	Elect	Tot	Gas (%)	Elect (%)	Tot (%)	
N31	3845	183 (3764)	4028 (7609)	3952	696	4648	−2.7	−116.6	−14.3	$r^2 = 0.78$
N15	3377	256 (2465)	3633 (5842)	3611	697	4308	−6.7	−92.6	−17.0	$r^2 = 0.83$
N33	7556	206 (2451)	7762 (10007)	5407	684	6091	33.2	−107.4	24.1	$r^2 = 0.74$
N06	4110	231 (2042)	4341 (6152)	5228	703	5931	−23.9	−101.2	−31.0	$r^2 = 0.79$

[a]The first number is only the electrical use for lighting, pumps and fans, i.e. to match SAP. The second number in parentheses includes the total of all electricity use for each flat

average UK residential buildings with low gas use (9000 kWh/year). Though N33 is using 33% more gas than modelled, the correlation with HDD is still strong. In contrast, flat N31 showed a higher overall annual consumption of electricity compared to the average value for medium consumption (3200 kWh/year). As noted previously, despite being the smallest flat with one occupant only, the occupant in flat N31 worked from home, using many office/electric appliances and was at home most of the time. These data thus confirm that electricity use is not proportional to total floor area, as normally assumed in energy calculations, but rather more related to occupant behaviour.

Following the total energy analysis, it appears that flat N33 may have a valid claim with a 33% EPG in gas consumption and would, based on the I-Life framework, be subject to deeper investigation. Though the others would not have a valid claim, all flats are used to demonstrate the BPE process.

Use of Electrical Appliances

Activity diaries coupled with DomEARM helped evaluate how electricity was used in the flats. All the flats have been provided with the same high-efficiency (class A or A+) electrical appliances for cooking and washing (hobs, oven, dishwasher and washing machine), thereby excluding appliance efficiency as a variable. The largest uses are in the categories of electronics, cooking and wet appliances. The use of laundry and dish washing accounts for significant variation in electricity use; the average number of loads per week varies from one (N33) to ten (N15) and it is not proportional to the number of occupants or guests (for example, in flat N06 (one occupant) there are normally six loads of laundrys per week). However, most of the difference in electricity use across flats is probably due to the use of electrical

Table 5 Percentage breakdown of electricity consumption in the flats

No.	Computer/ home office (%)	Electronics (%)	Cooking (%)	Refrigeration (%)	Wet appliances (%)	Other appliances (%)	Pumps and fans (%)	Lighting (%)
N31	7	57	17	6	8	0	3	2
N15	5	25	18	14	27	1	4	6
N33	0	23	33	18	16	2	4	4
N06	11	12	23	13	22	5	6	6

appliances for home office and home entertainment. Perhaps the cause of overconsumption of electricity in flat N31 is indeed the use of electronics, which accounts for almost 60% of total electricity use (Table 5).

Fabric Assessment

Air permeability was designed to be 4 $m^3/h\ m^2$ at 50 Pa, well below the maximum value admitted by UK building regulations (10 $m^3/h\ m^2$ at 50 Pa). The air permeability tests showed that all the flats were below the design target, apart from flat N31 where the test indicated a slightly higher value (N31: 4.64, N15: 3.03, N33: 3.95, N06: 3.31 $m^3/h\ m^2$ at 50 Pa). However, this slight difference in actual air permeability was not enough to significantly affect the heating demand. Air tightness was confirmed to be good in all five flats.

Thermal imaging surveys were conducted in the flats on the 17th and 30th of November 2017. In flat N31, the thermal performance of walls and roof was deemed good. The wall surface temperature varied between 18 °C and 20 °C, while the ceiling surface temperature was more consistently around 20 °C. In flat N33 the thermal performance of walls and roof was also deemed good. The wall indoor surface temperature varied between 18 °C and 19 °C. The coldest spot reached 16 °C under the window thresholds. In N15, the wall indoor surface temperature varied between 20 °C and 21 °C, falling to about 17 °C in the coldest spot under the window. In N06, infrared images highlighted the presence of some thermal bridges at the foot of the walls and at the French windows' thresholds, where the tenant reported frequent condensation in the winter. In the coldest spots, surface temperature was around 14–15 °C (Fig. 2).

Heating Control, Heating Patterns and Temperatures

The mean indoor temperatures in winter varied from 14.8 to 21.7 °C across flats.

The mean in all heated rooms varied from 18.4 to 21.7 °C. Monitoring of the radiator temperatures and heating schedules demonstrated differing heating

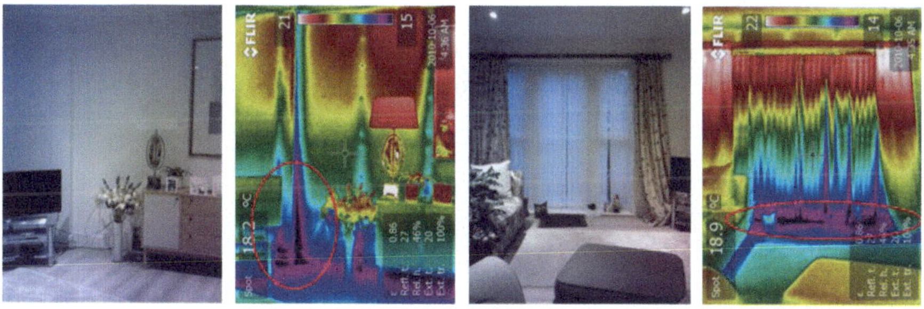

Fig. 2 Thermal imaging in the living room of flat N06. In the red circles, detection of thermal bridges at the foot of the external walls and window threshold

behaviours in the flats. N31 controls the heating by timer, set from 6:30 to 10:00 and from 17:00 to 23:00 in both bedroom and living room. The high maximum indoor temperatures were recorded in the bedroom of flat N31 (27.7 °C) and the highest maximum mean, 21.7 °C, was found in this bedroom. N15 controlled heating manually for use mainly in the evening in the living room; however, the heating patter was irregular. N15 had the second highest maximum temperature readings of 27.4 °C. N33 controlled the heating manually and separately for the living room and bedroom. Temperatures recorded in the bedroom of N33 were significantly lower than their living room and the bedrooms in any other flat. This result suggests a problem with the heating system of the room, which was also confirmed by the occupant, who reported that the radiators of the bedroom and the study room never warmed up properly, while those of the lounge/hallway, which are on another circuit, worked well. The fault was not fixed by bleeding them, suggesting a more serious, claim worthy fault, such as incorrect design of the circuit. The tenant of flat N06 reported in the survey to have the heating always on and to control it by thermostat temperatures and radiators valves. The heating pattern shows in fact a large use of heating in both rooms. N31 and N06 are close to the correspondent heating patterns, whereas those of N15 and N33 show other high peaks of gas use, denoting a significant use of DHW in addition to space heating.

Despite high external temperatures in the spring of 2018 (high of 26.1 °C reached), analysis of radiator temperatures showed that two flats N06 & N33, continued to use heating rather regularly in April. In N15 the use of heating was low all around, while in N31 it decreased consistently in April. This is probably the result of the increase of heat gains, being N31 a penthouse with southern exposition. The same beneficial effect is not evident in the other penthouse, N33, probably due to the northerly exposure of the flat.

Window Opening and Air Quality

The monitoring data on window opening (during February 2018) showed different patterns among the flats. Two (N06 and N15) opened windows regularly, while the other two (N33 and N31) almost never opened the windows. Even though windows opening patterns are different, the air quality was good for all the flats over the monitored period, which confirms a good design of the background ventilation and extraction. The mean CO_2 levels varied from 541 to 1194 ppm, which is well below the 2000 ppm threshold.

Discussion

The results showed that, in the high thermal performance building, occupants' behaviour has significant influence on electricity use and gas use for water heating, while space heating demand is more heavily determined by physical factors.

Gas Consumption

Regarding gas use for space heating, significant occupant related variables are the thermostat settings (heating demand temperatures) and the average number of hours per day the heating system is used. On the other hand, even if the building fabric performance is the same, other physical factors deeply influence space heating demands, such as total floor area and amount of exposed surface area. Results from this monitoring study suggest that physical factors have stronger impact than occupant factors on gas use for space heating, most probably due to the high thermal insulation level and air tightness of the building, which reduces the potential impact of occupant behaviour. This conclusion was drawn because, despite significant differences in heating control among the flats, the average indoor temperatures (weighted by room size) are homogenous across the flats (Table 6).

The desired comfort temperature is very similar among the flats, but the space heating demand to achieve it varies significantly due to different size, orientation and location of the flats within the building, indicating energy requirement is strongly determined by physical factors.

The highest gas use for space heating was found in N33, which is neither the one with the highest heating demand temperatures (N15) nor the one with the largest use of heating over the day (N06), but it is a large penthouse, with a lot of external surface and two north-facing facades.

The results of linear regression analysis between gas use for space heating and both total floor area and ratio of external surface to TFA showed positive relationships with strong coefficients of determination (r^2), equal to 0.81 and 0.85,

Table 6 Average indoor air temperature over the heating season weighted by room size

	N31	N15	N06	N33
Mean (°C)	20	20.2	20.2	20.6
Std. deviation	1	1	0.7	1.1

respectively. The sample is too small to consider the fitted equations as predictors of gas use; however, the high values of r^2 confirm a strong relationship between gas use and flat size, which is proportional to the volume of air to heat, as well as to the amount of external surfaces, which is proportional to the heating losses. The two penthouses (N31 and N33) and the ground floor flat (N06) have a higher amount of exposed surface area compared to N15, which explains the higher gas consumption per square meter.

DHW demand, instead, was not found to be proportional either to total floor area or to number of occupants, as opposite to what is normally assumed in energy models such as SAP. The highest actual consumption of DHW was estimated for N33, which is a large flat with one only occupant. However, a similar flat in terms of occupancy and size such as N06 is estimated to have much lower gas use for DHW. This happened because in flat N33 DHW is regularly used during the day also for other purposes apart from showering. This result also confirmed that DHW demand tends to increase with the age of occupant, as found in other studies (Guerra-Santin & Itard, 2010; Schleich & Hillenbrand, 2009), since the occupant in N33 is an elderly person. The results highlighted the need to carry out further investigations on the energy use for water heating, which accounts for an important share of domestic energy use in buildings designed to high thermal performance standards.

Electricity Consumption

Electricity consumption appears to be completely independent from physical factors such as flat size, as already outlined by other studies in the UK (Firth, Lomas, Wright, & Wall, 2008); the largest flat (N06) used the least electricity over the analysed period, while the flat with the highest electricity consumption (either absolute and per square meter) was the smallest flat housing only one occupant. The electricity use per square meter varied with a factor of 2.7 across the flats which had identical mechanical ventilation systems and efficiency of kitchen appliances and wet appliances. This variability must be thus determined by occupant behaviour. Further to this, regarding lighting, the relationship between lighting consumption and floor area was $r^2 = 0.05$, while the relationship between lighting consumption and average occupancy count was $r^2 = 0.85$. The strength in the latter suggests again the greater influence of occupant behaviour over physical factors on electricity use. To reduce the true performance gap, energy modelling tools should be improved regarding the estimation of non-regulated energy uses such as electrical appliances, which account for the biggest share of electricity use. This is challenging because overall, electricity use does not appear to be related to physical factors such as flat size or actual number of occupants but rather it seems to vary randomly with occupant behaviour.

Future Application

This study has shown that the availability of high-frequency data on energy use allow evaluators to derive useful insights on occupant behaviours, analysing, for example, the daily profiles of gas and electricity use. Therefore, effective procedures could be developed to derive typical profiles of electricity use from smart meters data, which are planned to be installed in every home in the UK by 2024. This kind of high-frequency information could be used also to adjust hot water estimations, which is the second energy end use strictly related to occupant behaviour. In this way, the uncertainties related to occupant variables would decrease and the buildings' energy performance certificates (EPCs) could be updated into 'in-use EPCs', that would be reliable indicators of actual energy use and carbon emission reduction achieved by buildings in-use. The study could be used to influence energy consumption with regard to DHW and electricity consumption using normative messages (Schultz, Nolan, Cialdini, Goldstein, & Griskevicius, 2007) and messages of approval or disapproval. Beyond smart meters' usefulness in making energy data collection simpler, providing smart meter feedback to occupants through the use of energy displays resulted in a 7% decrease in electricity consumption in Denmark (Grønhøj & Thøgersen, 2011) and an overall average energy use reduction of 10% in Australia (Anda & Temmen, 2014).

Conclusion

This paper has described a BPE study carried out on four flats located in the same building to empirically assess the influence of occupant factors and physical factors on the energy performance of a building in-use. The differentiation between occupant physical influence on energy consumption is important for the application of a new insurance-backed energy performance warranty for guaranteeing the in-use performance of new homes in the UK.

The BPE process is also important for identifying the root cause of issues and to potentially remediate problems.

Summary of Findings

- Heating demand temperatures in the living rooms were closer to 20 °C rather than 21 °C, as it was also found in a wider survey on the UK housing stock (Shipworth et al., 2010). Heating demand in the rest of the flats were higher than 18 °C in most of the cases.
- Measured internal temperature and occupant behaviour would have suggested certain flats were consuming more space heating energy then others; therefore,

building characteristics were more able to explain the difference in space heating results.

- Heating was on for fewer hours than SAP assumptions in all the flats except for N06 (weekdays and weekends).
- SAP overestimated (based on TFA) the number of occupants in three of the four flats. As a result, incorrect occupancy count results in incorrect DHW estimations. DHW use was found to be highly dependent on occupant behaviour and potentially the age of the occupant.
- Electricity use has been shown to be highly dependent on occupant behaviour; furthermore, energy from lighting use is strongly related to average number of occupants.
- The number and use of electrical appliances vary significantly across the flats but SAP does not account for this energy end use.

Recommendations

- Install sub-meters for DHW to disaggregate DHW from space heating to help investigate whether the cause of the EPG is in the building or in occupant behaviours.
- Models tend to overestimate appliance use. This limitation could be overcome in future monitoring studies by using non-intrusive electrical appliance monitoring devices (Agyeman, Han, & Han, 2015; Aladesanmi & Folly, 2015), which would provide a more detailed disaggregation of electricity use by each appliance.
- For the claims process to be less invasive, smart meter data can be used to identify the influence of physical or occupant factors.

Acknowledgments We are grateful to Innovate UK for sponsoring the I-life research project on Insurance-backed warranty for whole life housing energy performance (Application no: 55662-413534).

References

Agyeman, K., Han, S., & Han, S. (2015). Real-time recognition non-intrusive electrical appliance monitoring algorithm for a residential building energy management system. *Energies, 8*(9), 9029–9048.

Aladesanmi, E., & Folly, K. (2015). Overview of non-intrusive load monitoring and identification techniques. *IFAC-Papers On Line, 48*(30), 415–420.

Anda, M., & Temmen, J. (2014). Smart metering for residential energy efficiency: The use of community based social marketing for behavioural change and smart grid introduction. *Renewable Energy, 67*, 119–127.

BEIS. (2018). In Department for Business, E. a. I. S (Ed.), *Clean growth strategy: Leading the way to a low carbon future*. London: HM Government.

BEIS. (2019). Uk becomes first major economy to pass net zero emissions law. Retrieved Date Accessed, 2020 from https://www.gov.uk/government/news/uk-becomes-first-major-economy-to-pass-net-zero-emissions-law.

Bordass, B., & Leaman, A. (2005). Making feedback and post-occupancy evaluation routine 1: A portfolio of feedback techniques. *Building Research and Information, 33*(4), 347–352.

Firth, S., Lomas, K., Wright, A., & Wall, R. (2008). Identifying trends in the use of domestic appliances from household electricity consumption measurements. *Energy and Buildings, 40*(5), 926–936.

Gaze, C. (2014a). *AIMC4 Information paper 5: lessons from AIMC4 for cost-effective fabric-first low-energy housing Part 5: As-Built performance and Post Occupancy Evaluation.*

Gaze, C. (2014b). How did the homes perform? The data. *Proceedings of final AIMC4 conference.*

Gill, Z. M., Tierney, M. J., Pegg, I. M., & Allan, N. (2010). Low-energy dwellings: the contribution of behaviours to actual performance. *Building Research and Information, 38*(5), 491–508.

Gram-Hanssen, K. (2010). Residential heat comfort practices: understanding users. *Building Research and Information, 38*(2), 175–186.

Grønhøj, A., & Thøgersen, J. (2011). Feedback on household electricity consumption: learning and social influence processes. *International Journal of Consumer Studies, 35*(2), 138–145.

Guerra-Santin, O., & Itard, L. (2010). Occupants' behaviour: determinants and effects on residential heating consumption. *Building Research and Information, 38*(3), 318–338.

Gupta, R., & Kapsali, M. (2014). How effective are 'close to zero' carbon new dwellings in reducing actual energy demand: Insights from UK. *30th International PLEA Conference, 2014*, 16–18.

HM Government. (2011). *The carbon plan: Delivering our low carbon future.* London: Change, D. o. E. a. C.

Lowe, R., Wingfield, J., Bell, M., & Bell, J. (2007). Evidence for heat losses via party wall cavities in masonry construction. *Building Services Engineering Research and Technology, 28*(2), 161–181.

Met Office. (2018). *Averages tables. District: England SE & Central S. Climate period 1981–2010.* [Accessed].

Monahan, S., & Gemmell, A. (2011). How occupants behave and interact with their homes. *The impact on energy use, comfort, control and satisfaction. HIS-BRE press on behalf of the NHBC foundation, Milton Keynes.*

Ofgem. (2019). *Typical domestic consumption values* [Online]. Ofgem. Retrieved March 26, 2019, from https://www.ofgem.gov.uk/gas/retail-market/monitoring-data-and-statistics/typical-domestic-consumption-values.

Preiser, W., Rabinowitz, H., & White, E. (1988). *Post-occupancy evaluation.* New York: Van Nostrand Reinhold.

Preiser, W., & Vischer, J. (2006). The evolution of building performance evaluation: An introduction. In W. Preiser & J. Vischer (Eds.), *Assessing building performance*. London: Routledge.

Schleich, J., & Hillenbrand, T. (2009). Determinants of residential water demand in Germany. *Ecological Economics, 68*(6), 1756–1769.

Schultz, P. W., Nolan, J. M., Cialdini, R. B., Goldstein, N. J., & Griskevicius, V. (2007). The constructive, destructive, and reconstructive power of social norms. *Psychological Science, 18*(5), 429–434.

Shipworth, M., Firth, S. K., Gentry, M. I., Wright, A. J., Shipworth, D. T., & Lomas, K. J. (2010). Central heating thermostat settings and timing: building demographics. *Building Research and Information, 38*(1), 50–69.

Steemers, K., & Yun, G. Y. (2009). Household energy consumption: a study of the role of occupants. *Building Research and Information, 37*(5–6), 625–637.

Stevenson, F., & Leaman, A. (2010). Evaluating housing performance in relation to human behaviour: new challenges. *Building Research and Information, 38*(5), 437–441.

Stevenson, F., & Rijal, H. B. (2010). Developing occupancy feedback from a prototype to improve housing production. *Building Research and Information, 38*(5), 549–563.

Thompson, P., & Bootland, J. (2011). GHA monitoring programme 2009–2011: technical report results from Phase 1: post-construction testing of a sample of highly sustainable new homes, *Report, Good Homes Alliance*.

Williamson, T., Soebarto, V., & Radford, A. (2010). Comfort and energy use in five Australian award-winning houses: regulated, measured and perceived. *Building Research and Information, 38*(5), 509–529.

Wilson, W., Barton, C., & Smith, L. (2017). *Tackling the under-supply of housing in England, London*.

Wingfield, J., Bell, M., Miles-Shenton, D., & Seavers, J. (2011), *Elm tree mews field trial–Evaluation and monitoring of dwellings performance* (Final technical report).

Wingfield, J., Bell, M., Miles-Shenton, D., South, T., & Lowe, R. (2011). *Evaluating the impact of an enhanced energy performance standard on load-bearing masonry domestic construction: Understanding the gap between designed and real performance: lessons from Stamford Brook*. (Report) Retrieved from Department for Communities and Local Government: http://data.parliament.uk/DepositedPapers/Files/DEP2012-1435/StamfordBrook.pdf.

Zero Carbon Hub. (2014). *Closing the gap between design and as-built performance*. End of term report.

A Multidimensional Analysis of Smart Energy Systems: Towards Developing a Common Framework for Assessing the Sustainability of Small-Scale Renewables in Selected Societal Sectors

Peter Gudde, Justine Oakes, Nic Bury, Peter Cochrane, and Nicholas Caldwell

Introduction

In the last decade renewable generation technologies have experienced rapid uptake by new entrants onto the UK power network. This has primarily been driven by national decarbonisation policies and tariff-based instruments supporting small-scale generation. The market has continued to mature, and UK Government has now taken the step to remove state-sponsored support for new solar photovoltaic (solar PV) from April 2019, except for a small number of pre-registered schemes.

There are now over one million low and zero carbon power stations connected to the UK electricity grid, with solar PV accounting for 13 GW of installed capacity (BEIS, 2019). This compares to total UK generating capacity at around 106 GW (Statista, 2019) producing 339 TWh in 2017 (BEIS, 2018).

Moving towards a more diverse, decentralised model of power generation is seen by many as part of a smarter, technology-based solution. However, such Distributed Energy Resources on their own create problems which network operators and the sector more generally are trying to manage. The system must transform if it is to become sustainable (Fig. 1).

Some commentators (Table 1) consider that current UK Government policies are not addressing the energy trilemma: resilience, carbon reduction and affordability.

P. Gudde (✉) · N. Bury
School of Science, Technology and Engineering, University of Suffolk, Ipswich, UK

Suffolk Sustainability Institute, University of Suffolk, Ipswich, UK
e-mail: p.gudde@uos.ac.uk

J. Oakes
Suffolk Sustainability Institute, University of Suffolk, Ipswich, UK

P. Cochrane · N. Caldwell
School of Science, Technology and Engineering, University of Suffolk, Ipswich, UK

L. Scott et al. (eds.), *Sustainable Ecological Engineering Design*,
https://doi.org/10.1007/978-3-030-44381-8_27

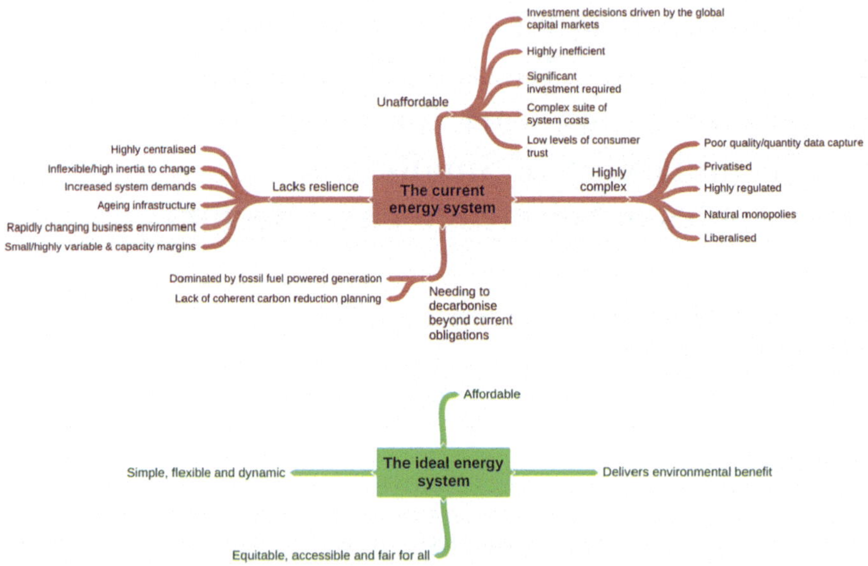

Fig. 1 Comparing the features of the current power system and a more sustainable model

Table 1 Example of comments on the energy trilemma

Energy trilemma	Commentary
Resilience	The Royal Academy of Engineering's report to the Council of Science and Technology commented because of a combination of factors the UK power system "*in the absence of intervention, would reduce the capacity margin during the time frame considered by this report, in a manner that would present an increasing risk to security of supply*" (Royal Academy of Engineering, 2015, p. 2)
Carbon reduction	Latest evidence from the IPCC stated the need to decarbonise at a far faster rate than the UK or other countries are achieving. Committee on Climate Change (2019) concluded that, "*Globally, current pledges of effort do not go far enough.*" (CCC, 2019, p. 16)
Affordability	The level of investment required to upgrade the transmission network alone was estimated in 2012 at around £8.8 Billion (Electricity Networks Strategy Group, 2012). The current highly regulated model is creating inertia at national infrastructure level. This is highlighted by the recent announcements from the commercial developers to abandon two planned nuclear power projects in the UK (The Guardian Online, 2019). Similar issues are evident at distribution network level

"*Each pillar of [this] trilemma is important—ignoring any one will result in failure across the system*", (Royal Academy of Engineering, 2015, p. 9).

This is all at a time when established UK energy companies are losing market share to new entrants (Ofgem, 2018) and the sector experiences declining public trust (Competition and Markets Authority, 2014). However, the energy trilemma is only part of a broader set of issues and a failure to account for these could undermine the securing of a smarter UK energy system.

This research has yet to identify a model which takes a multi-stakeholder perspective, sustainability-led approach to managing change in the UK energy system. A new assessment model is, therefore, required to ensure that policy and investment decisions move towards sustainable energy provision whilst embracing the implications for all stakeholders. It is important to ensure that optimal decisions are made throughout the network to secure a sustainable future for the UK.

Preparatory Review: Measuring Sustainable Development

The commonly used definition of Sustainable Development adopted by the United Nations is "*Development which meets the needs of the present without compromising the ability of future generations to meet their own needs*" (UN World Commission on Environment and Development, 1987). The UN General Assembly subsequently adopted 17 Sustainable Development Goals (SDGs) for achievement by 2030 (UN, 2015), summarised in Fig. 2.

The UN has derived 169 targets from these goals against which each national signatory is required to report its progress towards their achievement. In its 2018 report, the UK Stakeholders for Sustainable Development identified 143 of the targets as being relevant to the UK. The UK Government, through the Office of National Statistics, reports national progress against the relevant targets (ONS, 2019).

Currently, there are a wide range of tools and methods to assess progress towards some or all the factors considered part of Sustainable Development (Table 2).

Fig. 2 UN sustainable development goals (United Nations, 2019)

Nilsson, Griggs, Visbeck, and Ringler (2016) demonstrated the value of developing an integrated approach to measuring progress towards achieving the SDGs. In so doing, a key challenge of managing policy and delivery at global, national and institutional level across a range of policy agendas with often diverging time-horizons was highlighted,

> "*The aspirations of the 2030 Agenda are far reaching, and many will only be achieved in the long term (possibly beyond 2030). And leveraging policies on positive interactions may have longer pay back times and require longer planning horizons than one-goal approaches ... A barrier is that much of today's governmental actions are coloured by short-termism.*" (Nilsson, 2016, pp. 23–24)

Table 2 A summary of selected performance measurement tools linking the power system to the UN SDGs

Theme	Methodology	Summary
Sustainability indexes	The World Energy Council's Energy Trilemma Index tool (2019)	Index ranking countries on their ability to provide sustainable energy through three dimensions: Energy security, Energy equity (accessibility and affordability), Environmental Sustainability
	Environmental Sustainability Index (Yale and Columbia Universities, 2019) in collaboration with the World Economic Forum and the EU Joint Research Centre	Published through the Socioeconomic Data and Applications Center (SEDAC), the index provides a composite profile of national environmental stewardship based on a compilation of indicators derived from underlying datasets
Environmental/ ecological impact	Life cycle assessment or analysis	A process of evaluating the environmental impacts of a product at each stage of its life
	Ecological foot printing	Defined as the amount of land and resources required to sustain a way of living. Examples include: The Natural Step Framework (2019), Global Footprint Network (2019) and World Wide Fund for Nature (2019)
	Environmental Impact Assessment methodology within—UK Town and Country Planning	A statutory tool applied when major decisions are being considered within the UK planning system
	The UK Building Research Establishment Environmental Assessment Model (BRE, 2019)	A method for calculating the sustainability of a masterplan, development or individual building
	Green accounting methodologies	An approach for defining financial value to non-financial (environmental and social) factors to aid business decision-making. An example of this is the tool described by Stanojevića, Vranešal, and Gökalp (2010)

(continued)

Table 2 (continued)

Theme	Methodology	Summary
Economic impact	Return on Investment, Internal Rate of Return and Net Present Value, Pay back	An established suite of financial and economic metrics to predict and assess the performance of a proposal from an investment perspective at organisational level
	Gross Value Added, Gross Domestic Product as examples	Another established suite of established metrics, in this case, applied at regional or nation state level to measure economic growth
	Whole-life costing	A financial evaluation of a range of capital and revenue costs throughout the life of a project, product or service from raw material production to end of life
	Levelised cost of energy	A method for comparing the financial competitiveness of a various energy generating technologies. Similar conceptually to payback, it calculates the minimum cost per unit of energy to recover the capital outlay and running costs
	Marginal abatement cost (MAC) curves	MAC curves provide an economic estimate of abatement measures such as Greenhouse Gas emissions reduction technologies compared to Business as Usual. They can operate at various scales, an example being at the global in models developed by Mckinsey (2007) as well as being used in policy modelling by UK Government
Social impact	Social Return on Investment	These include work by the SROI Network (2012) based on material originally by published by UK Government Cabinet Office in 2008
	Community Energy project assessment	An Excel based tool published by Community Energy England to help a community groups considering energy projects to be able to measure and assess the benefits that they bring to their community and to the planet (O'Leary & Speciale, 2017)
	Various GIS-based modelling tools	Various tools (academic and commercial) using spatially defined data sets to map patterns or features at various geographical scales with a range of functionalities

Research Aims and Methods

By looking at how investment decisions are being made in organisations within three societal sectors (the public, private and third sectors) who are participating in a new and potentially disruptive ways, this research will consider the following questions:

(a) What drives stakeholders to invest in renewable energy technologies like solar PV?

Table 3 Data and capture methods

Area	Proposed data and capture methods
Technical	Literature review of performance data for deployed technologies and impact on energy profile Predicted and actual electricity generation data from the deployed technology Predicted and actual energy consumption from the selected organisations
Economic	Literature review of the economics of the deployed technology Analysis of supply chain costs and energy market costs
Environmental	Use of published industry, site and network-generated data along the supply chain Analysis of the impact on greenhouse gas emissions Assessment against a range of possible scenarios (e.g. National Grid Futures)
Social	Literature review of the health on stakeholders up and downstream Collection and evaluation of perceptions of key stakeholder groups

(b) How do we measure the "sustainability" impact of renewable energy deployment?
(c) What are the resulting synergies, conflicts and trade-offs across the key Sustainable Development themes?
(d) How do stakeholders beyond those who are making the investment decisions influence technology deployment—do they benefit or lose out and in what ways?

The following pre-curser questions will also be considered:

(e) What assumptions are being made?
(f) What outcomes or scenarios should be considered based on these assumptions?
(g) If the UN SDGs are inappropriate, what are the alternatives at sub-nation state level?

A multidimensional view of a smarter electricity system is, therefore, needed. This research will consider economic, environmental, social and organisational values and their inter-relationships upstream and downstream to understand how they deliver against both institutional drivers and the UN SDGs. Quantitative and qualitative data from a variety of sources including scientific and grey literature will be considered. It is intended to utilise a range of research methods as a basis for further definition and refinement (Table 3).

Research Results

Preparatory research shows that defining the impact of change in the energy system using the conventional energy trilemma model does not provide confidence that a sustainable approach will be followed (Fig. 3). It is proposed, therefore, to develop a framework to assess and measure the performance of energy-related interventions against defined sustainable development criteria. It is anticipated that any ensuing model or tool will require spatial and time-variable functionality.

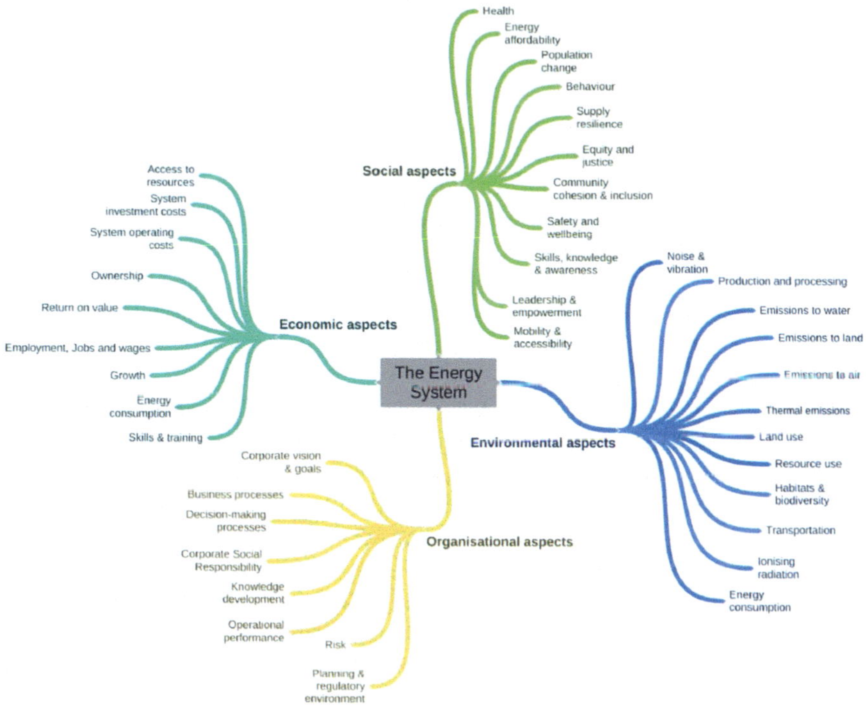

Fig. 3 Factors affecting investment decisions in the UK Energy System

Initial research suggests that although the SDGs provide a comprehensive platform, there is very little available published material to show how the SDGs and the relevant targets can be linked to institutional activity in the UK, more specifically to the changing nature of the power system and investment decisions made by new entrants investing in renewable generation.

Building on the work of Nilsson et al. (2016), the International Council for Science (ICSU, 2017) explored the interlinkages between and within some of the SDGs to evaluate an initial set of four SDGs. When applied to the aspects model in Fig. 3, this causal and functional relationship framework helps to show how positive or negative feedbacks could be measured, for example, between ownership (Economic aspect), energy affordability and stakeholder behaviours (Social aspects) in the context of small-scale renewable energy generation. However, although the methodology for scoring put forward by Nilsson et al. may be appropriate for this research, it will be necessary to translate the intent of the SDGs to see if they apply to the UK institutional sectors that will be evaluated.

Discussion

The mind map shown in Fig. 3 indicates that there are significant factors that require consideration as we transition towards a smarter energy system. Initial research suggests that the focus on the energy trilemma fails to consider the impact of deploying Distributed Energy Resources upstream or downstream of the UK energy system.

Currently, the energy industry takes a predominantly investment and shareholder return-led approach based on technical solutions operating within a complex regulatory environment. Government policy and the way that the principal power sector participants behave, for example, have been shown to contribute to higher than expected customer costs and slower technology deployment to deal with the energy trilemma (Helm, 2017, pp. xi–xiii).

Scientific research has tended to take a single aspect or dimensional approach whether within the disciplines of engineering, economics or behavioural science. Examples include Wolsink (2012), Blazqueza, Bracamontesa, Bollinob, and Nezamuddina (2018), and Shove and Walker (2014). Espinoza, Samaniego, Jara-Alvear, and Ochoa (2017) considered the interface between the technology and the citizen to explore the possibility of a "power ecosystem".

This suggests that in order to understand the impact on society and key stakeholders of a smarter energy system, we must first model the various scenarios and be able to measure the system's total value taking account of all aspects, weighted according to their relative importance to each stakeholder. From this, we can then derive an understanding of the relative values in time and space from each stakeholder perspective. This means that we need to understand how to compare aspects which may exhibit quite different quantitative and qualitative characteristics at different spatial and temporal scales.

The initial literature research undertaken to date has highlighted that it is necessary to broaden the viewpoint of stakeholders in a rapidly expanding and complex part of the UK's key infrastructure. For example, it is important to have a comprehensive dialogue with all those involved from system design through to investors, operators and consumers. Currently, there are few shared commonalities of understanding.

Conclusions and Future Work

So far, this research has established that the energy trilemma provides a narrow basis on which to assess the value of deploying renewable energy technologies like solar PV. The current power system demonstrates significant policy, regulatory and financial inertial forces. These inertial forces hamper progress to the ideal system state shown in Fig. 1.

There are a wide range of tools and methods available to model the impact of change (Table 2) which are applicable to specific aspects of the energy system.

Although there is common agreement amongst the key participants who perform the energy policy, regulatory and operational functions in the UK that change is needed to address the energy trilemma, there is a lack of an effective mechanism, including an agreed and shared model, to assess the impact of change in the power sector at different scales and across time. Exploring a Geographical Information System (GIS) based approach, among others, could have significant merit given the need to manage and model spatial information.

The model presented at Fig. 3 indicates that there are at least forty factors that could play a part when an organisation is planning to make an investment decision affecting their interaction with the energy system, whether that is installing solar PV on a single building or making decisions that affect national infrastructure. At this time, no framework has been identified which looks at the relative importance of these aspects from different stakeholder perspectives to derive an optimal investment position.

Furthermore, there is a gap between how nations, using the UN Sustainable Development Goals as the basis and individual organisations using a range of environmental, social and economic metrics, set and track progress to achieving sustainability.

The proposed research will, therefore, aim to verify these initial conclusions and, if shown to be valid, will try to put forward a model to address the gap that this initial analysis suggests.

References

BEIS. (2018). Digest of UK Energy Statistics (DUKES) 2018—Chapter 5: Electricity. *BEIS*. Retrieved April 27, 2019, from assets.publishing.service.gov.uk/government/uploads/system/uploads/attachment_data/file/736152/Ch5.pdf.

BEIS. (2019). Solar photovoltaics deployment in the UK. *BEIS*. Retrieved March 31, 2019, from www.gov.uk/government/statistics/solar-photovoltaics-deployment.

Blazqueza, J., Bracamontesa, F., Bollinob, C., & Nezamuddina, N. (2018). The renewable energy policy paradox. *Renewable and Sustainable Energy Reviews, 82*(1), 1–5.

Building Research Establishment. (2019). The Building Research Establishment Environmental Assessment Method. *Building Research Establishment*. Retrieved May 2, 2019, from www.breeam.com/.

Committee on Climate Change. (2019). Net Zero—The UK's contribution to stopping global warming. Retrieved May 2, 2019, from www.theccc.org.uk/wp-content/uploads/2019/05/Net-Zero-The-UKs-contribution-to-stopping-global-warming.pdf.

Competition and Markets Authority. (2014). Energy market investigation. *Competition and Markets Authority*. Retrieved June 14, 2018, from www.gov.uk/cma-cases/energy-market-investigation#final-report.

Electricity Networks Strategy Group. (2012). Our electricity transmission network: A vision for 2020—A Summary of an Updated Report to the Electricity Networks Strategy Group, *URN 11D/955*. Retrieved May 2, 2019, from https://assets.publishing.service.gov.uk/government/uploads/system/uploads/attachment_data/file/48275/4264-ensg-summary.pdf.

Espinoza, J. L., Samaniego, E., Jara-Alvear, J., & Ochoa, T. D. (2017). Smart grids: A multi-scale framework of analysis. *Industrial Electronics (ISIE)*, IEEE 26th International Symposium.

Global Footprint Network. (2019). Retrieved May 2, 2019, from www.footprintnetwork.org/our-work/ecological-footprint/.

Helm, D. (2017). Cost of Energy Review, Department for Business, Energy & Industrial Strategy. Retrieved June 17, 2019, from www.gov.uk/government/publications/cost-of-energy-independent-review.

International Council for Science. (2017). A guide to SDG interactions: from science to implementation. *International Council for Science*. pp. 18–30. Retrieved June 17, 2019, from https://council.science/cms/2017/05/SDGs-Guide-to-Interactions.pdf.

McKinsey. (2007). A cost curve for greenhouse gas reduction. Retrieved May 4, 2019, from www.mckinsey.com/business-functions/sustainability/our-insights/a-cost-curve-for-greenhouse-gas-reduction.

Nilsson, M. (2016). Understanding and mapping important interactions among SDGs—Background paper for Expert meeting in preparation for HLPF 2017. *Readying institutions and policies for integrated approaches to implementation of the 2030 Agenda, Vienna*.

Nilsson, M., Griggs, D., Visbeck, M., & Ringler, C. (2016). *A draft framework for understanding the SDG interactions, ICSU Working Paper*. Paris: International Council for Science.

O'Leary, A., & Speciale, G. (2017). Monitoring and evaluating community energy: Reviewing existing tools and outlining a pilot standardised tool. Retrieved May 10, 2018, from www.pureleapfrog.org/monitoring-and-evaluation.

Office of National Statistics. (2019). UK data for Sustainable Development Goal Indicators. *Office of National Statistics*. Retrieved April 25, 2019, from https://sustainabledevelopment-uk.github.io/.

Ofgem. (2018). Electricity supply market shares by company: Domestic, GB, *Ofgem*. Retrieved June 13, 2018, from www.ofgem.gov.uk/data-portal/electricity-supply-market-shares-company-domestic-gb.

Royal Academy of Engineering. (2015). A critical time for UK energy policy what must be done now to deliver the UK's future energy system—A report for the Council for Science and Technology. *Royal Academy of Engineering*. Retrieved May 2, 2019, from www.raeng.org.uk/publications/reports/a-critical-time-for-uk-energy-policy.

Shove, E., & Walker, G. (2014). What is energy for? *Social Practice and Energy Demand, Theory, Culture and Society, 31*, 41–58.

Stanojevića, M., Vranešal, S., & Gökalp, I. (2010). Green accounting for greener energy. *Renewable and Sustainable Energy Reviews, 14*(9), 2473–2491.

Statista. (2019). Total electricity generation capacity projections of major power producers in the United Kingdom (UK) from 2018 to 2035 (in Gigawat*t*), *Statista*. Retrieved April 27, 2019, from www.statista.com/statistics/496283/total-electricity-generation-capacity-uk/.

The Guardian Online. (2019). *Hitachi scraps £16bn nuclear power station in Wales*. Retrieved April 27, 2019 from www.theguardian.com/business/2019/jan/17/hitachi-set-to-scrap-16bn-nuclear-project-anglesey-wales.

The Natural Step. (2019). Retrieved May 2, 2019, from https://thenaturalstep.org/.

The SROI Network. (2012). A guide to Social Return on Investment. Retrieved May 1, 2019, from www.socialvalueuk.org/app/uploads/2016/03/The%20Guide%20to%20Social%20Return%20on%20Investment%202015.pdf.

UN World Commission on Environment and Development. (1987). Report of the World Commission on Environment and Development: Our Common Future, *United Nations*. Retrieved May 2, 2019, from www.un-documents.net/wced-ocf.htm.

United Nations. (2015). Resolution adopted by the General Assembly on 25 September 2015, 70/1. Transforming our world: the 2030 Agenda for Sustainable Development, 70th Session 21st October 2015, *United Nations*. Retrieved May 9, 2019, from www.un.org/en/development/desa/population/migration/generalassembly/docs/globalcompact/A_RES_70_1_E.pdf.

United Nations. (2019). Sustainable Development Goals, *United Nations*. Retrieved April 29, 2019, from www.un.org/sustainabledevelopment/sustainable-development-goals/.

United Nations World Commission on Environment and Development. (1987). Our Common future A/42/427, *United Nations*. Retrieved April 29, 2019, from www.un-documents.net/wced-ocf.htm.

Wolsink, M. (2012). The research agenda on social acceptance of distributed generation in smart grids: Renewable as common pool resources. *Renewable and Sustainable Energy Reviews, 16*, 822–835.

World Energy Council. (2019). Energy Trilemma Index. *World Energy Council*. Retrieved May 2, 2019, from https://trilemma.worldenergy.org/#!/energy-index.

Worldwide Fund for Nature. (2019). Ecological foot printing—A measure of sustainability. *Worldwide Fund for Nature*. Retrieved May 2, 2019, from wwf.panda.org/knowledge_hub/teacher_resources/webfieldtrips/ecological_balance/eco_footprint/.

Yale University. (2019). Environmental Performance Index Global metrics for the environment. *Yale University*. Retrieved April 21, 2019, from https://epi.envirocenter.yale.edu/.

Evaluating Solar Prediction Methods to Improve PV Micro-grid Effectiveness Using Nonlinear Autoregressive Exogenous Neural Network (NARX NN)

Norbert Uche Aningo, Adam Hardy, and David Glew

Introduction

The growth in digital technologies helps to improve energy efficiency and smooth operation of smart grids but often creates potential challenges such as rapid increase in energy demand and needs for decarbonization (WEO, 2019). Renewable energy resources are essential to mitigate these challenges.

In recent years, renewable energy resources have played the leading role in meeting the increasing energy demands as well as contributing to the fight against global Green House Gas (GHG) emissions (IRENA, 2017). The integration of renewable energy resources into the electricity grid also reduces environmental pollution (Zou, Zhao, Zhang, & Xiong, 2016). Moreover, renewable power sources (especially solar power) are increasing their share of the global energy market. Indeed, the renewable energy contribution to global energy installations was about 160 GW in 2016 (Tanti, 2018). The installed Solar PV represented the biggest addition, by providing 43.3% (about half) of the overall newly installed global renewable power generators that year (GlobalData, 2017). The contribution of solar power to global electricity generation can be clearly observed with the trend in electricity generation as given in Fig. 1, where solar PV, hydro and wind electricity generations have each recorded substantial increase from 1995 to 2016 (IEA, 2018). Hence, solar power can help in meeting the global power demand if properly harnessed. However, the amount of solar power that can be harnessed each day fluctuates according to various factors such as solar irradiance, cloud cover and temperature. Predicting the yield from solar micro-grids can therefore be imprecise.

Solar power is a technological innovation that has the benefits of availability and cost effectiveness. These benefits, coupled with the benefits of flexibility and

N. U. Aningo (✉) · A. Hardy · D. Glew
Leeds Sustainability Institute, School of the Built Environment and Engineering, Leeds Beckett University, Leeds, UK

L. Scott et al. (eds.), *Sustainable Ecological Engineering Design*, https://doi.org/10.1007/978-3-030-44381-8_28

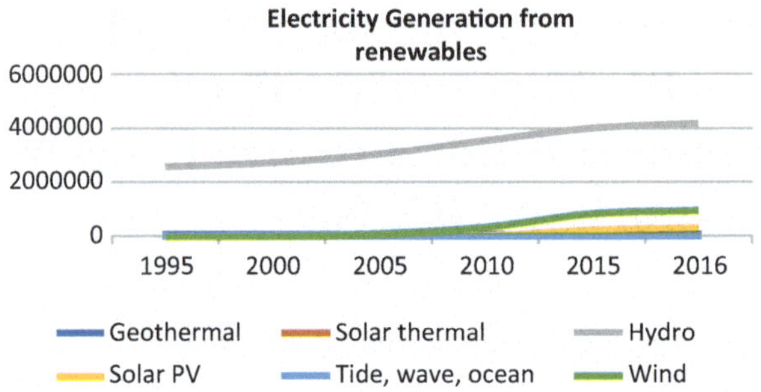

Fig. 1 Electricity generation from renewable by source (IEA, 2018)

potential as both small-scale and large-scale power generating source, has resulted in a proliferation and wide distribution of PVs (IEA, 2015; IRENA, 2019). However, solar power is highly imprecise and variable by nature (Wu, Wen, Lou, & Xin, 2015). The variable and intermittent nature of solar power sources introduce significant challenges in power system planning and scheduling operations of smart grids (Pelland, Galanis, & Kallos, 2011; Raza, Mithulananthan, & Summerfield, 2018). For smart grids to perform at optimum levels, accurate prediction is required to reduce intermittency and variability. Solar power prediction is essentially known to improve integration of solar power resources into smart grid (Wan et al., 2015).

Solar power prediction is a requirement in power system operation and maintenance because it enhances the effectiveness of power system balance, power system stability, frequency response, power factor correction and reactive power compensation (Nuchhi, Sali, & Ankaliki, 2013; Ueckerdt, Brecha, & Luderer, 2015). For instance, the accurate short-term prediction of PV output ensures smoothing of transmitted voltage by reducing voltage ramps and frequency fluctuations in smart grids (Chen, Du, Xiao, & Lu, 2017; Chu, Urquhart, Gohari, & Pedro, 2015).

This paper presents an Artificial Neural Network (ANN) model for solar power prediction, comparing how weather input variables from Leeds, UK, affect the prediction accuracy. Following this, the Nonlinear Autoregressive Exogenous Neural Network (NARX NN) model performance is compared with Nonlinear Autoregressive Neural Network (NAR NN) model using time series modelling approach. The result shows that NARX NN model outperformed NAR NN model for selected variables from the geographical location. Three algorithms used for NN training were also examined and compared using best weights and Neuron parameters. This research paper is part of wider scope of renewable energy micro-grid design for a local hospital and residential housing. The main contribution of this research is development of prediction model that accounts for selection of best input variables and the effect of those variables on power peak.

Methods

In this research study, two methods—NARX and NAR—are compared to determine which is more powerful in predicting solar power generation. The proposed methodology aims at improving 24-h solar power prediction using comparison of the statistical methods. Hence, this section presents a brief explanation of those methods for clarity, the structure of network model and the descriptive analysis of datasets selected. The NARX and NAR models for time series prediction are based purely on machine learning techniques. The following subsection specifically describe these models.

NARX Neural Network (NARX NN) Method

Artificial neural networks (ANNs) are black-box techniques that can learn, store and retrieve information based on the data fed into it (Gurney, 1997; Haykin, 2005). ANNs have been used in solving a wide range of issues such as prediction, optimization, pattern recognition and classification (Araújo, Oliveira, & Meira, 2017; Gong, Fan, Guo, & Cai, 2017; Haykin, 2005). More specific examples of ANNs include feedforward neural networks (FFNN) and recurrent neural networks (RNN): FFNNs have been widely applied to study nonlinear behaviour by feeding the networks with external data. RNNs, meanwhile, can make use of their own output as an input to improve model accuracy. It is assumed that all inputs and outputs are independent of each other in FFNN, which is not typically a good assumption. But in RNN, the output depends on previous computations and performs the same task for every element in a sequence. Hence, RNNs make use of their internal memory to process an arbitrary sequence and RNNs can thus make use of sequential information.

Neural networks for time series prediction can either be Nonlinear Autoregressive (NAR) without exogenous input or Nonlinear Autoregressive with exogenous input (NARX) model. NARX NN models allow the use of extra variables (such as temperature, humidity and cloud cover) that can enhance prediction accuracy. Figure 2 shows typical open loop and closed loop NARX neural network.

ANN models have been widely applied to forecast time series for different purposes and applications (Siegelman & Sontag, 1992). These studies applied different input variables, time horizon and structure to conduct the time series prediction. For example, Hourly Solar Radiation prediction was carried by Mohammed, Hamdan, and Abdelhafez (2013) using NARX NN in Jordan. The study recommended the use of NARX models for hourly solar prediction in Jordan and nearby regions (Mohammed et al., 2013). A general review of power system applications using artificial neural networks can be seen in Kalogirou (2000).

The benefits of NARX NNs include the often superior prediction caused by the extra inputs. NARX NNs are also relatively easy to understand and use, compared

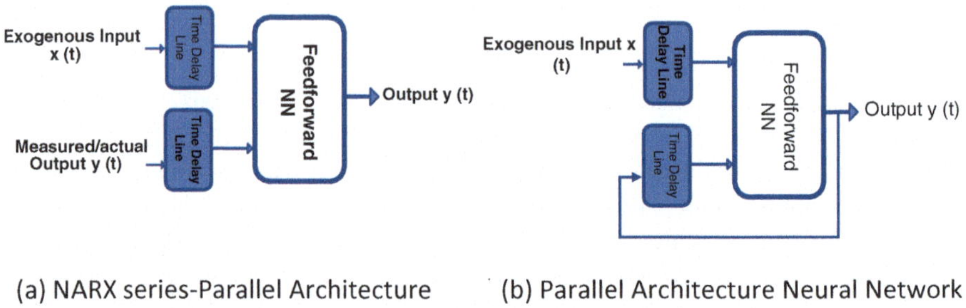

Fig. 2 (**a**) NARX series-parallel architecture, (**b**) parallel architecture neural network

with some other statistical or parametric methods that require prior knowledge of statistics. Furthermore, they can utilize Input-Output (IO) mapping—this means that it can learn from previous examples by creating IO mapping to solve problems (Haykin, 2005).

However, one of the major limitations of NARX Neural network is its black-box learning approach and interactions (or relationships) between underlying models do not reflect real physics (Haykin, 2005; Tu, 1996). This is oppose to other statistical methods (such as fuzzy logic algorithm) which can interpret relationships using logic rules (Haykin, 2005; Tu, 1996).

Data Collection

Study Sites and Data

The study is conducted on data from Leeds, UK. The hourly data for 1-year period were collected from a weather station for this study. The deployed remote sensing devices were used to acquire a dataset on an hourly basis from the weather station, which is managed by Leeds-Bradford airport. Twelve ground weather parameters useful for solar irradiance prediction were selected from the weather station which is equipped with a set of sensors to take record of radiation, temperature, pressure, wind speed, visibility and sky cover.

Data Selection and Analysis

Data selection is important in time series modelling. The preliminary pre-processing of the dataset from Leeds weather station was carried out. After obtaining the datasets to be used for modelling, the data with most predictive power was determined.

The data was normalized to have the same range of values for each of the inputs to be modelled. Data normalization is done to ensure equal attributes and stable convergence of the neural network weights and bias. During this process, the data with zero values which corresponds to night period was removed.

Methods Adopted

This section explains the methods adopted and the process for modelling of Nonlinear Autoregressive Exogenous (NARX) Neural Network for solar power prediction. The process of modelling time series (for both NAR and NARX) using MATLAB toolkit is shown in flowchart of Fig. 3.

In the model flowchart, the datasets were divided into two different time series sets for training and testing.

The process involves split of the 2017 datasets into train set (80%) and test set (20%) in order to create the time series k-fold rolling windows for cross validation. The training process also includes adjusting the number of weights and neurons in each of the nodes using a predefined error tolerance value such that the measured and predicted values are as close as possible. After the train/test stage and the consequent acceptance of the model, then the validation of the network would be carried out to obtain the overall best performance. We applied statistical analysis, which include coefficient of determination (R^2), mean squared error (MSE) and root mean square error (RMSE), to determine the performance of the model.

Evaluating the Success of the Models

Generally, regression problems deal with continuous data and have the aim at predicting a value that is as close as possible to the actual value. Supervised learning is used to decrease the errors between the prediction and actual target of the neural network (Mitchell, 1997). It is essential to evaluate the models that are trained using supervised learning by choosing the right evaluation metrics. The results of the NARX models used in this study will be analysed by comparing the MSE and R^2 performances of the training stage, and then comparing the MSE, R^2 and RMSE of predicted data against the observed data in the testing stage of NARX model. The mean squared error (MSE), coefficient of determination (R^2) and root mean square error (RMSE) are discussed below (Nagelkerke, 1991; Vernier, 2001):

Data Pre-processing

Due to difference in magnitude of the data used, the data pre-processing was done by normalizing between 0 and 1. Normalizing or down-scaling data is necessary to adjust the measured values that have different scales such that they are converted to

Fig. 3 Flowchart for NN time series modelling training and testing of the datasets

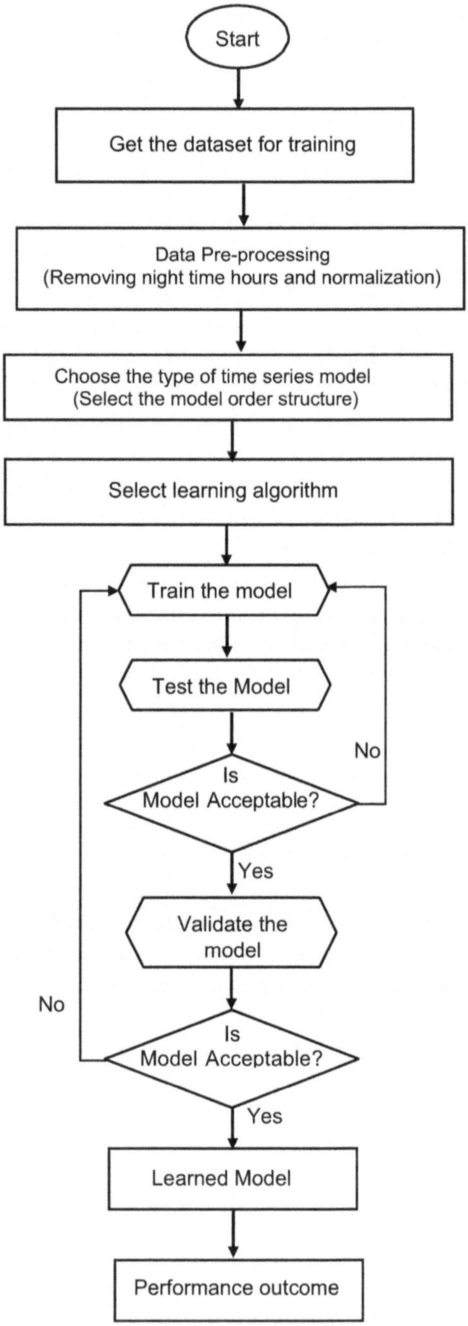

common size. The normalization of the data was achieved with the formula in Eq. (1):

$$X_{normalized} = (X - X_{min}) / (X_{max} - X_{min}) \quad (1)$$

Results and Discussion

Training the NARX Model

An experiment was conducted using different numbers of hidden layers, tapped delay lines (d) (or delay parameter), number of neurons and transfer functions. The delay parameter is described as the amount of time taken to perform the prediction. To obtain accurate prediction model, the delay parameter is adjusted repeatedly for each model during training.

The experiment is performed to obtain the best network structure that could be used to determine best performance of our proposed NARX model. This is achieved (using trail-and-error method) by adjusting delay parameter whereas other parameters remain fixed. The MSE and R^2 obtained with the delay parameters for NARX model after several executions are given in Table 1. The best MSE obtained with the delay parameter is marked with shaded (yellow) box. From the average MSE values obtained, it is observed that minimum delay parameter required for accurate prediction is 8. Delay parameters between 8 and 18 are needed to get good prediction using our proposed model. Another neural network parameter that requires determination is number of neuron. Table 2 shows the best MSE using a fixed delay with variable number of neurons for five different executions. The average MSE obtained from the experiment showed that between 2 and 6 neurons are required for NARX model. This indicates that large number of neurons could lead to bad prediction or worst outcomes such as overtraining (or network overfitting). The training process

Table 1 The MSE and R^2 of delay parameter obtained for each algorithm

Delay	1		4		8		12		18	
Algorithm	MSE	R^2	MSE	R^2	MSE	R^2	MSE	R^2	MSE	R^2
Levenberg	0.0033	0.98	0.0038	0.98	0.0034	0.98	0.0032	0.98	0.0023	0.98
Bayesian R	0.0047	0.98	0.0045	0.98	0.0027	0.98	0.0029	0.98	0.0037	0.98
Scale conjugate	0.0069	0.96	0.0078	0.98	0.0049	0.96	0.0042	0.98	0.0059	0.96
Average	0.0049		0.0054		0.0037		0.0034		0.0039	

Table 2 The MSE and R^2 of neuron parameter obtained for each algorithm

Neurons	2		4		6		8		10	
Algorithm	MSE	R^2	MSE	R^2	MSE	R^2	MSE	R^2	MSE	R^2
Levenberg *M*	0.0034	0.98	0.0028	0.98	0.0037	0.98	0.0039	0.98	0.0047	0.98
Bayesian *R*	0.0027	0.98	0.0039	0.98	0.0049	0.96	0.0058	0.98	0.0049	0.96
Scale conjugate	0.0049	0.96	0.0048	0.98	0.0034	0.98	0.0048	0.98	0.0064	0.97
Average	0.0037		0.0038		0.0040		0.0048		0.0053	

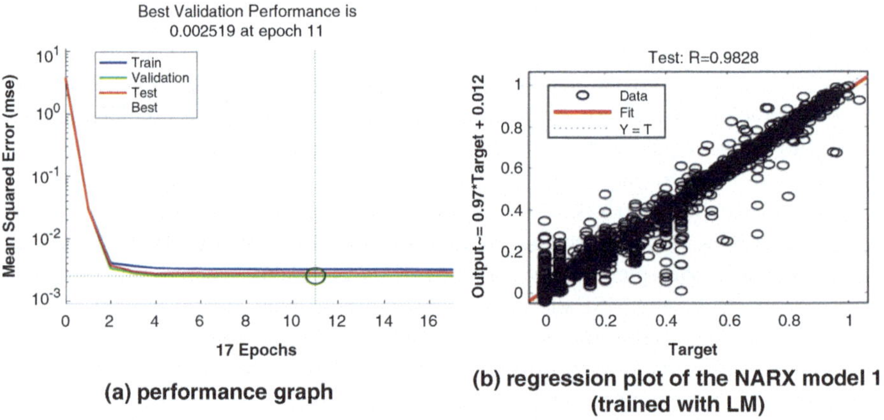

Fig. 4 (**a**) Performance graph and (**b**) regression plot of the NARX model 1 (trained with LM)

involves several iterations that gives different level of accuracy and the weights/bias values (i.e. the process can give different outputs using the same inputs for better performance).

NARX Model Trained with Levenberg Marquardt (LM) Results

When the NARX model is trained for the first time using the LM algorithm, it is observed that network completed and stopped at 17 epochs. The network is retrained several times (which increased slightly the number of epochs), and the best results recorded. The best performance obtained from the NARX LM1 model is 0.0025 as shown in Fig. 4. The best MSE and R^2 of the network for testing sets after several retraining processes are shown in Table 3:

Table 3 Performance of NARX models

Model type	Epoch	Regression (R^2) (%)	MSE	Gradient
NARX LM1	17	98	0.0025	0.0013
NARX LM2	20	97	0.0048	0.0059
NARX LM3	20	98	0.0028	0.0047

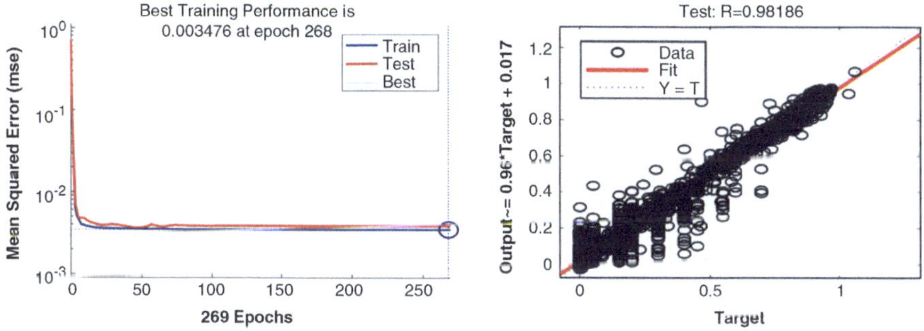

Fig. 5 (**a**) Performance graph and (**b**) regression plot of the NARX model 1 (trained with BR)

ANN NARX Model Trained with Bayesian Regularization (BR) Results

When a second NARX model is trained using the BR algorithm, it is observed that the model completed and stopped at 269 epochs. We apply the usual 'early stop' to get the best performance and avoid overfitting. The best performance of the NARX BR1 is 0.0035 as seen in Fig. 5. The MSE and R^2 of the network for testing after several retraining are shown in Table 3:

ANN NARX Model Trained with Scaled Conjugate Gradient (SCG) Results

The NARX model was then trained with a SCG algorithm and it is observed that the network completed in 136 iterations and the best performance is obtained at epoch 130 as shown in Fig. 6. The best performance of the NARX SCGs is 0.0051. The MSE and R^2 of the network for testing after several retraining steps of the model are shown in Table 3:

From the performance Tables 3, 4 and 5, two features are observed to be common for the three examined NARX models:

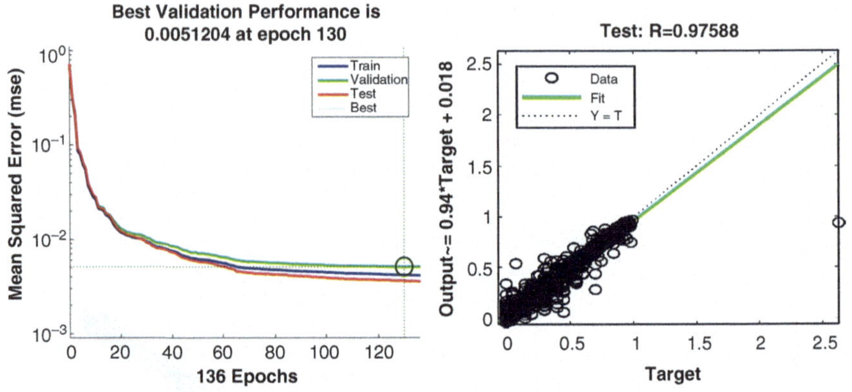

(a) performance graph

(b) regression plot of the NARX model 1 (trained with SCG)

Fig. 6 (**a**) Performance graph and (**b**) regression plot of the NARX model 1 (trained with SCG)

Table 4 Performance of NARX models

Model type	Epoch	Regression (R^2) (%)	MSE	Gradient
NARX BR1	269	98	0.0035	0.0013
NARX BR2	254	98	0.0049	0.0059
NARX BR3	199	95	0.0038	0.0047

Table 5 Performance of NARX models

Model type	Epoch	Regression (R^2) (%)	MSE	Gradient
NARX SCG1	136	97	0.0051	0.0017
NARX SCG2	157	95	0.0059	0.0019
NARX SCG3	179	96	0.0063	0.0016

(a) As the number of epochs increase, the MSE decreases. This shows that their error characteristics for testing and validation are comparable. Hence, no overlapping is observed from the best performance plots.
(b) All values of R^2 are observed to be above or equal to 95%. This indicates the correlations between the outputs and target.

Two contrasting features can as well be observed from performance plots of the three NARX models: (a) the training of NARX LM takes below 20 epochs to complete, whereas NARX BRs and NARX SCGs take above 50 epochs in all the cases examined. This shows that the NARX model with LM algorithms trains and converges faster than NARX models with BR and SCG algorithms. (b) The MSE of NARX LMs is within the acceptable range while the MSE of NARX SCGs and

Table 6 The NARX model parameter summary

Feature	Choice
Training algorithm	Levenberg Marquardt
Number of hidden neurons	6
Normalized datasets intervals	[0.005–1]
Delay parameter	8
Error metrics	MSE

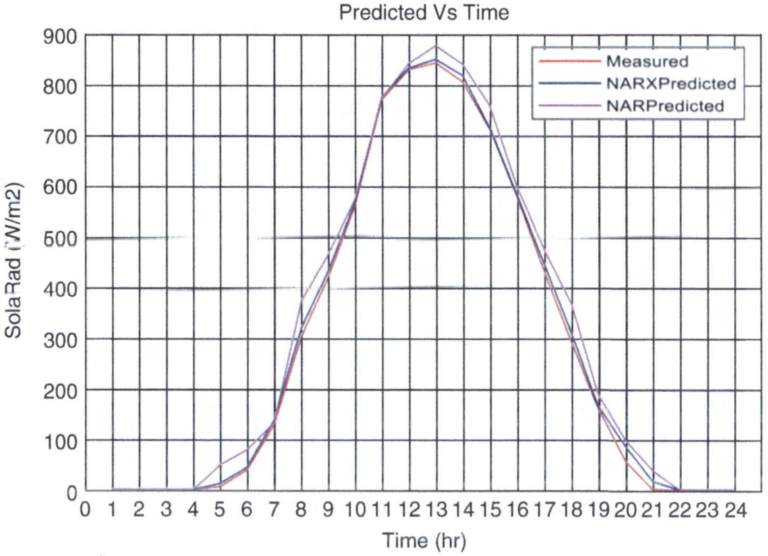

Fig. 7 Comparison of actual and predicted values of solar irradiance of models

NARX BRs are very high and very low, respectively. The gradient of NARX LMs and NARX BRs are similar and within the acceptable range, while the gradient of NARX SGCs are very high, and therefore above the acceptable range. This experiment further shows that higher numbers of neurons or a lower delay parameter may lead to bad prediction due to overfitting or network getting stuck in local maxima. Hence, the summary of the parameters for this study is presented in Table 6.

The Actual Values and Predicted Values for NARX Models

Both NAR and NARX models are applied to predict values for 24-h ahead after training. To demonstrate the merits of the proposed NARX model, we compare with NAR.

The plot of comparison for models are shown in Fig. 7. The red curve represents actual data while the blue and magenta curves represent predicted values for NARX

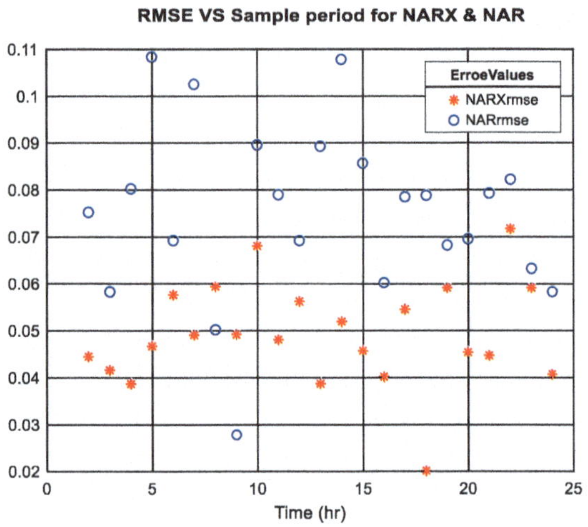

Fig. 8 Comparison of RMSE for different prediction models

and NAR, respectively. The plot demonstrates that NARX prediction outcomes are better than NAR prediction outcomes. From the plot, we can see that the actual curve is similar to the NARX predicted curve but differ slightly for NAR model. It is observed that both NARX and NAR curves follow the same path with the actual values, but there are deviations for NAR from the actual values especially between 5, 8, 13 and 18 h ahead as seen from the plots. Generally, there are slight fluctuation from NARX and NAR predicted values but it can be observed that NARX predicted curve is much closer to actual values. This indicates that our proposed NARX model shows more accuracy prediction which suggests that NARX model is more accurate for short-term predictions. The proposed model can be more useful for daily solar radiation prediction. It also shows that using exogenous input (such as Temperature) can help improve day-ahead prediction accuracy. The next subsection is the residual plot of the models.

Error Analysis

There are different metrics to evaluate the accuracy of the NAR and NARX models. On this stage of the study, RMSE is used to assess the accuracy of the models. Generally, the NARX model show good accuracy in short-term prediction. The RMSE values obtained from the NARX model is typically less than RMSE of NAR model. For example, the error in NARX models never exceeds 0.08 and shows only slight fluctuation, whereas the NAR model RMSE is inconsistent with big fluctuations (Fig. 8).

Conclusion

In this study, two time series forecasting methods, NARX and NAR models, were compared using 2017 datasets from Leeds, UK. During the training stage of the NARX, initial simulations were carried out using different combinations of variables from the Leeds UK dataset, and with consideration of different values of neural network parameters. Different delay parameters, neuron parameters and three different algorithms were tested during the network training stage using two evaluation criteria (MSE and R^2). The findings showed that models trained with the Levenberg-Marquardt (LM) and Bayesian regularization (BR) algorithms outperformed models trained with Scaled Conjugate gradient algorithm. However, it was observed that the model with the LM algorithm (with best performance of 0.02519 MSE and 0.981 R^2) trained faster than other models but the model trained with Bayesian regularization still achieved good performance (with best MSE of 0.0347 and R of 0.981). Models trained with SCG have the highest MSE values. Furthermore, it was observed that NARX model with LMs trained faster and were able to converge at much lower number of iterations (less than 20 epochs). From the plot curve and RMSEs obtained, NARX NN outperformed NAR using the same parameters. This indicates that the use of exogenous inputs helps to improve accuracy of short-term prediction. The future research work will be targeted at developing different time series models that have comparable advantages using additional dataset from different region.

References

Araújo, R. A., Oliveira, A. L. I., & Meira, S. (2017). A morphological neural network for binary classification problems. *Engineering Applications of Artificial Intelligence, 65*, 12–28.

Chen, X., Du, Y., Xiao, W., & Lu, S., 2017. *Power ramp-rate control based on power forecasting for PV grid-tied systems with minimum energy storage*. IEEE Conference 2017

Chu, Y., Urquhart, B., Gohari, S. M., & Pedro, H. T. C. (2015). Short-term reforecasting of power output from a 48 MWe solar PV plant. *Solar Energy, 112*, 68–77.

GlobalData. (2017). *Global PV capacity is expected to reach 969GW by 2025*. Power Technology

Gong, T., Fan, T., Guo, J., & Cai, Z. (2017). GPU-based parallel optimization of immune convolutional neural network and embedded system. *Engineering Applications of Artificial Intelligence, 62*, 384–395.

Gurney, K. (1997). *An introduction to neural networks*. ROUTLEDGE, UCL Press Limited 11 New Fetter Lane London EC4P 4EE, UCL Press Limited is an imprint of the Taylor & Francis Group, Record Number: 163. https://www.inf.ed.ac.uk/teaching/courses/nlu/assets/reading/Gurney_et_al.pdf.

Haykin, S. (2005). *Neural Network: A Comprehensive Foundation*, Pearson Education (Singapore) Pte. Ltd., Indian Branch, 482 F. I. E. Patparganj Delhi 110092, India, Pearson Prentice Hall.

IEA. (2015). *Energy from the desert: Very large scale PV power plants for shifting to renewable energy future*. International Energy Agency Photovoltaic Power Systems Program.

IEA. (2018). *Electricity generation from renewables by source World 1990–2016*. International Energy Agency. Key World Energy Statistics 2018

IRENA. (2017). Renewable energy: A key climate solution. *International Renewable Energy Agency.*

IRENA. (2019). *Innovation landscape for a renewable-powered future: Solutions to integrate variable renewables.* International Renewable Energy Agency, Abu Dhabi.

Kalogirou. (2000). Applications of artificial neural-networks for energy systems. *Applied Energy, 67*(1–2), 17–35.

Mitchell, T. M. (1997). *Machine learning, McGraw-Hill Science/Engineering/Math.* Retrieved from https://www.cs.ubbcluj.ro/~gabis/ml/ml-books/McGrawHill%20-%20Machine%20 Learning%20-Tom%20Mitchell.pdf.

Mohammed, L. B., Hamdan, M. A., & Abdelhafez, E. A. (2013). Hourly solar radiation prediction based on Nonlinear Autoregressive Exogenous (Narx) neural network. *Jordan Journal of Mechanical and Industrial Engineering, 7,* 11–18.

Nagelkerke, N. J. D. (1991). *A note on a general definition of the coefficient of determination.* Retrieved from http://www.cesarzamudio.com/uploads/1/7/9/1/17916581/ nagelkerke_n.j.d._1991_-_a_note_on_a_general_definition_of_the_coefficient_of_determination.pdf. *Biometrika trust* http://links.jstor.org/sici?sici=0006-3444%28199109%2978%3A3% 3C691%3AANOAGD%3E2.0.CO%3B2-V.

Nuchhi, S. S., Sali, R. B., & Ankaliki, S. G. (2013). Effect of reactive power compensation on voltage profile. *International Journal of Engineering Research and Technology, 2*(6), 2627.

Pelland, S., Galanis, G., & Kallos, G. (2011). Solar and photovoltaic forecasting through post- processing of the Global Environmental Multiscale numerical weather prediction model. *Progress in Photovoltaics: Research and Applications, 21*(3). https://doi.org/10.1002/pip.1180

Raza, M. Q., Mithulananthan, N., & Summerfield, A. (2018). Solar output power forecast using an ensemble framework with neural predictors and Bayesian adaptive combination. *Solar Energy, 166*, 226–241.

Siegelman & Sontag. (1992). On the computational power of neural nets. In *Proceedings of the fifth annual workshop on Computational learning theory.* ACM conference Proceedings, 440–449.

Tanti, T. (2018). The key trends that will shape renewable energy in 2018 and beyond. *World Economic Forum.*

Tu, J. V. (1996). Advantages and disadvantages of using artificial neural networks versus logistic regression for predicting medical outcomes. *Journal of Clinical Epidemiology, 49*, 1225–1231.

Ueckerdt, F., Brecha, R., & Luderer, G. (2015). Analyzing major challenges of wind and solar variability in power systems. *Renewable Energy, 81*, 1–10.

Vernier. (2001). *What are mean squared error and root mean squared error?* Beaverton, OR: Vernier Software & Technology. Retrieved from https://www.vernier.com/til/1014/.

Wan, C., Zhao, J., Song, Y., Xu, Z., Lin, J., & Hu, Z. (2015). Photovoltaic and solar power forecasting for smart grid energy management. *CSEE Journal of Power and Energy Systems, 1*, 38–46.

WEO. (2019). World Energy Outlook 2017: A world in transformation. *IEA.*

Wu, C., Wen, F., Lou, Y., & Xin, F. (2015). Probabilistic load flow analysis of photovoltaic generation system with plug-in electric vehicles. *International Journal of Electrical Power & Energy Systems, 64*, 1221–1228.

Zou, C., Zhao, Q., Zhang, G., & Xiong, B. (2016). Energy revolution: From a fossil energy era to a new energy era. *Natural Gas Industry B, 3*, 1–11.

Part V
Sustainable Retrofit and Lifecycle Assessment

Responsible Retrofit Measures for Traditional Listed Dwellings: An Energy Simulation Validation Strategy

Michela Menconi, Noel Painting, and Poorang Piroozfar

Introduction

Research Background

The use of dynamic Building Energy Simulation (BES) tools, as a means to optimise the energy performance of buildings, has sensibly increased since they first emerged in the 70s. This has been driven by more stringent requirements for tighter energy conservation measures. However, several studies to date have highlighted significant discrepancies between simulated performance and measured data when BES tools were deployed to assess the energy performance of existing buildings or to foresee that of new ones (BRE, 2014; Dall'o, Sarto, Sanna, & Martucci, 2012; Heath, Pearson, Barnham, & Atkins, 2010; Hubbard, 2011; Ingram & Jenkins, 2013; Jenkins, 2008; Moran, 2013; STBA, 2012; Thompson & Bootland, 2011; Wingfield, Bell, Miles-Shenton, & Seavers, 2011). Therefore, such instruments are yet frequently criticised for their lack of precision (Coakley, 2014). Concerns are even more profound when it comes to simulation of traditional (and more so for traditional listed) buildings because of the complexities of processes and synergies that characterise their thermal behaviour. The challenges in reproducing such complex behaviour using BES for traditional heritage buildings are mainly due to the wide range of assumptions concerning the materials build-up of their thermal envelopes. Frequently, constructions and materials, not available within the default options provided by the software library, need to be created ex novo by the

M. Menconi (✉) · N. Painting
School of Environment and Technology, University of Brighton, Brighton, UK
e-mail: m.menconi3@brighton.ac.uk

P. Piroozfar
School of Environment and Technology, University of Brighton, Brighton, UK

Digital Construction Lab, University of Brighton, Brighton, UK

L. Scott et al. (eds.), *Sustainable Ecological Engineering Design*,
https://doi.org/10.1007/978-3-030-44381-8_29

researchers, based on the visual and measured survey of the building investigated, as well as on the knowledge gained about local traditional methods of construction and materials properties. This increases the difficulties in simulating the actual energy consumption of traditional buildings correctly (BRE, 2014; STBA, 2012).

Therefore, the need for thoroughly devised strategies for validating the output of energy simulations is pressing, to ensure that the created models can generate realistic results. When BES is deployed for the performance analysis of the status-quo of a real building and/or for testing possible solutions to improve such performance, the comparison between the results generated by simulation and empirical data provides a powerful validation tool (Baranowski & Ferdyn-Grygierek, 2009; Maile, Bazjanac, & Fischer, 2012; Raftery, Keane, & O'Donnell, 2011; Ryan & Sanquist, 2012).

Aim of This Paper

This paper sets out to describe the validation strategy specifically devised for an ongoing study that aims to propose responsive and effective retrofit interventions for C19th traditional listed dwellings (TLDs) in South-East England.

Overview of the Research Methodology

The research methodology deploys a sequential mixed method approach. Qualitative methods have been used for the choice of eight representative case studies (CSs)—carefully selected in the city of Brighton and Hove (UK), using stringent inclusion/exclusion criteria set in this project—and for collecting data to be used for model creation, validation and normalisation. The research uses then mostly quantitative methods performing BES using IES-VE. BES results will be deployed to carry out a sensitivity analysis to demonstrate statistically how the interventions impact the energy performance in the selected cases. This paper addresses the validation process which stems out of two phases of data collection (as indicated in blue in Fig. 1).

The first phase of data collection deployed multiple methods to gather a wide range of input data, necessary for the creation of each energy model, as follows:

- A critical review of literature established the gaps in knowledge with reference to: methodology, methods, data collection/generation and analysis instruments. This was followed and complemented by secondary data collection and expert consultation with local conservation experts, to help with necessary assumptions concerning the construction methods and materials build-up of the envelope as well as air leakage values.

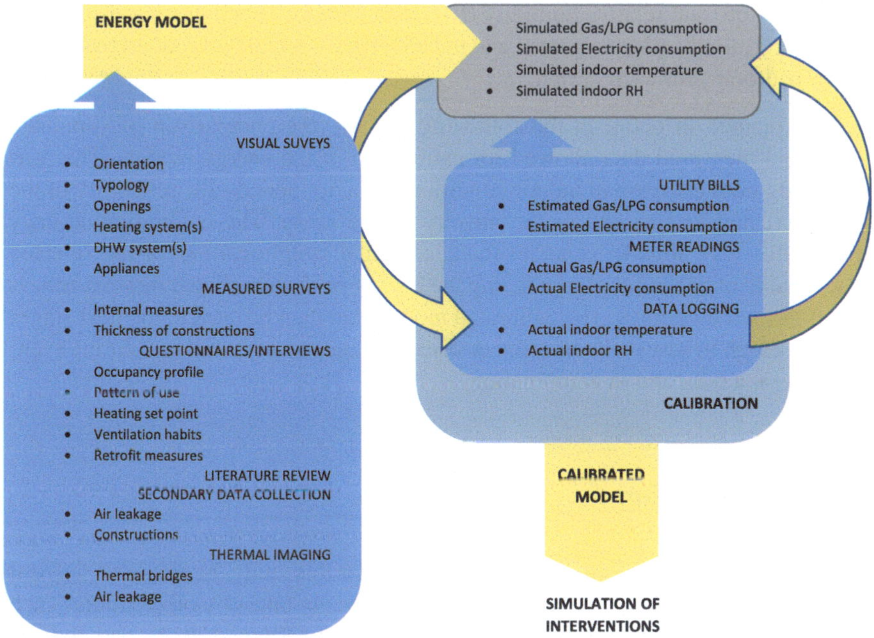

Fig. 1 The methodological framework for the generation of calibrated models

- Visual and measured surveys provided data concerning location, typology, orientation, layout and measures, openings (size, typology, age and condition), traditional features, appliances and lighting fixtures.
- Questionnaires and interviews complemented the data previously gathered and added data about occupancy profiles, pattern of use of the heating and domestic hot water systems and appliances as well as windows operation.
- Thermographic surveys enriched the understanding of the composition of the thermal envelope and aided in identifying possible thermal bridges, or areas of air leakage.

The second phase of data collection started in parallel and was carried out over a period of 1 year, aimed at gathering energy consumption and indoor conditions data for each CS, to be used in the following stage of calibration of the results of energy simulations, involving the following steps:

- Utility bills relative to the previous year's energy consumption were collected from the participants.
- Gas/LPG and electricity meter readings were performed as spot measurements over one full year.

- Temperature and relative humidity (RH) data were collected using sensors, over at least 3 months (during heated and unheated periods), in two rooms (living area and bedroom) for each period.

As explained in detail in the following section, the comparison of simulated results and measured data allowed for calibration of the energy models. This was done using both energy and indoor conditions data, to increase the reliability of the models created and to add to the novelty of the study. The calibration strategy devised this way aims to generate models, virtual simplifications of the selected CSs, capable of accurately reproducing the energy performance and thermal behaviour of the real dwellings. The calibrated models can then be used in the next stages of the research to assess the benefits or disadvantages of potential retrofit interventions, one at a time and in combination.

Models Calibration

The validation procedure specifically conceived for this study, based on the critical literature review, develops further the approach already taken by other studies such as Bertagnolio (2012), Mustafaraj, Marini, Costa, and Keane (2014), and Raftery et al. (2011). It constitutes of three subsequent stages of calibration, as follows:

- Stage one, based on input data obtained from visual and measured surveys, questionnaires, interviews, thermographic surveys, literature review and secondary data collection.
- Stage two, where monitored energy consumption data is compared to simulated results and the input data are fine-tuned accordingly.
- Stage three, where monitored temperature and RH data are compared to the simulation results from the model to complete the calibration.

A detailed description of the three subsequent stages of calibration follows:

Stage 1

At this stage, the energy models were created using the input data gathered during the first phase of data collection (see Fig. 1). The energy simulations were then run using the average weather file for Brighton, as provided by Meteo-Norm.[1]

[1] IES-VE uses site data that contain values for latitude, longitude and altitude of a wide range of sites throughout the world, taken from standard tables published by CIBSE and ASHRAE. Brighton is not included in such locations; therefore, local weather data were requested to Meteo-Norm.

Stage 2

An initial screening of the validity of the results generated by the energy simulations was carried out using the estimated annual energy consumption data, as evinced from the energy bills, in order to assess the capability of the models to predict the energy consumption of the dwellings as investigated in their status-quo, as well as energy and CO_2 emission savings in the later design stages.

Meanwhile, the energy meter readings of each CS were carried out. This allowed for a calibration to be performed using actual figures where a whole year of energy consumption data was accounted for. The simulation results were compared to actual monitored data in two main categories of building energy usage: electricity and gas (or LPG wherever applicable, which was limited to one CS). Percentage differences (PD) between simulation results and measured data were calculated for each energy consumption category over the annual period of investigation. This was done using the following equation, based on the work of Reeves, Olbina, and Issa (2012):

PD (%) = [(Simulated Results – Measured Results)/Measured Results] × 100

where positive values of the PD show that the simulation overestimated the annual energy consumption, and negative values indicate that it underestimated such consumption. Values of the PD in the range of ±10% were considered acceptable. This is in line with previous research (Ogando, Cid, & Fernández, 2017) and more challenging than the tolerances adopted by other researches (Reeves et al., 2012, following Maamari et al., 2006). The input values used in the first set of simulations for each CS were therefore fine-tuned, where needed, to calibrate the simulation outcome with the metered data and obtain results within the acceptable range.

At this stage, to increase the reliability of the models, a further calibration was performed using sub-annual monitored and simulated energy data. This calibration phase was more challenging than the one operated on annual data. In fact, the achievement of similar values for simulated output and monitored data relative to a shorter period was much more affected by changes in a multitude of behavioural factors and occupancy patterns. In the case of dwellings in use, such patterns may not be consistent, during shorter periods of investigation, with the general profiles relative to the whole year, produced according to the questionnaires and interviews with the occupants. Therefore, on the day of each meter reading, a few more questions were asked from the occupants and note was taken of any changes that may have happened in the general profiles of use generated from the first interviews, during the specific sub-period of investigation. The energy consumption figures produced by the simulations were estimated by the software, given the input data concerning envelope constructions, heating schedule (s), heating set-point(s), natural ventilation habits and appliances use pattern provided by the surveys and interviews. Such data were updated at this stage, in each sub-annual period between the meter readings, according to any special condition that the occupants were aware of.

At this stage, the applicability of ASHRAE (2002) criteria for calibration was considered as frequently adopted in previous research (Pan, Huang, & Wu, 2007;

Parker, Cropper, & Shao, 2012; Raftery et al., 2011; Yang & Becerik-Gerber, 2015; Yoon, Lee, & Claridge, 2003) when validating BES models using energy data. ASHRAE (2002) recommends the use of statistical indices, namely the Normalised Mean Bias Error[2] (NMBE) and the Coefficient of Variation of the Root Mean Square Error[3] (CV(RMSE)), applying them to hourly or monthly energy use data collected over 1 year. However, due to the constraints of this research, energy use data for a year was collected at longer intervals that spanned from 2 to 5 months. Nevertheless, it needs to be noted that the ASHRAE procedure (2002) only aims at achieving an acceptable correspondence between predicted and actual energy consumption; hence, the statistical indices proposed by ASHRAE only relate to energy in the context of the Guidelines. This validation approach, as noted by Coakley (2014) and Fabrizio and Monetti (2015), does not take into account other influential parameters, such as indoor conditions, i.e. temperature and RH patterns, at the risk of producing models that do not really correspond to the actual building or of potentially producing more than one namely calibrated solution.

This is even more true when studying traditional dwellings, whose materials and constructions are different from modern ones; hence it is challenging to reproduce their complex thermal behaviour accurately adopting the same methods utilised for modern buildings (BRE, 2014). Therefore, the strategy developed for this research aims at balancing accuracy in energy consumption outputs as well as thermal behaviour data. This was done using a more flexible approach than the one suggested by ASHRAE when considering energy data for the calibration. This stage, in fact, was aimed at achieving ±15% PD between each sub-annual metered and simulated energy consumption. Because of the challenges imposed by the implications of varied patterns of use during short periods of time, such limits were not possible to achieve for all the CSs resulting in values slightly in excess of ±15% for one or more periods of investigation in a few cases. Therefore, NMBE and CV (RMSE) were finally also calculated over 1 year on all the sub-annual periods. For the reasons explained before, these indexes were allowed more flexible limits than the ones imposed by ASHRAE for monthly data.

Therefore, the models were iteratively refined, and the outputs were compared to measured data until:

- The PDs calculated for annual energy data were within the limits of ±10%.
- The PDs calculated for sub-annual energy data were within the limits of ±15% wherever possible and/or the NMBE and CV(RMSE) calculated over 1 year of sub-annual data were, respectively, within ±10% and below 30%.

Such level of accuracy concerning energy data was considered enough at this stage to proceed with the following stage of calibration. It has to be stressed, in fact, that the main purpose of the calibration process described in this paper is to serve

[2] NMBE (%) = [Σ(aD-sD)/ΣaD] × 100 where: aD = actual data; sD = simulated data.

[3] CV(RMSE) (%) = (RMSE/μ) × 100 = {$\sqrt{}$ [Σ(aD-sD)²/n] /μ}× 100 where: aD = actual data; sD = simulated data; n = number of data ; μ = mean of actual data.

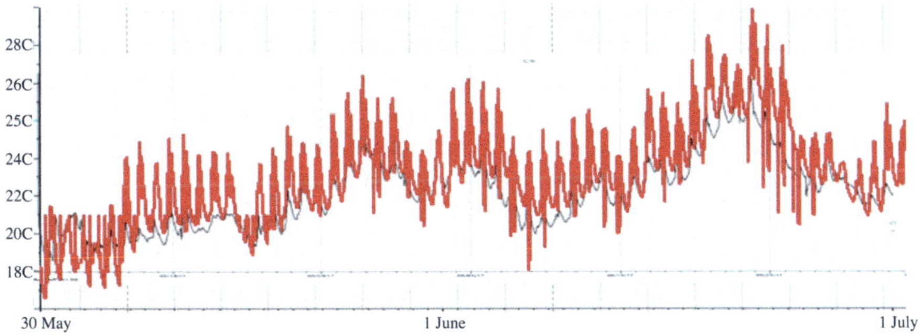

Fig. 2 CS2: Graphic analysis of simulated (in red) and monitored (in black) temperature data for the living room over 2 summer months

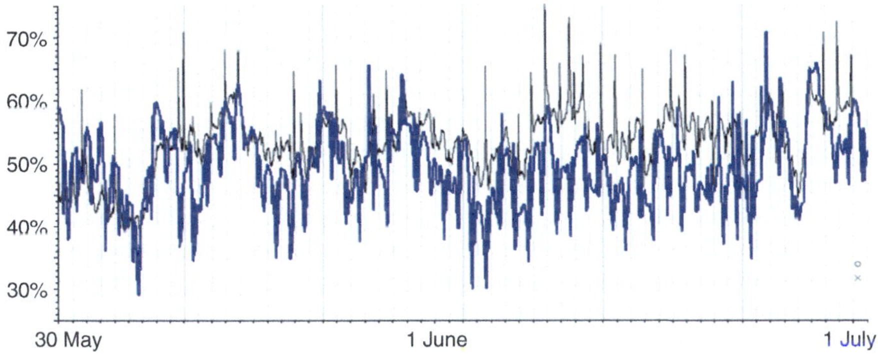

Fig. 3 CS2: Graphic analysis of simulated (in blue) and monitored (in black) RH data for the living room over 2 summer months

the final aim of this research, which is to test the relative effects of potential retrofit interventions in the CS selected, in order to propose responsive and effective combinations of retrofit measures applicable to TLDs. This is done assessing the changes in energy consumption between the base case dwellings and the variations with interventions. Therefore, whereas it is desirable to produce models capable of predicting the actual energy consumption of the dwellings investigated as accurately as possible, it is the change in energy consumption between the base case dwellings and the variations with interventions that this research aims to investigate.

Stage 3

When the calibration process was successful to this point, a further stage of calibration was performed using indoor temperature and RH data. This was done initially using graphic analysis (see Figs. 2 and 3), when the winter and summer cycles of data logging were completed. The sub-hourly temperature and RH data acquired by the sensors were compared with the ones outputted from the dynamic simulations for the same periods and for the same rooms. At this stage, the objective was to validate the thermal behaviour of the models, establishing if the graphs presented significant discrepancies or were otherwise reliable, while also aiding in the understanding of the building envelope characteristics and the behaviour of its thermal mass. Furthermore, the graphic analysis provides a straightforward visual comparison of simulated and monitored data, aiding in identifying where the most evident discrepancies between such data exist, therefore where it is most likely that errors occur. Hence, complementary to the manual iterative calibration method—used in stage 1 and 2—this further stage of calibration was conducted using visual analysis of the graphs to facilitate the decision concerning the specific periods of investigation needing further checks.

This stage of calibration was performed using the average Brighton weather file for all the data collected during 2018. For the data collected during 2017, a parallel simulation of the same models was run using the weather file generated via Weather Analytics for Brighton, relative to that specific year (as recommended by ASHRAE Guideline, 2002) acquired from IES. This further simulation was aimed at excluding the variables potentially causing discrepancies, as a result of weather, from the fine-tuning process. Having excluded differences between the weather file used in simulation and the actual weather, other factors—mainly pertaining to the pattern of use, heating system, building fabric, as indicated in the literature as the principal sources of errors in simulation—were examined more confidently to find the source of discrepancies.

When such analysis produced satisfactory results, simulated and monitored sub-hourly recorded temperature and RH data (time intervals of 10 and 30 min) were used in the final stage of calibration, comparing them according to ASHRAE (2002) statistical indices for hourly data. All the CSs were calibrated calculating NMBE and CV(RMSE) for temperature and RH data for at least two (heated and unheated) periods, of 1–2 months each, in two different rooms (living areas and bedrooms, respectively).

The iterative calibration process, illustrated in Fig. 4, was repeated for each CS, until:

- PD between simulated and measured annual energy data was within 10%.
- PDs between simulated and measured sub-annual energy data were within 15% and/or NMBE and CV(RMSE) between simulated and measured sub-annual energy data relating to a whole year were within ±10% and <30%, respectively.
- A good similarity between the graphs of simulated and monitored temperature and RH data was reached for all the periods of data collection and all the rooms monitored.
- NMBE and CV(RMSE) between simulated and monitored temperature and RH sub-hourly data achieved values within ±10% and <30%, respectively, for all the periods of data collection and all the rooms.

When all these criteria were fulfilled, the models were considered calibrated.

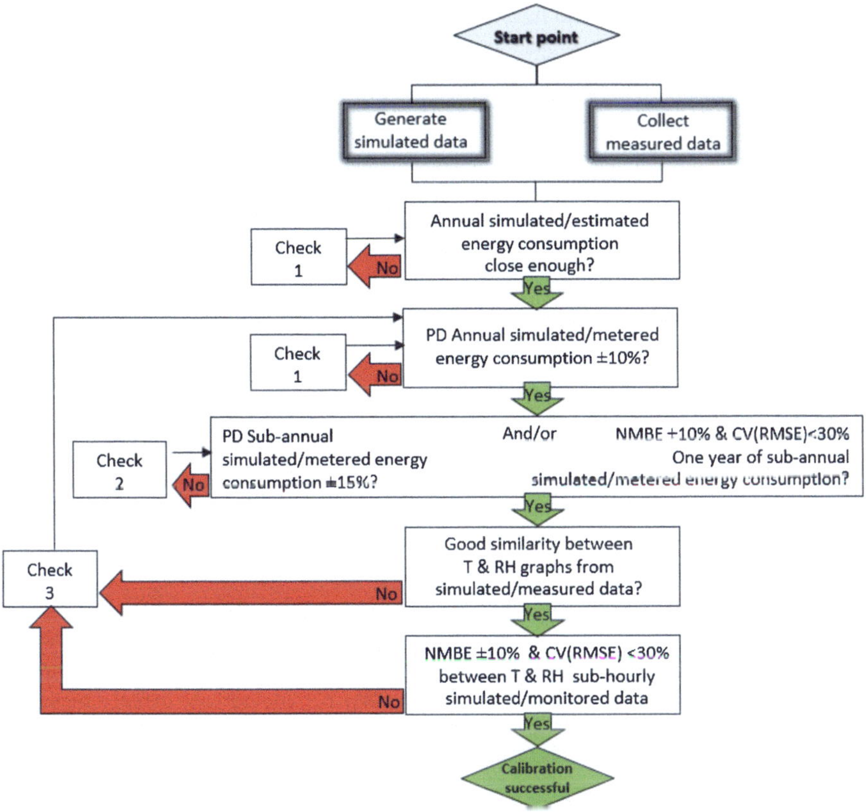

Fig. 4 The calibration process

Input Data Fine Tuning

Hierarchy of Input Data

Previous research has pointed out how the calibration process can be extremely dependent on the researcher's personal judgement of the individual relevance of the multitude of parameters that need to be inputted in the simulation software (Maile et al., 2012; Raftery et al., 2011). Therefore, to avoid such subjective approach and to improve the validity and reliability of the calibration and the study, a strategy has been applied, developing the one used in previous studies such as Gines Cooke (2018), Parker et al. (2012), and Raftery et al. (2011). First of all, a hierarchy was devised for the wide range of input data, depending on the source used to obtain them. The data sourcing strategies capable of providing on-site measured data concerning each specific CS under investigation were considered the most reliable ones in the context of this study. Therefore, to ensure consistency throughout the calibration process, input data have been sourced and checked according to the following hierarchy:

- Data from direct observation (mainly concerning envelope constructions -when visible-, openings, heating/DHW systems, appliances, lights) and measured data, recorded during the site surveys.
- Data from questionnaires and interviews with the occupants concerning occupancy profile and pattern of use (related to heating schedules and temperature set-points, lighting schedules, DHW use, natural ventilation habits and frequency of use of the equipment).
- Data from benchmark studies, codes and legislations, best practice guides, standards and guidelines, operation manuals (CIBSE, 2015; EST, 2008; IES, 2009; Wood, Bordass, & Baker, 2009) to cover areas where it was not possible to secure data from other sources.

Data Quality Checks

During the subsequent stages of calibration, three levels of data quality checks (see Fig. 4) were carried out, based on the proposed hierarchy of sources, and the literature review concerning the sources of uncertainty in calibration (including, inter-alia, Blecich, Franković, & Kristl, 2016; De Wit & Augenbroe, 2002; Gucyeter, 2018; Hoes, Hensen, Loomans, De Vries, & Bourgeois, 2009; Marini, Webb, Diamantis, & Buswell, 2014; Ogando et al., 2017; Parker et al., 2012; Reeves et al., 2012; Ryan & Sanquist, 2012).

Conclusion

The process of calibration specifically devised for this study has been carried out until the results were falling within the specified error margin. This iterative process can be described as:

- Running the simulations of the models created.
- Comparing the outputs with the monitored data.
- Identifying discrepancies to spot potential and relevant source(s) of error.
- Fine-tuning relevant parameters.
- Running the simulations of the newly modified models.

Numerous simulation runs were needed to obtain acceptable calibration levels. The performed calibration process was successful for all the CSs investigated, adding to the validity, reliability and hence generalisability of future findings of the research project and, equally importantly, to its novelty. In fact, most of the precedent studies about traditional dwellings that aspired to complement the use of BES with a calibration strategy (Ingram, 2013; Mohammadpourkarbasi, 2015; Moran, 2013) were only based on a limited number of—sometimes one—CSs. This study, by contrast, uses a multiple CSs approach to be able to increase its reach and validity.

All the models were calibrated using a full year record of energy consumption data as well as temperature and RH data. The strategy adopted is meant to ensure the validity of the created models by proving them capable of reproducing the real thermal behaviour and energy performance of the dwellings under investigation. Such strategy was deemed even more necessary in a research that utilises CS dwellings in use, where the flexibility requested for the collection of energy consumption data (that relies on the availability and ease of contact with the occupants) can be counteracted by the application of an in-depth use of long periods of indoor conditions monitoring.

It needs to be noted that the periods of data logging and meter readings are slightly different for each CS, but this is not affecting the quality of the calibration as each CS has been validated independently and individually within the period of data monitoring associated with it.

The calibration process devised in this way moves one step further from previous research that deployed both energy and indoor conditions data for calibration of BES using CSs of public buildings with heritage value (Ogando et al., 2017; Şahin et al., 2015), dwellings (Georgiou, 2015) or traditional dwellings (Flores, 2013). In fact, in this study, temperature data was monitored over a longer period of 3–6 months (depending on the actual participation of the occupants). Notably, the use of two monitoring stages allowed for considering heated and unheated periods for calibration instead of just one, which offer higher degree of accuracy and increase the validity and reliability of the findings. Furthermore, the monitored temperature data was collected at intervals shorter than 1 h (½ h and 10 min) and calibrated using graphic analysis as well as ASHRAE statistical indexes for all the data (vs. exclusive use of graphic analysis, or satisfactory PD for most of data, as done by previous researches concerning dwellings). Importantly, this research also makes use of RH data for calibration adopting graphic analysis as well as ASHRAE criteria, adding to the novelty of the study. In fact, the few precedent calibrations of BES applied to dwellings, which attempted the use of temperature as well as RH data (Bozonnet, Doya, & Allard, 2011; Drissi Lamrhari & Benhamou, 2018), adopted PD and graphic analysis or maximum deviation, on indoor conditions data only, and did not involve energy consumption data at all. The novelty of this calibration approach is even more evident when compared to what has previously been done in the studies concerning traditional dwellings in the UK (Ingram, 2013; Mohammadpourkarbasi, 2015; Moran, 2013), where the calibration was only performed on energy (electricity and/or gas) data comparing annual total or monthly individual data—simulated and measured—using graphic analysis and/or PD.

The unique way in which the calibration process was devised in this study contributes to the novelty of the study and helps to ensure that the created models are accurate virtual reproductions of the real dwellings. To date, all the models are fully calibrated and ready to be used in the following stages of research to simulate the effects of potential retrofit interventions on the selected TLDs.

References

ASHRAE. (2002). *ASHRAE Guideline 14–2002 for measurement of energy and demand savings*. Atlanta, GA: American Society of Heating, Refrigerating and Air-conditioning Engineers (ASHRAE).

Baranowski, A., & Ferdyn-Grygierek, J. (2009). Heat demand and air exchange in a multifamily building—Simulation with elements of validation. *Building Services Engineering Research and Technology, 30*, 227–240.

Bertagnolio, S. (2012). *Evidence-based model calibration for efficient building energy services*. PhD, University of Liège.

Blecich, P., Franković, M., & Kristl, Ž. (2016). Energy retrofit of the Krsan Castle: From sustainable to responsible design—A case study. *Energy and Buildings, 122*, 23–33.

Bozonnet, E., Doya, M., & Allard, F. (2011). Cool roofs impact on building thermal response: A French case study. *Energy and Buildings, 43*, 3006–3012.

BRE. (2014). *Solid walls heat losses and the potential for energy saving*.

CIBSE. (2015). *CIBSE guide A: Environmental Design*. London: Chartered Institution of Building Services Engineers.

Coakley, D. (2014). *Calibration of detailed building energy simulation models to measured data using uncertainty analysis*. Galway: NUI.

Dall'o, G., Sarto, L., Sanna, N., & Martucci, A. (2012). Comparison between predicted and actual energy performance for summer cooling in high-performance residential buildings in the Lombardy region. *Energy and Buildings, 54*, 234.

De Wit, S., & Augenbroe, G. (2002). Analysis of uncertainty in building design evaluations and its implications. *Energy and Buildings, 34*, 951–958.

Drissi Lamrhari, E.-H., & Benhamou, B. (2018). Thermal behavior and energy saving analysis of a flat with different energy efficiency measures in six climates. *Building Simulation, 11*, 1123–1144.

EST. (2008). *Measurement of domestic hot water consumption in dwellings*. Energy Saving Trust.

Fabrizio, E., & Monetti, V. (2015). Methodologies and advancements in the calibration of building energy models. *Energies, 8*, 2548–2574.

Flores, J. A. M. (2013). *The investigation of energy efficiency measures in traditional buildings in the Oporto World Heritage Site*. PhD, Oxford Brookes University.

Georgiou, G. (2015). *Assessing energy and thermal comfort of domestic buildings in the mediterranean region*. PhD, Loughborough University.

Gines Cooke, M. (2018). *Analysis of implication and value of the process of calibration of building energy models (BEMs)*. IES.

Gucyeter, B. (2018). *Calibration of a building energy performance simulation model via monitoring data*. Building performance analysis conference and SimBuild co-organized by ASHRAE and IBPSA-USA. Chicago, IL.

Heath, N., Pearson, G., Barnham, B., & Atkins, R. (2010). *Historic Scotland Technical Paper 8: Energy modelling of the Garden Bothy, Dumfries House*. Edinburgh: Historic Scotland.

Hoes, P., Hensen, J. L. M., Loomans, M. G. L. C., De Vries, B., & Bourgeois, D. (2009). User behavior in whole building simulation. *Energy and Buildings, 41*, 295–302.

Hubbard, D. (2011). Ventilation, Infiltration and Air Permeability of Traditional UK Dwellings. *Journal of Architectural Conservation, 17*, 59–74.

IES. (2009). *Historic Scotland Technical Paper 5: Energy modelling of a mid 19th century villa—Baseline performance and improvement options*. Edinburgh: Historic Scotland.

Ingram, V. (2013). *Energy performance of traditionally constructed dwellings in Scotland*. PhD, Heriot-Watt University.

Ingram, V., & Jenkins, D. (2013). *Historic Scotland Technical Paper 18: Evaluating energy modelling for traditionally constructed dwellings*. Edinburgh: Historic Scotland.

Jenkins, D. (2008). *Historic Scotland Technical Paper 4: Energy modelling in traditional scottish houses*. Edinburgh: Historic Scotland.

Maamari, F., Andersen, M., De Boer, J., Carroll, W. L., Dumortier, D., & Greenup, P. (2006). Experimental validation of simulation methods for bi-directional transmission properties at the daylighting performance level. *Energy and Buildings, 38*, 878–889.

Maile, T., Bazjanac, V., & Fischer, M. (2012). A method to compare simulated and measured data to assess building energy performance. *Building and Environment, 56*, 241.

Marini, D., Webb, L. H., Diamantis, G., & Buswell, R. A. (2014). *Exploring the impact of model calibration on estimating energy savings through better space heating control, 2014*. London: Building Simulation and Optimization Conference.

Mohammadpourkarbasi, M. (2015). *The Eco-Refurbishment of a 19th Century Terraced House*. PhD, University of Liverpool.

Moran, F. (2013). *Benchmarking the energy use of historic dwellings in Bath and the role for retrofit and LZC technologies to reduce CO2 emissions*. Doctor of Philosophy, University of Bath.

Mustafaraj, G., Marini, D., Costa, A., & Keane, M. (2014). Model calibration for building energy efficiency simulation. *Applied Energy, 130*, 72–85.

Ogando, A., Cid, N., & Fernández, M. (2017). Energy modelling and automated calibrations of ancient building simulations: A case study of a School in the Northwest of Spain. *Energies, 10*, 807.

Pan, Y., Huang, Z., & Wu, G. (2007). Calibrated building energy simulation and its application in a high-rise commercial building in Shanghai. *Energy and Buildings, 39*, 651–657.

Parker, J., Cropper, P., & Shao, L. (2012). *A calibrated whole building simulation approach to assessing retrofit options for Birmingham Airport*. Loughborough, UK: First Building Simulation and Optimization Conference.

Raftery, P., Keane, M., & O'Donnell, J. (2011). Calibrating whole building energy models: An evidence-based methodology. *Energy and Buildings, 43*, 2356–2364.

Reeves, T., Olbina, S., & Issa, R. (2012). *Validation of building energy modeling tools: Ecotect, green building studio and IES. Winter simulation conference*, IEEE, pp. 1–12.

Ryan, E. M., & Sanquist, T. F. (2012). Validation of building energy modeling tools under idealized and realistic conditions. *Energy and Buildings, 47*, 375–382.

Şahin, C. D., Arsan, Z. D., Tunçoku, S. S., Broström, T., Akkurt, G. G., Humanistisk-Samhällsvetenskapliga, V., … Konstvetenskapliga, I. (2015). A transdisciplinary approach on the energy efficient retrofitting of a historic building in the Aegean Region of Turkey. *Energy and Buildings, 96*, 128–139.

STBA. (2012). *Performance and energy efficiency of traditional buildings: Gap analysis study*. London: STBA.

Thompson, P., & Bootland, J. (2011). *Good home alliance monitoring programme 2009–11: Technical Report*. *In:* Alliance, G. H. (ed.).

Wingfield, J., Bell, M., Miles-Shenton, D., & Seavers, J. (2011). *Elm tree mews field trial Evaluation and monitoring of dwellings performance: Final technical report*. *In:* University, L. M. (ed.).

Wood, C., Bordass, B., & Baker, P. (2009). *Research into the thermal performance of traditional windows: timber sash windows*. English Heritage.

Yang, Z., & Becerik-Gerber, B. (2015). A model calibration framework for simultaneous multi-level building energy simulation. *Applied Energy, 149*, 415–431.

Yoon, J., Lee, E. J., & Claridge, D. E. (2003). Calibration procedure for energy performance simulation of a commercial building. *Journal of Solar Energy Engineering, 125*, 251–257.

Understanding Factors Influencing Overheating: The UK's First Large-Scale Domestic Passivhaus Retrofit

Dean Myers, Christopher Gorse, and David Johnston

Literature Review

Introduction

In recent years, the issue of overheating in dwellings has been gaining momentum in Europe and beyond. The impacts of global warming and climate change are affecting weather patterns around the world, leading to longer periods of hot weather in the summer and warmer winters (IPCC, 2014). In the UK, summers have become significantly hotter and dryer and winters are warmer and wetter (Lowe et al., 2018). UK climate data shows that summer extremes are becoming more common, 9 of the 10 warmest years for the UK have occurred since 2002 and all the top 10 warmest years have occurred since 1990 (Kendon, McCarthy, Jevrejeva, Matthews, & Legg, 2018). In order to reduce global CO_2 emissions, improvements are being made to the energy efficiency of our buildings, by increasing levels of thermal insulation and improving airtightness to reduce the amount of energy needed during the heating seasons (Building Regulations, 2010 approved document). While such measures are beneficial during the heating season, the improvement to the building's ability to retain heat can be problematic during periods of high solar gain, as temperatures in the building increase. Furthermore, the use of large glazed windows used to increase passive solar gain also increases the overheating risk when the solar intensity is high, especially in those buildings designed to be low energy with high levels of

D. Myers (✉) · C. Gorse · D. Johnston
Leeds Sustainability Institute, School of the Built Environment and Engineering, Leeds Beckett University, Leeds, UK
e-mail: Dean.Myers@incommunities.co.uk

L. Scott et al. (eds.), *Sustainable Ecological Engineering Design*,
https://doi.org/10.1007/978-3-030-44381-8_30

building fabric thermal performance (Passivhaus Trust, 2012). The combination of these factors affects the internal environment, causing temperatures to rise and this, together with inadequate ventilation, can lead to discomfort for the occupants and potentially serious health risks for the most vulnerable in society (Zero Carbon Hub, 2014a).

The 2003 heat wave cost the lives of some 35,000 people across Europe and in London temperatures of 27 °C were experienced for 12 successive days (Zero Carbon Hub, 2014b). While the adverse health implications due to overheating buildings should not be overlooked (House of Commons, 2018), in the UK, excess winter mortality exceeds the death rate during the summer in the UK (Office for National Statistics, 2018) meaning that insulation of homes is a priority and will help protect against extreme winter deaths (Healy, 2002; Wilkinson, Armstrong, & Landon, 2001). Nevertheless, the extreme temperatures experienced in the summer also result in excess mortality (Public Health England, 2018).

The problem faced by designers is to provide energy-efficient dwellings that consider future climate scenarios, which provide adequate levels of comfort in both summer and winter. The current estimates of heat wave patterns would make the 2003 heatwave "*a one in two year event by 2050. In 2070 those deadly conditions of 2003 will be considered an unusually cool summer.*" (John P. Holdren, 17th January 2008, The John H. Chafee Memorial Lecture). Therefore, it is probable that the risk of summertime overheating in dwellings will be exacerbated in future years, particularly as countries with temperate climates experience more regular and intense heat waves (Meehl & Tebaldi, 2004).

Therefore, the aim the research reported here is to present an initial review of the UK's first large-scale Passivhaus retrofit, with particular emphasis on the factors affecting overheating in a number of the dwellings. Quantitative data was collected from temperature, humidity and CO_2 sensors installed in seven flats. Qualitative data was collected by semi-structured interviews with the residents of the same seven flats. The intention is that the results will influence the future design of large-scale low-energy retrofitted buildings to further mitigate the effects of overheating.

The study focuses on a retrofit of two duplex maisonette blocks in Manchester, which has one of the highest areas of deprivation in the country (DCLG, 2010; MCC, 2015). The primary drivers for the project were reducing energy costs for the tenants, improving the thermal efficiency of the buildings as part of the wider Decent Homes programme (DCLG, 2006) and reducing carbon emissions in the context of the local, national and global climate change agenda.

The research reported first defines overheating, together with relevant thermal comfort standards. Evidence of overheating in dwellings is then presented, followed by an explanation of the health impacts. Overheating in low-energy buildings and Passivhaus dwellings are also discussed, before considering occupant behaviour. Finally, the research methodology is presented along with the results of the data analysis.

Overheating in Dwellings

Overheating Definition

Providing a clear definition of overheating is difficult, as there is no accepted or agreed classification that can be consistently applied (Zero Carbon Hub, 2016). The factors that influence overheating constantly change, e.g. weather and human behaviour. Thus, such variable conditions make modelling a challenge and assumptions regarding human activity are made and historic data are used to represent dynamic environmental conditions. Due to the complexity, the simplifications to models are often made; for example, the PHPP/SAP metrics are generated from steady-state assessment tools, which do not cater for the dynamic changes in occupants, building and environmental conditions (Lomas & Porritt, 2017). Furthermore, the internal environment of a building is not simply a condition that results from the fabric, services and occupants but is also affected by the surrounding physical and natural environment; for example, phenomena such as the urban heat island effect, which are influenced by density of urban infrastructure, vegetation, water and other landscape conditions, affect internal conditions of the building (Lauwaet et al., 2015). Moreover, human reactions to heat are different, with some groups particularly vulnerable to overheating (Hansen et al., 2011); research on heat-susceptibility shows that occupant needs can be grouped, but personal preferences towards heat and perceptions overheating need to be considered as well as the conditions in and around the building.

Prior research has centred on outdoor rather than indoor temperature and therefore it is has not been possible to say with any degree of certainty what temperature actually constitutes as overheating (DCLG, 2012). CIBSE (2006) proposed a definition of overheating in living areas as *"1% annual occupied hours over 28 °C"* (CIBSE, 2006). Additionally, the Building Regulations provide no guidance on a maximum internal temperature, although CIBSE recommends 25 °C as "warm", 28 °C as "hot" and 35 °C as the temperature at which there is a serious risk to health (Gething, 2010).

Furthermore, Public Health England (PHE, 2018) advise that an external temperature of 30 °C is the threshold for "*alert and readiness*" in their five-stage heatwave plan. DCLG (2012) state that in the UK there are around 2000 heat-related deaths in a typical UK year and predictions for future climate scenarios suggest that this could be as high as 5000 per year by 2080. In TM52 (2013) a CIBSE taskforce recognised that a new approach to assessing overheating was necessary. Utilising the adaptive approach, this resulted in TM59 (2017) which provides a design methodology for the assessment of overheating risk in homes.

In the document "Next Steps in defining Overheating", The Zero Carbon Hub (2016) proposed to seek the guidance of professionals from throughout the construction sector to provide a definition of overheating and consider the future overheating potential of buildings. However, it is worth noting that the research will be aimed at new rather than existing dwellings, which still account for a large portion of the UK housing stock.

Overheating Standards

The following table shows the most commonly used current standards and guidance for UK buildings. However, the Passivhaus Planning Package (PHPP) thermal comfort parameters not contained in the table is over 25 °C for 10% of the year. The CIBSE Guide A (2006) standard prior to the current standard contained in the table for living rooms was "*1% annual occupied hours over 28 °C*" and for bedrooms, "*1% annual occupied hours over 26 °C*" (CIBSE, 2006, p. 12) (Table 1).

Table 1 Overheating standards

(a) DfES and HH5RS overheating standards

Standard	Building type	Peak summer temp.	Durations of time	Permitted exceedance
DfES BB87	Schools	28 °C	Occupied hours/ year	80 h
Housing Health & Safety Rating System (HHSRS)	Dwellings	25 °C	Occupied hours/ year (inferred)	Unspecified

(b) CSSSE overheating standards

Standard	Building type	Acceptable range	PMV[a]	Max. daily temp. summer	Duration of time	Occupied hours exceeding θ_{max}	Assessment of daily overheating severity
CIBSE Guide A (section 1.5.3.2) based upon BS EN 15251 (Category II)	Dwelling	±3 K	±0.5	26 °C	Occupied hours/ year	<3% when ΔT >1 K between May and Sept.	Weighted exceedance <6
	Offices	±3 K	±0.5	26 °C			
	Retail	±3 K	±0.5	25 °C			
	Schools	±3 K	±0.5	25.5 °C			

For more information on these standards refer to CIBSE Guide A
[a]Predicted mean vote

(c) SAP overheating standards

Standard	Building type	Calculated peak temperature	Evaluation
SAP[a] Appendix P (Table P2)	Dwellings	<20.5 °C	Not significant
		≥20.5 °C & <22 °C	Sight
		≥22 °C & <23.5 °C	Medium
		≥23.5 °C	High

[a]It should be noted that the Standard Assessment Procedure (SAP) is a regulatory tool, not a design tool
Source: Passivhaus Trust (2012))

Evidence of Overheating in Dwellings

Until recently, the majority of research has concentrated on outdoor temperature and as of 2012 there was limited data regarding measured indoor temperatures. However, there is now a growing body of literature addressing the issue of overheating in dwellings (DCLG, 2012; NHBC, 2012; Beizze, Lomas, & Firth, 2013; Liu & Coley, 2015; Saleem Gul et al., 2015; Tabatabaei Sameni, Gatrell, Montazami, & Ahmaed, 2015). The Good Homes Alliance (2014) (GHA) report identified that 30% of cases of overheating was in new-build flats and care homes constructed after the year 2000. Contributory factors were found to include insufficient ventilation and large glazed areas. Senior Policy Advisor James Hulme from the House Builders Association claimed that the high levels of airtightness specified by Part L of the Building Regulations were causing new-build properties to overheat (Building.co.uk, 2014). Ascione, Böttcher, Kaltenbrunner, and Vanoli (2015) studied a new-build multi-storey building in Berlin and found that overheating events had been recorded from excess solar gain from large windows. Using the EPBD Recast approach and BS EN 15251, they proposed white external curtains as a low-cost intervention to mitigate overheating.

Health and Well-Being: The Impacts of Overheating in Dwellings

Overheating in dwellings can cause serious health problems for building occupants (Vardoulakis et al., 2015), and a growing population coupled with urbanisation and a loss of green space will require us to consider future mitigation strategies (Santamouris & Kolokotsa, 2015). Heat illness places a significant drain on NHS resources, and research has shown that Doctor and Accident & Emergency appointments rose for the most vulnerable in society in one particular UK emergency department during the mild heat wave of 2013 (Smith et al., 2016). Additionally, overheating events are particularly more dangerous for older people who spend most of their time indoors. Van Loenhout et al. (2016) found that the link between heat related health problems and internal temperature is most prevalent in the living room and bedroom. They concluded by recommending that this should be considered when investigating how to mitigate heat illness in older people.

Overheating in Low-Energy Buildings (LEB's)

Low-energy buildings play an important role in reducing energy demand and CO_2 emissions. However, there is an increasing amount of literature that shows that low-energy homes are prone to overheating events, which will become much worse in future climate scenarios (Rodrigues et al., 2012), as global mean temperatures rise in response to climate change (Jentsch et al., 2014). Furthermore, McLeod et al. (2013) carried out an investigation into overheating risk in homes built to the Passivhaus standard. Simulation modelling was carried out using weather data and the findings suggested that by 2015, internal temperatures could go beyond 25 °C for 5–10% annually.

There is limited data available for new-build flats built to the Passivhaus standard in the UK and none for multi-residential large-scale Passivhaus retrofit. However, Tabatabaei Sameni et al. (2015) carried out an investigation on 25 social housing flats built to the PH standard over three cooling seasons. The properties were assessed using PH criteria and it was found that two thirds of the flats were at risk of overheating. Significantly, they concluded that occupant behaviour was the most significant factor in increasing or decreasing overheating risk. Furthermore, they found occurrences of overheating when applying both adaptive and a static 25 °C exceedance method. They also found a difference applying the threshold for vulnerable and non-vulnerable occupants. They concluded that "the *results not only have implications for the evaluation of overheating risk but also for the way in which social landlords place tenants of differing vulnerabilities*" (Tabatabaei Sameni et al., 2015, p. 222). Building geometry can also increase or decrease cooling demand. Lavafpour and Sharples (2014) suggest that the large, south facing glazed areas often employed in Passivhaus design to utilise solar gain can lead to overheating. They propose that "*inclined south facing façade geometries*" could mitigate overheating, although this would be difficult to achieve with retrofit. Morten (2015) proposed modelling the Passivhaus standards with a selection of future climate scenarios explaining that this may enable us to better understand what will happen, which measures could mitigate overheating and which measures are the most appropriate. It is worth noting that Morton suggests that care must be taken with multi-residential buildings, as individual results may not be representative of the building as a whole.

Foster et al. (2016) collected temperature, humidity and CO_2 data from five Scottish Passivhaus dwellings and found problems with overheating. They suggest that further research is required regarding the internal environment, occupant health and well-being to fill a knowledge gap in this area. Material science could also play an important role in mitigating overheating. Mavrogianni, Wilkinson, Davies, Biddulph, and Oikonomou (2011) suggested that knowledge of how insulation performs post-retrofit is necessary to distinguish which buildings are at risk of overheating.

Furthermore, summertime overheating risk tends to be greatest in dense urban areas, rather than rural environments (Gartland, 2012; Mavrogianni et al., 2011; Oikonomou et al., 2012; Taylor et al., 2014). The development of urban and green areas leads to higher temperatures and the heat island effect. Therefore, cities need to take a strategic approach to counteract future climate scenarios by using the latest in materials technology, e.g. phase change materials to mitigate the impact of solar radiation (Santamouris & Kolokotsa, 2015).

Occupant Behaviour: The Unintended Consequences

Notwithstanding the variety of factors which can influence overheating, such as weather and climate, building orientation, insulation levels (Mavrogianni et al., 2011), and inadequate ventilation (Taylor et al., 2014; Abdul Hamid et al., 2020), little research has been carried out regarding the relationship between human behaviour and overheating. Morgan et al. (2016), found that, when investigating overheating in Passive House dwellings of identical form, construction and layout in Scotland, occupant behaviour appeared to be the biggest variable in determining the likelihood of overheating. This was a result of both increased IHG from high users of consumer electronics and lower than anticipated levels of ventilation.

In the document "*Investigation into Overheating in Homes Analysis of Gaps and Recommendations*" DCLG (2012) state that there are five main gaps and recommendations in research as follows:

1. *"A good understanding of the relationship between health and indoor temperatures.*
2. *A good understanding of the impact of behaviour on this relationship and the interaction with building characteristics. This should include a better understanding of how much impact behaviour has in its different forms has on overheating and how it can be influenced.*
3. *Validated models for individual dwellings to understand how buildings modify external environmental conditions for a range of occupancy patterns and behaviours.*
4. *Validated dwelling models to understand the variation of internal conditions—impact of orientation, location within a building, local factors & regional climatic factors.*
5. *Validated dwelling model to understand the impact of a range of interventions to reduce overheating risks—technical & behavioural*". (DCLG, 2012, p. 7)

They go further and suggest that more evidence is needed with regard to human behaviour in dwellings and the role of building occupants is pivotal in determining the conditions for overheating to exist. Additionally, this has previously been modelled by assuming what activities occupants perform, rather than what the actual "in use" conditions are in reality. In the USA, Phillips & Levin, 2015 support this and raised concerns about overheating in low-energy homes in the California area and state that "*the major drivers and research gaps were related to human behaviour, pollutant sources and integrated design*" (Phillips & Levin, 2015, p. 80). In temperate climate zones, reliance is placed on the occupants to cool Passivhaus dwellings using natural ventilation strategies, such as opening windows and night-time purging. It is therefore essential that occupants understand these strategies. However, there may be reasons why occupants are reluctant to use these strategies, such as noise, pollution or for security reasons. Palmer, Godoy-Shimizu, Tillson, and Mawditt (2016) reported that a Passivhaus home on the Dormont Park estate exceeded 28 °C for 9% of the summer, and exceeded 25 °C for around a fifth of the summer, with the likely cause being that the windows were left closed for much of the time.

Thermal Comfort

Introduction

In order to maintain a stable "core" temperature of 37 °C, the human body must regulate itself by moving blood to and from the surface of the skin. This is done by the physiological processes of evaporation of sweat when we are too hot and shivering when we are too cold (Nicol et al., 2012). Following the invention of modern heating systems and air conditioning, people became aware that we could now control the internal environment to satisfy our requirements (Fabbri, 2015). It was the Danish Physiologist Per Ole Fangers work in the 1960s that provided the basis for what we now know as the adaptive thermal comfort standard (Fanger, 1973). This literature review will introduce the subject of thermal comfort as being vital to understanding overheating risk in low-energy buildings and, in particular, Passivhaus.

Defining Thermal Comfort

The American Society of Heating, Refrigeration and Air-Conditioning Engineers (ASHRAE) provide a definition of thermal comfort as follows:

> *"… that condition of mind which expresses satisfaction with the thermal environment and is assessed by subjective evaluation."* (ASHRAE, 2017, p. 2)

Thermal comfort can be divided into three categories, outdoor, indoor and semi-indoor. This research will focus on indoor thermal comfort only; outdoor and semi outdoor will not be discussed in detail.

Thermal Comfort Standards

Modern indoor comfort standards can be traced back to the 1970s and these were widely accepted as the industry standard until the turn of the century. The industry began to realise that the previous standards were not appropriate and proposed a revised approach called BS EN 15251 (BSI, 2007) using the "adaptive thermal comfort" model. The adaptive approach recognised the direct correlation between the indoor and outdoor temperature. Within the adaptive temperature range, the building occupants adapt to the internal environment by removing clothing and opening windows, etc. (BSI, 2007). The result being that thermal comfort is maintained.

The adaptive model was adopted by ASHRAE and Europe, and is based on a "linear relationship" of the internal and prevailing external temperatures. Furthermore, the adaptive model assumes that the range of acceptable temperatures are larger in naturally ventilated (NV) buildings than air conditioned (AC) buildings (Vellei et al., 2017). In recent years, the UK and European standards for assessing thermal comfort are BS EN 15251 (BSI, 2007) and TM59 (CIBSE, 2017) which are adaptive thermal comfort models and CIBSE Environmental Design Guide A—7th revision (CIBSE, 2006) and PHPP (PHI, 2015) which are static aggregate exceedance methods.

Following the introduction of new European standards aimed at improving the energy efficiency of buildings, there has been an increased focus on assessing thermal comfort in buildings (Antoniadou & Papadopoulous, 2017). Other thermal comfort models are proposed such as a Bayesian thermal comfort model (Wong et al., 2014), a 13-value comfort scale (Buratti et al., 2016) and Buyak et al. (2017) proposed a model that considers exergy as part of the model with exergy being the energy available when a system is in balance with its environment. In terms of thermal comfort, exergy and energy can be used to analyse and compare the qualities of various energy types in order to ensure that the internal environment is sufficient (Tsatsaronis, 2007).

Others have suggested using infrared thermography of the human face for predicting thermal comfort (Ghahramni et al., 2016). Choi and Yeom (2017) propose using advanced sensing technology with data from standard comfort surveys to estimate thermal comfort. However, it could be argued that this will merely support models rather than replace them. Vilches et al. (2017) suggest that thermal comfort and monthly income should be used as the main driver for selecting the most suitable energy-efficient retrofit methodology. However, this would be difficult to achieve in the social housing sector. This is due to the turnover of residents and one retrofit measure may be suitable for one person or family and not another.

Thermal Experience: The Nuances of Thermal Comfort

Current research is also investigating the nuances of thermal comfort. Mishra et al. (2016) suggest that variations in internal temperature can be agreeable to building occupants, while at the same time improving comfort perception and reduced energy use. They go further to discuss personalised comfort systems and a holistic approach to controlling thermal comfort with building design by choice of internal décor, fixtures and finishes. Liu et al. (2017) state that research into thermal adaptation in transitional seasons is inadequate. They suggest that thermal experience influences thermal comfort through behavioural, physiological and psychological paths. This supports one of the aims of this research, which is to study behavioural aspects of overheating risk. Geng et al. (2017), on the other hand, discovered that variations in thermal comfort could affect an individual's perception of other factors associated with Internal Environmental Quality (IEQ), such as air, lighting and ventilation.

In summary, this literature review has introduced the topic of overheating and thermal comfort, which is vital to understanding the cause and impacts of overheating in dwellings. A definition of thermal comfort has been provided, along with current international and domestic standards and how they have developed. Current research, which attempts to improve and modify standards, has also been explained. Finally, the review closed with thermal experience and how the nuances of how we as individuals perceive thermal comfort can assist in reducing energy use, while at the same time improving thermal comfort.

Research Methodology

The research methodology being used on this project is "Grounded Theory" (Glaser, Strauss, 1967). Birks & Mills (2012) consider Grounded Theory Method (GTM) *"one of the most popular research designs in the world"* (Birks & Mills, 2012, p. 1). It involves a process of generating theory from the research, as opposed to testing existing theories. The researcher collects data and categorises it into groups, before examining it to see if it fits into any of the groups (Davies & Hughes, 2014). The emerging theory is therefore firmly "grounded" in the data with a thorough examination of the data producing the optimum results.

The method for collecting the data is a mixed methods approach. This is a relatively recent method in the social and human sciences spheres. It involves gathering quantitative and qualitative data, with the consensus being that this approach will allow us to understand the research area more comprehensively rather than using either method on its own (Cresswell, 2014). The intended outcome is that this will provide sufficient data saturation from all available data sources. In their seminal work, "On the Discovery of Grounded Theory: Strategies for Qualitative Research" (Glaser & Strauss, 1967), they suggest that both quantitative and qualitative data can be used together for mixed methods research. However, they go further to explain that they should not be used together to test each method, but should supplement one another to provide a joint verification.

Due to research constraints, the installation of the monitoring equipment was restricted to seven flats, which were purposely selected for the research. Four flats were selected from the larger block of 20 units and 3 flats from the smaller block of 12 units. The participants were selected due to their demographic, with the view being that a wider spread of results would be generated. The sample contained a mixture of older, disabled, middle aged, younger people with some participants being single, co-habiting or co-habiting with children and pets.

Qualitative data was collected via two semi-structured interviews with the occupant(s) of each property. The first round of interviews was conducted in October 2016. The semi-structured interview gives the interviewer the flexibility to "probe" the beliefs of the participant in order to develop their response and explore new areas, which may not have been previously considered by the researcher (Gray, 2014). Brinkmann & Kvale (2015) explain that *"the qualitative research interview attempts to understand the world from the subjects point of view"* (Brinkmann & Kvale, 2015, p. 3). The second stage of qualitative data gathering will utilise ethnography. Ethnography is a technique which involves the researcher placing him or herself in the research setting, facilitating direct observation. The researcher establishes a relationship with the participants in their natural environment, through interaction, in order to describe their actions and understand the meaning of those actions (Silverman, 2011).

Three types of quantitative data are used in this study. The first was collected from monitoring instruments installed in each property over a 14-month period. Temperature and humidity sensors were installed in the lounge, main bedroom and hallway and a CO_2 sensor was also placed in the lounge. The sensors were positioned away from direct sunlight, which would affect the readings.

Research Results

This section will present an overview of the first and second round of qualitative and quantitative data analysis to date. The interviews were digitally recorded and transcribed with the participants anonymised and the data stored securely in line with the University code of ethics. The data from the first interviews was transposed on to an Excel spreadsheet and divided into categories.

The results from the first planned semi structured interviews indicated that summer overheating was a concern for most of the occupants. Line by line in vivo coding generated an emerging theme showing that five out of the seven participants expressed some concern regarding temperature. Comments included, ***"it does get too hot" (flat C)*** and ***"it can get too warm" (flat A)***.

The housing provider had given a great deal of consideration to training and education. Each resident attended practical training sessions in a show flat and were shown how to use the technology. A useful home user guide was also provided for the residents to refer to following occupation of the property and all expressed some recollection of the training. However, opinion was divided, with four of the participants providing positive remarks like, ***"I was given a booklet and they told us about a vent (MVHR) and how it works and things like that" (flat A)*** and ***"we've still got the book there" (flat D)***.

The two sub-categories to emerge from data regarding control/behaviour were "windows and doors" and the "MVHR system". Although both are sub-categories individually, there is a crossover between both sets of data. The consensus was that subjects prefer to open the windows or doors and that it is a behaviour pattern they have always performed. It can be reactive with comments like ***"if we're too warm…. we can open a window or door" (flat D)*** and ***"if it gets too warm, we just open the back door and the front door" (flat F)***. It can also be a pro-active behaviour with some participants preferring to open windows and doors as a programmed behaviour with ***"I like my doors open most of the time" (flat A)*** and ***"it's something I've done in every single house" (flat E).***

The most significant outcome in terms of health and well-being is that three of participating properties explained that asthma or respiratory infections had completely disappeared. One individual stated ***"I used to have asthma very, very badly when I lived here (pre-retrofit)" (flat E)***. Two participants from the seven properties expressed no change in their general health. However, they both admitted to existing health conditions.

Three of the participants acknowledged that appliances emit heat. They also realised that appliance use contributes to increased internal temperatures. Flat A explained that **"it (amplifier) gets warmer"** and **"that's another thing about heat, you know, I've got the TV amp, I've got the two televisions, you've got your router, I've got my sky box, all these things"**. Another participant displayed concern about the heat stating, **"if we're at home on the weekends, it's usually washing machine, dryer, the cooker. Then the heat spikes!" (Flat G)**.

Results from the qualitative analysis showed that overheating is an issue in this particular large-scale Passivhaus retrofit. Therefore, the quantitative data was analysed to investigate if internal temperatures are high enough to constitute overheating. Initial examination of the quantitative temperature data supports qualitative data but not to the extent that one would expect. The preliminary data analysis focussed on the week commencing 18th of July 2016 with 19th of July 2016, being the hottest day of the year and the highest outdoor air temperature was 30 °C (rp5.co.uk, 2018). Four out of the seven flats showed internal temperatures of approximately 27 °C. The other three flats were around 26 °C and these flats were either unoccupied in the day or had the windows open. These temperatures are not what one would expect when discussing overheating in the context of Passivhaus but are important in the context of future climate scenarios and when considering night-time temperatures. The methodology involved filtering the raw data for the main lounge in each property and obtaining the external temperature for the period of 18th of July to 24th July 2016. Two sources of weather data were used to verify the temperatures from metoffice.co.uk (2018) and www.rp5.co.uk (2018). The following graph displays the temperatures (Fig. 1):

An overheating analysis was then carried out for the period when any property was outside of the PHPP parameters. The temperature data was assessed for the whole year and two out of the seven participating properties showed overheating outside of the PHPP parameters of over 25 °C for more than 10% of the year. The methodology for arriving at this figure involved filtering the raw temperature data in Excel. The results were that, all flats were close to the overheating parameters with only two of the flats falling outside of PHPP parameters. The results were surprising with the hallway of flat A over 25 °C for 61 days and the lounge of flat C displayed signs of overheating for most of the year.

The second round of qualitative data gathering involved producing an interview guide, which focussed on the three main themes of thermal comfort, control/behaviour and communication. The research takes an ethnographic approach to study the

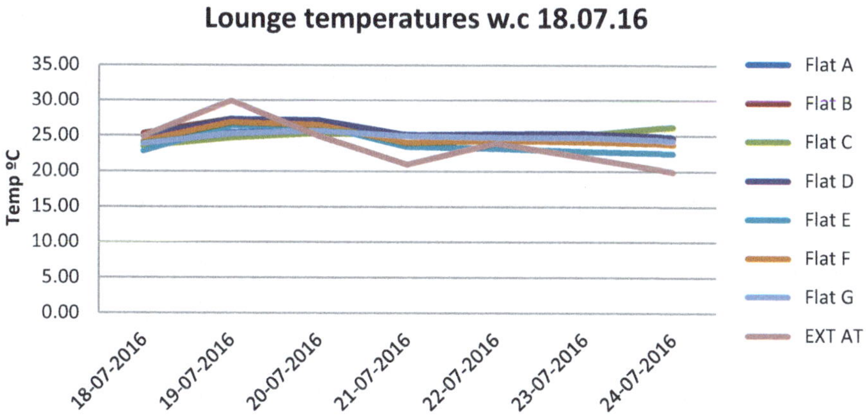

Fig. 1 Lounge temperatures for participating properties including hottest day of the year

participants experiencing high summer temperatures. Due to the high summer temperatures and subsequent heatwave of 2018, the researcher interviewed and observed the participants in their natural environment in late July and early August 2018. The period of June to July 2018 experienced temperatures regularly above 30 °C during a 15-day period with some weather stations recording 50 dry days into the latter part of July (Met Office, 2018). The three main categories to emerge from the interviews are discomfort from heat, window opening behaviour and user experience of MVHR and internal air quality (AQ).

A QUEST temperature sensor was set up in the lounge of each property 1 h prior to the interview. The participants were asked to close all of their windows and entrance doors for the hour preceding the interview. They were advised that if it became too uncomfortable, they could open the windows and doors. The interview was then carried out. The preliminary findings from the data are that lounge temperatures in three flats actually reduced by up to 4 °C with the windows closed. The results were spread across the blocks with two flats from the smaller block and one flat from the larger block. Lounge temperatures in the remaining three flats increased with the windows closed. The occupant of Flat A refused to close the windows and doors due to discomfort. It is worth noting that flat A is an older disabled person, who is sedentary for a large portion of the day and is a high user of heat-emitting appliances.

Discussion

This paper has introduced factors influencing overheating risk in the UK's first large-scale domestic Passivhaus retrofit. Seven flats and the occupants have been studied and two flats have shown evidence of overheating outside of the PHPP parameters. Furthermore, window opening behaviour, appliance ownership, MVHR design and use, building orientation and mis-communication have been proposed as indicators of overheating risk in periods of high day-time summer temperatures.

Fletcher et al. (2017) and Bundle et al. (2018) support this research that older, more vulnerable people are at risk from the impacts of overheating in well-insulated, airtight buildings including Passivhaus. Furthermore, Lomas and Porritt (2017) and Baborska-Naroźny et al. (2017) support that understanding human behaviour and how people learn and adapt to incidents of extreme heat is vital to understanding overheating in dwellings. MVHR design, maintenance and user behaviour adds another layer of complexity to the issue. This research has shown that temperatures can reduce with windows closed and a properly designed and maintained MVHR system is important to consider in the context of overheating in low-energy dwellings. Recent research has explained that particulate matter can form on the heat exchanger coil within the MVHR units (Abdul Hamid et al., 2020). Additionally, Taylor et al. (2014) explained that window opening behaviour can affect the level of outdoor pollutants entering the property. This could also affect the performance of the MVHR system and the internal air quality. Therefore, communication and education to residents are extremely important considerations for communicating advice about living in Passivhaus and low-energy buildings in general.

Conclusion

There is no UK monitoring data for large-scale Passivhaus retrofit to compare the findings in this paper. The only similar building, which is known to this research, is Wilmcote House in Portsmouth, which has yet to achieve EnerPHit status. However, Southampton University have carried out post-occupancy monitoring of Wilmcote House, although it has not been possible to obtain any data. The results of this research certainly present a new perspective given that the buildings are the first fully certified EnerPHit in the UK. The results from this study indicate that large-scale Passivhaus developments are prone to overheating and that temperatures appear to reduce with closed windows in some flats in periods of extreme hot weather. However, further research will investigate how and why this occurs. PHPP provides design criteria, which requires the dwelling to achieve internal temperatures of 25 °C for less than 10% of the year. This can be spread out over a long period or it could be condensed into a shorter period. Short periods of extreme heat could lead to serious health impacts for the most vulnerable in society residing in low-energy buildings. Although there are many factors, which can influence overheating, this research will further investigate how human behaviour affects overheating in large-scale Passivhaus retrofits.

References

Abdul Hamid, A., Johansson, D., & Lempart, M. (2020). Determining the impact of air-side cleaning for heat exchangers in ventilation systems. *Building Services Engineering Research and Technology, 41*(1), 46–59.

American Society of Heating Refrigeration Air Conditioning Engineers (ASHRAE). (2017). *Thermal environmental conditions for human occupancy*. Atlanta, GA: American Society of Heating and Refrigeration Engineers.

Antoniadou, P., & Papadopoulous, A. (2017). Occupants thermal comfort: State of the art and the prospects of personalized assessment in office buildings. *Energy and Buildings, 153*, 136–149.

Ascione, F., Böttcher, O., Kaltenbrunner, R., & Vanoli, G. P. (2015). Summer overheating in a new multi-story building in Berlin: Numerical study for improving the indoor microclimate. *Energy Procedia, 75*, 1305–1314.

Baborska-Narożnya, M., Stevenson, F., & Grudzińskac, M. (2017). Overheating in retrofitted flats:occupant practices, learning and interventions. *Building and Environment, 45*(1–2), pp. 40-59, Taylor & Francis Group.

Bennet, M. (2018). Excess winter mortality in England and Wales: 2017 to 2018 (provisional) and 2016 to 2017 (final). Office for National Statistics, London.

Beizze, A., Lomas, K. J., & Firth, S. K. (2013). National survey of summertime temperatures and overheating risk in English homes. *Building and Environment, 65*, 1–17.

Birks, M., & Mills, J. (2012). *Grounded theory: A practical guide*. London: Sage.

Brinkmann, S., & Kvale, S. (2015). *Interviews: Learning the craft of qualitative research interviewing*. London: Sage.

British Standards Institute. (2007). *BS EN 15251: 2007 indoor environmental input parameters for design and assessment of energy performance of buildings addressing indoor air quality, thermal environment, lighting and acoustics*. London: BSI.

Building Regulations. (2010). *SI. NBS*, online version.

Bundle, N., O'Connell, E., O'Connor, N., & Bone, A. (2018). A public health needs assessment for domestic indoor overheating. *Public Health, 161*, 147–153.

Buratti, C., Palladino, P., & Ricciardi, P. (2016). Application of a new 13 value comfort scale to moderate environments. *Applied Energy, 180*, 859–866.

Buyak, N. A., Deshko, V. I., & Sukhodub, I. O. (2017). Buildings energy use and human thermal comfort according to energy and exergy approach. *Energy and Buildings, 146*, 172–181.

Chartered Institution of Building Services Engineers. (2006). *Environmental design CIBSE guide A*. London: CIBSE.

Chartered Institution of Building Services Engineers. (2013). *TM52: The limits of thermal comfort: Avoiding overheating in European buildings*. London: CIBSE.

Chartered Institution of Building Services Engineers. (2017). *TM59: Design methodology for the assessment of overheating risk in homes*. London: CIBSE.

Choi, J. H., & Yeom, Y. (2017). Study of data driven thermal sensation prediction model as a function of local body skin temperatures in a built environment. *Building and Environment, 121*, 130–147.

Cresswell, J. W. (2014). *Research Design qualitative quantitative, and mixed methods approaches*. London: Sage.

Davies, M., & Huges, N. (2014). *Doing a successful research project*. London: Palgrave Macmillan.

Department for Communities and Local Government. (2006). *A decent home: Definition and guidance for implementation*. Wetherby: DCLG Publications.

Department for Communities and Local Government. (2010). *The English indices of deprivation*. London: Department for Communities and Local Government.

Department for Communities and Local Government. (2012). *Investigation into overheating in homes analysis of gaps and recommendations. Her Majesty's Stationery Office Environmental Design CIBSE Guide A*. London: CIBSE.

Fabbri, K. (2015). A brief history of thermal comfort: From effective temperature to adaptive thermal comfort. In: *Indoor thermal comfort perception*. Springer, Chambers

Fanger, P. O. (1973). Assessment of man's thermal comfort in practice. *British Journal of Industrial Medicine, 30*, 313–324.

Foster, J., et al. (2016). Scottish passive house: Insights into environmental conditions in monitored passive houses. *Sustainability, 8*, 412.

Fletcher, M. J., Johnston, D., Glew, G. W., & Parker, J. M. (2017). An empirical evaluation of temporal overheating. *Building and Environment, 121*, 106–118.

Gartland, L. (2012). Heat Islands: Understanding and Mitigating Heat in Urban Areas. Earthscan, London.

Gething, B. (2010). Design for future climate: opportunities for adaptation in the built environment. Technology Strategy Board, Swindon.

Geng, Y., Wenjie, J., Lin, B., & Zhu. (2017). The impact of thermal environment on occupant IEQ perception and productivity. *Building and Environment, 121*, 158–133.

Ghahramni, A., Castro, G., Becerik-Gerber, B., & Yu, Y. (2016). Infrared thermography of human face for monitoring thermoregulation performance and estimating personal thermal comfort. *Building and Environment, 109*, 1–11.

Glaser, B., & Strauss, A. (1967). *The discovery of grounded theory*. New York: de Gruyter.

Gray, D. E. (2014). *Doing research in the real world* (3rd ed.). London: Sage Publications.

Hansen, A., Bi, P., Nitschke, M., Pisaniello, D., Newbury, J., & Kitson, A. (2011). Perceptions of heat-susceptibility in older persons: Barriers to adaption. *International Journal of Environmental Research and Public Health, 8*(12), 4214–4728.

Healy, J. D. (2002). Excess winter mortality in Europe: A cross country analysis identifying key risk factor. *Journal of Epidemiology & Community Health, 57*(10), 784–789. Retrieved from https://www.jrf.org.uk/report/cold-comfort-social-and-environmental-determinants-excess-winter-deaths-england-1986-1996

Holdren, J. P. (2008). *Meeting the climate change challenge. Eighth Annual Joan H. Chafee Memorial Lecture on Science and the Environment*. Washington: National Council for Science and the Environment.

House of Commons. (2018). *Heatwaves: Adapting to climate change, ninth report of session, 2017–2019*. House of Commons Environmental Audit Committee, HC 826. Retrieved from https://publications.parliament.uk/pa/cm201719/cmselect/cmenvaud/826/826.pdf.

IPCC. (2014). Climate change 2014: Synthesis report. In R. K. Pachauri & L. A. Meyer (Eds.), *Contribution of working groups I, II and III to the fifth assessment report of the intergovernmental panel on climate change. Core Writing Team*. Geneva: IPCC.

Jentsch, M. F. et al. (2014) Limitations of the CIBSE design summer year approach for delivering representative near-extreme summer weather conditions. *Building Services Engineering Research & Technology, 35*(2), 155–169.

Jones, R. V., et al. (2016). *Measured indoor temperatures thermal comfort and overheating risk: Post occupancy evaluation of low energy houses in the UK*.

Kendon, M., McCarthy, M., Jevrejeva, S., Matthews, A., & Legg, T. (2018). State of the UK climate 2017. *International Journal of Climatology, 38*(S2), 1–35.

Lavafpour, Y., & Sharples, S. (2014). Impact of the envelope geometry on cooling demand in very airtight UK dwellings under current and future weather projections. *Energy Procedia, 62*, 421–430.

Lauwaet, D., Hooyberghs, H., Maiheu, B., Lefebvre, W., Driesen, G., Van Looy, S., De Ridder, K. (2015). Detailed urban heat Island projections for cities worldwide: dynamical downscaling CMIP5 global climate models. *Climate, 3*, 391–415.

Liu, C., & Coley, D. (2015). Overheating risk of UK dwellings under a changing climate. *Energy Procedia, 78*, 2796–2801.

Liu, L., Wu, Y., Baizan, L., Cheng, Y., & Yao, R. (2017). Seasonal variation of thermal sensations in residential buildings in the hot summer and cold winter zone of China. *Energy and Buildings, 140*, 9–18.

Lomas, K. J., & Porritt, S. M. (2017). Overheating in buildings: lessons from research. *Building Research & Information, 45*,(1–2), 1–18.

Lowe, J. A. Bernie, D., Bett, P., Bricheno, L., Brown, S., Calvert, D., Clark, R., Eagle, K., Edwards, T., Fosser, G., Fung, F., Gohar, L., Good, P., Gregory, J., Harris, G., Howard, T., Kaye, N., Kendon, E., Krijnen, J., Maisey, P., McDonald, R., McInnes, R., McSweeney, C., Mitchell, J.F.B., Murphy, J., Palmer, M., Roberts, C., Rostron, J., Sexton. D., Thornton, H., Tinker, J., Tucker, S., Yamazaki, K. & Belcher, S. (2018). *UKCP 18 science overview report. Met Office, November 2018 (updated March 2019).* Retrieved from https://www.metoffice.gov.uk/pub/data/weather/uk/ukcp18/science-reports/UKCP18-Overview-report.pdf.

Manchester City Council. (2015). Indices of deprivation 2015. Manchester City Council.

Mavrogianni, A., Wilkinson, P., Davies, M., Biddulph, P., & Oikonomou, E. (2011). Building characteristics as determinants of propensity to high indoor summer temperatures in London dwellings. *Building and Environment, 55*(2011), 117–130.

McLeod, R. S., Hopfe, C. J., & Kwan, A. (2013). An investigation into future performance and overheating risks in Passivhaus dwellings. *Building and Environment, 70*, 189–209.

Meehl, G. A., & Tebaldi, C. (2004). More intense, more frequent, and longer lasting heat waves in the 21st century. *Science, 305*(5686), 994–997.

Met Office. (2018). *Summer 2018*. Retrieved from https://www.metoffice.gov.uk/climate/uk/interesting/summer_2018.

Mishra, K., Loomans, M. G. L. C., & Hensen, J. L. M. (2016). Thermal comfort of heterogenous and dynamic indoor conditions—An overview. *Building and Environment, 109*, 82–100.

Morten, W. (2015). *Strategies for mitigating the risk of overheating in current and future climate scenarios*. Leamington Spa: Encraft.

Morgan, C., Foster, J., Poston, A. & Sharpe, T. (2016). Overheating in Scotland: contributing factors in occupied homes. *Building Research & Information. 45*, 1–14.

National House Building Council. (2012). *Understanding overheating–where to start*. London: Richards Partington Architects.

Nicol, F., Humphreys, M., & Roaf, S. (2012). *Adaptive thermal comfort–principles and practice*. London: Routledge.

Oikonomou, E., Davies, M., Mavrogianni, A., Biddulph, P., Wilkinson, P. & Kolokotroni, M. (2012). Modelling the relative importance of the urban heat island and the thermal quality of dwellings for overheating in London. *Building and Environment*, (57), 223–238.

Palmer, J., Godoy-Shimizu, D., Tillson, A., & Mawditt, I. (2016). *Building performance evaluation programme: Findings from domestic projects–making reality match design*. Swindon: Innovate UK.

Passive House Institute. (2015). *Criteria for the passive house EnerPHit and PHI low energy building standard*. Darmstadt: Passive House Institute.

Passivhaus Trust. (2012). *Passsivhaus–an introduction*. London: Passivhaus Trust.

Phillips, T., & Levin, H. (2015). Indoor environmental quality research needs for low energy homes. *Science and Technology for the Built Environment, 21*, 80–90.

Public Health England. (2018). *PHE heatwave mortality monitoring, Summer 2018*. Retrieved from https://assets.publishing.service.gov.uk/government/uploads/system/uploads/attachment_data/file/771819/PHE_heatwave_mortality_monitoring_report_2018.pdf.

rp5.co.uk. (2018). *Weather in Manchester (airport), UK [Online]*. Retrieved 2018, from http://rp5.co.uk/.

Rodrigues, L. T., et al. (2012). Summer overheating potential in a low energy steel framed house in future climate scenarios. *Sustainable Cities and Society, 7*, 1–15.

Saleem Gul, M., Jenkins, D. P., Patidar, S., Menzies, G. F., Banfill, P. F. G., & Gibson, G. J. (2015). Communicating future overheating risks to building design practitioners: Using the low carbon futures tool. *Building Services Engineering Research and Technology, 2*, 182–195.

Santamouris, M., & Kolokotsa, D. (2015). On the impact of urban overheating and extreme climatic conditions on housing, energy, comfort and environmental quality of vulnerable population in Europe. *Energy and Buildings, 98*, 125–133.

Silverman, D. (2011). *Qualitative research* (3rd ed.). Sage Publications: London.
Smith, S., Elliot, A. J., Hajat, S., Bone, A., Smith, G. E., & Kovats, S. (2016). Estimating the burden of heat illness in England during the 2013 summer heatwave using syndromic surveillance. *Journal of Epidemiology and Community Health, 70*, 459–465.
Tabatabaei Sameni, S. M., Gatrell, M., Montazami, A., & Ahmaed, A. (2015). Overheating investigation into UK social housing flats built to the Passivhaus standard. *Building and Environment, 92*, 222–235.
Taylor, M (2014). Preventing Overheating. London: Good Homes Alliance.
Taylor, J., Davies, M., Mavrogianni, A., Chalabi, Z., Biddulph, P., Oikonomoiu, E., ... Jones, B. (2014). The relative importance of input weather data for indoor overheating risk assessment in dwellings. *Building and Environment, 76*, 81–91.
Tsatsaronis, G. (2007). Definitions and nomenclature in exergy analysis and exergoeconomics. *Energy, 32*(4), 249–253.
Vardoulakis, S., Dimitroulopoulou, C., Thornes, J., Ka-Man Lai., Taylor, J., Myers, I., Heaviside, C., Mavrogianni, A., Shrubsole, C., Chalabi, Z., Davies, M. & Wilkinson, P. (2015). Impact of climate change on the domestic indoor environment and associated health risks in the UK. *Environment International 85*, 299–313.
van Loenhout, J. A. F., et al. (2016). The effect of high indoor temperatures on self-perceived health of elderly persons. *Environmental Research, 146*, 27–34.
Vellei, M., Herrera, D. F., & Natarajan, S. (2017). The influence of relative humidity on adaptive thermal comfort. *Building and Environment, 124*, 171–185.
Vilches, A., Padura, A. B., & Huelva, M. M. (2017). Retrofitting of homes for people in fuel poverty: Approach based on household thermal comfort. *Energy Policy, 100*, 283–291.
Wilkinson, P., Armstrong, B., & Landon, M. (2001). *Cold comfort: The social and environmental determinants of excess winter deaths in England, 1986–1996*. York: Joseph Rowntree Foundation.
Wong, L. T., Mui, K. W. & Cheung, C, T. (2014). Bayesian thermal comfort model. *Building and Environment, 82*, 171–179.
Zero Carbon Hub. (2014a). *Closing the gap between design & as-built performance evidence review report*. London: Zero Carbon Hub.
Zero Carbon Hub. (2014b). *Overheating in homes drivers of change*. London: Zero Carbon Hub.
Zero Carbon Hub. (2016). *Next steps in defining overheating*. London: Zero Carbon Hub.

A South African Experience of Building Energy Retrofit Project Challenges and Solutions

Chikezirim Okorafor, Fidelis Emuze, and Dillip Kumar Das

Introduction

According to Miller and Buys (2011), retrofitting existing buildings is more challenging, than designing a new sustainable building from scratch. Sustainable retrofit is viewed as an expensive disruptive process. Building occupants also show resistance to change and the disruptive processes (Miller & Buys, 2011). Retrofit processes often include addition of existing walls, opening walling elements, removal and installation of Heating, Ventilation and Air Conditioning (HVAC) elements, and strengthening of frames (Wilkinson, 2011). This leads to costly effect such as demolishing, lengthy construction time and occupant relocation. Such disruptions associated with these processes deter building owners from retrofitting their buildings (Cheung, Foo, & Granadino, 2000; Wilkinson, 2012). Dixon (2014) stated that the most important challenge to retrofitting is related to economic factors, organisational issues and lease structure. Oppong and Masahudu (2014) also postulate that legislation, programming and funding have been hindering the retrofitting of existing buildings. However, despite these challenges, the potentials of energy retrofit of existing buildings remain untapped. This can be attributable to multidisciplinary processes where several factors are involved (Godwin, 2011). These factors make each building energy retrofit project a unique and complex optimisation effort. To

C. Okorafor
Unit for Lean Construction and Sustainability, Central University of Technology Free State, Bloemfontein, South Africa

F. Emuze (✉)
Department of Built Environment, Central University of Technology Free State, Bloemfontein, South Africa
e-mail: femuze@cut.ac.za

D. K. Das
Department of Civil Engineering, Faculty of Engineering and Information Technology, Central University of Technology Free State, Bloemfontein, South Africa

L. Scott et al. (eds.), *Sustainable Ecological Engineering Design*,
https://doi.org/10.1007/978-3-030-44381-8_31

this end, this particular paper highlights the challenges and solutions that are appropriate in combating this problem in the South African built environment sector.

Literature Review

The phenomenon of climate change is responsible for the current focus on carbon management as one must be mindful of the wider implications of CO_2 emission for sustainable development, and the role that the built environment plays in this interaction. In spite of all the attention on carbon management in recent years, the fact remains that global CO_2 gas emissions and carbon intensity (measured as CO_2 emissions per unit of economic output) have continued to rise (Pielke, 2010). The world emitted twice as much CO_2 per marginal unit of economic activity in the decade leading to 2008 than in the previous decade (Diakaki, Grigoroudis, & Kolokotsa, 2008). It seems that global economic output is unable to extricate itself from carbon dependency (99% of the variations in carbon emissions can be explained by the changes in the approximately USD 50 trillion global economy—Pielke, 2010), and the trend is unlikely to reverse. This is made clear by the "Kaya Identity" (Kaya, 1990; Peters et al., 2017), which is composed of two primary factors: economic growth and technology changes.

1. CO_2 emissions = population × per capita GDP × energy intensity × carbon intensity
2. P = total population
3. GDP/P = per capita GDP
4. GDP = economic growth (contraction) = P × GDP/P = GDP
5. Energy intensity (EI) = TE/GDP = total energy (TE) consumption/GDP carbon intensity (CI) = C/TE = CO_2/total energy consumption
6. EI × CI = "carbon intensity of the economy" = TE/GDP × C/TE = C/GDP

Thus, according to the logic of these relationships, carbon accumulating in the atmosphere can be reduced only by reducing one or more of the following: population, per capita GDP, energy intensity, or the carbon intensity of the economy. It is at this point that the wider importance of "sustainable development" comes into play. The definition of "sustainable development" is by now well known: the Brundtland Commission Report (WCED, 1987) defines sustainable development as "development that meets the needs of the present without compromising the ability of future generations to meet their own needs". It contains within it two key concepts:

1. The concept of "needs", in particular the essential needs of the world's poor, to which overriding priority should be given.
2. The idea of limitations imposed by the state of technology and social organisation on the environment's ability to meet present and future needs (WCED, 1987).

The built environment is critical to both of these concepts of "sustainable development", and therefore the management of carbon in the built environment is central to our efforts to bequeath a "sustainable" world to future generations. Buildings (especially housing, but also other infrastructure) contribute to fulfilling the need for sustainable development, especially for the poor; the state of technology in the built environment provides a quick win for the world to achieve a low-carbon (and therefore sustainable) future. The built environment is a major consumer of energy, and it is thus a significant contributor of CO_2. The United Nations estimates that buildings consume 30–40% of the total energy used worldwide (United Nations Environment Programme, 2007). If we take into account cities, up to 90% of energy use occurs in and/or for cities (Svirejeva-Hopkins, Schellnhuber, & Pomaz, 2004). Given the rapid urbanisation and associated development in built infrastructure in both developed and developing nations, the role of the built environment in energy use, and therefore CO_2 emissions, is likely to be considerable (Ma, Cooper, Daly, & Ledo, 2012; Svirejeva-Hopkins et al., 2004). This is especially the case in Asia, but also in Latin America and, to a lesser extent, in sub-Saharan Africa. At the same time, the technical know-how needed to achieve substantial savings in energy use in the built environment (and therefore large reductions in CO_2 emissions) is largely well known. Therefore, in theory at least, the expected boom in built infrastructure in the world could potentially offer huge opportunities to reduce emissions and wean the world away from its carbon-intensive ways (Wang, Yan, & Xiao, 2012). Even the Intergovernmental Panel on Climate Change (IPCC, 2007) has asserted that the built environment sector is not only the most technically feasible sector but also the most cost-effective sector for reducing carbon emissions (Rivers, 2010). The question is "What, then, is preventing such huge and cost-effective potential from being realised?" This necessitates the need for this study in South Africa.

Research Method

Research is an organised way of exploring a problem to get the best solution (Gray, 2014; Sutrisna, 2009). A researcher's chosen research methodology provides the foundation for utilisation of a particular research philosophy, approach, strategy and choice. The study was situated in a pragmatist paradigm; a mixed-methods research design was adopted. Pragmatism is founded on the assumption of finding a solution to a research problem. Creswell (2013) contends that pragmatist researchers mainly focus on the "what" and the "how" of the research problem, by applying all the methods based on the criterion they think will work best in answering their research questions, utilising both qualitative and quantitative approaches. This study has adopted the research philosophy of pragmatism, for three main reasons, namely the nature of the research problem, the data and the methods of collecting this data, and the purpose of the research. The research problem, as discussed above, entails answering a number of research questions in order to fulfil the aim and the objectives of the research. This includes answering questions relating to "what" and "how", and, as such; it means that no single approach can be used to answer those questions. This informed the decision to use a method that combines both the qualitative and the quantitative research strategies. In realisation of this objective, themes, questions, and gaps were identified in the review of the literature. Information from the interviewees was gathered through the use of a digital recording device, which was clearly identified to them prior to the start of the interview, to confirm their consent to be recorded. The recording device allowed the researcher the opportunity to concentrate fully on the responses, although notes were also taken during and after the recording. These notes were taken into account during data analysis. Each interview was transcribed and underwent a series of coding exercises related to themes, relationships, and differences regarding the subject matter for the observation stage of the data collection and analysis. Extensive reading on the research topic was done to ensure understanding of the context and to capture perspectives from the participants. This included revisiting and reviewing the interview transcripts to ensure that the emphasis on the topic was accurately captured.

Research Results

The participation of professionals from building energy retrofit projects (BERPs) in this interview is paramount for obtaining in-depth knowledge on the study. The sample of interviewees comes from various organisation, background and experience. This ensured diverse views on the subject matter. A total of 13 purposive interviewee were sampled, with an average of 15.6 years of experience in BER projects. These skilled professionals are not only able to highlight the challenges for managing BER projects, but also to demonstrate with their expertise the complexity of the profession itself, and to inform on gaps which need to be filled in order to deliver projects more successfully. The data collection was also gathered through live case studies; this was done in order to obtain insight into the actual running of a building energy retrofit project. For clarity purposes, the findings of the structured interview are presented in Table 1.

Structured Interview

Challenges from Structured Interview

The interviewees were requested to outline and explain the challenges encountered when retrofitting buildings. In their individual responses, several challenges were highlighted. The challenges that top the list range from insufficient communication to insufficient consultation. They all stated that the activities of stakeholders are

Table 1 Matrix of challenges and solution from the interview

Challenges	Solution
Insufficient communication and consultation	Improvement in the area of communication and consultation with all stakeholders
Activities of stakeholders are insufficient and inadequate	They concurred that managing the different parties involved in the BERP delivery process will help
Lack of stakeholder agreement, the piecemeal fashion of doing things, and the lack of social data incorporated in the project were also highlighted	They stated that information sessions are needed to receive and communicate details of activities needed in the project
Lack of flexibility/adaptability	They advocated that experts dealing in BER projects should obtain substantial benefits from being flexible
Lack of technical know-how	The interviewees reiterate that training provision for contractors, subcontractors and installers is of the utmost importance
Capital cost of such investments, uncertainty about the payback period, and limited access to finance	Provision of grants and low-interest loans for such retrofitting schemes
Lack of building science/energy efficiency awareness among contractors, subcontractors etc.	Use of suitably qualified and experienced retrofit technicians can help in improving the delivery
Human behavioural factors	Incorporation of human behavioural pattern is needful
Governmental policy change in terms of funding the project	A comprehensive business case should be deployed in this case
Selection of retrofit technologies	The selection should be based on the need, availability, ease of work or installations and user friendly
Operational challenges (interruption of operations)	Operational activities for the energy efficiency drive should be communicated pre and during the project delivery
Resistance from building owners	Targeting owners of buildings in dire need of retrofit and pointing out how such retrofit works can improve comfort levels in a building
Specific nature and characteristics of the building	Flexibility and innovative drive from the contractors can help in this case
Project budget and target	Investment decisions for energy efficiency should be properly carried out against the project deliveries before the commencement of the project
Comprehensive and concrete weighted evaluation of the social, cultural and economic benefits of retrofitting	Educational level can impact how receptive the potential adopter will be towards retrofitting an existing building

Source: Researchers (2019)

insufficient and inadequate. A lack of stakeholder agreement, the piecemeal fashion of doing things, and the lack of social data incorporated in the project were also highlighted. These were closely followed by a lack of collaboration and cooperation, which revolves around a lack of stakeholder activities to obtain buy-in into the project, and experts operating randomly, which makes work move haphazardly. Lack of flexibility/adaptability was also outlined as another serious challenge in building energy retrofit projects. This ability in times of unpredictability and complexity requires a trade-off between adherence to a process and adaptability/flexibility. They advocated that experts dealing in BER projects should obtain substantial benefits from being flexible, by applying learning through continuous improvement and administrative innovation, because no two BER projects are the same.

In the same argument, the interviewees posited that lack of technical know-how is equally a challenge. The absence of suitably qualified and experienced retrofit technicians can result in poorly installed systems. This can then lead to project failures and disappointment, making the client sceptical of the benefits of such initiatives. Such failures are damaging to the retrofit sector and the construction industry in general. Therefore, it becomes vital that contractors make use of engineers and installers that understand the importance of getting it right, and the implications of not following procedures and standards, which often results in failed projects. Other challenges are such as capital cost of such investments, uncertainty about the payback period, and limited access to finance. This can be linked to the income of the potential investor; most clients (building owners) prefer a shorter payback time, so when the cost of investing in retrofit works is seen as one that will take over 5 years to recoup, many withdraw from such schemes, as the interviewees suggest. Education level can impact how receptive the potential adopter will be towards retrofitting an existing building. Low levels of education contribute to low awareness of the direct and indirect benefits of such systems to the individual and society at large; homeowners also fear disruption or damage to the building. The interviewees added that location can place certain constraints on the type and the nature of technologies that can be applied. Some interviewees went further to say that lack of building science/energy efficiency awareness among contractors, subcontractors, buyers, and others and lack of awareness seem to be a challenge, because some of them cannot differentiate between renovation and energy retrofit of buildings. The worst of it all is that contractors, subcontractors, and others lack the requisite experience to carry out the task, thereby creating more problems for the building after the retrofit. Which might be in terms of technical know-how of some issues, such as air-tightness, ducting, etc. of the dwellings? Uncertainties such as weather conditions (climate change), human behaviour factors, government policy change in terms of funding the project, selection of retrofit technologies, operational challenges (interruption of operations), resistance from building owners and investors, specific nature and characteristics of the building, and the project budget and targets all present huge challenges.

In addition, the interviewees put it that economic, technical, environmental and social dynamics of energy retrofit are challenging. Even the investment decisions for energy efficiency on retrofits are quite complex in nature. Other challenges relate to end users' awareness, attitudes and behaviours in relation to energy use, and clients' requirements and experience (in most cases they have limited knowledge) were also highlighted. Comprehensive and concrete weighted evaluation of the social, cultural and economic benefits of retrofitting an existing building can be a daunting task. There is always an element of uncertainty regarding the technical, financial and operational benefits of implementing BERPs, as suggested by the interviewees. Finally, most of the interviewees concurred that the success of a retrofit depends firstly on understanding the building and its context in sufficient detail and depth. The professionals also need to understand that some of the formal standards and methods used by government and industry are incorrect or incomplete. Finally, it is important to understand the interactions between all these different elements and how different aims for retrofit may conflict with each other.

Solution from Structured Interview

When asked to consider how best to overcome these retrofitting challenges and what was needed to improve the delivery of BERPs, most of the interviewees suggested improvement in the area of communication and consultation with all stakeholders, especially in affected buildings. They stated that information sessions are needed to receive and communicate details of activities, in order to deliver what is actually needed in the BERP. Collaboration and coordination are elements that must be improved upon. They concurred that it takes more time, money and effort to make decisions collaboratively; the potential gains of doing so outweigh the costs involved. Collaboration and coordination happen on the assumption that the resulting decisions will be superior to decisions made individually regarding BERPs. The interviewees reiterate that training provision for contractors, subcontractors and installers is of the utmost importance. Other potential solutions offered were provision of grants and low-interest loans for such retrofitting schemes, to serve as incentives, and awareness creation through targeted education, so that homeowners and occupants can see the energy and cost-saving potential of such retrofit practices. Other solutions offered were targeting owners of buildings in dire need of retrofit and pointing out how such retrofit works can improve comfort levels in a building, emphasising the importance of environmental and climate protection, and increased awareness through special campaigns by the government and professional bodies on issues relating to energy efficiency in buildings. Also, training, in the form of certification for contractors and subcontractors, and post-retrofit assessment of the building in order to detect issues with the retrofit will equally add value, as suggested by the interviewees.

They concurred that managing the different parties involved in the BERP delivery process in order to streamline implementation will also prevent or at least minimise fragmentation of the retrofit sector. The interviewees further emphasised that this can be achieved by better customer service provision, through changes to the level of marketing and an increase in the size of the target group, which will result in more awareness creation, preventing the rebound effect, i.e. a situation where savings from retrofitting lead to increased demand for other energy-consuming goods and services, thereby offsetting the initial gains. Where this happens, the client may dispute the needs and savings.

Live Case Study

The live case studies were conducted in South Africa, where the researcher undertook visits to the projects. The building was predominantly high rise government office building. The case studies were undertaken with significant input from the project team members. Live case studies provide a unique opportunity to obtain insight into the actual running of a BER project, revealing the challenges experienced by the project team. Consequently, six respondents were interviewed across 14 on-going retrofitted projects in South Africa. The line of inquiry for the live case studies focused on how the themes identified earlier were applied across the projects, and the relevance of issues highlighted through the structured interviews. The findings are presented in Table 2.

Table 2 Matrix of challenges and solution from the live case studies

Challenges	Solution
Difficulties in decanting staff	A detailed communication strategy should be developed detailing each activity that is going to be undertaken
Stakeholders are reluctant to cooperate with energy retrofit initiatives	Detailed communication strategy will ensure total buy-in among stakeholders
Stakeholder management is difficult with respect to collaboration and cooperation	That stakeholder's management initiative should be created before energy retrofit activities and should be sustained after these activities as suggested by the interviewees
Government projects do not have funding for retrofit	A well written business case proposal will suffice in this case, detailing direct and indirect benefit of such EE drives in the building
No existing drawings to work from	This calls for being flexible and adaptable
The issue of changes in technology	Proper planning of the retrofitting technology needs to be done, so that it does not affect or hamper the working of other building components
Challenges in building up rates	This include a legal framework informed by policy is required to promote competitive pricing in this trade

Source: Researchers (2019)

Challenges from Live Case Studies

Many challenges were mentioned by the interviewees. These include the difficulties in decanting staff, especially when the building is a multipurpose building, which is often the case with government buildings. They stated that stakeholders are reluctant to cooperate with energy retrofit initiatives. Four of the six interviewees stated emphatically that stakeholder management is difficult with respect to collaboration and cooperation. For instance, they mentioned that they always face a huge challenge when they are making a business case with the sponsors and the end user. According to the interviewees, most government projects do not have enough funding for retrofit. They said that the government always approaches ESCOs to finance government projects, and that this is a daunting task. In addition, retrofit projects are poorly funded, especially those ones that are delivered by the government. The interviewees claim that in few cases there are no existing drawings to work from, which makes energy retrofit projects a tedious task. The issue of changes in technology and challenges in building up rates in retrofit projects were also highlighted by the interviewees.

Solution from Live Case Studies

According to the interviewees, a detailed communication strategy should be developed and must be practically driven, to inform the occupants on step-by-step procedures and the benefits of each activity that is going to be undertaken. This will ensure total buy-in among stakeholders. In addition, they assert that raising awareness and promoting behaviour change through communication and education must be a push-up approach against a pull approach to the stakeholders. That this initiative should be created before energy retrofit activities and should be sustained after these activities as suggested by the interviewees. They added that there is a need to develop behaviour change initiatives in parallel with technological change, in order to maximise the benefits of both. Those occupants with genuine complaints as a result of post retrofit activities should be acknowledged, and the issues highlighted must be addressed, according to the interviewees. In the same argument, the interviewees reported that experience has shown that behaviour change can achieve energy savings of up to 5–15%, and that such initiatives must be intentional. For example, facilitating conditions for behaviour change is at least as important as trying to influence it directly. They concurred that, behavioural change is most effective when a number of levers are pulled, in a coherent, coordinated and systematic way. The interviewees reiterate that policies aimed at changing behaviour need to be simple and transparent. That the issue of awareness raising and information provision alone does not work effectively, but does work better when tailored to the target audience. The installation of digital feedback, for example, the displays on smart meters, needs to be designed to prompt action. For instance, most occupants will be more receptive to financial savings than to energy or emissions savings, and substitute behaviour needs to be more attractive than the default. The interviewees went on to conclude that consultation, collaboration, and cooperation should be improved, as energy retrofit is a multifaceted activity, where different trades complement each other.

Discussion

The results of the research indicate that improvement in the area of communication and consultation with all stakeholders, especially in affected buildings, will go a long way to addressing the challenge of lack of communication and consultation. The participants stated that information sessions are needed to receive and communicate details of activities, in order to deliver what is actually needed in the BERP (Oppong & Masahudu, 2014). Collaboration and coordination between stakeholders is also a factor that can be improved upon. The study also found that training provision for contractors, subcontractors, and installers is of the utmost importance. Provision of grants and low-interest loans for such retrofitting schemes, to serve as incentives, is also crucial (Oppong & Masahudu, 2014). Awareness creation through targeted education to homeowners and occupants, so that they can see the energy and cost-saving potential for such retrofit practices, is also needful. Proper planning of the retrofitting needs to be done, so that it does not affect or hamper the working of the building operations (Godwin, 2011). Simultaneous operation of old and new energy systems should be ensured. There is also the limitation of lack of information and availability of energy-efficient products. To overcome this problem, EE products should be bought through good vendors, after doing thorough market research (Sullivan, Mark, & Parnell, 2006). A well thought-out plan to manage different parties involved, in order to streamline implementation, is also crucial, in order to prevent, or at least minimise, fragmentation of the retrofit activities. The study suggested that experts dealing in BER projects should obtain substantial benefits from being flexible, by applying learning through continuous improvement and administrative innovation, because no two BER projects are the same.

Concluding Remarks

The major challenges mentioned in this study pertain to communication and consultation. The participants stated that the activities of stakeholders are insufficient and inadequate, there is a lack of stakeholder agreement, the piecemeal fashion of doing things is a challenge, and there is a lack of social data incorporated in projects, among other things. This is closely followed by a lack of collaboration and cooperation, which revolves around a lack of stakeholder activities to obtain buy-in into the project, and experts operating randomly, which makes work move haphazardly. The issue of flexibility/adaptability is another serious challenge in BER projects. The study also established that technical know-how is a key challenge. Other challenges, such as the capital cost of such investments, uncertainty about the payback period, and limited access to finance, are equally significant.

The following recommendations for policy and industry practice are made based on the findings and the conclusions. A legal framework informed by policy is required to promote sustainability practices, especially as they concern existing government buildings. This will create a platform for standardised operation for BERP implementation processes. There has to be training and certification given to retrofit professionals, where this can serve as a criterion for the awarding of contracts by government and its agencies. Contractors should strategically build their competitive capabilities and competencies, through acquisition of the relevant training, qualifications, experience and professionalism, which is a prerequisite for developing the skills needed for transformational change in the trade. Finally, South African construction industry should intensify construction protocols, procedures and activities that will encourage synergy towards strategic alliances and subcontracting partnerships, so as to facilitate skills and knowledge transfer between contractors, as the retrofit trade is relatively new to the country.

References

Cheung, M., Foo, S., & Granadino, J. (2000). *Seismic retrofit of existing buildings: Innovative alternatives* (pp. 1–10). Canada: Public Works and Government Services.

Creswell, J. W. (2013). *Research design: Qualitative, quantitative, and mixed methods approach* (4th ed.). Los Angeles: SAGE.

Diakaki, C., Grigoroudis, E., & Kolokotsa, D. (2008). Towards a multi-objective optimization approach for improving energy efficiency in buildings. *Energy and Buildings, 40*(9), 1747–1754.

Dixon, T. (2014). Commercial property retrofitting. *Journal of Property Investment & Finance, 32*, 443–452.

Godwin, P. (2011). Building conservation and sustainability in the United Kingdom. *Procedia Engineering, 20*, 12–21.

Gray, D. E. (2014). *Doing research in the real World* (3rd ed.). London: SAGE Publishing Ltd.

IPCC. (2007). *Climate change 2007: Synthesis report (AR4). IPCC Fourth Assessment Report.* New York: Cambridge University Press.

Kaya, Y. (1990). *Impact of carbon dioxide emission control on GNP growth: interpretation of proposed scenarios*. Paper presented to the IPCC Energy and Industry Subgroup, Response Strategies Working Group, Paris. (Mimeo).

Ma, Z., Cooper, P., Daly, D., & Ledo, L. (2012). Existing building retrofits: Methodology and state-of-the-art. *Energy and Buildings, 55*, 889–902.

Miller, E., & Buys, L. (2011). Retrofitting commercial office buildings for sustainability: Tenants expectations and experiences. *Management and Innovation for a Sustainable Built Environment, 3*, 33–50.

Oppong, R. A., & Masahudu, M. (2014). Exploration of building adaptation and retrofitting challenges in Ghana: The case study of selected rural bank projects. *Structural Survey, 32*, 5–11.

Peters, G. P., Andrew, R. M., Canadell, J. G., Fuss, S., Jackson, R. B., Korsbakken, J. I., … Nakicenovic, N. (2017). Key indicators to track current progress and future ambition of the Paris Agreement. *Nature Climate Change, 7*, 118–122.

Pielke, R. A. (2010). The carbon economy and climate mitigation: An editorial essay. *WIREs Climate Change, 1*, 770–772.

Rivers, N. (2010). Impacts of climate policy on the competitiveness of Canadian industry: How big and how to mitigate? *Energy Economics, 32*, 1092–1104. https://doi.org/10.1016/j.eneco.2010.01.003, ISSN: 0140–9883.

Sullivan, L., Mark, B. and Parnell, T. (2006). *Lessons for the application of renewable energy technologies in high density urban locations*. Proceedings of the 23rd Conference on Passive and Low Energy Architecture (PLEA 2006), Geneva, Switzerland, 6–8 September 2006

Sutrisna, M. (2009). *Research methodology in doctoral research: Understanding the meaning of conducting qualitative research*. Working paper presented in ARCOM Doctorial Workshop. John Moores University, Liverpool, 12 May 2009.

Svirejeva-Hopkins, A., Schellnhuber, H. J., & Pomaz, V. L. (2004). Urbanised territories as a specific component of the Global Carbon Cycle. *Ecological Modelling, 173*(2–3), 295–312.

Wang, S., Yan, C., & Xiao, F. (2012). Quantitative energy performance assessment methods for existing buildings. *Energy and Buildings, 55*, 873–888.

Wilkinson, S. (2011). *The sustainable retrofit potential in lower quality office stock in the central business district*. Amsterdam: Management and Innovation for a Sustainable Built Environment.

Wilkinson, S. (2012). Analysing sustainable retrofit potential in premium office buildings. *Structural Survey, 30*, 398–410.

World Commission on Environment and Development. (1987). *Our common future: Report of the World commission on environment and development*. New York: United Nations. Retrieved July 23, 2018, from http://www.un-documents.net/ocf-02.htm.

Part VI
Ecology and General Sustainability

Serious Games for the Built Environment: Eco Material Trumps

John Lester Clarke

Introduction: Environmental Impact of Buildings

The design, construction, operation and use of buildings has a significant impact on the natural environment in terms of energy use, the burning of fossil fuels, CO_2 emissions, the depletion of finite resources, air, water and ground pollution, biodiversity and the production of waste (UKGBC, 2019).

The built environment in the UK accounts for 40–50% of all CO_2 emissions amounting to 360 million tonnes of CO_2 released into the atmosphere per year and produces 120 million tonnes of waste (WRAP, 2007), and studies have shown (Horvath, 2004) that up to 29% of all solid waste going to landfill originates from construction and demolition. During the process of decomposition this can result in the release of methane, a greenhouse gas, with four times the global warming potential of CO_2. There is an ever-increasing demand for the provision of built environments as the global population is projected to reach 9.8 billion by 2050 and there is strong scientific consensus that human endeavours, such as construction, will contribute exponentially to global warming with resultant effects on climate change and sea level rises, with potentially disastrous effects for life on earth.

Legislation (Climate Change Act, 2008) sets legally binding national greenhouse gas reduction targets of 34% by 2020 and at least 80% by 2050 compared to 1990 levels. The recognition that building practices need to change is evidenced by revisions in legislation with targets for zero carbon and low water usage new-build housing by as early as 2016 and new build non-domestic buildings by 2019 for England and Wales (DCLG, 2007).

As well as global, national and regional environmental benefits of a more sustainable built environment, there are social advantages in terms of healthier

J. L. Clarke (✉)
School of the Built Environment, University College of Estate Management Horizons, Reading, UK
e-mail: a.williams@ucem.ac.uk

L. Scott et al. (eds.), *Sustainable Ecological Engineering Design*,
https://doi.org/10.1007/978-3-030-44381-8_32

buildings and economic benefits by optimising the productivity of occupants and by investing now in a more sustainable future to offset future costs, as indicated in the Stern Report (Stern, 2006).

Sustainable buildings and the materials from which they are constituted also have social benefits in providing healthier indoor environments by minimising volatile organic compounds (VOCs) contained within and released/off-gassed from synthetic materials, which contributes to 'sick building syndrome'. Healthier indoor environments also have economic benefits encouraging greater productivity through the health and well-being of occupants and ultimately general populations, reducing expenditure on health care.

Environmental Impact of Construction Materials

The need for a more sustainable built environment is widely accepted and construction materials have significant potential throughout their life cycle to have positive benefits for the environment by reducing our reliance on finite resources, eliminating damaging production processes and long-distance transportation of materials. It is easier to recycle and reuse natural materials and specify non-toxic materials with low embodied energy from renewable sources for more durable and adaptable buildings. Waste can be designed-out avoiding increasingly expensive disposal costs and cutting out costly sourcing of virgin materials (UKGBC, 2019).

Sustainable building materials are more efficient, less toxic and they respect the health, safety and welfare of building occupants and the community in general. If sourced locally, sustainable construction materials can encourage local economic development and supply chains.

The range of green building materials and products that are currently available has grown exponentially in response to the growth in awareness of environmental issues and offer a range of aesthetic options, perform well and are increasingly cost-competitive (Greenspec, 2018), and the emergence of sustainable building rating systems such as the Building Research Establishment Environmental Assessment Method (BREEAM) and Leadership in Energy and Environmental Design (LEED) has given high importance to the sustainability of construction materials.

Four hundred and twenty million tonnes of materials are used in the construction of buildings in the UK each year (Lazarus, 2009) which accounts for 30–50% by volume of all manufactured goods, excluding food production (Roaf, 2004). One hundred and twenty million tonnes end up as waste from construction, demolition refurbishment and excavation processes and it has been estimated that 20 million tonnes of *unused* materials end up in landfill each year (WRAP, 2007). Over their entire life cycle, materials used in construction contribute a significant amount to the environmental impact of the construction sector in terms of the extraction of finite raw materials, their processing, transportation, manufacture into building products, packaging, installation on-site and their destination after primary use, such as recycling, reuse, incineration or landfill disposal.

In order to reduce the environmental impact of building materials, those responsible for creating and maintaining our built environment must engage with sustainable material issues at each stage of procurement, design, construction and beyond the design life of the materials. For example, during the construction phase alone, research has shown that CO_2 emissions can be reduced by as much as 30% through a careful selection of low environmental impact materials (Gonzalez & Navarro, 2006).

A range of analytical techniques have been developed, collectively called environmental life cycle assessment (LCA). There are also a number of guides, resources and initiatives aimed at tackling the environmental impact of construction materials, including the Building Research Establishment (BRE) 'Green Guide to Specification' and the 'Waste & Resources Action Plan' (WRAP) 'halving waste to landfill' initiative.

However, research has shown (Greenspec, 2018) that there is a quite disparate body of existing data related to the sustainability credentials and environmental impact of construction materials due in part to the complexity of the market with a wide variety of materials and alternatives available, lack of comprehensive research in this field and often unsubstantiated manufacturers' claims. Therefore, it is difficult for built environment students and professionals alike to access clear, concise and impartial data without undertaking time-consuming information-gathering and research activities. Hence, this resource has been developed using serious game theory.

Serious Game Theory

The need to understand and apply sustainable building practices is continually increasing. The ability to absorb technical information can be daunting for built environment students and professionals alike and the value of games, used to support training and learning, has been widely recognised for many years, and as gaming has become increasingly popular it is more widely considered (Owen & Dewick, 2015). Lujan and Di Carlo (2005, p. 19) argue that 'the packed curriculum leaves little time for students to acquire a deep understanding of the subject or to develop life-long skills such as critical thinking, problem solving, and communication'. Research has shown that didactic methods, such as lecturing, merely expose students to content, and exposure is not sufficient for learning. Active processing of information, not passive reception of information, leads to learning (Bolles, 1988).

Serious games have been defined as 'a specific interaction of students with others (or individually) using specific game mechanics and dynamics, oriented towards specific outcomes' (Beetham, 2008). It is important to map learning outcomes to such activities highlighted by Bloom's taxonomy of learning outcomes, including remembering, understanding, applying, analysis, evaluating and creating and how they organise ideas, opinions and thoughts.

The use of innovative educational games can increase enthusiasm and reinforce previously presented didactic information. Games create a challenging, constructively competitive atmosphere that facilitates interaction among learners in a friendly and fun environment.

Active and experiential learning strategies reach all types of learners in the visual, auditory and tactile learning styles. Research by Fleming and Mills (1992) suggests that visual learners have a preference for seeing, auditory learner's best learn through listening to debates and discussions and tactile learners prefer to learn via experience—moving, touching, and doing.

Motivation is a key element in the effectiveness of serious games and the focus on tasks and challenges can be contextualised and explained through the flow framework in applied game theory. 'Flow' is described as a state of '*deep concentration in which thoughts, intentions and feelings are focused on the same goal*' (Csikszentmihalyi, 1990) and is widely discussed in Schell (2015). The flow channel (see Fig. 1) illustrates where the participant is in an enjoyable state of flow, where the application of and challenge to the learner's skills and abilities are neither too difficult, resulting in stress and anxiety and de-motivate, nor too easy, resulting in boredom and distraction.

Meaningful feedback is also key for helping students to achieve embedded learning goals and encourages knowledge construction and reflection. Whitelock (2011) argues that feedback forms an essential component of any assessment task, which is described as 'advice for learning'.

From a game design perspective, features that necessitate careful planning of educator's different role types need to balance both learning and fun. Types of roles can be as designer, player, facilitator, motivator, observer and evaluator.

In a recent report it has been highlighted that the key challenge for effective learning with games is for the experience to be undertaken in relation to clear learning outcomes as well as being made relevant to real-world contexts and practice.

The success of game-based learning strategies owes to active participation and interaction being at the centre of the experience, and signals that current educational

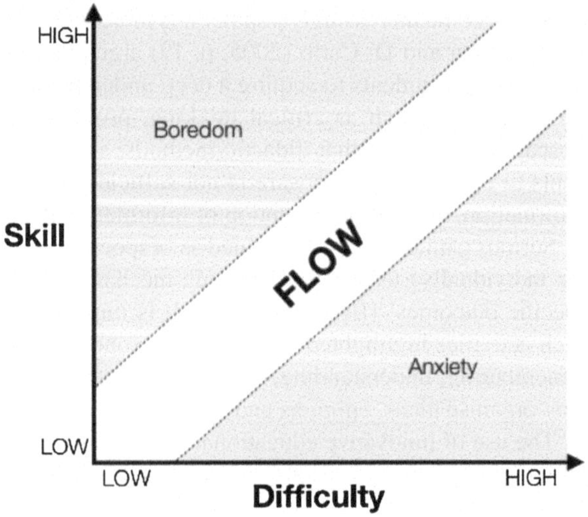

Fig. 1 The flow channel (Source: Csikszentmihalyi, 1990)

methods are not engaging students sufficiently. Experience with and affinity for games as learning tools is an increasingly universal characteristic among those entering higher education and the workforce. Based on research from a number of models, the National Training Laboratories (NTL) in the 1960s produced the 'learning pyramid' (see Fig. 2). This illustrates average student retention rates for differing teaching and learning strategies. The game-based approach lies within the 50–80% range. These types of activity are also ones that encourage intellectual effort at the higher levels of Bloom's Taxonomy (Bloom et al., 1956).

The 'learning pyramid' has had some criticism of its empirical and methodological diligence (Letrud, 2012), despite being referenced widely, and therefore there is some need to establish its evidential derivation.

This project brings the concept of gaming into the classroom of built environment students and aims to establish to what extent the impact of gaming has on teaching and learning in applied and action-based settings.

Development of Eco Material Trumps

The development of 'Eco Material Trumps' is an attempt to collate a wide body of sustainability data on construction materials into a single resource in order to assist in the selection of sustainable materials but also to act as an educational catalyst to encourage debate and discussion on a complex topic. Rather than focusing on one material's environmental effects, which may not provide much of an explanation out of context, it is often more valuable to perform a comparative assessment of construc-

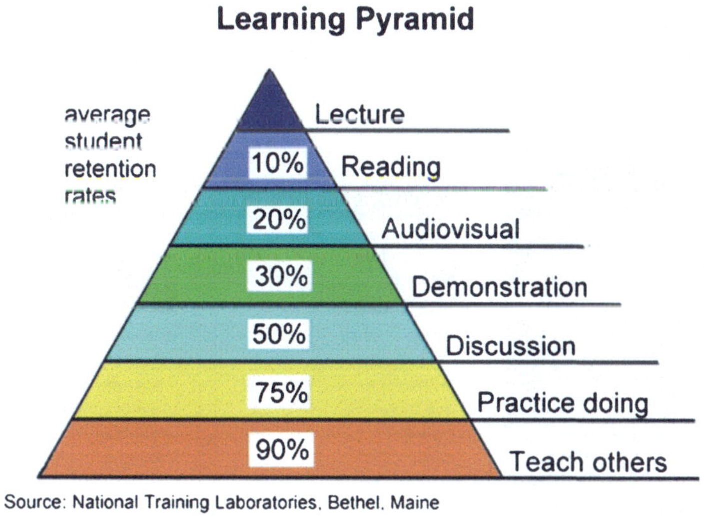

Fig. 2 The learning pyramid. (Source: National Training Laboratories, Bethel, Maine)

tion materials. Franklin, Peat, and Lewin (2003, p. 81) describe the use of such aids to 'stimulate group discussion, by bringing together information and concepts from a number of sources to assist learners in the pulling together and linking of material'.

The aim is to bring the game within the learning environment, acting as an aid to support the wider built environment curriculum to stimulate debate on how, in practice, decision makers need to balance the differing criteria that are used to establish the environmental impact of construction materials.

The game can be used across disciplines, age groups and educational levels. The potential beneficiaries will therefore comprise anyone concerned with the environmental impact of building materials such as teachers, students, academics, existing and future built environment professionals.

The cards focus on six key sustainability criteria identified and selected for their widely recognised relevance to the built environment, their impact on sustainability and the availability of valid and reliable data for the chosen materials. The criteria are: embodied energy, global warming potential, recyclability, landfill decomposition, durability and thermal conductivity. Forty-four common construction materials have been selected based on their common usage in general building construction. Therefore, the resource brings together a wide variety of otherwise disparate sets of data into a coherent and accessible teaching, learning and reference resource (see Fig. 3) allowing for direct comparison, analysis and evaluation through experiential game-based learning.

Included in the pack are instructions on how to play the game, clear definitions of each of the sustainability criteria and further information of the environmental impact of key materials which can be used to increase knowledge and understanding prior to playing the game or follow-up activities.

Fig. 3 Example of eco-construction trumps format and content (Source: Author)

Research Study Method

One of the main barriers to the uptake of games in a learning context is the lack of empirical data to support the hypothesis for its effectiveness, as well as a lack of understanding about how these games might be used most appropriately in practice.

The study is based on the use of the card game during several workshops involving UCEM staff, students and EAUC conference delegates, all with an interest in the built environment and/or pedagogy, totalling 17 in number. Feedback was elicited using a tailored semi-structured questionnaire using a variety of question types, both open and closed response approaches to investigate participants' perceptions of the value and importance of the active learning and cooperative activities they undertook during the workshops. Two of the workshops were held at UCEM, the first with NUS Responsible Futures participants and the second at a staff 'Show M' session. The other workshop was held at EAUC conference 2018 (see Fig. 4).

Data was analysed using a mixed methodology approach. An overall thematic analysis of data was used for open-ended qualitative responses leading to an analytic or theoretical coding. The closed, structured questions enabled full quantitative analysis.

Analysis and interpretation of the responses to the questionnaire follows.

Fig. 4 EAUC conference workshop (Source: Author)

Findings from Workshops

The questionnaire elicited a great deal of useful data regarding the perceptions of the respondents to engaging with the game. The participants/respondents were from either an academic or built environment professional background.

Figure 5 shows responses to the question 'what 3 words come to mind when you think of the cards?' represented in a 'word cloud' which gives greater prominence to words that appear more frequently in the source data. Clearly the most popular answer was 'fun' which has been highlighted as a key driver for using game-based learning. The second most popular response was 'informative' again reinforcing previous research findings that learning can occur through game-based initiatives, and thirdly, 'engaging' indicating that interest was maintained. In descending order of significant frequency came 'interesting', 'play', 'thought-provoking', 'useful', 'materials', 'competition' and 'education'.

When asked 'how did you find playing the game?' 35% of the respondents found the game 'very easy' and 65% found the game to be 'relatively easy' to play whilst none found the game 'difficult' or 'very difficult'. Additional comments suggested that 'it's a good, fun icebreaker to engage people' and initially the participants took a short time to familiarise themselves with the game play 'it took a while to get going but then we flew!' Some of the participants found the definitions and concepts a little difficult to grasp but this encouraged discussion and debate based on the definitions on the cards and required participants to refer to the definitions and supplementary information initially, to fully benefit from the resource.

In terms of usefulness in their field of work 35% found the game 'very useful', 41% found the game 'moderately useful', 18% found it to be 'quite useful' and 6% thought it not to be useful. Many of the participants related the subject matter to elements of their teaching content or directly with work they were undertaking, stating that the game offers the data on these materials in a very accessible and interac-

Fig. 5 A 'word cloud' indicating feedback from workshops (Source: wordle.net)

tive way and that the cards reinforced the success of 'play' in information sharing. Some stated it would only be useful if it was available online or in a digital format and others suggested it could be developed for more bespoke specialisms such as cost data.

When asked 'how has your understanding of sustainable material selection improved as a result of playing the game?' all respondents indicated some improvement with 6% stating their understanding had improved 'completely', 35% 'significantly' and 59%, moderately'. Additional comments included that playing for longer or a few more times would increase understanding. One participant stated that the game helped the understanding of the whole process of evaluating material impact and the relative environmental credentials of different materials.

In terms of engagement with the game 65% were 'very interested' and 35% were 'interested' with no respondents stating they were either 'bored' or 'slightly bored'. One respondent stated that 'competing always excites people!', another stated that 'once you're short of cards interest wanes' and another 'just loves playing Trumps'.

Fifty-three percent of respondents would 'definitely' recommend the cards to colleagues and 47% would be 'likely' to recommend the cards whilst no respondents were 'unlikely' or explicitly wouldn't recommend them to colleagues.

Other questions related to understanding the data which many stated became clearer after reading the definitions and discussing their meanings during the game. Some of the data was not available which frustrated some players and interrupted the flow of the game whilst some were surprised at materials they had considered to be of high environmental impact that proved to have an overall lower impact, e.g. concrete.

There were many suggestions on how the cards could be used in different or innovative ways including as a research resource 'It makes finding information really easy instead of trawling through the internet!' and as an educational tool on training courses or to engage members of the public and non-specialists. There were also several ideas for developing and expanding the resource to enable comparison of specific materials, for example, a whole set dedicated to insulation materials and related topics such as renewable energy versus non-renewable energy sources, sustainable communities and lifestyle habits.

Criticisms of the cards included the need for clearer definitions and that some of the data was too general. Some participants questioned the validity of the data, illustrating the difficulty of quantifying sustainability criteria from a number of data sets and product champions who wanted their material to be shown in the best environmental light possible, even if in contradiction to some of the research findings. Another criticism was the inability of the cards to be updated and that they could quickly become obsolete in the light of future research and construction material developments. This is a good case for having an online version that could be easily updated and either played remotely online or printed, for use as a physical resource.

Conclusions

It has been shown that game-based learning has a place in contributing to educational provision where these complex issues can be tackled in a fun, informative and engaging way. The ability to absorb technical information can be daunting for built environment students and professionals alike and the value of games, used to support training and learning, has been widely recognised for many years. The packed curriculum leaves little time for students to acquire a deep understanding of the subject or to develop life-long skills such as critical thinking, problem solving, and communication.

Combining environmental impact criteria into a card game encourages a non-lecturing educational experience and provides an opportunity to engage students, staff and professionals in discussion and creates a source of reference that will assist in sustainable construction material selection and will impact the built environment sector through awareness and application of best and future practice.

Results from workshops and participants' feedback show that the cards help students process and retain technical information. Games create a challenging, constructively competitive atmosphere that facilitates interaction among learners in a collegiate and enjoyable environment. Findings from the feedback elicited from workshops show that this resource enables the processing of complex sets of data, brings together data from disparate sources, encourages interaction and discussion, promotes learning through visual and tangible presentation of data and encourages sustainable thinking about and beyond the subject matter.

Feedback indicates that the curriculum requires some adjustment to incorporate active methods that provide educational experiences designed to develop life-long learners and students who can solve novel problems and challenges. It is hoped that use of the cards will lead to reconfigurations of some module teaching and learning strategies.

Initial indications from the workshops suggest the hypothesis that students have better retention and understanding of knowledge when taught by active as opposed to passive methods holds some truth but requires further empirical study to establish causation between engaging with the game and levels of knowledge retention.

It is hoped to undertake further research and develop this concept further for a professional audience allowing an even wider dissemination and potential impact beyond built environment undergraduates, and into practice and industry such as the comparison of exemplar sustainable buildings on their sustainability credentials.

References

Beetham, H.. (2008). *Review: Design for learning programme phase 2. Review of learning design as part of the JISC's Design for Learning Programme*. Retrieved May 18, 2018, from http://www.jisc.ac.uk/whatwedo/programmes/elearningpedagogy/designlearn.aspx.

Bloom, B., Engelhart, S., Furst, M. D., Hill, E. J., Krathwohl, W. H., & D. R. (1956). *Taxonomy of educational objectives: The classification of educational goals. Handbook I: Cognitive domain*. New York: David McKay Company.

Bolles, E. (1988). *Remembering & forgetting: Inquiries into the nature of memory*. New York: Walker.

Csikszentmihalyi, M. (1990). *Flow: The psychology of optimal experience*. New York, NY: Harper & Row.

DCLG Department for Communities & Local Government. (2007). *Building a greener future: Policy statement*. London: DCLG.

Fleming, N. D., & Mills, C. (1992). *VARK a guide to learning styles* [online]. Retrieved April 20, 2018, from https://serc.carleton.edu/resources/13902.html.

Franklin, S., Peat, M., & Lewin, A. (2003). Non-traditional interventions to stimulate discussion: The use of games and puzzles. *Journal of Biological Education, 37*(2), 79–84.

Gonzalez, J. M., & Navarro, J. G. (2006). Assessment of the decrease of CO_2 emissions in the construction field through the selection of materials: A practical case study of three houses of low environmental impact. *Journal of Building and Environment, 41*(7), 902–909.

Greenspec. (2018). *Materials guides*. Retrieved April 20, 2018, from http://www.greenspec.co.uk/.

Horvath, A. (2004). Construction materials and the environment. *Annual Review of Environment and Resources, 29*, 181–204.

Lazarus, N. (2009). *Construction materials report: Toolkit for carbon neutral developments–part 1*. London: Bioregional.

Lujan, H. L.& Di Carlo, S. E. (2005). Too much teaching, not enough learning: what is the solution? Advances in Physiology Education, *30*(1), 17–22. Retrieved April 21, 2011 from http://www.advan.physiology.org

Legislation gov uk, Climate Change Act. (2008). *Chapter 27. Part 1 Carbon targeting and budgeting*. London: HMSO.

Letrud, K. (2012). A rebuttal of NTL institutes learning pyramid. *Education, 133*(1), 117–124.

Owen, P., & Dewick, P. (2015). *How effective is a games-centric approach in changing student eco behaviours?* Manchester: University of Manchester, UK Eco Action Games.

Roaf, S. (2004). *Closing the loop: Benchmarks for sustainable buildings*. London: RIBA Publishing.

Schell, J. (2015). *The art of game design: A book of lenses* (2nd ed.). Boca Raton, FL: CRC Press, Taylor & Francis.

Stern, N. (2006). *Stern review: The economics of climate change*. Cambridge: Cambridge University Press.

UKGBC. (2019). *Climate change*. www.ukgbc.org/climate-change/. UK green building council Retrieved from April 17, 2019.

Whitelock, D. (2011). Activating assessment for learning: Are we on the way with Web 2.0? In M. J. W. Lee & C. McLoughlin (Eds.), *Web 2.0-based-E-learning: Applying social informatics for tertiary teaching* (pp. 319–342). Hershey, PA: IGI Global.

WRAP UK—Waste & Resources Action Programme. (2007). *Halving construction waste to landfill by 2012: Construction briefing note*. Retrieved April 20, 2018, from www.wrap.org.uk/construction.

Sustainable Development of Mankind Through Ecological Services: A Literature Review

Mpho Ndou and Clinton Aigbavboa

Introduction

Sustainable development requires a focal point between the three pillars of economic development, which namely are social value, economic systems and environmental preservation, which can be complicated at times to unite (Kivati & Onyango, 2015). Following the Brundtland report and the Rio Summit in 1987 and 1992, respectively, governments and associations have taken up sustainable development as an attractive objective and created measurements for sustainable development, yet execution has been problematic. Matthews and Hammill (2010) mentioned that the fundamental issue after the Rio Summit has been in outlining the move of sustainability from hypothesis to reality regarding the industrious hold of mechanical, political and different requirements of sustainable policies.

It is yet one illustration of how sustainable development has been stretched in different directions over the last two decades. Characterizing and actualizing sustainable development has required the strains between the three pillars, and the predominant 'impacts' to work harmoniously. For this reason, more research is required to investigate alternative methods of achieving sustainable development in developing countries in view of mankind. The paper was structured on secondary based data, the adopted research methodology and the lessons learnt through research findings of the study. Subsequently, conclusions and recommendations relating to the sustainable development of mankind are drawn based on the findings of this paper.

M. Ndou (✉) · C. Aigbavboa
SARChi in Sustainable Construction Management and Leadership in the Built Environment, University of Johannesburg, Auckland Park, South Africa

L. Scott et al. (eds.), *Sustainable Ecological Engineering Design*,
https://doi.org/10.1007/978-3-030-44381-8_33

Sustainable Development: An Ecological Aspect

According to Goh (2014), there is no standard definition for sustainable development to date despite its popularity. An extensive variety of ideas of sustainability is still open, and various new considerations and thoughts have advanced after some time. The WCED definition to understand sustainable development has been generally acknowledged to be the cornerstone of ecological, financial and social interfaces in the dedication of reasonable improvement.

The term 'green' is regularly associated with sustainability, but in this context, it is linked with sustainable development. Yanarella, Levine, and Lancaster (2009) explained that green separates itself from sustainability since green is more concerned about the product of the three pillars, while sustainability focuses on each pillar, which mainly is the environment, economy and society. Similarly, Goh (2014) further illustrated through Fig. 1 below that green development is the focal point in sustainability. This results from the harmony between social consideration, economic welfare and environmental security.

Sustainable development focuses on the interrelation of the ecological and economic systems and how they function within the environment (Kivati & Onyango, 2015). Connecting biophysical elements of ecological systems with human development through the idea of Ecological Services is vital given the evaluation of the exchange offs (ecosystem, socio-cultural and monetary) included in the loss of ecological systems and biodiversity unmistakably and reliably. Similarly, the push to coordinate ecological and economic aspects which enhances natural and monetary administration and guarantee long-term sustainability has allowed more profound comprehension of the ecological functions and qualities.

Fig. 1 Three pillars of sustainability. (Source: Goh, 2014)

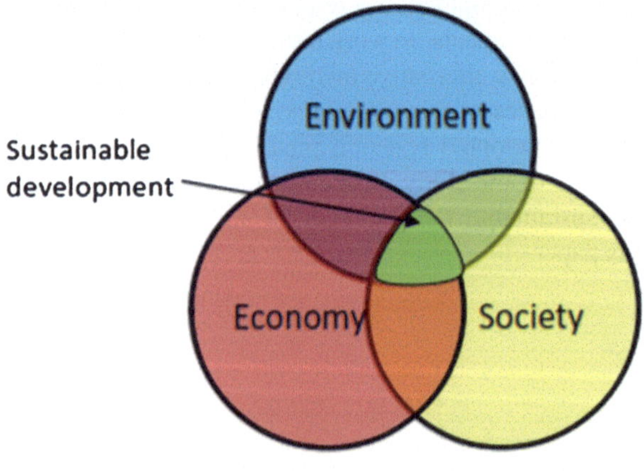

Ecological Services: A Development for Mankind

Ecological services (ES) are vital in the evaluation of mankind's development as environmental aspects are assessed over the long-term socio-economic and ecological expense frameworks which are beneficial towards various day-to-day activities of mankind. In turn, this long-term framework could be compared with the private short-term expenses framework by determining the main drivers of decision-making relating to sustainable development. According to Heal (2005), ecological systems are rare; material commitments to financial welfare cannot be underestimated and can be influenced by cognizant decisions. Figure 2 below shows the dependency of relations.

The outline as indicated by Kivati and Onyango (2015) underscores that the presence of mankind and society relies on upon the life support of the physical nature. Human culture has become reliant on the physically managed of the ecosystem. Nature, in any case, can proceed capably without mankind. Similarly, nature can sustain itself if mankind would become extinct. Again, if nature got to be terminated, or decline past a specific point, mankind would likewise suffer from the imbalance of the ecosystem. Nature, however, constitutes the external environment of mankind. Maiteny and Parker (2002) concluded an inevitable principle that makes mention of the organisation that bankrupts its surroundings bankrupts itself.

Ecological settings are inclined to degrade if they are overused or abused since the deterioration of nature might be unchangeable, or the system may take quite a while to recoup (Dasgupta, 2008). Similarly, Dasgupta (2008) added that this could cause the ecosystem to break down without earlier cautioning as the environment cannot be supplemented once exhausted or corrupted. Moreover, Kivati and Onyango (2015) stated that as the activities of mankind debilitate ES, it is vital to take note of long-term environmental well-being and its part in empowering the habitation and economic movement of mankind. The economic significance of ES to humanity legitimises the need to comprehend their value.

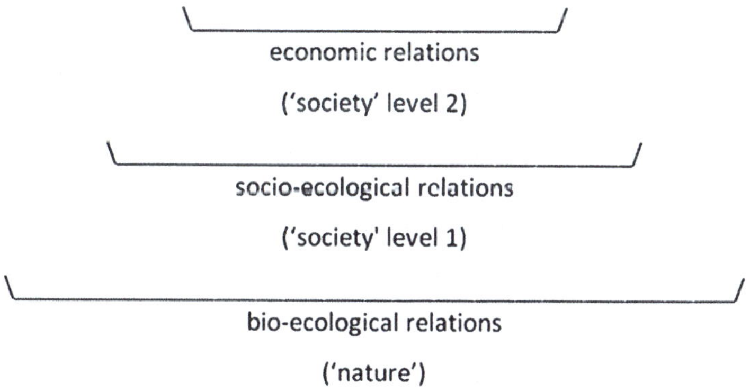

Fig. 2 The existence of dependency. (Source: Kivati & Onyango, 2015)

The clearest threat of disregarding the element of nature in economic matters is that the coexistence of the economy is due to nature, and by overlooking it we may unintentionally harm its self-recuperation function. Van der Heide (2013) opined that concerning the economy, ecosystems might be viewed as an extraordinary type of capital resources. The functions of ES and the reserves of natural resources that create them are working for each other rather than against one another. Kivati and Onyango (2015) indicated that ES contribute altogether towards human welfare, both directly and indirectly, and in this way speaks to a critical path of value in the planet economically. In addition, they realised that since these services are not taken note of in the business sectors or accurately measured in wording identically to economic development and services, they will be frequently given too little weight in policy decisions.

Besides, it was observed that the majority of economists analyse natural and human developed capital as substitutes instead of complementary; thus, neither one of the two elements can be restricting, but in complementary elements either might be restricting (Kivati & Onyango, 2015). Similarly, ecological economists view natural and human developed capital as complementary items, and along these lines stress the significance of restricting factors and the use of alternatives when it comes to scarcity patterns. Summarily, to comprehend the association between ES and the development of mankind as a component that supports policy decisions, different approaches can be implemented towards the remuneration of ES.

Remuneration for Ecological Services

According to the Millennium Ecosystem Assessment (MEA, 2005), ES can be widely described as the payment or remuneration people gain from conserving and developing ecosystems. On the other hand, alterations to the ecosystems and the debasement of ES are expanding at a disturbing rate. Jenkins, Murray, Kramer, and Faulkner (2010) explained that from an economic viewpoint, debasement takes place when multiple ES portray the attributes of public goods, bringing about externalities. Concerning public goods, ES has been undersupplied because of their absence of quality in the commercial setting. Owing to this, Pattanayak, Wunder, and Ferraro (2010) concluded that society neglects to set up establishments that internalise the value of services made available by ES.

Remuneration for ecological services (RES) is described by Engel, Pagiola, and Wunder (2008) to be a peculiar application that encourages the prevention of wasteful utilisation of resource post the 1992 Rio Summit as it envisaged to conquer the issue of externalities. Similarly, the concept form of RES can be associated within the ecological economic framework. Environmental debasement occurs within the framework due to the ceaseless business sector's repeated failure to internalise ecological externalities. Similarly, this results in an impaired public's view of ES as being centred around a 'free riding' state (Jack, Kousky, & Sims, 2008). The pseudo market together with other market developments through environmental externalities internalise the whole RES concept (Schomers & Matzdorf, 2013).

In the most recent years, the conceptual development of both the ES and RES ideas received considerable attention from researchers. The attention was created by the possibility of integrating these ideas into various remuneration plans and markets (Gomez-Baggethun, de Groot, Lomas, & Montes, 2010). The RES phenomenon continues to be a versatile term with numerous definitions that interchange (Schomers & Matzdorf, 2013). Recent studies revealed that numerous RES cases relatively include governmental mediation and public remuneration schemes (Muradian, Corbera, Pascual, Kosoy, & May, 2010; Vatn, 2010).

The intent of RES seeks to be the formation of incentives for the procurement of such goods, resulting in changing individual or joint departments that generally would prompt unnecessary deterioration of the ecosystem and the environment at large (Schomers & Matzdorf, 2013). According to Muradian et al. (2010), it might be suitable to define RES as an exchange of assets between social participants, which seeks to generate incentives to adjust individual and joint land use choices with the social enthusiasm for the proper utilisation of natural resources. Moreover, the definition also encompassed well-known Pigouvian concept under the RES scheme, generally referred to the governmental remuneration scheme (Vatn, 2010).

The Market Approach

The market approach aims to maintain a minimum to a no exchange costs including prescribed legislative laws about non-legislative or state power. These legislations are supposed to iron out the impact of externalities within the market (Schomers & Matzdorf, 2013). In addition, they added that preferably private or independent markets can be set amid public participants prompting sufficient distribution of natural assets despite what was allocated in the beginning, as the consumer will repay the supplier for the externalities. Similarly, governmental mediation will yield better efficient results than relying on the market sector for the dispersion of natural assets (Coase, 1960). Moreover, Coase (1960) mentioned that the underlying designation of property rights are somewhat limited to the government's responsibility amid justifying a lawful domain that enforces property legislative rights.

Engel et al. (2008) maintained that this approach advocates the emphasis of constructive outcomes that relates environmental efficiency together with economic efficiency when contrasted with governmental approaches. The market approach to RES is commonly referred to as the Coasean scheme. The adequacy of this approach is within its ability to provide a value-added service directly to the participation. Similarly, it also motivates or incentivise participation by guaranteeing the functionality of the components, monitoring specifically whether service delivery is being conveyed, and re-negotiating or cancelling the agreement if necessary. Genuine market approach case studies are not really depicted in literature. Most RES examples relating to the market approach involving the remuneration from the administration of ES is made available at a local level.

Similarly, Wunder and Alban (2008) investigated that in France, the water bottling company, Vittel, has been operating a RES programme with more than 25 agriculturists in the Vosges Mountains since the early 90s. Agriculturists are incentivized for their efforts of preserving the high-water qualities through cultivation processes in their farming activities. Subsequently, these actions drive the project to become perplexing with basic market transactions. Moreover, there are promotions of wildlife preservation through other studies conducted on the RES programme for the ecosystems in Africa. Tourism administrators contract regions between the borders of Tanzania and Kenya through a preservation concession contract that ensure monetary instalments are being paid out to the occupants of the regions on a yearly basis (Schomers & Matzdorf, 2013). Similarly, this contract can be further identified as a land lease that ensure contractual compliancy from the remunerator.

In a similar study, it was discovered that documented agreements suspend agrarian development, permanent residency, and the consumption of charcoal and unlicensed hunting within various areas (Schomers & Matzdorf, 2013). The objective of the remuneration seeks to prevent and re-address the considerable increment in the acquisition of land for livestock cultivating. Nonetheless, the loss of wildlife habitat and poaching is deduced to have brought about a considerable reduction in the overall wildlife habitat of those regions (Nelson et al., 2010). Moreover, Fisher, Kulindwa, Mwanyoka, Turner, and Burgess (2010) realised that various articles published portray concentration on assessing the likelihood of adopting the market-based RES programme in the near future. The possible adoption could infer that the market-based RES concept is liable to pick up insignificance in the near future.

The Governmental Approach

The governmental or state approach to RES is generally introduced as the Pigouvian scheme and follows the ecological tariff and standard procedures (Vatn, 2010). In another study, Van Hecken and Bastiaensen (2010) stated that the state approach is centred on the Pigouvian concept of financing current product markets using positive externalities. Similarly, the governmental approach depends on the notion (total remuneration = the total benefit) which should be yielded. Schomers and Matzdorf (2013) emphasised that the ecological tariff and standard procedures conversely commence with a prior arrangement of standards for ecological quality and after that enforced subsidies are required to accomplish these standards. Likewise, it is stressed that legislative RES programme from the traditional Pigouvian concept fails, as remunerations schemes are not connected to a product which is presumed to cater for the rewarding or benefit of the externality (Van Hecken & Bastiaensen, 2010).

Preferably, ES themselves are then changed over into a tradable product. This is evident when the government is regarded as a mediator following up for the benefit of service customers (Schomers & Matzdorf, 2013). Moreover, it was indicated that the primary contrast between the market and governmental approach to RES programme is, therefore, the candid quality of exchange, where previously the direct end-user remunerated the service supplier, which lead end-users being disregarded as the direct users. Subsequently, the depiction between these two approaches to the RES programme is regularly described by various exclusion cost structures. Also, the market approach relating to the RES programme often as possible remunerates land officials for the procurement of ES that is portrayed as club products (Vatn, 2010). Grantees of these ES are found mainly within the primary level which makes the identification process much easier.

The state approach to the RES programme is centred on the supply of public goods. The state approach to the RES programme is explored under the RES programme for China, the European Union (EU) and the United States (US). Although Australia might have similar national programmes of the same nature as those in the US and EU, there is limited literature linking Australia to direct RES programmes. Moreover, the governmental preservation tools use in China have direct RES titles were their state attributes are different from other national RES programmes. In another study, it was mentioned that there were no RES programmes particularly in Brazil (Schomers & Matzdorf, 2013). Nonetheless, in a recent development, it was noted that a proposal relating to the RES programmes are being prepared for submission to the environmental stakeholder in Brazil's governance. In addition, it was further indicated that South Africa has a legislative project which is slightly in line with the RES programme scope, the Working for Water programme (WfW), which concentrates basically on reducing the individual rates of unemployment and poverty. The preservation of the mountain catchments relating to its hydrology and ecology is a secondary act.

Research Methodology

The study was conducted based on a research objective of a larger research project. The research objective of this literature review paper was to assess the methods that can encourage the adoption of ecological services as a sustainable approach. The objective was achieved using existing theoretical frameworks and concepts in key literature on sustainable development and ecological economics. The study is a theoretical review which systematically prepared a conceptual foundation based on the various school of thoughts relating to ecological services and how they can aid towards the sustainability of mankind. The current methodology falls within a quantitative research approach that is adopted in the larger research project. A quantitative strategy was adopted based on its rigorous scientific and sampling properties together with its generalizability influence. Moreover, the current study adds to the body of knowledge relating to ecological economics as there are only a few studies on various remuneration schemes of ecological services in developed and developing countries.

Ecological Services in Developed and Developing Countries: A Case Study

A case study on RES programmes in various nations is portrayed in more detail below:

Ecological Services in the European Union

The EU has been practising the RES programme as an instrument to internalise externalities prior to the usage in Latin America (Schomers & Matzdorf, 2013). It is noteworthy that the RES programmes in the EU were enacted at a singular state membership. In another statement, Baylis, Peplow, Rausser, and Simon (2006) asserted that the RES programmes were converted into a pioneer state membership in terms of which the policies were now coordinated by one body. The Agri-environmental programmes (AEPs) were another branch of the RES programme which administered remuneration to farmers who contributed to the preservation endeavours that enhanced and maintained the environment voluntarily. It is nonetheless observed that the implementation of the AEPs likewise actuated a disputable discourse. The disputable argument questioned whether AEPs are masked production endowments giving a more worthy method that remunerates farmers, or an empowerment tool for externalities, whether negative or positive (Baylis et al., 2006).

It was further noted that farmers in the EU were demanding the first pillar remuneration, which was the first instalment received, a subsidy to continue farming (Baylis et al., 2006). These farmers were required to comply with Good Farm Practice (GFP), and any additional remuneration past the GFP standard in the form of a RES payment could be awarded on a voluntary basis. AEPs comprise an assortment of various Agri-environmental plans and measures. Similarly, it was noted that adoption was contingent upon the plan, whether a decrease of the externalities that negatively impact the plan (such as lessening the contamination of nitrate and pesticide and change of acceleration to comprehensive arable cultivating land, among others) or the procurement of positive externalities which attracts remuneration (Baylis et al., 2006). Around 20% of arable land in the EU makes use of some AEPs to decrease the detrimental effects currently experienced in the agribusiness, which subsequently affects the ecology. It is for this reason that Uthes, Matzdorf, Mueller, and Kaechele (2010) concluded that AEPs regularly needs focusing on critical areas, but unacceptable results are achieved regularly.

Ecological Services in the United States of America

The historical backdrop of the governmental approach to incentives for ES that encouraged ecological preservation in the US have been in practice for a longer period when compared to the EU. On the same note, Baylis et al. (2006) highlighted that in the pioneers of the Conservation Reserve Program (CRP) in ensured soils and brought about a major decrease in certain crop harvests to prevent an oversupply in the 1930s (Baylis et al., 2006). In the late 80s, the Farm Bill widened the US Agri-environmental strategy to incorporate environment and farm revenue concerns. Similarly, this revealed that Swampbuster and Sodbuster were alternative programmes which were incorporated in the Farm Bill to end conversion of wetland and very erodible land to farmland (Baylis et al., 2006). In addition, erodible areas were brought out of generations with the formation of the Conservation Reserve Program (CRP) (Dobbs, 2006).

During the late 90s, the Environmental Quality Incentives Program (EQIP) was included from the Farm Bill, and it was constantly modified during early 2000 into the Farm Bill, with extended financing and the development of the Conservation Security Program (CSP) (Schomers & Matzdorf, 2013). EQUIP and CSP are AEPs for operational lands and are fundamental programmes for the state to attain ES from farming. Moreover, the CSP is the nearest programme to what versatility is in the UN.

Ecological Services in China

In China, RES programmes are mostly referred to as eco-compensation (EC). There is no universal definition for EC to date. It can be interpreted as a financial instrument intended for the provision of public goods. The programme is public legislation that utilises the fiscal exchange instruments to internalise externalities and to accordingly adjust the damage among private and social interest (Xiong & Wang, 2010). This conforms to the RES definition of Muradian et al. (2010), and Xiong and Wang (2010) added that either a fee is levied to reduce negative externalities or remunerations in various schemes are allocated for the procurement of positive externalities. Moreover, it consents to the governmental approach of the RES programme. Be that as it may, the EC repudiates the RES since remuneration is actually a compensation for legitimate land use curb and, consequently, not a monetary incentive to cultivate land use transformation (Mullan, Kontoleon, Swanson, & Zhnag, 2011). Subsequently, it is simply a project rather than legislation.

The most important EC legislation is the forest ecological benefit compensation scheme (Xiong and Wang, 2010). Similarly, the Natural Forest Protection Program (NFCP) was introduced as a test programme in the late 90s and was implemented in more than ten provinces and self-governing districts. It was reported that by the beginning of the twenty-first century, the NFCP was extended to 18 regions and districts and, in this way, became one of the biggest forest protection programmes on the planet (Mullan et al., 2011). Also, forests were to be re-established and secured using bans on logging to conserve ES that included soil disintegration, and water maintenance and flood control (Liu, Li, Ouyang, Tam, & Chen, 2008).

In support of the above statement, it was mentioned that payments made up for financial misfortunes because of the legislative confinements on logging and remuneration for reforestation and the sustainability of forest lands (Muradian et al., 2010). In addition, subsidies of up to 81.5% for the programme were given by the central government with the local state providing 18.5% towards the programme (Liu et al., 2008). This reveals that the approach to the RES which is used in China is focused more towards a remuneration policy for legislative restrictions.

Ecological Services in South Africa

The introduction of the South African Working for Water Programme (WfW) as a state programme or initiative in the mid-90s lead to poverty alleviation communal project which is still operational today. The WfW was incorporated into the RES scheme in light of the hydrological capacities and the biodiversity of mountain catchments which the initiative mainly focused on (Schomers & Matzdorf, 2013). In addition, the WfW scheme does not pay duty fees to land officials for land utilisation and changes that are expected to give or save certain ES. Rather, unemployed people are contracted to destroy alien plant species and to re-establish natural fire systems in private, communal mountain catchments and riverine areas.

Moreover, Swallow, Leimona, Yatich, and Velarde (2010) realised that financing for the WfW scheme was generally generated from communal poverty projects and water levies. In light of the explained RES definitions, it was noted that the WfW scheme is not a financial component to internalise externalities by relegating monetary qualities to ES (Schomers & Matzdorf, 2013). In context, it is a communal employment project and still uses monetary links, which remunerates conducts that protect ES. Similarly, the Comprehensive Agricultural Support Programme (CASP) was initiated in 2004, which was aimed at reducing poverty and unemployment and bring about food security. This initiative relied more on fiscal policy through a loan structure which was aimed at fast-tracking the implementation of a then vision 2021 of the National Development Plan.

Ecological Services in Brazil

Brazil presently has neither a national RES programme nor does it practice the lawful idea of ES. Be that as it may, Farley and Costanza (2010) indicated that both a governmental approach for ES and a national RES programme are under debate. If the programme is endorsed, the Brazilian RES scheme will depend on how Proambiente concept is defined as within the ES title. Under the Proambiente concept, remuneration plans were created to compensate farmers for the procurement of ES like the decrease or evasion of deforestation, carbon sequestration, recovery of ecological hydrological capacities, soil protection, the conservation of the ecosystem, the lessening of forest wildfires and diminishing the loss of ES caused by agriculture (Costenbader, 2009). In addition, for Brazil to have a fully fledged state RES programme, they need to incorporate reduced emission from deforestation and carbon sequestration programmes.

Discussion

The appraisal of the study reveals that there are two main approaches which are generally followed by developed and developing countries. This excludes the Proambiente concept which has not gained much interest from researchers and other publications under the RES theme. The case study shows that RES programmes in developing countries receive less attention, often pertaining to the need for innovation and absence of the distribution of remuneration linked with opportunity costs when compared to developed countries. Also, the reviews revealed how other forms of financial incentives were currently categorised as RES by field professional, governments and researchers. Similarly, this categorisation reveals a more politically driven initiative than an implementation scheme, which was its original mandate. It was observed that most of the literature published under the theme of 'RES' adopted a governmental approach as opposed to a market approach (Wunder, 2005). Schomers and Matzdorf (2013) revealed that most RES programmes adopted a governmental approach in incentivising ES in most parts of the world. On the other hand, the classification of the incentive scheme as a RES programme or just an incentive-driven programme was dependent on how the government implemented its policies. In another study, Latin America as a developing country was a party to using various fiscal and market-related policy instrument. Nonetheless, Latin America classifies the method of RES as being more policy inclined, than the intended incentive scheme when compared to industrialised nations.

In other similar studies, both developed and developing countries adopting ES such as the agricultural and environmental based governmental RES schemes (Hajkowicz, 2009; Uthes et al., 2010). Countries such as China, US, Mexico, European nations and Costa Rica revealed in their practical RES schemes that the market approach was an insignificant role player in their request for sustainable development. On the other hand, The South African WfW initiative is more economic orientated as opposed to being ecological bias. This means that it is more focused on communal employment rather than a regulatory remuneration for ecological services. Similarly, the CASP initiative is also driven to address the National Development Plan (NDP) of 2030, which seeks to alleviate poverty, job creation, and food security which still lacks the incentives to drive more ecological practices within NDP policies. Various authors have cited the relevance of market-based approach as a more theoretical model than a practical model as it received most of its attention from academics and researchers (Pascual, Muradian, Rodriguez, & Duraiappah, 2010; Wunder & Alban, 2008). In addition, the market approach created a platform for the theme of economic efficiency and development to be discussed in academic forums and sessions (Pascual et al., 2010).

From the case study, it is evident that the market approach adopts a more economical model as opposed to the governmental approach which adopts a more sustainable and ecological incentive scheme. The distinction between the two approaches does not mimic the current policy settings diversity in various nations. In industrial driven nations such as the US and the EU, there have been multiple incentive schemes that rewarded land conservation practices of stewards which were not classified as ecological practices, for which literature publications were undocumented. Similarly, there have been other market-related approaches which adopted a RES model but was categorised outside the theme of RES, which resulted in multiple publications being regarded as mere environmental conservation schemes rather than ES (Carroll, Fox, & Bayon, 2008; Hallwood, 2010). This brings to light that the RES has been misunderstood as a mere incentive scheme rather than a constituent of a nation's sustainable policy instrument which seeks to transform and develop ecological habits practice by mankind (Schomers & Matzdorf, 2013; Van Hecken & Bastiaensen, 2010). A functional RES scheme serves as an 'add-on' in terms of environmental conservation linked to ecological services and economic efficiency of existing policy regulations as opposed to replacing these regulations. Nonetheless, complete RES adoption worldwide will require full-time participation from the governance of these nations.

Conclusion and Recommendations

The conceptual foundation laid by the theoretical analysis of the paper appraised the sustainability of the two RES schemes adopted in both developed and developing countries. The intercountry case study realised that more efforts of RES adoption were needed in developing countries who were more geared towards economic policy instruments rather than ecological practice. The packaging of RES schemes in developing countries needs to be more incentive orientated rather than programmed based in order to ensure adoption. Moreover, the theoretical value of the paper revealed that, although RES programmes are in motion in some developed and industrialised nations, the 'add-on' functionality of these programmes still needs to be fused into an existing sustainable policy which seeks to develop mankind.

It is therefore recommended that developing countries adopt similar schemes such as the WfW and CASP initiative implemented in South Africa. In countries that have existing agricultural schemes and initiatives, these can be modified to include a policy instrument were environmental conservation stewards are remunerated for any ES practised. This, in turn, will drive the sustainable development of mankind due to the remuneration behind the motive. Also, developed as well as industrialized nations can consider more policy measures that will drive the ecological 'add-on' functionality in a more practical manner as opposed to just being the desired legislation. An extension to the study can be conducted on how the market approach could be introduced in conjunction with the governmental approach in the effort of creating a hybrid approach. The hybridisation will not only introduce an economic instrument but open a platform for more stakeholders such as the private sector participants to contribute to the ecological movement nations need to adopt.

References

Baylis, K., Peplow, S., Rausser, G., & Simon, L. (2006). Agri-environmental policy in the European Union: Who's in charge? Commissioned Paper. *Canadian Agricultural Trade Policy Research Network (CATPRN), 4*, 1–23.

Carroll, N., Fox, J., & Bayon, R. (2008). *Conservation and biodiversity banking: A guide to setting up and running biodiversity credit trading systems*. London: Earthscan.

Coase, R. H. (1960). The problem of social cost. *Journal of Law and Economics, 3*, 1–44.

Costenbader, J. (2009). Legal frameworks for REDD. Design and implementation at the National Level, IUCN. *Environmental Policy and Law*, Paper No. 77.

Dasgupta, P. (2008). Nature in economics. *Environmental and Resource Economics, 39*, 1–7.

Dobbs, T. (2006). Working lands agri-environmental policy options and issues for the next United States Fram Bill, economic staff paper, 3. *Department of Economics, South Dakota State University, 2006*, 1–37.

Engel, S., Pagiola, S., & Wunder, S. (2008). Designing payments for environmental services in theory and practice: An overview of the issue. *Ecological Economics, 65*, 663–674.

Farley, J., & Costanza, R. (2010). Payments for ecosystem services: From local to global. *Ecological Economics, 69*, 2060–2068.

Fisher, B., Kulindwa, K., Mwanyoka, I., Turner, R. K., & Burgess, N. D. (2010). Common pool resource management and PES: Lessons and constraints for water PES in Tanzania. *Ecological Economics, 69*, 1253–1261.

Goh, C.S. (2014) *Development of a capability maturity model for sustainable construction*. HKU Theses Online (HKUTO), Hong Kong.

Gómez-Baggethun, E., de Groot, R., Lomas, P. L., & Montes, C. (2010). The history of ecosystem services in economic theory and practice: From early notions to markets and payment schemes. *Ecological Economics, 69*, 1209–1218.

Hajkowicz, S (2009). The evolution of Australia's natural resource management programs: Towards improved targeting and evaluation of investments. *Land Use Policy, 26*(2):471–478.

Hallwood, P. (2010). Contractual difficulties in environmental management: The case of wetland mitigation banking. *Ecological Economics, 63*, 446–451.

Heal, G. (2005). Valuing ecosystem services. *Ecosystems, 3*(1), 24–30.

Jack, B. K., Kousky, C., & Sims, K. R. E. (2008). Designing payments for ecosystem services: Lessons from previous experience with incentive-based mechanisms. *Proceedings of the National Academy of Sciences, 105*, 9465–9470.

Jenkins, W. A., Murray, B. C., Kramer, R. A., & Faulkner, S. P. (2010). Valuing ecosystem services from wetland restoration in the Mississippi Alluvial Valley. *Ecological Economics, 69*, 1051 1061.

Kivati, G.N., & Onyango, J. (2015). Conventional vs. ecological economics perspectives in the valuation of ecosystem services in East Africa. *African Centre for Technology Studies*, paper 010.

Liu, J. G., Li, S. X., Ouyang, Z. Y., Tam, C., & Chen, X. D. (2008). Ecological and socioeconomic effects of China's policies for ecosystem services. *Proceedings of the National Academy of Sciences of the United States of America, 105*, 9477–9482.

Maiteny, P., & Parker, J. (2002). *Unit 6 study guide-science and culture in education for sustainability*. London: Distance Learning Centre, London South Bank University, London.

Matthews, R. A., & Hammill, A. (2010). Sustainable development and climate change. *International Affairs, 85*(6), 1117–1128.

Millennium Ecosystem Assessment (MEA). (2005). *Ecosystems and human wellbeing: Synthesis*. Washington, DC: Island Press.

Mullan, K., Kontoleon, A., Swanson, T., & Zhnag, S. Q. (2011). When should household be compensated for land-use restriction? A decision-making framework for Chinese forest policy. *Land Use Policy, 28*, 402–412.

Muradian, R., Corbera, E., Pascual, U., Kosoy, N., & May, P. H. (2010). Reconciling theory and practice: An alternative conceptual framework for understanding payments for environmental services. *Ecological Economics, 69*, 1202–1208.

Nelson, F., Foley, C., Foley, L. S., Leposo, A., Loure, E., Peterson, D., … Williams, A. (2010). Payments for ecosystem services as a framework for community-based conservation in Northern Tanzania. *Conservation Biology, 24*, 78–85.

Pascual, U., Muradian, R., Rodriguez, L. C., & Duraiappah. (2010). Exploring the links between equity and efficiency in payments for environmental services: A conceptual approach. *Ecological Economics, 69*, 1237–1244.

Pattanayak, S. K., Wunder, S., & Ferraro, P. J. (2010). Show me the money: Do payments supply environmental services in developing countries? *Review of Environmental Economics and Policy, 4*, 254–274.

Schomers, S., & Matzdorf, B. (2013). Payments for ecosystem services: A review and comparison of developing and industrialized countries. *Ecosystem Services, 6*, 16–30.

Swallow, B. M., Leimona, B., Yatich, T., & Velarde, S. J. (2010). The conditions for functional mechanisms of compensation and rewards for environmental services. *Ecology and Society, 15*(4), 6.

Uthes, S., Matzdorf, B., Mueller, K., & Kaechele, H. (2010). Spatial targeting of Agri-environmental measures: Cost-effectiveness and distributional consequences. *Environmental Management, 46*, 494–509.

Van Hecken, G., & Bastiaensen, J. (2010). Payments for ecosystem services: Justified or not? A political view. *Environmental Science and Policy, 13*, 785–792.

Van der Heide, S. H. (2013). Economic viewpoints on ecosystem services, Wageningen, Statutory Research Tasks Unit for Nature and the Environment (WOT Namur & Milieu). *WOt-rapport 123*, 68, 7.

Vatn, A. (2010). An institutional analysis of payments for environmental services. *Ecological Economics, 69*, 1245–1252.

Wunder, S. (2005). *Payments for environmental services: Some nuts and bolts*. CIFOR Occasional Paper 42.

Wunder, S., & Alban, M. (2008). Decentralized payments for environmental services: The cases of Pimampiro and PROFAFOR in Ecuador. *Ecological Economics, 65*, 685–698.

Xiong, Y., & Wang, K. L. (2010). Eco-compensation effects of the wetland recovery in Dongting Lake area. *Journal of Geographical Sciences, 20*, 389–405.

Yanarella, E. J., Levine, R. S., & Lancaster, R. W. (2009). Research and solutions: "Green" vs. sustainability: From semantics to enlightenment. *Sustainability: The Journal of Record, 2*(5), 296–302.

Waste Management: The Case of Construction and Demolition Waste in Port Elizabeth

Siyabonga Jikeka, John Smallwood, Fidelis Emuze, and Eric Simpeh

Introduction

The construction industry is one of the most significant industries on a global scale, with respect to its economic, technological, and environmental impact. Its rapid growth over the recent decades has resulted in an increase in the amount of CDW, thus burdening the environment (Banias, Achillas, Vlachokostas, Moussiopoulos, & Tarsenis, 2010). CDW is produced from the construction, renovation, repair and demolition of technical structures such as residential and commercial buildings, roads, and bridges (Fatta et al., 2003). The composition of CDW varies for these different activities and structures. CDW often contains bulky and heavy materials, including bricks, wood, reinforced concrete, lime, concrete, mixed earth, electrical wiring, steel bars, plumbing fixtures, and earth and rock arising from clearing site. More importantly, CDW waste also consists of hazardous substances such as lead and asbestos (Chatterjee, Ghosh, Ghosh, & Haldar, 2016). All these varied materials must be managed in an environmentally sound and economic feasible manner (Fatta et al., 2003). According to Ekanayake and Ofori (2004), this waste is problematic, not so much for its hazardous nature as for the sheer volume generated. However,

S. Jikeka · J. Smallwood (✉)
Department of Construction Management, Nelson Mandela University, Port Elizabeth, South Africa
e-mail: john.smallwood@mandela.ac.za

F. Emuze
Department of Construction Management, Nelson Mandela University, Port Elizabeth, South Africa

Department of Built Environment, Central University of Technology, Bloemfontein, South Africa

E. Simpeh
Department of Construction Management and Quantity Surveying, Cape Peninsula University of Technology, Cape Town, South Africa

L. Scott et al. (eds.), *Sustainable Ecological Engineering Design*,
https://doi.org/10.1007/978-3-030-44381-8_34

between 50% and 80% of construction waste is reusable or recycled. In an increasingly urban world, the growth in waste generation, particularly in CDW, has led to serious management problems in cities and countries (Rodriguez, Alegre, & Martinez, 2007). Environmental problems resulting from CDW disposal are a cause for concern because of the impact that illegal disposal sites, which are common, have on cities and their surroundings (Duran, Lenihan, & O'Regan, 2006). CDW represents a significant portion of municipal solid waste and is a problem to be managed in most municipalities due to the lack of information with respect to generation and disposal in urban areas (Pinto & Gonzalez, 2007). Nunes (2004) states that CDW is part of municipal solid waste and has a constitution that varies, depending on the generating source, phase of work site, constructive technology, and nature of work site. The rapid growth of large and mid-sized cities has increased the generation of CDW. This increase and lack of municipal policies has aggravated public problems related to the collection, transport, and disposal of urban solid waste, with illegal disposal of C&D waste being frequently observed in places such as streets, pavements, wastelands, hillsides, and stream and river beds.

Techniques for the safe disposal of solid waste are available. However, it is believed that these techniques will be insufficient to meet future needs and that advanced technology needs to be developed to utilise waste as a resource in the future, by recovering materials and perhaps energy from waste for recycling and synthesising of new products from waste material. Hence, this research is aimed at addressing key issues and current activities related to CDW management in the Nelson Mandela Bay metropolitan area, and specifically to investigate whether: CDW is indiscriminately disposed of; certain CDW is hazardous; there is a disproportionate amount of CDW, and generated CDW is often re-wasted. Thus, the management of CDW is a diverse challenge. Policy making, planning, and management roles and responsibilities are shared by every tier of government and sectors, both private and public, which provide CDW management services. Ownership and operation of CDW management systems and facilities, which includes storage, collection, transport, recycling, combustion, and land filling, are shared by local government service providers, commercial enterprises, industrial organisations, and private solid waste management service providers.

Review of the Literature

Waste Emanating from Construction and Demolition Sites

The construction industry as a sector generates a significant amount of CDW, approximately 40% of the waste stream of a country, which excludes mining and power station waste, and about 15–30% of the waste that ends up in landfill sites (Macozoma, 2001). At construction site level, it is estimated that up to 15% of materials purchased end up as waste, while at demolition sites, up to 100% of waste generated can end up as waste destined for landfill after the usual stripping of basic high-value products. Demolition sites usually generate the highest proportion of CDW, followed by renovations, while construction sites generate the least (Macozoma, 2001). For instance, The United States of America (USA) produced 534 million tons of CDW in 2014, 166 million tons emanating from buildings (US Environmental Protection Agency, 2016). More so, Hong Kong produces 57,547 tons of CDW and landfills 3942 tons of CDW per day (Hong Kong Environmental Protection Department, 2015). In the South African context, CDW generated by the construction industry in 2011 was 4,725,542 tons, and of that, only 756,087 tons was recycled. The C&D waste generated, accounted for 20% of all the general waste produced in the country (Department of Environmental Affairs, 2012). For example, in the Gauteng province, 25% of the waste landfilled is C&D waste and contributed to the diminishing air space in landfills (Gauteng Department of Agriculture Conversation and Environment, 2009). Although it may appear as though construction activities present the least problem in terms of CDW generation, closer examination will indicate that it is still significant. For example, it has been determined that up to 80% of a homebuilder's waste stream is recyclable (Macozoma, 2001). If recovered for reuse and recycling, this waste could reduce the cost of waste disposal, reduce procurement costs of virgin materials through the reuse of materials on site where appropriate, and also possibly generate revenue through the reuse of materials on site where appropriate, and through the sale of such material to the public. Although the disposal costs of construction site waste form as little as 0.5% of the total budget of a typical home, contractors know that this can affect their profits since they generally operate within a tight profit margin.

Certain Construction and Demolition Waste is Hazardous

Most construction waste can be classified as inert, such as bricks, solid concrete, stone, sand, gravel, and other arid materials. However, a small portion of construction waste is hazardous such as asbestos, with particularly fine fibres, solvents, and resins, while other types of non hazardous waste may also be present, such as rubbish, cardboard, and glass (Rodriguez et al., 2007). One environmental aspect requiring special attention in construction firms is the generation of waste, due not so much to their hazardous nature, but as to the sheer volume produced. If this waste is deposited in landfills, the useful life of the landfill is reduced due to its limited capacity, resulting in a substantial environmental impact (Ekanayake & Ofori, 2004). Moreover, CDW may contain harmful substances which can seep into the ground and cause air, water, and land pollution within the natural environment (Hareli, Portnov, & Seror, 2014). The interpretation Guide of Standard ISO 14001: 96, which serves to facilitate implementation of environmental management systems by companies in the construction sector, places special stress on this environmental aspect, and establishes criteria for the specific management of different types of waste generated on the construction site, namely used oil, hazardous waste,

waste derived from containers and packaging, and inert waste (Rodriguez et al., 2007). Consequently, construction firms should apply different methods for the management of solid, hazardous and inert waste on sites, to observe the specific legal requirements regulating each type of waste. It should be stressed that one of the main preventive instruments currently applicable for environmental management of construction projects is environmental impact evaluation.

Minimising the Amount of CDW Through Materials Management and Construction Management

Reclamation of material from the waste stream is not considered recycling *per se*. Recycling, on the other hand, is the reprocessing of a reclaimed material and converting it into a new material or use. Recycling CDW is hardly a new concept for the construction industry. Concrete and paving materials have been reused as fill material or roadbed for many years. In fact, the form of concrete used by the Romans included 'recycled' aggregate from older stone structures. There are several reasons besides the cost of disposal for the increased interest in recycling CDW. The decreasing availability of a quality resource base for the manufacture of building materials is an important consideration. In many cases, the sources for 'virgin' materials are great distances from installation/building projects and transportation costs warrant contractors to seek a local replacement. For example, aggregate from demolition rubble can be an abundant and prevalent local source of road base and fill. An additional benefit to the use of demolition aggregate is that the virgin quarried aggregate will not have to be used in such low value applications such as backfill. In this case, a low cost, but equivalent recycled product such as demolition aggregate is an acceptable substitute. There are other products such as recycled gypsum board and lumber that are being used in new construction projects.

Research

Research Methodology

The quantitative research approach was adopted to explore, investigate, and understand how solid waste in the form of CDW is managed in Port Elizabeth.

A custom-designed questionnaire was used for the collection of the primary data. The questionnaire included Likert-type scale type questions—scales of 1–5, and 1–4 were used. The questionnaire consisted of two sections. Section A addressed the demographic details of respondents, such as gender, age, profession, qualification(s), function in the firm, size of the firm, and the sector in which the firm operates. Section B addressed whether: CDW is indiscriminately disposed of;

Table 1 Survey response rate

Category	Approached (No.)	Responded (No.)	Response (%)
Contractor	10	7	70.0
Site agents	8	6	75.0
Municipal officials	6	4	66.7
Officials working on landfills	1	1	100.0
Total	**25**	**18**	**72.0**

certain CDW is hazardous; there is a disproportionate amount of CDW, and generated CDW is often re-wasted.

A non-probability sampling technique was adopted, as the whole population was not accessible, and the distribution and collection of questionnaires was subject to the availability of officials. The non-probability sampling technique can be judgmental, opportunistic, and snowball. This research entailed a balance of judgmental and snowball. For the purposes of this study, the population is defined as all full-time employees, employed by construction and demolition firms, government departments assigned to waste management duties, and service providers dealing with CDW. The empirical study included the following: contractors; municipal waste management officials, and officials working on landfills around Port Elizabeth.

Participants were requested to complete the questionnaire and submit to the lead-researcher by means of a facsimile, hand delivery, or by means of an e-mail. A total of 25 questionnaires were sent to the sample, of which 18 questionnaires were received by the cut-off date for receipt of completed questionnaires (Table 1). The average response rate to surveys is 55.7%; therefore, a response of 72.0% (Table 1) can be considered as good, compared with the average of 55.7%. Moreover, a response rate of at least 50% can be considered as adequate for conducting analysis.

Results

The demographic data includes gender; age group; highest formal qualification; length of service in the construction industry; firm/practice size; function in firm/practice, and sector in which the firm/practice operates.

48.4% of respondents were in the age range 40–49 years, which may be the more experienced employees. 35.5% of the respondents were in the age range 30–39 years. A further 8.7% were in the age range '50–59' years, and 1% in the 60 years plus range. Therefore, 54.1% were 40 years and older, which constitutes 'older workers', which also reflects maturity and experience, and thus gives credibility to the responses.

In terms of highest academic qualifications of the respondents, matric (32%) predominates, followed by diploma (24%), postgraduate diploma (15%), and Bachelor's degree (9%). Only 4% have an Honours degree, and 2% a Master's degree. Therefore, 68% have a tertiary qualification.

In terms of the length of service of respondents within the industry, 32% have been working for 5–9 years, 24% have been working for 10–19 years, 20% have been working for 20 years or more, 16% 4 years and less, and 8% have been in the industry with different firms. Therefore, 52% have been working in the industry for 10 or more years.

55% of respondents work for firms that employ between 21 and 30 people, 25% 30 or more people, 11% between 11 and 20 people, and 9% between 1 and 10 people. In summary, 20% work for firms that employ between 1 and 20, and 80% between 21 and more.

50% of the respondents are site managers, who are responsible for the physical construction process and should have intimate knowledge of site practices. 30% are Health and Safety Officers, 17% are contract managers, and only 3% are directors. 45% of the respondents work for a public sector client, 30% work for the private sector, and the remaining 25% work for private-public partnership entities.

Table 1 indicates the respondents' ratings of aspects relative to CDW disposal in the Nelson Mandela Bay Metropolitan area in terms of percentage responses to a scale of very poor to very good, and a mean score (MS) ranging between 1.00 and 4.00, the midpoint being 2.50. MSs > 2.50 indicate a rating of good, as opposed to poor as in the case of MSs ≤ 2.50. However, given that the MSs relative to all the statements are >1.75 ≤ 2.50, the ratings can be deemed to be between very poor to poor/poor. In general, the responses indicate that there is very poor management of CDW disposal within the municipality (Table 2).

Table 3 indicates the degree of concurrence with statements related to illegal dumping of CDW in terms of percentage responses to a scale of strongly disagree to strongly agree, and a MS ranging between 1.00 and 5.00, the midpoint being 3.00 (Neutral). It is notable from Table 3 that only 2/6 (33.33%) MSs are >3.00, which indicates that in general the participants agree, as opposed to disagree with the two statements.

The MSs (>2.60 ≤ 3.40) relative to statements 'There is insufficient space on landfill sites to cater for construction and demolition waste', 'The limited number of

Table 2 Rating of aspects relative to CDW disposal in the Nelson Mandela Bay metropolitan area

	Response (%)						
Aspect	Unsure	Very poor	Poor	Good	Very good	MS	Rank
Is the disposal of CDW organised and monitored in the municipality?	0.0	27.8	27.8	33.3	11.1	2.28	1
What is your perception of landfill management in the municipal area?	0.0	38.9	33.3	11.1	16.7	2.05	2
How effective is the municipality in terms of landfill management?	0.0	44.4	33.3	16.7	5.6	1.95	3
What is your perception of CDW disposal in the municipality?	0.0	55.6	16.7	16.7	11.1	1.84	4
How effective is the municipality in terms of waste disposal monitoring?	0.0	44.4	38.9	11.1	5.6	1.78	5

Table 3 Degree of concurrence with statements related to illegal dumping of CDW

Statement	Response (%)						MS
	Unsure	Strongly disagree	Disagree	Neutral	Agree	Strongly agree	
There is insufficient space on landfill sites to cater for CDW	0.0	3.2	22.6	29.0	32.3	12.9	3.32
The location of landfill sites is too far from construction sites which leads to illegal dumping	3.2	19.4	41.9	25.8	9.7	0.0	2.30
The limited number of landfill sites around town leads to illegal dumping	6.5	9.7	32.3	35.5	12.9	3.2	2.66
Disposing of waste properly to landfill sites is not always affordable due to the high rate of disposal trucks per m^3	6.5	16.1	38.7	29.0	9.7	0.0	2.42
Illegal dumping of waste is due to the lack of understanding of laws	0.0	3.2	9.7	29.0	38.7	19.4	3.61
Municipality does provide waste management services (skips) to sites	0.0	9.7	41.9	25.8	19.4	3.2	2.71

landfill sites around town leads to illegal dumping', and 'Municipality does provide waste management services (skips) to sites' indicate that the concurrence is between disagree to neutral/neutral. The MSs ($>1.80 \leq 2.60$) relative to the statements 'The location of landfill sites is too far from construction sites which leads to illegal dumping' and 'Disposing of waste properly to landfill sites is not always affordable due to the high rate of disposal trucks per m^3' indicate that the concurrence is between strongly disagree to disagree/disagree. The MS ($>3.40 \leq 4.20$) relative to the statement 'Illegal dumping of waste is due to the lack of understanding of laws' indicates the concurrence is between neutral to agree/agree. In general, the responses indicate that the number and location of landfill sites has nothing to do with illegal dumping.

Table 4 indicates the degree of concurrence with statements related to hazards and CDW in terms of percentage responses to a scale of strongly disagree to strongly agree, and a MS ranging between 1.00 and 5.00, the midpoint being 3.00 (Neutral). 4/5 (80.0%) MSs are >3.00, which indicates that in general the respondents agree, as opposed to disagree with the statements.

The MSs ($>2.60 \leq 3.40$) relative to 'There is insufficient space on the landfill sites to cater for CDW', 'There are procedures in place for disposal and treatment of hazardous waste', 'There are management procedures for hazardous waste on your site', and 'Hazardous waste is disposed on specific specialised landfill site' indicate that the concurrence is between disagree to neutral/neutral. The MS ($>3.40 \leq 4.20$) relative to 'The illegal dumping of waste effects the environment by contaminating the soil and can impact proper draining, and makes areas more sus-

Table 4 Degree of concurrence with statements related to hazards and CDW

Statement	Response (%)						MS
	Unsure	Strongly disagree	Disagree	Neutral	Agree	Strongly agree	
There is insufficient space on the landfill sites to cater for CDW	0.0	3.2	19.4	25.8	45.2	6.5	3.23
There are procedures in place for disposal and treatment of hazardous waste	0.0	6.5	12.9	25.8	48.4	6.5	3.35
The illegal dumping of waste effects the environment by contaminating the soil and can impact proper draining and make areas more susceptible to flooding	0.0	3.2	12.9	19.4	54.8	9.7	3.55
There are management procedures for hazardous waste on your sites	3.2	0.0	25.8	32.3	32.3	6.5	3.20
Hazardous waste is disposed on specific specialised landfill sites	0.0	3.2	45.2	25.8	22.6	3.3	2.81

ceptible to flooding' indicates that the consensus is between neutral to agree/agree. In general, the responses indicate that the insufficient space on landfill sites to dispose hazardous waste effects the environment and poses a threat to the public.

Table 5 indicates the degree of concurrence with statements relating to materials management and recycling in terms of percentage responses to a scale of strongly disagree to strongly agree, and a MS ranging between 1.00 and 5.00, the midpoint being 3.00 (Neutral). It is notable that all the MSs are >3.00, which indicates that in general the respondents agree, as opposed to disagree with the statements. Given that the MS of the statement 'The schedule of material ordered is always as accurate as possible', the concurrence can be deemed to be between disagree to neutral/neutral. The MSs (>3.40 ≤ 4.20) relative to 'Packaging used for material delivered on site is always recycled', 'Delivered material is stored in a safe, secure and waterproof storage area', and 'Suitable spoil, demolition materials, prunings, and surplus construction material arising from the works is recycled to avoid the need to transport to landfills' indicate that the consensus is between neutral to agree/agree. The MS (>4.20 ≤ 5.00) relative to 'Damaged or incomplete material delivered to site is isolated for recycling purposes' indicates that the consensus is between agree to strongly agree/strongly agree. In general, it may be assumed that contractors are managing their schedule of materials ordered, packaging of material, and recycling of damaged or incomplete material.

Table 5 Degree of concurrence with statements related to materials management and recycling

Statement	Response (%)						MS
	Unsure	Strongly disagree	Disagree	Neutral	Agree	Strongly agree	
The schedule of material ordered is always as accurate as possible	0.0	3.2	29.0	29.0	35.5	3.2	3.06
Packaging used for material delivered on site is always recycled	0.0	0.0	0.0	9.7	83.9	6.5	3.97
Damaged or incomplete material delivered to site is isolated for recycling purposes	0.0	0.0	0.0	6.5	54.8	38.7	4.32
Delivered material is stored in a safe, secure and waterproof storage area	0.0	0.0	6.5	19.4	64.5	9.7	3.77
Suitable spoil, demolition materials, pruning's and surplus construction material arising from the works is recycled to avoid the need to transport to landfills	0.0	0.0	3.2	25.8	67.7	3.2	3.71

Discussion

According to the literature, the construction industry contributes significantly to the amount of waste that is generated globally, South Africa included. More so, empirical evidence seems to suggest that the uptake of CDW management concepts in South Africa is rather slow. Reasons for this vary, but generally they are not unconnected with insufficient regulatory interventions, low levels of awareness and education, underdeveloped secondary markets for construction materials, negative perceptions, and perceived limited incentives. The study determined that the management of CDW disposal within the municipality is generally poor. Many countries are concerned with respect to how the construction industry treats CDW, as well as the perceived diminishing landfill disposal capacity. These apparent concerns thus encourage the evolution of approaches that are synonymous with sustainable practices. CDW management is therefore one of the approaches that is continuously proffered. It is therefore argued that consistency should be achieved at least nationally in terms of the definition and classification of CDW.

The municipality, within the context of this study, is viewed with respect to the management of CDW to protect the environment and the public. For instance:

- There is a gradual clamp-down on CDW disposal on landfill sites.
- There is a shift towards improved site management and waste prioritisation with the focus on waste prevention, source control, and separate waste collection.
- Secondary markets are starting to play a pivotal role in the management of CDW.

- Awareness creation through information dissemination, apart from enforcing applicable legislation.

However, the findings from the study indicate that the insufficient space on landfill sites to dispose hazardous waste effects the environment and poses a threat to the public. It is notable that the MSs relative to the majority of statements are $>2.60 \leq 3.40$, which indicates that the consensus with respect to 'There is insufficient space on the landfill sites to cater for construction and demolition waste', 'There are procedures in place for disposal and treatment of hazardous waste', 'there are management procedures of hazardous waste on your site', and 'hazardous waste is disposed on specific specialised landfill site' is between disagree to neutral/neutral. This is because current inconsistencies could account for the difficulties associated with the generation, reuse, recycling, disposal, and illegal dumping of waste. There is a need for accurate information regarding CDW, its characteristics, and origin. The absence of such information could lead to difficulties in terms of the assessment of the CDW problem, planning for the management of the problem, which could possibly include the development of the secondary construction materials market.

Conclusions

The organising and monitoring of the disposal of CDW, and the management of landfill sites in Port Elizabeth is poor. It can be concluded that planning and monitoring relative to landfill sites is inadequate.

There is a lack of understanding of laws relating to the dumping of waste, and there is insufficient space on landfill sites. It can be concluded that endeavours to create awareness and that planning relative to the provision of landfill sites are inadequate.

Contractors do have materials management systems and processes in place and are contributing to recycling endeavours. The problem appears to be isolated to CDW, which is to a large extent related to the traditional construction methods and materials adopted and used in South Africa, which include, inter alia, the use of in situ reinforced concrete, and masonry construction, which in turn requires cement plaster rendering.

Recommendations

To be succinct, recommendations are related to the management of CDW, the role of the construction industry, and solid waste in generic terms.

In terms of the management of CDW, it is recommended that the concerned parties, particularly the municipality, should undertake a study to characterise CDW in their area by:

- Improving the administrative systems for information related to CDW.
- Implementing legislative instruments to support the prevention and redirection of CDW from disposal to extended use applications, reuse, and recycling.
- Promote, support, plan, and showcase CDW prevention, reduction, recovery, reuse, and recycling strategies.

The construction industry should endeavour to:

- Promote the application of the waste hierarchy in dealing with waste produced on construction, demolition, and renovation sites such as waste avoidance before minimisation, then treatment before disposal.
- Develop waste specifications to ensure that waste management planning is part of tender specifications.
- Insist on the inclusion of waste management plans in tender documents.
- Promote recycling of buildings as a preferred approach, instead of demolition, and ensure that its principles are applied throughout the life cycles of buildings.
- Promote waste recovery for reuse and recycling in accordance with waste hierarchy before waste disposal at demolition sites.

In terms of waste generation, the following recommendations are proffered:

- In terms of waste at source: separation of waste at source should be initiated on a small scale, and piloted in different areas, and assorted colour waste bins should be provided for waste separation at construction sites for pilot purposes. Lowering taxation for recycled products to encourage the value of these products and strengthening legislation that supports recycling and reuse of materials.
- With respect to the collection of waste: the municipality should provide containers on construction sites for waste storage, and encouraging an incentive scheme facilitated by the municipality to reward firms that embrace recycling and reuse of construction CDW.
- Relative to the dumping of waste: improved management and utilisation of open spaces within the metropolitan area should be supported by all concerned parties; there should be rewards for individuals who report the illegal dumping of waste, and a policy must be developed to avoid illegal disposal sites.
- Concerning landfill sites: the lifespan of landfill sites needs to be prolonged; landfill sites should be rehabilitated at regular intervals, or when needed, and people should be discouraged from residing close to waste disposal sites or landfill sites.

References

Banias, G., Achillas, C., Vlachokostas, C., Moussiopoulos, N., & Tarsenis, S. (2010). Assessing multiple criteria for the optimal location of a construction and demolition waste management facility. *Building and Environment, 45*, 2317–2326.

Chatterjee, S., Ghosh, P., Ghosh, S. K., & Haldar, H. S. (2016). An optimization model on construction and demolition waste quantification from building. *Procedia Environmental Sciences, 35*(279), 288.

Department of Environmental Affairs. (2012). *National waste information baseline report.* Pretoria: Department of Environmental Affairs.

Duran, X., Lenihan, H., & O'Regan, B. A. (2006). A model for assessing the economic viability of construction and demolition waste recycling—The case of Ireland. *Resources, Conservation and Recycling, 46*(3), 302–320.

Ekanayake, L. L., & Ofori, G. (2004). Building waste assessment score: Design-based tool. *Building and Environment, 39*(7), 851–861.

Fatta, D., Papadopoulos, A., Avramikos, E., Sgourou, E., Moustakas, K., & Kourmoussis, F. (2003). Generation and management of construction and demolition waste in Greece—An existing challenge. *Resources, Conservation and Recycling, 40*(1), 81–91.

Gauteng Department of Agriculture Conservation and Environment. (2009). *Gauteng Provincial Building and Demolition Waste Guidelines.* Johannesburg: Gauteng Department of Agriculture Conservation and Environment.

Hareli, S., Portnov, B. A., & Seror, N. (2014). Evaluating the effect of vehicle impoundment policy on illegal construction and demolition waste dumping: Israel as a case study. *Waste Management, 34*, 1436–1445.

Hong Kong Environmental Protection Department. (2015). *Monitoring of solid waste in Hong Kong-Waste Statistics for 2014.* Hong Kong.

Macozoma, D. S. (2001). *Building Deconstruction International Council for Research and Innovation in Building and Construction (CIB) Report No. BOU/C358.* Rotterdam: CIB.

Nunes, K.R.A. (2004). *Evaluation of investments in and the performance of civil construction and demolition waste recycling centres: A study case for Municipality of Rio de Janeiro.*

Pinto, T.P., & Gonzalez, J.L.R. (2007). Construction waste handling and management: Guidance manual: How to deploy a system for handling and management in municipalities. Vol. 1. Portugal.

Rodriguez, G., Alegre, F. J., & Martinez, G. (2007). The contribution of environmental management systems to the management of construction and demolition waste: The case of Autonomous Community of Madrid (Spain). *Resources, Conservation and Recycling, 50*(3), 334–349.

United States of America Environmental Protection Agency (EPA). (2016). *Advancing Sustainable materials management: 2014 Fact sheet.* Washington, DC: EPA.

Performance Evaluation of Five Sediment Barriers Using a Full-Scale Testing Apparatus

Alan Bugg, Wesley Donald, and Wesley Zech

Introduction

According to the 2000 National Water Quality Inventory conducted by the United States Environmental Protection Agency (USEPA) (2000), sedimentation is one of the most widespread pollutants affecting rivers and streams, second only to bacterial pathogens. Sources of sedimentation include agriculture, urban runoff, construction, and forestry. However, sedimentation from construction sites is typically 10–20 times greater than from agriculture lands, and 1000–2000 times greater than from forestlands. In a short time, sedimentation from construction sites can contribute more sediment to waterways than would occur naturally over several decades from the same land. The resulting siltation clogs natural waterways and damages aquatic habitats (U.S. Environmental Protection Agency (USEPA), 2018). The United States (US) Federal Government recognized the detrimental effects caused by stormwater runoff and sediment discharge from construction sites. The US Congress passed the Clean Water Act in 1972 and the Water Quality Act of 1987 in response to these concerns, resulting in significant changes regarding environmental management methods used in the construction industry (U.S. Congress, 1972, 1987).

Construction site boundaries are typically encircled by perimeter control practices (i.e., silt fences, wattles, brush barriers), also known as sediment barriers, which act as the final control to capture and contain transported sediment prior to off-site discharge. Devices used as perimeter controls treat sheet flow by removing

A. Bugg (✉)
McWhorter School of Building Science, Auburn University, Auburn, AL, USA
e-mail: rab0018@tigermail.auburn.edu

W. Donald
Civil Engineering Department, Auburn University, Auburn, AL, USA

W. Zech
Department of Civil, Construction, and Environmental Engineering,
The University of Alabama at Birmingham, Birmingham, AL, USA

L. Scott et al. (eds.), *Sustainable Ecological Engineering Design*,
https://doi.org/10.1007/978-3-030-44381-8_35

sediment primarily through sedimentation and, to a minor degree, trapping of soil particles via filtration. As ponding occurs upstream of a perimeter control, particles fall out of suspension and are retained on-site. The filtration efficiency of the perimeter control material is limited by small soil particles passing through the void spaces within the filtering medium (Thompson et al., 2016). In addition, the flow through capacity of silt fence material has the potential to degrade over time as pores in the material become clogged with sediment.

Current Sediment Barrier Testing Methods and Protocols

ASTM recognizes two standards for testing sediment barrier performance: (1) ASTM D5141, *Standard Test Method for Determining Filtering Efficiency and Flow Rate of the Filtration Component for a Sediment Retention Device (SRD)* (ASTM International, 2011) and (2) ASTM D7351, *Standard Test Method for Determination of Sediment Retention Device (SRD) Effectiveness in Sheet Flow Applications* (ASTM International, 2013). Tests performed conforming to the procedures contained in ASTM D5141 are small-scale and conducted in a laboratory setting. The test apparatus consists of a 125 cm (49.2 in.) long by 85 cm (33.5 in.) wide flume and a 75 L (19.8 gal) container with a mechanical stirrer used to introduce sediment-laden flow into the flume. Test results are limited to determining the tested sediment barrier material properties, such as filtering efficiency and flow-through rate. This test procedure is not designed to evaluate installation methods and procedures, structural integrity, or full-scale field performance.

The ASTM D7351 standard test method introduces sediment-laden flow by mixing 2270 kg (5005 lbs) of water and 136 kg (300 lbs) of sediment prior to testing with a tank equipped with an internal agitator. The tank is positioned on a scale and the weight of the tank is monitored at regular intervals while discharging sediment-laden water at a constant flow rate of 90 kg/min (198.4 lbs/min) during a 30-min test. Test conditions are designed to simulate the peak 30 min of 10-year, 6-h storm event in the mid-Atlantic region that produces 10.1 cm (4 in.) of rainfall. The flow and sediment load are determined by assuming that 25% of the rainfall from the 10-year, 6-h storm occurs in the peak 30 min of the storm event and that 50% of the precipitation infiltrates into the ground. The associated sediment load resulting from erosion is calculated using the Modified Universal Soil Loss Equation (MUSLE) (Williams & Berndt, 1977), which allows for the calculation of a storm-specific quantity of sediment yield. The sediment-laden flow is directed down an impervious $3H{:}1V$ slope to the 6 m (20 ft.) wide impervious test area where the SRD is installed. The flow passing through the sediment is collected and directed toward a tank where effluent weight is measured using a scale. Though the collection tanks provide a measurement of the amount of sediment-laden runoff that is discharged and collected, the flow rate for the 30-min test is limited by the capacity of the tank. In addition, the scales only provide the total weight of sediment-laden water and do not have the ability to differentiate between the composition of sediment or water.

Other researchers have used rainfall simulators to generate rainfall-induced erosion on earth embankments while also being able to simulate different rainfall intensities. In this test procedure, rainfall simulation is used to generate sediment-laden runoff emanating from a slope to evaluate the installation, structural integrity, and sediment containment capabilities of sediment barriers. Simulated rainfall, applied to a $3H{:}1V$ constructed embankment plot, is used to simulate the natural erosion process to introduce sediment-laden flow to the SRD. The plot is 2.4 m (8 ft.) wide by 8.2 m (27 ft.) long with the sediment barriers installed at the toe of the embankment. A collection system is used to channel flow passing through the sediment barrier into a collection tank. This method introduces a series of variables that are difficult to control: the sediment load generated by the erosion process is dependent upon the preparation of the earthen test bed prior to testing, the simulated rainfall intensity, and the speed and direction of the wind. Factors such as the moisture content of the test bed, compaction, and surface roughness prior to testing can also impact the amount of soil erosion and sediment transport resulting from simulated rainfall. Preparing the test bed so that moisture content and soil compaction remain consistent over a large number of testing cycles requires considerable effort and is critical to producing meaningful test results that are repeatable and comparable.

The sediment barrier research described above using rainfall simulators uses a fixed slope, which limits the size and slope of the drainage area that can be used to subject the sediment to field-like runoff conditions. Some researchers have overcome the slope limitations by using a 2.4 m (8 ft.) wide by 9.1 m (30 ft.) long tilting test bed for sediment barrier testing (Gogo-Abite & Chopra, 2013). However, these apparatuses share the same limited test bed size as the apparatuses that use a fixed slope. Therefore, it is not possible to test sediment barriers using currently devised rainfall simulators under realistic, worst-case field conditions using the currently accepted silt fence design criteria due to the drainage area limitations inherent to these methodologies and apparatuses.

Zech et al. (2009) conducted field evaluations of silt fence tieback systems on an active construction site. The purpose of the silt fence tieback system is to provide sufficient capacity for sediment capture and to create intermediate check structures to decrease water velocity, thereby reducing erosive forces. Monitoring the tieback systems over four rainfall events showed that the tieback system captured sediment and minimized undercutting compared to the control system (no tieback) despite the fourth storm exceeding the design criteria for silt fence. The field evaluation of the tieback system was valid because the evaluated system and the control system were installed on mirrored slopes that were adjacent to each other and, therefore, had the same slope and soil type and were subjected to the same amount of rainfall. However, this type of field testing is difficult to replicate because it may be difficult to locate identical slopes with similar soil types. In addition, rainfall is a highly stochastic variable resulting in highly variable rainfall volumes and intensities and makes comparison testing of various sediment barriers difficult to perform.

The most widely recognized design criteria for unreinforced silt fence is 0.10 ha (0.25 acre) drainage area per 30.5 m (100 ft.) of installed fence. Using this criterion, the length of the drainage area upstream of the installed fence is 33.2 m (108.9 ft.)

(Bugg et al., 2017). None of the current methodologies or protocols listed above can simulate this size drainage area while at the same test sediment barriers under field conditions.

In order to evaluate the performance of sediment barriers under field conditions, researchers at the Auburn University-Erosion and Sediment Control Testing Facility (AU-ESCTF) were tasked by the Alabama Department of Transportation (ALDOT) to test different sediment barriers to evaluate overall performance and make recommendations for possible design and installation improvements. As a result, a full-scale testing apparatus was designed and constructed at the AU-ESCTF to evaluate sediment barriers simulating flow conditions experienced by sediment barriers installed on construction sites. Full-scale testing allows perimeter controls to be subjected to field conditions and is a better predictor of actual performance than laboratory or other small-scale testing techniques. This study focuses on the evaluation of five sediment barriers using the full-scale test apparatus at the AU-ESCTF.

Testing Methodology

The testing methodology and procedures used to evaluate five different silt fence installations are those detailed in Bugg et al. (2017). All tests were performed in the full-scale testing apparatus shown in Fig. 1 at the AU-ESCTF. Performance evaluations of the five sediment barrier practices tested included structural integrity, sediment retention, and effect on water quality.

Simulated flow is introduced to the system via a trash pump that draws water from a supply pond into an 1136 L (300 gal.) water tank. The water tank uses a series of valves and orifices to regulate flow through a calibrated weir into a mixing trough where sediment is introduced at a controlled rate and mixed with the flowing, highly turbulent water. Each test uses a stockpile of soil native to the state of Alabama with a soil texture classification as a loam soil (46.9% sand, 28.1% silt, 25.0% clay) according to the Natural Resources Conservation Services (NRCS) soil texture method. This soil is used to create the sediment-laden flow as well as construct the earthen test area. Sheet flow is generated using slotted diversion vanes mounted to the impervious slope. A $3H{:}1V$ test slope conveys flow to the 6.1 m (20 ft.) wide, 1% longitudinal sloped earthen test area. The test area is equipped with water-tight, removable access door sections that are 2.4 m (8 ft.) wide. The access doors can be removed to allow a silt fence to be installed using a tractor-drawn slicing machine. Any flow passing through the perimeter control discharges into a collection tank that is 2.4 m (8 ft.) wide by 1.8 m (6 ft.) long by 1.5 m (4.7 ft.) deep, downstream of the test area.

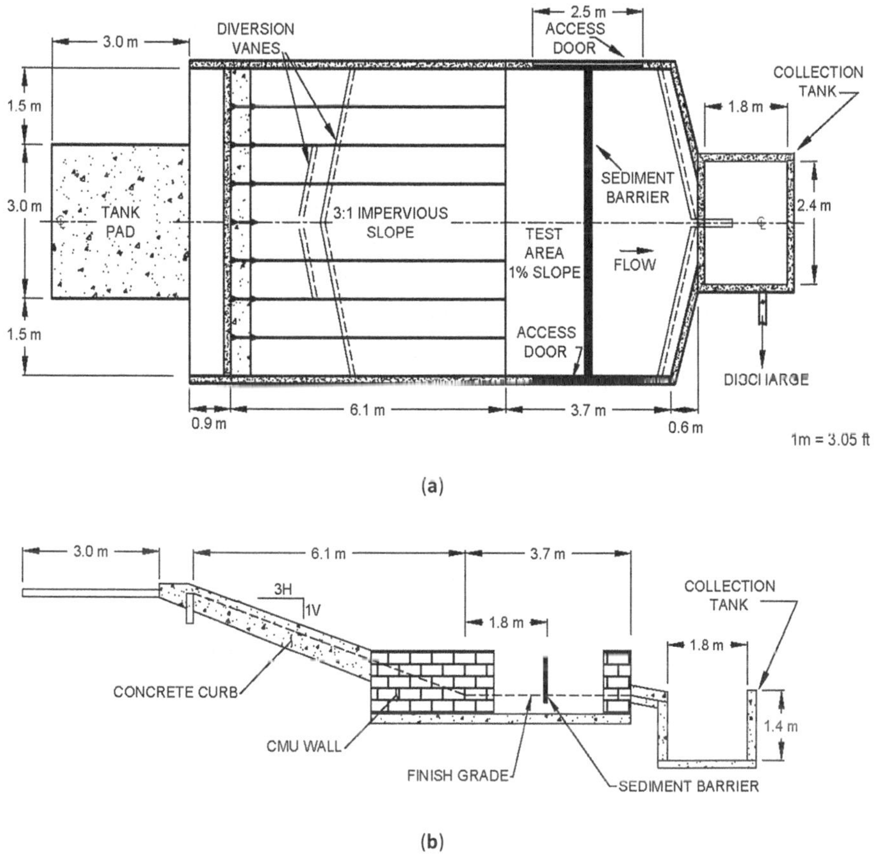

Fig. 1 Plan and profile views of the test apparatus (Bugg et al., 2017). (**a**) Plan view, (**b**) pro file view

Calculation of Test Flow Rate and Sediment Quantity

The test flow rate was based on the current design requirement in Alabama (Alabama Department of Environmental Management (ADEM), 2016) that sediment barriers retain eroded sediment onsite resulting from a 2-year, 24-h rainfall event. A 2-year, 24-h rain event was selected because sediment barriers on construction are required to withstand and not fail during this storm event. This results in a design criterion allowing a maximum drainage area of 0.2 ha (0.5 ac) per 30.5 m (100 ft.) of wire reinforced silt fence.

The flow rate for testing was calculated to mimic the average 2-year, 24-h rainfall event for Alabama that has an average precipitation depth of 11.7 cm (4.43 in.). Using a curve number (CN) of 88.5, the average CN for the state based upon GIS analysis for newly graded areas (Perez et al., 2015), and the average flow rate for the peak 30 min of the rainfall event, a standardized flow rate was developed. The representative drainage area is scaled down from 30.5 m (100 ft.) wide to 6.1 m (20 ft.) wide to match the width of the test area. Assuming a flow length of 66.4 m (217.8 ft.), the standardized flow rate was calculated to be 6.2 L/s (0.22 ft.^{3}/s).

The quantity of sediment required to be introduced for silt fence testing was calculated using the Modified Universal Soil Loss Equation (MUSLE) (Williams & Berndt, 1977). Based upon the flow calculations, the soil type used for sediment introduction, and the theoretical drainage area, the total sediment load for a 30-min test is 507 kg (1116 lbs) of soil that is introduced at a constant rate of 16.9 kg/min (37.2 lbs/min).

Testing Regime

A series of full-scale experiments introducing sediment-laden flow at a constant rate for 30 min are conducted to evaluate the performance of each sediment barrier installation. Each sediment barrier was installed three times (I-1, I-2, and I-3) with each installation undergoing three performance evaluations (P-1, P-2, and P-3), each simulating a 2-year, 24-h storm event to determine performance repeatability and longevity.

Evaluated Sediment Barriers

The five evaluated sediment barriers were as follows: (1) *ALDOT Trenched Silt Fence*, (2) *ALDOT Sliced Silt Fence*, (3) *Alabama Soil and Water Conservation Commission (AL-SWCC) Trenched Silt Fence*, (4) *ALDOT Standard Sediment Retention Barrier*, and (5) *ALDOT Standard Wattle Perimeter Control.* Details of the tested silt fence practices are shown in Fig. 2a–e.

Results and Discussion

The following is a summary of test results for the five evaluated sediment barriers based on the testing methodology described above. The evaluated performance areas include effects on structural performance, sediment retention, and water quality.

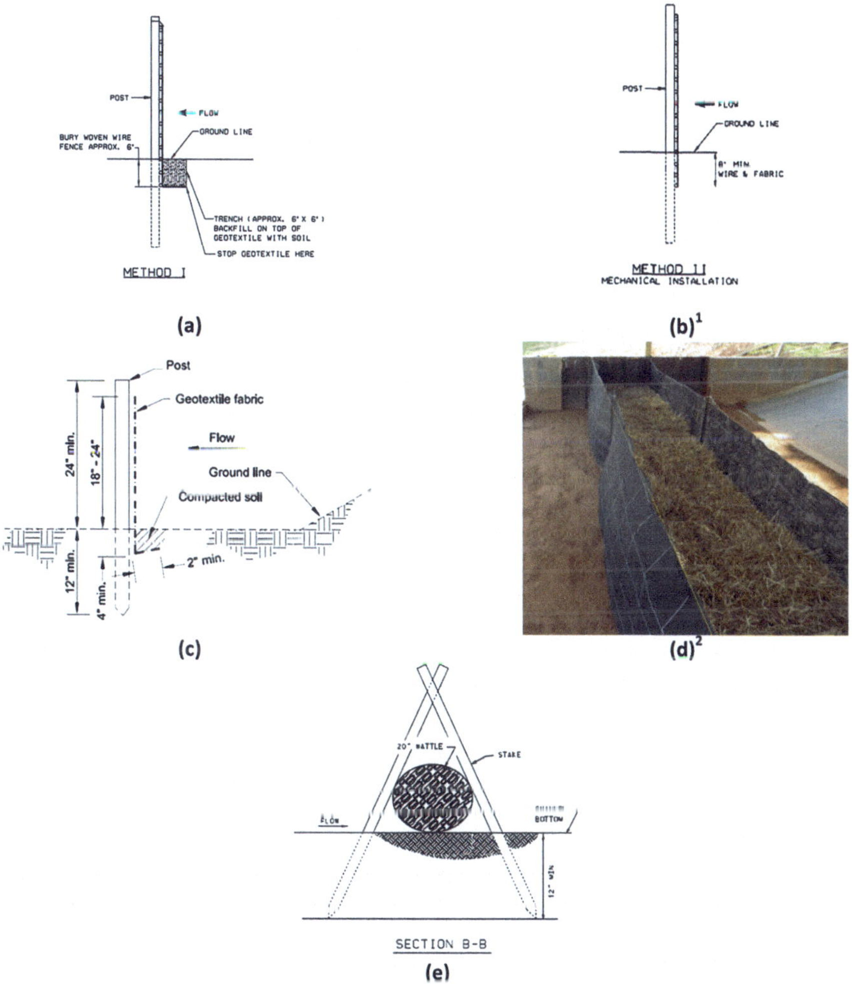

Fig. 2 Silt fence installation details. (**a**) *ALDOT Trenched Silt Fence* (2016), (**b**) *ALDOT Sliced Silt Fence* (2016), (**c**) *AL-SWCC Trenched Silt Fence* (2014), (**d**) *ALDOT Standard Sediment Retention Barrier* (2015), and (**e**) *ALDOT Standard Wattle Perimeter Control* (2006). Notes: (1) The *ALDOT Sliced Silt Fence* is installed with a slicing machine towed behind a farm tractor; (2) The silt fences for the *ALDOT Standard Sediment Retention Barrier* are installed in the same; manner as the *ALDOT Trenched Silt Fence* with a row of hay bales installed between the two rows of silt fence

Structural Performance

All five sediment barriers were installed 1.8 m (6 ft.) downstream of the toe of the impervious slope to allow adequate space for an impoundment to form. As previously discussed, impoundment is an important factor in improving water quality by promoting sedimentation. The ability of silt fence to form and maintain an impoundment is dependent on its structural performance. During testing, it was observed that the impoundment increased after each storm event due to the clogging of the fabric pore passages, restricting the flow-through rate. This placed increasing strain on the sediment barriers as the impoundment depth and hydrostatic pressure increased.

The following is a summary of the structural performance of each evaluated sediment barrier:

- *ALDOT Trenched Silt Fence*: For all three installations of the ALDOT Trenched Silt Fence, the steel t-posts failed on either the second or third performance test for each installation as a result of the geotextile fabric becoming increasingly less porous, creating larger impoundments in shorter time periods. As the height of the impoundment increased, the steel t-posts began to deflect. This deflection continued until water overtopped the silt fence fabric, resulting in a failure of the silt fence installation. The test results indicate that this configuration would not perform adequately from a structural standpoint when exposed to a 2-year, 24-h rainfall event in the field unless revisions are made to the spacing of posts or the posts are constructed out of heavier material.
- *ALDOT Sliced Silt Fence*: All three installations of the ALDOT Sliced Silt Fence were undermined and failed during the first performance test. The failure of all three installations of this silt fence configuration were very similar in nature and occurred between 8 and 12 min after the introduction of sediment-laden flow during the first performance test. The test results indicate that this configuration would not perform adequately from a structural standpoint when exposed to a 2-year, 24-h rainfall event in the field.
- *AL-SWCC Trenched Silt Fence*: The only significant structural deficiency noted during any of the performance tests for the AL-SWCC Trenched Silt Fence was undermining around one of the six posts as the impoundment reached full height 28 min into the first performance test (P-1) of the second installation. However, the area that undermined eventually sealed due to sediment deposition as the impoundment drained and did not reappear during subsequent tests P-2 and P-3. The structural performance of the AL-SWCC Trenched Silt Fence during testing indicates that this configuration would be adequate to perform from a structural standpoint when exposed to multiple 2-year, 24-h storm events.
- *ALDOT Standard Sediment Barrier*: The only significant structural deficiency noted during performance testing for the *ALDOT Standard Sediment Barrier* was that the first row of silt fence overtopped during the second and third performance tests for all three installations. At the end of the third performance test of the second installation, the downstream silt fence failed due to the center post deflecting in a similar manner to the *ALDOT Trenched Silt Fence*. However, since the upstream silt fence did not collapse, the impoundment was not lost. Consideration should be given to making the posts out of heavier material so they can better resist bending.
- *ALDOT Standard Wattle Perimeter Control*: All three installations of the *ALDOT Standard Wattle Perimeter Control* failed during the first performance test due to undermining. As the impoundment formed during testing, the straw wattle started to float, allowing the sediment-laden flow to pass underneath it. The test

results indicate that this configuration would not perform adequately from a structural standpoint when exposed to a 2-year, 24-h rainfall event in the field.

Sediment Retention

Complete topographic surveys of the test area using a robotic total station were conducted pretest and posttest to quantify sediment deposition and erosion both upstream and downstream of the silt fence practice. The topographic data from the surveys were then analyzed using computer-aided design software. This software converted the raw data into a triangulated irregular network for a three-dimensional representation of the test area surface and allowed for a comparison of the pretest and posttest channel topography (Thompson et al., 2016). Since the amount of sediment introduced was a known volume, the amount retained was compared to the amount introduced, allowing for the determination of percentage retained by the system. Table 1 summarizes the performance for each silt fence installation in retaining sediment.

The *AL-SWCC Trenched Silt Fence* and *AL-SWCC Trenched Silt Fence* both had a sediment retention rate of 90.5%, outperforming the *ALDOT Trenched Silt Fence* and the *ALDOT Sliced Silt Fence* that had sediment retention rates of 82.7% and 66.9%, respectively. Unfortunately, the survey data for the *ALDOT Standard Wattle Perimeter Control* was lost and could not be recovered. However, given its failure mode was similar to that of the *ALDOT Sliced Silt Fence*, its sediment retention rate can be assumed to be similar.

Table 1 Summary of silt fence practice sediment retention data

Sediment barrier	Description	Installation	% Sediment retained by installation	Avg. % retained
1	*ALDOT trenched silt fence*	I-1	86.6	82.7
		I-2	86.7	
		I-3	74.8	
2	*ALDOT sliced silt fence*	I-1	59.5	66.9
		I-2	68.2	
		I-3	73.1	
3	*AL-SWCC trenched silt fence*	I-1	90.5	90.5
		I-2	91.0	
		I-3	90.0	
4	*ALDOT standard sediment retention barrier*	I-1	89.6	90.5
		I-2	92.1	
		I-3	89.9	
5	*ALDOT standard wattle perimeter control*	I-1	No Data	No Data
		I-2	No Data	
		I-3	No Data	

The *AL-SWCC Trenched Silt Fence* and the *AL-SWCC Trenched Silt Fence* were both effective at creating and withstanding a full impoundment upstream without overtopping or structurally failing, unlike the other three sediment barriers. When the steel t-posts on the *ALDOT Trenched Silt Fence* deflected to the point where the posts were nearly parallel to the ground allowing the practice to be overtopped, the impoundment upstream was lost and decreased the ability of the sediment barrier to retain sediment. When the *ALDOT Sliced Silt Fence* and the *ALDOT Standard Wattle Perimeter Control* were undermined, water passing underneath caused scouring that resulted in erosion both upstream and downstream of the fence. This also limited the upstream impoundment. Due to the high level of undercutting and loss of impoundment, failure was considered to have occurred for each installation of these practices after the first performance test and resulted in the installations not being subjected to tests P-2 and P-3. As previously mentioned, each silt fence was offset from the toe of the slope 1.8 m (6 ft.) to provide extra storage for impoundment. This 1.8 m (6 ft.) transition has only a 1% gradient slope, likely helping to slow runoff emanating from the 3:1 (*H*:*V*) slope. This transition likely provides enough energy dissipation to help the larger sand particles to fall out of suspension, even with minimal impoundment from the silt fence. This further emphasizes the benefit of not installing sediment barriers directly at the toe of the slope.

Water Quality

Turbidity is a metric often used by regulatory agencies to quantify the quality of stormwater runoff discharge from construction sites. To evaluate the effect each silt fence practice had on water quality, water samples were taken at four locations: (1) on the impervious slope upstream of the impoundment formed by the silt fence, (2) in the impoundment immediately upstream of the silt fence, (3) immediately downstream of the silt fence, and (4) at the discharge pipe where the outflow enters the collection tank (Bugg et al., 2017). Figure 3 contains graphs depicting the average turbidity for the duration of all performance tests conducted on the five sediment barriers. These test results indicate that none of the evaluated sediment barriers provided a significant improvement in turbidity between the upstream and downstream sampling points. In fact, the average turbidity readings were consistently slightly higher for the samples taken downstream of the tested sediment barriers when compared to the readings for upstream samples. This may be attributable to the fact that some of the suspended sediment had started to settle out in the upstream impoundment. As the flow passed through the barrier, remixing of the sediment occurred, thereby increasing turbidity.

The exception to this condition occurs between during the performance testing for the *ALDOT Sliced Silt Fence* and the *ALDOT Standard Wattle Perimeter Control*. As can be seen in Fig. 3b, e, the downstream readings for turbidity were significantly higher than the upstream turbidity readings after the sediment barriers failed. This is attributable to the fact that both sediment barriers were undercut, allowing

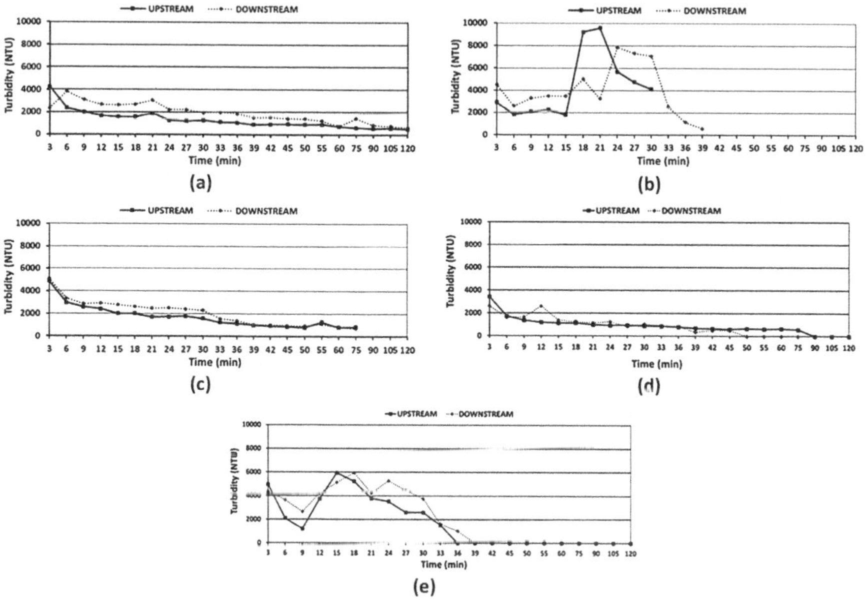

Fig. 3 Average turbidity over time for each silt fence installation tested. (**a**) *ALDOT Trenched Silt Fence*, (**b**) ALDOT Sliced Silt Fence, (**c**) AL-SWCC Trenched Silt Fence, (**d**) ALDOT Standard Sediment Retention Barrier, (**e**) ALDOT Standard Wattle Perimeter Control

the sediment-laden flow to pass unimpeded under the barrier material. This resulted in little or no upstream impoundment. In addition, the unimpeded sediment-laden flow underneath these sediment barriers eroded the test bed, which increased turbidity downstream.

Conclusions

The research study evaluated the performance of five different types of sediment barriers under full-scale performance testing that simulated conditions in the field. This investigation provided researchers at the AU-ESCTF the ability to evaluate the performance of standard sediment barrier practices and effects of common failure modes.

The test results indicate that the structural performance of sediment barriers is the most important component in capturing sediment, thereby improving water quality. Sediment barriers that structurally perform well allow a larger quantity of suspended soil particles to settle out of suspension in the upstream impoundment upstream of the barrier, capturing them before they are transported off-site, thereby having a direct effect in improving water quality. The test results also reflect the need for regular sediment barrier maintenance practices since the structural performance and the ability to maintain upstream impoundment is directly affected by sediment accumulation resulting from multiple storm events. Furthermore, it should be noted that these practices are not standalone sediment controls. As they are often used as perimeter controls, they may act as the last line of defense against elicit discharges. Therefore, it is important that the practice be complimentary to other erosion and sediment controls used throughout the site. This will help decrease the chance of failure with less sediment accumulation.

Performance testing indicated that the *ALDOT Sliced Silt Fence and the ALDOT Standard Wattle Perimeter Control* are not acceptable to use as perimeter sediment controls due to their failure to stand up structurally when exposed to conditions simulating a 2-year storm. In addition, the *ALDOT Trenched Silt Fence* and the *ALDOT Standard Sediment Retention Barrier* require modifications to their post spacing and/or material in order to improve their structural performance. The *AL-SWCC Trenched Silt Fence* was the only sediment barrier that performed flawlessly during performance testing. As shown in the test results, the sediment retention rate for the tested sediment barriers increased based on their structural performance. Therefore, structural performance is the most important factor when determining which sediment barrier to use in the field.

The results indicate that the barrier geotextile materials (i.e., nonwoven vs. woven or straw) had little or no effect on water quality when measuring turbidity. Neither geotextile material showed increased effectiveness in filtering based upon water quality data. The higher turbidity recorded downstream when compared to upstream, however, may have been due to the upstream samples being taken at the top of the water column of the impoundment, whereas the downstream samples were taken from the flow exiting the bottom of the silt fence, which would likely have a higher concentration of suspended sediment fines as concentration levels increase with sedimentation occurring. Further evaluation of this possibility is warranted to better understand sediment barrier performance. However, the test results did indicate that the impoundment upstream of silt fence is effective in decreasing turbidity and improving water quality via the process of sedimentation.

The testing methodology used to evaluate the five sediment barriers proved to be functional and allowed for their assessment under simulated field conditions. The information obtained through this study will be beneficial to designers when specifying sediment barriers used as perimeter controls on construction sites. Understanding the potential failure modes of sediment barriers is a critical component for formulating design, installation, and maintenance methods to enhance both initial and sustained performance in order to prevent environmental damage and regulatory violations.

Acknowledgments This paper is based on a study sponsored by ALDOT and the Auburn University Highway Research Center. The authors gratefully acknowledge this financial support. The findings, opinions, and conclusions expressed in this paper are those of the authors and do not necessarily reflect the view of the sponsors.

References

Alabama Department of Environmental Management (ADEM). (2016). *National pollutant discharge elimination system permit, general permit*. Montgomery, AL: Alabama Department of Environmental Management.

Alabama Department of Transportation (ALDOT). (2006). *Standard Drawing ESC-300-4*. Montgomery, AL: Alabama Department of Transportation.

Alabama Department of Transportation (ALDOT). (2015). *Standard Drawing ESC-300-5*. Montgomery, AL: Alabama Department of Transportation.

Alabama Department of Transportation (ALDOT). (2016). *Standard Drawing ESC-200 (Sheet 4 of 5)*. Montgomery, AL: Alabama Department of Transportation.

Alabama Soil and Water Conservation Committee (AL-SWCC). (2014). *Alabama Handbook for Erosion Control, Sediment Control and Stormwater Management on Construction Sites and Urban Areas* (Vol. 1). Montgomery, AL: Alabama Soil and Water Conservation Committee.

ASTM International. (2011). *Standard test method for determining filtering efficiency and flow rate of the filtration component for a sediment retention device*. West Conshohocken, PA: ASTM D541–11.

ASTM International. (2013). *Standard test method for determination of sediment retention device effectiveness in sheet flow applications*. ASTM D7351–13. West Conshohocken, PA: ASTM International.

Bugg, R. A., Zech, W. C., Donald, W. N., & Perez, M. A. (2017). Improvements in standardized testing for evaluating sediment barrier performance: Design of a full-scale testing apparatus. *Journal of Irrigation and Drainage Engineering, 143*, 1–11. https://doi.org/10.1061/(ASCE)IR.1943-4774.0001194

Gogo-Abite, I., & Chopra, M. (2013). Performance evaluation of two silt fence geotextiles using a tilting test-bed with simulated rainfall. *Geotextiles and Geomembranes, 39*, 30–38.

Perez, M. A., Zech, W. C., Donald, W. N., & Fang, X. (2015). Methodology for evaluating inlet protection practices using large-scale testing techniques. *Journal of Hydrologic Engineering, 20*, 1–9. https://doi.org/10.1061/(ASCE)HE.1943-5584.0001019

Thompson, J., Sattar, A., Gharabaghi, B., & Warner, R. (2016). Event based total suspended sediment particle size distribution model. *Journal of Hydrology, 536*, 236–246. https://doi.org/10.1016/j.jhydrol.2016.02.056

U.S. Congress. (1972). *Clean Water Act. In Federal water pollution control Act, 33 U.S.C. §1251 et seq*. Washington, DC: U.S. Congress.

U.S. Congress. (1987). *Water Quality Act of 1987. In Public Law 100-4*. Washington, DC: U.S. Congress.

U.S. Environmental Protection Agency (USEPA). (2000). *Water Quality Inventory*. Washington, DC: U.S. Environmental Protection Agency.

U.S. Environmental Protection Agency (USEPA). (2018). *Construction site runoff control minimum control measures*. Washington DC: U.S. Environmental Protection Agency.

Williams, J. R., & Berndt, H. D. (1977). Sediment yield prediction based on watershed hydrology. *Transactions of ASAE, 20*(6), 1100–1104.

Zech, W. C., McDonald, J. S., & Clement, T. P. (2009). Field evaluation of silt fence tieback systems at a highway construction site. *Practice Periodical on Structural Design and Construction, ASCE, 14*, 3. https://doi.org/10.1061/(ASCE)1084-0680(2009)14:3(105)

Contractor's Readiness for LEED Requirements in Ireland: Factors for Consideration

Cian Ryan and John P. Spillane

Introduction

While the construction industry contributes hugely to society through economic and social means, it has a huge impact on the environment. The industry is accountable for 40% of global resources, which unquestionably contributes significantly to global warming, air pollution, water pollution and the depletion of natural resources (Suzer, 2015). This issue is becoming increasingly important in the industry, as concern for the environment has become a high priority in modern-day society. To combat these significant effects, more and more buildings are being designed and procured with environmentally friendly principles in mind. The carbon footprint of a building is not easy to alter once it has been built, highlighting the need to rethink the building processes and systems in the industry. Similarly, a building's life cycle must be considered throughout the design and construction stages to ensure lasting sustainability is achieved (Fenner & Ryce, 2008).

This new building approach requires a level of measurement and acknowledgement to ensure a universal standard of green construction is met where necessary. Therefore, green building rating systems were developed. These rating systems ensure the building contributes to the overall reputation of the building, consumes less energy and provides a better living environment (Doan et al., 2017). There are several popular green building system ratings in use globally, such as LEED (Leadership in Energy and Environmental Design), BREEAM (Building Research Establishment Environmental Assessment Method), CASBEE (Comprehensive Assessment System for Built Environment Efficiency) which is in use in Japan, Green Star of Australia and New Zealand and BEAM (Building Energy Assessment

C. Ryan · J. P. Spillane (✉)
Construction Management and Engineering, School of Engineering, University of Limerick, Limerick, Ireland
e-mail: john.spillane@ul.ie

L. Scott et al. (eds.), *Sustainable Ecological Engineering Design*,
https://doi.org/10.1007/978-3-030-44381-8_36

method) in Hong Kong. While these certifications differ in parts, they strive towards a common goal. These initiatives were initially introduced by government bodies in the hope of changing the industry for the better, and while they have been successful, there is still a lot of progress to be made.

This paper focuses on the LEED (Leadership in Energy and Environmental Design) certification, which is a rating system used to certify buildings with regard to their sustainability and environmental impact. It was first used by the United States Green Building Council in 1998. Subsequently, it has become increasingly prevalent in the Irish construction industry and across Europe in recent years, due to clients demands for environmentally friendly building. At the time of writing, there are currently 213 LEED registered projects in Ireland with this number constantly rising (Irish Green Building Council, 2018). While the LEED certification is advantageous for the environment, it also poses its own disadvantages from a building contractor's perspective as it provides another obstacle in the tender process.

There are many publications available discussing the requirements of the LEED certification without delving into the Irish contractor's perspective or the level of preparedness of Irish contractors. This research provides a useful tool for Irish contractors to enable them to clearly see their readiness with regard to obtaining LEED certification, what is required to receive LEED certification while also providing a clear comparison between the standard of foreign procedures and Irish. Based on this, the aim of this paper is to identify and document the requirements from a contractor's perspective in Ireland to achieve LEED certification and investigate their current readiness.

The Growing Importance of Leed and Green Building: A Literature Review

With the continued global focus on sustainability and green building, the construction industry must also come into focus. Currently, the construction industry is accountable for using 40% of the world's natural resources (Doan et al., 2017; Suzer, 2015; Tam, Tam, & Tsui, 2004). This is an alarming figure that highlights the need for green building rating systems. The LEED certification is the most widely used green building rating system in the world, being used on over 94,000 projects in over 165 countries and territories worldwide (U.S Green Building Council, 2018a). Also, the square footage of LEED registered projects has grown rapidly, from 0.15 billion to 15 billion between 2008 and 2016 (Doan et al., 2017). The 2015 Paris agreement, as outlined in the world green building council publication (Victoria Burrows, 2017), states that all new buildings from 2030 must operate at net zero carbon and all buildings by 2050. These statistics show the need for Irish contractors to prepare for the inevitable growth of green building across the construction industry, both at home and abroad. The LEED certification is currently in

LEED version 4 (U.S Green Building Council, 2018b). This is the 5th updated version of LEED, which further demonstrates the required growth and evaluation the certification body is undertaking, which ensures the system will have a significant impact on the Irish construction Industry for years to come.

With 213 LEED projects in Ireland (Irish Green Building Council, 2018), this number can be expected to rise quickly, as outlined in the World Green building trends report, stating that growth in sustainable building was anticipated to double from 2017 to 2018 (Dodge Data & Analytics, 2016). This provides further confirmation of the importance of Irish contractor readiness for building rating systems such as LEED in Ireland. Other factors highlighted in the World Green building trends report showing the importance of the implementation of green building rating systems include client demand, which has risen from 35% to 40% from 2012 to 2015, and also higher standards of environmental regulations (Dodge Data & Analytics, 2016).

The reason for this growing interest in green building is due to its clear advantages over traditional building. These advantages come in various forms throughout the building's life cycle and are advantageous to both the developer/owner of the building and its tenants and not only benefit financially but in other ways also such as living quality and public image. The Canada Green Building trends report (McGraw Hill Construction, 2014) highlights the benefits of green building as: higher building quality, future proofing assets and public demonstration of corporate sustainability, while also having financial benefits, such as lower operating costs, higher return on overall investment and higher rental rates. Similar benefits can be seen in a study by Scofield (2013), which shows that LEED Gold buildings in New York City consumed 20% less energy than others. Also, Newsham, Mancini, and Birt (2009) and Matisoff, Noonan, and Mazzolini (2014) both show the energy consumption advantage LEED certified buildings have when compared with uncertified buildings. While there is an underlying assumption that green building is more expensive than traditional methods, this is not always true. For example, the additional cost of building green in the United States averages at only 2% (Kats, 2003). Gan, Zuo, Ye, Skitmore, and Xiong (2015) and Bartlett & Howard (2000) in addition to Ambec and Lanoie (2008) also argue that green building has a financial advantage over non-certified counterparts. There is insufficient data to support these statistics in Ireland at present, but these advantages can be expected to transfer to the Irish construction industry also, reinforcing the need to prepare for LEED in Ireland as these advantages are realised.

However, there have only been 55 LEED certified projects in Ireland since 2010 (U.S Green Building Council, 2018c). Compare this with the number of planning permissions granted in Ireland in the same period (155,322) (Central Statistics Office, 2018) and the standard of the Irish green building industry becomes clear. While there is clear improvement seen in the Irish Green building council Activity report for 2017 in green building involvement and training (Irish Green Building Council, 2018), there is much more to be done.

Research Design

Following on from the demonstrated need for Irish contractors to be LEED ready, it is necessary to ascertain the factors from those within the sector. To do so, a qualitative approach was adopted as the construction industry in Ireland currently does not have sufficient information regarding LEED due to it being in its infancy in the country (Irish Green Building Council, 2018). To complement this, a case study method was used to only gather rich and useful information from those within the industry. To aid in the identification of potential participants, a dual criterion sampling strategy was adopted. Firstly, interviewees had to have worked on a LEED certificated project, and secondly, would have had done so in the past 3 years.

Case Study Overview

To gauge insight into actuality while also garnering the viewpoints of those working within industry, a case study approach is adopted. Similar studies also considered this approach including Spillane and Oyedele (2017), while Zanial (2007) argue that it is one of the more beneficial approaches where 'in-depth explanations of a social behaviour are sought', as in this instance. The case studies were undertaken to investigate the processes, obstacles faced and various other factors in the construction of a LEED certified project. To aid triangulation, three case studies were identified, of three distinctly separate completed LEED certified projects in Ireland. Conducting three separate project case studies provides an in-depth and detailed view of the overall preparedness of the contractors for LEED. Interviews were undertaken with several key members of the project team from each of the respective projects, to gain valuable insight into the current situation regarding LEED in Irish construction projects. These project members were the Project manager, the Site manager and the site engineer for the companies involved and their respective LEED certified project. These roles are all key in the project delivery; therefore, making these interviews invaluable information sources as their various perspectives on the project can be considered.

The first case study was an industrial warehouse building with attached offices located in industrial estate. A LEED Gold requirement was specified with an approximate project value of €8 million. The second case study was a three-storey office building located in an industrial estate. A LEED Silver Requirement was specified in this instance and had an approximate project value of €10 million. The third and final case study was a four-storey office building located in an industrial Estate, where a LEED Gold requirement was specified. In each context, four participants provided their insight into their respective case studies.

The interviews consisted of an unstructured open-ended interview approach, to ensure that each interviewee is asked the same questions, in the same order, thus providing uniformity in the results, while also eliminating interview bias. This interview

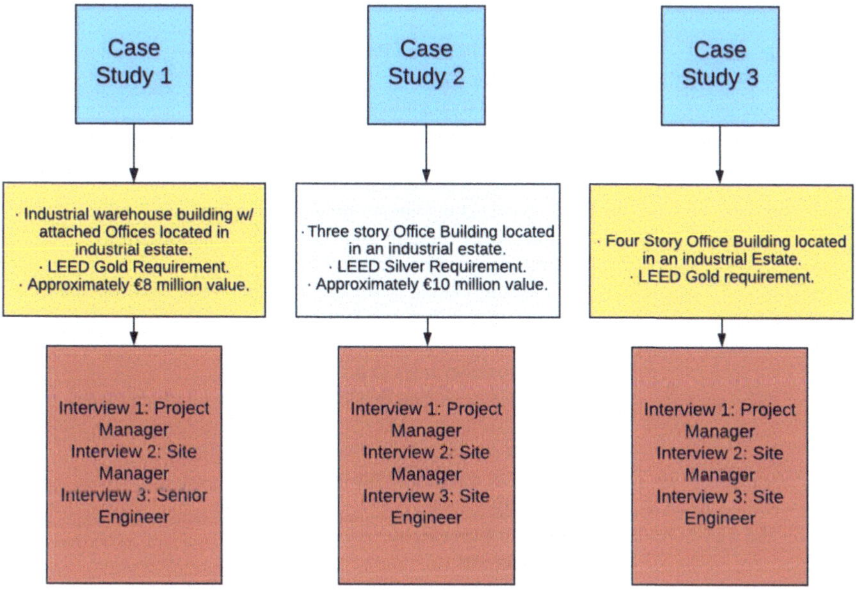

Fig. 1 Case study overview

method also allows for the interviewee's opinion to be considered, in addition to any additional relevant information they may provide.

Figure 1 illustrates the case study approach adopted.

The interviews were analysed using a thematic content coding and analysis. This method of analysis provides an overview of the occurring themes that can be seen across the gathered interviews information. Firstly, the interviews were read over several times to ensure familiarisation with the results. The main themes/points are then noted from each separate interview to provide a clear view of the interviewee's view on the topic. Reoccurring themes/points between the interviews are then analysed to provide useful data of overall opinions. The main points found are then written up using flow charts to provide a clear representation of the data. This analysis method was used for each project to provide a detailed case study into LEED for these projects.

Case Study Analysis

Of the nine interviews from three separate case studies, a total of six themes and forty-seven factors were identified. The following three tables document the themes and their underlying factors for each of the respective case studies.

Theme	Factors
Sustainable sites	Storm Water Pollution Prevention Plan Sedimentation Tank Sand Bags at road gullies Refuelling Station Concrete wash-out Station Surrounding trees maintained Difficulty reaching landscaping requirement Pumping water through foul sewer system and filtering Dust filter at site boundary Tyre cleaning station at site entrance/exit—Full-time operator Wheel washer difficulty Spill kits No storage under trees to protect roots Lack of experience on behalf of subcontractors with regard to SWPPP LEED not considered in some pricing elements Difficulty in road upkeep/wheel washing UV reflecting roof
Location and transportation	Public transport nearby Close to airport Bicycle stands/paths constructed Electric car charging stations Services put in place to cater for future developments nearby
Water efficiency	Rainwater Harvesting Tank Efficient water usage systems—Cisterns/taps/Urinals Water meters at all connections Design team specifies Difficulty with subcontractors despite LEED training
Energy and atmosphere	PV Panels LED lights—reduced light pollution Extra programme for LEED commissioning Excessive paperwork
Materials and sources	Segregation of waste plan Waste management companies unfamiliar with LEED Subcontractor not used to waste segregation Difficulty sourcing some specific materials Budget factor—Results in lower LEED points Sourcing local materials can result in higher cost/wait time FSC compliancy difficulty for doors etc. Quantity surveyor not focused on LEED Trouble sourcing from abroad—Product History BPDO difficulty—Suppliers not experienced Difficulty keeping track of materials used by subcontractors
Indoor environmental quality	All ducting/plumbing/ventilation systems sealed Clean as you go system to reduce dust build up Filters changed + breathing period before handover Segregated smoking area away from building—Difficulty with smokers Lack of experience from subcontractors

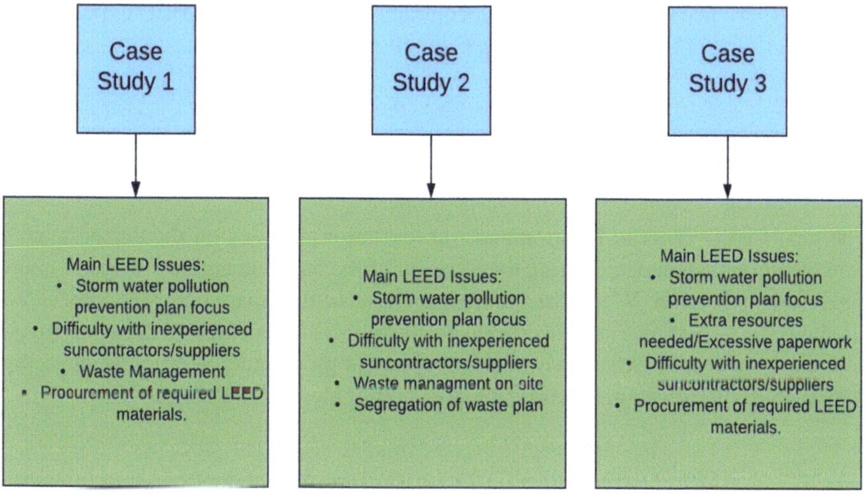

Fig. 2 Main LEED issues by case study

In summary, Fig. 2 highlights the main LEED issues that arose in each case study.

Discussion

Using the six themes identified from the factors listed, each is taken in turn, as follows.

Sustainable Sites

In the sustainable sites section of the LEED criteria, the focus is on the environment surrounding the building and preserving the natural systems in place nearby. Taking the view of a contractor, this is done by preventing construction activity pollution as much as possible and by implementing the correct management systems for this aspect of the project. One of the most prevalent issues in the data gathered is the need to focus on the storm water pollution prevention plan for the site. This was referred to by all nine interviewees. The interview subjects all agree that this is an area where problems may arise for the contractor and resources must be allocated to ensure this is managed appropriately, a sentiment echoed by the Environmental Protection Agency (2017). Furthermore, there is a strong agreement across all interviews to ensure the usage of spill kits, sandbags at road gullies and also to ensure refuelling stations are planned out. Several interviewees also state that there is a lack of experience on behalf of the subcontractor to manage the LEED expectations in these criteria (U.S. Green Building Council, n.d.).

Location and Transportation

The next LEED criteria, location and transportation, focuses on the design team's considerations when deciding on the location of the building and the usage of alternative transportation. While the contractors have little impact here, the interviews all show the importance of this criteria as all projects are located with public transport easily accessed. Also, two of the projects implemented electric car charging stations into their design, an approach also highlighted by Litman (2015).

Water Efficiency

Similarly, in this section, the design team has the greater responsibility. This is evident in the data gathered, as each project incorporates a rainwater harvesting tank into their design. However, a recurring issue from the contractor's perspective is the lack of experience on behalf of the subcontractor, despite the LEED training undertaken. The U.S. Green Building Council (2019a) also argues that this poses difficulties for the contractor to reach the required LEED standards.

Energy and Atmosphere

The energy and atmosphere criteria consider energy-efficient designs used in the project's construction and also the use of renewable energy sources. All three projects studied include solar-powered roof panels. According to the gathered data, this section creates difficulty for the contractor in the form of excessive paperwork and extra commissioning work. Several interviewees agree that resources and plans must be put in place to ensure these difficulties are properly managed as highlighted in Green Building Elements (2014) and the U.S. Green Building Council (2019b).

Materials and Sources

The materials and resources criteria focus on minimising the impacts of material procurement throughout the supply chain and also the management of waste on-site. The contractor has a large role to play in this section of the LEED rating process. The issue of a lack of experience on sub-contractor's behalf is evident here yet again as is mentioned by seven of the interviewees. Similarly, the waste management companies in the project areas were not adequately trained in LEED processes and lacked experience which caused delays and difficulties. Also, several interviewees agree that there is difficulty in sourcing certain specific materials required to obtain LEED points; this results in higher cost and delays to the project. An interesting point made by two site managers is that the quantity surveyor must be more focused on the LEED process as they do not consider the required materials when pricing (Corning, n.d.).

Indoor Environmental Quality

This section focuses on the indoor air quality of the building, not only in design but also in the construction of the building. This is achieved by the contractors using a construction indoor air quality management plan. There is a clear agreement across all interviews that extra resources must be assigned to address this. Each case study project kept the ventilation systems clean by sealing each duct/opening. As can be seen in previous sections, the lack of experience on the sub-contractor's behalf is again an issue when dealing with indoor environmental air quality, an aspect that Kubba (2017) also refers to.

Conclusion and Recommendations

As highlighted, LEED is continuing to grow in prominence, globally, but particularly in Ireland. To aid Irish contractor readiness for LEED certification, this paper aims to highlight contractor readiness factors for consideration. To achieve this aim, three distinction LEED Certified case studies were identified and three participants from each interviewed.

The key findings resonate around six core themes: Sustainable Sites, Location and Transportation, Water Efficiency, Energy and Atmosphere, Materials and Sources, and Indoor Environmental Quality. From this, a core trait that arises throughout the interviews and associated literature is the reoccurring issue of a lack of experience on behalf of many people involved in these projects across all LEED aspects. The data clearly shows that sub-contractors and suppliers in Ireland are not adequately prepared for LEED certified projects, resulting in delays and wasted resources. It is evident that all parties involved are not receiving the required level of training to be able to succeed in achieving LEED certification. This is a fundamental and concerning finding from the study, highlighting the inadequacy of both training and development of core skillsets in the context of LEED and its successful delivery. Also, the need for LEED planning and resource allocation is also prevalent in the case studies completed. Irish contractors are not allocating the required amount of resources to implementing LEED procedures and management systems and this must be considered. These issues all solidify the hypothesis that the current readiness of Irish contractors with regard to LEED certification is wholly inadequate.

As previously discussed, there is an urgent need for not only improvement in training but also the development of core material and routes to market for delivery to those that need it most—our practitioners in industry. This must follow with an increased focus on LEED for an improvement to be seen. The research undertaken would suggest that the attitude towards LEED is currently less than desirable, as contractors are not allocating enough resources to deal with the required workload. Also, some interviewees are of the opinion that LEED does not take priority on-site and can be overlooked. In conclusion, it is recommended that Irish contractors engage in LEED training over the coming years, to ensure the growth continues in LEED Certified projects and that more sustainable projects become the norm within Irish society. At a more local level and from the perspective of contactors, the research conducted has the potential for strong implications in the construction industry. The research clearly shows the areas in which Irish construction contractors must focus to achieve LEED certification. This can prove to be invaluable to contractors as it allows them to avoid waste and focus resources accordingly, in the attainment of LEED certification on their respective projects.

References

Ambec, S., & Lanoie, P. (2008). Does it pay to be green? A systematic overview. *Academy of Management Perspectives, 22*(4), 45–62.

Bartlett, E. N. H. & Howard, N. (2000). Informing the decision makers on the cost and value of green building, *Building, Research & Information, 28*(5–6), 315–324.

Central Statistics Office. (2018). *Planning permissions granted by type of construction, quarter and statistic*. Retrieved March 6, 2019, from https://www.cso.ie/px/pxeirestat/Statire/SelectVarVal/saveselections.asp.

Corning, O. (n.d.). *4 Strategies for achieving LEED v4 materials and resources expertise*. Retrieved March 6, 2019, from https://www.architectmagazine.com/design/buildings/4-strategies-for-achieving-leed-v4-materials-and-resources-expertise.

Doan, A. G., Ghaffarianhoseini, A., Naismith, N., Zhang, T., Ghaffarianhoseini, A., & Tookey, H. (2017). A critical comparison of green building rating systems. *Building and Environment, 123*(1), 243–260.

Dodge Data & Analytics. (2016). *World Green Building Trends 2016*. Bedford, MA: Dodge Data & Analytics.

Environmental Protection Agency. (2017). *Developing a Stormwater Pollution Prevention Plan (SWPPP)*. Retrieved March 6, 2019, from https://www.epa.gov/npdes/developing-stormwater-pollution-prevention-plan-swppp.

Fenner, R. A., & Ryce, T. (2008). A comparative analysis of two building rating systems. Part 1: Evaluation. *Engineering Sustainability, 161*(1), 55–63.

Gan, X., Zuo, J., Ye, K., Skitmore, M., & Xiong, B. (2015). Why sustainable construction? Why not? An owner's perspective. *Habitat International, 47*(1), 61–68.

Green Building Elements. (2014). *What is LEED?—Energy and atmosphere*. Retrieved March 6, 2019, from https://greenbuildingelements.com/2014/09/23/leed-energy-atmosphere/.

Irish Green Building Council. (2018). *IGBC activity report 2017*. Dublin: Irish Green Building Council.

Kats, G. H. (2003). *Green building costs and financial benefits*. Westborough, MA: Massachusetts Technology Collaborative.

Kubba, S. (2017). Chapter 7: Indoor environmental quality. In *Handbook of green building design and construction* (2nd ed., pp. 353–412). Amsterdam: Elsevier.

Litman, T. (2015). *Recommendations for improving LEED transportation*. Todd Alexander Litman: Victoria Transport Policy Institute.

Matisoff, D. S., Noonan, D. S., & Mazzolini, A. M. (2014). Performance or marketing benefits? The case of LEED certification. *Environmental Science & Technology, 48*(3), 2001–2007.

McGraw Hill Construction. (2014). *Canada green building trends: Benefits driving the new and retrofit market*. Canada: Canada Green Building Council.

Newsham, G. R., Mancini, S., & Birt, B. J. (2009). Do LEED-certified buildings save energy? Yes, but .... *Energy and Buildings, 41*(8), 897–905.

Scofield, J. H. (2013). Efficacy of LEED-certification in reducing energy consumption and greenhouse gas emission for large New York City office buildings. *Energy and Buildings, 67*(1), 517–524.

Spillane, J. P., & Oyedele, L. O. (2017). Effective material logistics in urban construction sites: a structural equation model. *Construction Innovation, 17*(4), 406 428.

Suzer, O. (2015). A comparative review of environmental concern prioritization: LEED vs other major certification systems. *Journal of Environmental Management, 154*(2015), 266–283.

Tam, C. M., Tam, V. W. Y., & Tsui, W. S. (2004). Green construction assessment for environmental management in Hong Kong. *International Journal of Project Management, 22*(7), 563–571.

U.S Green Building Council. (2018a). *LEED is green building*. Retrieved March 9, 2019, from https://new.usgbc.org/leed.

U.S Green Building Council. (2018b). *The next step for LEED is LEED v4.1*. Retrieved March 9, 2019, from https://new.usgbc.org/leed-v41.

U.S Green Building Council. (2018c). *Country market brief: Ireland*. Retrieved March 9, 2019, from https://www.usgbc.org/advocacy/country-market-brief.

U.S Green Building Council. (2019a). *Water efficiency*. Retrieved March 9, 2019, from https://www.usgbc.org/credits/healthcare/v4/water-efficiency.

U.S Green Building Council. (2019b). *Energy & atmosphere*. . Retrieved March 9, 2019, from https://www.usgbc.org/credits/new construction/v4/energy-%26-atmosphere.

U.S. Green Building Council. (n.d.). *LEED credits: Sustainable sites*. Retrieved March 9, 2019, from https://www.usgbc.org/credits/new-construction/v4/sustainable-sites.

Victoria Burrows, J. L. (2017). From Thousands to Billions - Coordinated Action towards 100% Net Zero Carbon Buildings by 2050. World Green Building Council.

Zanial, Z. (2007). Case study as a research method. *Jurnal Kemanusiaan, 5*(1), 1–6.

Index

L. Scott et al. (eds.), *Sustainable Ecological Engineering Design*,
https://doi.org/10.1007/978-3-030-44381-8

The manufacturer's authorised representative in the EU is Springer Nature Customer Service Centre GmbH, Europaplatz 3, 69115 Heidelberg, Germany. If you have any concerns regarding our products, please contact ProductSafety@springernature.com

Printed and bound by CPI Group (UK) Ltd, Croydon, CR0 4YY
30/09/2026
02225697-0015